Transport

Contemporary Foundations of Space and Place
Series Editor: John Agnew

Titles in the Series:

Politics
John Agnew and Virginie Mamadouh

Environment
Kay Anderson and Bruce Braun

Development
Stuart Corbridge

Regions
J. Nicholas Entrikin

Transport
Susan Hanson and Mei-Po Kwan

Culture and Society
Nuala C. Johnson

The City
Jacques Lévy

Economy
Ron Martin

The Rural
Richard Munton

Theory and Methods
Chris Philo

Transport

Critical Essays in Human Geography

Edited by

Susan Hanson

Clark University, USA

and

Mei-Po Kwan

Ohio State University, USA

Routledge
Taylor & Francis Group
LONDON AND NEW YORK

First published 2008 by Ashgate Publishing

2 Park Square, Milton Park, Abingdon, Oxfordshire OX14 4RN
711 Third Avenue, New York, NY 10017

Routledge is an imprint of the Taylor & Francis Group, an informa business

First issued in paperback 2018

British Library Cataloguing in Publication Data
Transport : critical essays in human geography. –
(Contemporary foundations of space and place)
1. Transportation geography 2. Human geography
3. Geographical perception
I. Hanson, Susan, 1943– II. Kwan, Mei-po
388

Library of Congress Control Number: 2007941049

ISBN 978-0-7546-2703-6 (hbk)
ISBN 978-1-138-37846-9 (pbk)

Contents

Acknowledgements

The editors and publishers wish to thank the following for permission to use copyright material.

American Geographical Society for the essay: Edward J. Taaffe, Richard L. Morrill and Peter R. Gould (1963), 'Transport Expansion in Underdeveloped Countries: A Comparative Analysis', *The Geographical Review*, **53**, pp. 503–29. Copyright © 1963 AGS.

Bellwether Publishing Limited for the essays: David C. Hodge (1990), 'Geography and the Political Economy of Urban Transportation', *Urban Geography*, **11**, pp. 87–100. Copyright © 1990 Bellwether Publishing Ltd; Susan Hanson (1982), 'The Determinants of Daily Travel–Activity Patterns: Relative Location and Sociodemographic Factors', *Urban Geography*, **3**, pp. 179–202. Copyright © 1982 Bellwether Publishing Ltd.

Blackwell Publishing Limited for the essays: Torsten Hägerstrand (1970), 'What about People in Regional Science?', *Papers of the Regional Science Association*, **24**, pp. 7–21; Patricia L. Mokhtarian (2003), 'Telecommunications and Travel: The Case for Complementarity', *Journal of Industrial Ecology*, **6**, pp. 43–57. Copyright © 2003 Massachusetts Institute of Technology and Yale University.

Copyright Clearance Center for the essays: Melvin M. Webber (1976), 'The BART Experience – What Have We Learned?', *The Public Interest*, **45**, pp. 79–108. Copyright © 1976 National Affairs Inc.; David Levinson (2002), 'Identifying Winners and Losers in Transportation', *Transportation Research Record*, **1812**, pp. 179–85. Copyright © 2002 Transportation Research Board; Mark Garrett and Brian Taylor (1999), 'Reconsidering Social Equity in Public Transit', *Berkeley Planning Journal*, **13**, pp. 6–27. Copyright © 1999 University of California Press; Genevieve Giuliano (1999), 'Land Use and Transportation: Why We Won't Get There From Here', *Transportation Research Circular*, **492**, pp. 179–98. Copyright © 1999 Transportation Research Board; Elizabeth Deakin (2002), 'Sustainable Transportation: U.S. Dilemmas and European Experiences', *Transportation Research Record*, **1792**, pp. 1–11. Copyright © 2002 Transportation Research Board.

Elsevier Limited for the essays: Patricia L. Mohktarian and Ilan Salomon (2001), 'How Derived is the Demand for Travel? Some Conceptual and Measurement Considerations', *Transportation Research Part A*, **35**, pp. 695–719. Copyright © 2001 Elsevier; T.R. Leinbach (2000), 'Mobility in Development Context: Changing Perspectives, New Interpretations, and the Real Issues', *Journal of Transport Geography*, **8**, pp. 1–9. Copyright © 2000 Elsevier; Morton E. O'Kelly (1998), 'A Geographer's Analysis of Hub-and-Spoke Networks', *Journal of Transport Geography*, **6**, pp. 171–86. Copyright © 1998 Elsevier; Eduardo A. Vasconcellos (1997), 'The Demand for Cars in Developing Countries', *Transportation Research Part A*, **31**, pp. 245–58. Copyright © 1997 Elsevier; Ralph Gakenheimer (1999), 'Urban Mobility in

Series Preface

This series collects together some of the most significant articles from the major fields of human geography published over the past 40 years. During this time something of a renaissance has occurred in the thinking that explores facets of human society using the concepts of space and place (and associated ones such as territory, geopolitics, mobility, diffusion, and locality). This reflects both the rediscovery of cultural difference and local knowledge as claims to universal, objective knowledge have come under critical scrutiny, and the exhaustion of historicist narratives owing to the failure of various projects to end global economic and political inequality. Thinking in terms of fixed territories of statehood has also become problematic owing to increased human mobility in a technologically-driven world in which religious and other, often non-national identities, appear to be in the political ascendancy. The period from 1970 until today has been one of particularly dramatic worldwide geopolitical change. Consider, for example, the impact of the political collapse of the Soviet Union, the economic rise of China, and the spread of globalization around the world. Many of the central tenets of Western social thought on nationalism, the primacy of economic change, and the centrality of states that were established during the Cold War are now subject to revision.

Of course, concepts of space and place as they are applied to human society have old philosophical roots, although these were undoubtedly obscured, particularly in the United States, for many years. Since the seventeenth century 'space' has meant the plane on which events and objects are located, while 'place', the older of the two terms, dates back to the ancient Greeks and refers to the 'lived' or 'occupied' space. What is new in recent years is the varied ways in which these concepts and related ones are now being used. This involves blending older understandings, drawing from the three classic geographic traditions that emphasize the environmental (physical-human), spatial (distributional), and regional (clustering) approaches to geographic sameness and difference, with new sensibilities and concerns about social and political divisions (from social class, ethnicity, disability, and gender to the new political divisions of a post-Cold War world). This sense of older ideas adapting to new circumstances accounts for the seemingly paradoxical title of the series: Contemporary Foundations of Space and Place. If the foundations of recent thinking are rooted in the past, such thinking also mirrors recent imperatives. What is also new is a more self-consciously 'critical' tenor to use of concepts compared to a past when much research was driven less by theoretical or methodological considerations and much more by just an interest in a given phenomenon in itself (landscape features, settlement types, borderlands, etc.) as if the conditions for its study were self-evident.

The ten volumes are divided, using fairly conventional distinctions, between those that are more general or cover large parts of the field (theory and methods, regions) and those that have relatively more constrained empirical subject matter (rural, economic, etc.). Unsurprisingly, there is considerable potential overlap across the volumes. Within the field as a whole, there is no simple division of labor between articles that are just 'theoretical' or 'methodological' and those that employ concepts in more mundanely 'empirical' ways. Nevertheless, there are definitely some articles that have had greater theoretical or methodological influence across the field as a whole. Within their respective areas, however, editors do attempt to cover the main conceptual differences and controversies of the past forty years, paying particular attention to the scope of the influence in question. Each volume is comprised of articles in English. Of course, English dominates global academic production today, but this does not mean that only work written by native English-speakers is significant. In choosing volume editors and volume content, an active effort has been made to ensure the representation of standpoints and perspectives from all over the world. The series as a whole should be particularly helpful to those looking for substantial overviews of the course of human geography between the 1970s and the early 2000s but who lack access to large library collections of academic journals. The individual volumes are crafted by specialists who have expansive rather than narrow conceptions of the purpose of the overall enterprise. The volumes are thus also designed to appeal to those looking for artful reviews of recent developments in different substantive areas of human geography.

JOHN AGNEW
General Editor
UCLA

Introduction

Mobility, accessibility, networks and interactions across space are at the heart of how spaces and places are brought into being and continue to change. A series on the contemporary foundations of space and place without a volume on transportation is, therefore, hard to imagine. In conceptualizing this book, we have sought to strike a balance between including essays that have become classics in the field and ensuring coverage of topics that are considered foundational to transportation geography, even if some of the essays on those topics are too recent to be considered classics. All of the organizing concepts, then, are indeed classics in the sense that they have long structured thinking about the field of transportation. In addition to work that charts the development of thinking about transportation, space and place, we have sought to include essays that point towards a research agenda for the next generation.

Perhaps the heyday of transportation studies in geography was in the 1960s and early 1970s – the time when quantitative approaches held sway in the discipline. Certainly, transportation issues were at the core of the thinking of those who instigated and led the Quantitative Revolution in geography at the University of Washington (UW) in the late 1950s. A notable case in point is the collection of studies by William L. Garrison *et al.* (1959) and his graduate students at UW, which examined the impacts of highway improvements on land-use change. Brian Berry's (1967) empirical studies of central place concepts, Duane Marble's (1959) studies of individual travel behaviour and Michael Dacey's (1966) work on settlement spacing were all rooted in transportation concepts and all had their origins in the authors' graduate work at Washington with Garrison.

As the work of these quantitative 'space cadets' illustrates, geographical studies in transportation have traditionally emphasized spatial separation, networks, spatial interaction, the friction of distance and the relationship between spatial interaction and place. Changes over the past 30 to 40 years in the ways in which transportation scholars in geography have conceptualized space and place have been more subtle than in other corners of human geography. In the late 1950s and the 1960s space was viewed relatively unproblematically as a pre-given plane on which a 'space economy' emerges as rational decision-makers locate households, firms, transport networks and the like. Scholars' goals at that time were to learn the general principles that gave rise to the spatial patterns so evident in maps of human activity on the landscape. Nevertheless there was the recognition that human decisions were creating the spatial configurations that defined places. It was also recognized, even in those early days, that spatial context, then thought of primarily in terms of the density of activity sites such as shops or settlements, affects people's decisions – primarily, at that time, their decisions about spatial behaviour, such as where to shop (see, for example, Rushton, 1969).

Since then, the shifts in thinking about space and place have entailed a deepening recognition that, instead of being pre-given, space is created through human action and that the geographic context of spaces and places shapes subsequent human decisions and activity. Conceptualizations of spatial or geographic context have broadened considerably since the late 1960s to encompass far more than just density, but also, for example, meanings, institutions and social relations. In addition, geographers now recognize that 'local context' is created through interactions with distant, not just spatially proximate, people and places. The idea that 'space' includes the biophysical environment has entered transportation studies mainly through the concerns surrounding the environmental impacts of transportation investments.

As John Agnew notes in the Preface to this series, one of the 'new sensibilities' to shape thinking about space and place has been that surrounding social difference, conceived in terms of gender, stage in the life course, ethnicity, class and so on. Mainly because of their long-ago intellectual ties to behavioural geography and because of manifest intergroup differences in mobility and access, transport scholars were among the first to incorporate social difference into not only their thinking about human agency and the elements of geographic context, but also their work on the relationship between people and place. Some transport geographers have also come to understand that just as social identity and geographic context shape human decision-making, so too they influence the construction of knowledge through, for example, the questions a scholar chooses to pose and all aspects of research design. In recognition of this point, we have selected essays for this volume written by authors whose roots lie in a variety of social identities and geographic contexts.

Transportation geography diverges from other branches of human geography in that since the 1970s it has had strong links to both civil engineering and policy. The link to civil engineering reflects the traditional dominance of that field in transportation studies writ large and in transport planning; this strong connection has, in our view, tended to direct attention away from theory and towards techniques. The civil engineering connection also helps to explain the enduring popularity of the utility-maximizing framework in transport research – a framework that not only oversimplifies the nature of human agency, but also seems to have limited conceptual development in the field. The strong link to policy speaks to the very pressing and immediate nature of transportation problems such as congestion and the need for access; it also speaks to the centrality of transportation to a range of broader policy issues, such as economic development, inequality, environmental degradation and energy consumption.

Transportation research has a long tradition of detailed, largely quantitative, empirical work. This strong empirical tradition does not mean that the field is devoid of theory; rather, it reflects the policy connection and the reality that describing, understanding and planning for transportation patterns and processes require the analysis of large and complex data sets. As a result, a great deal of attention in the transportation literature has been devoted to analytical techniques. As in other empirically-oriented social sciences, the analytical approaches taken in transportation research have been tied to technologies of data collection and analysis; as computing power and sophistication have grown, modelling approaches have shifted and geographic information systems (GIS) have become an essential part of transportation analyses. Current approaches involve blending general principles with local particularities. As you read the essays here, pay close attention to how authors are designing studies, measuring concepts and analysing data, and think about how their results are dependent on their research designs.

We have organized the essays in this book into four Parts. Part I, 'Fundamental Concepts', addresses issues that transcend time and geographic scale; Parts II and III, 'Individual Behaviour in Urban Spatial Context' and 'Interregional Transport', are loosely defined by scale. Finally, Part IV, 'Policy Issues', draws together themes raised in Parts I to III and underlines both the strong policy orientation of transport studies and the importance of transport-related issues for societal decision-making. In the remainder of this Introduction we seek to contextualize the essays in each section of the book by relating them to the periods during which they were published and, where applicable, to the other themes of the book.

Fundamental Concepts

Access and Mobility

Access and mobility are at the core of transportation concepts. Traditionally, access has referred to the number of opportunities (or activity sites) within a given distance or travel time of an individual or an area, although information and communication technologies (ICTs) are complicating this conventional definition. Mobility refers to the ability to move between different activity sites – for example, from home to a bank or workplace (see Hanson, 1986, p. 4). In thinking about access and mobility in tandem, all four of the essays in this section address issues of great contemporary significance – issues that lie at the heart of such current catchphrases as 'smart growth', 'walkable neighbourhoods' and 'urban sustainability'. For example, if access to opportunity is what transportation is all about, how can people achieve access with a minimum of motorized mobility?

In his classic essay 'What about People in Regional Science?' (Chapter 1), Torsten Hägerstrand outlines his influential ideas about time geography. By combining access and mobility, Hägerstrand provides a conceptual framework for understanding how accessible places really are to individual travellers. His emphasis on individuals rather than on areal zones as the preferred units of analysis reflects a shift in thinking that was underway within human geography at the time (1970): witness the publication of Cox's and Golledge's *Behavioral Problems in Geography* in 1969. But, in his claim that individuals should not be subdivided into 'dividuals', Hägerstrand was indeed ahead of his time. Here, he emphasizes the wholeness of a person's experience, her time–space path, which should not be carved up into variables that lose sight of their interrelation in the person. In insisting that time and space likewise are not separable and that the integrity of the individual's life path needs to be kept in tact, Hägerstrand prefigures contemporary qualitative approaches that underscore the benefits of retaining the wholeness of the individual's experience.

Hägerstrand's time geography concentrates on the movements of individuals through the course of their everyday lives at a micro-scale; these movements are shaped not only by choice, but also by constraints on access and mobility, thereby presaging later frameworks, such as Giddens's (1986) structuration theory, which combine agency and structure. The concept of opportunity (a potential destination) was abroad at the time, but Hägerstrand deepens and extends this concept by noting that opportunity for interaction (for meeting needs of access) involves more than just an individual's *location* vis-à-vis potential activity sites, together with available modes of travel, but also entails the hours when an establishment (for example, a store) is open, rules that permit the person to enter the place (for example, a library) and the time needed in the person's activity schedule to visit it. Hägerstrand's time geography is a powerful and influential conceptual framework. Although few have operationalized the framework in empirical studies, echoes of this essay appear in many other essays in this volume.

In 'Accessibility and Intraurban Travel' (Chapter 2) Susan Hanson and Margo Schwab ask a number of interrelated questions that have long fascinated, and continue to engage, scholars of urban areas. How does access affect mobility? How does an individual's location vis-à-vis potential destinations within an urban area affect the frequency and length of their trips (and therefore their vehicle miles travelled (VMT), energy consumption and contribution to greenhouse gases)? Do people who live in low-density settings travel greater distances (consume more travel) than those living in higher-density settings? Are people who live in low-density environments more likely to use motorized modes of transportation and less likely to walk or bike to their destinations than those living in high-density settings? The empirical portion of this study is based on data that were collected in 1970–71, which not only indicates the durability of these questions, but also was the time when Hägerstrand was first writing about his time geography. In this essay, note how access is conceptualized and measured.

In 'Individual Accessibility Revisited: Implications for Geographical Analysis in the Twenty-first Century' (Chapter 3), Mei-Po Kwan and Joe Weber argue that traditional measures of access (like those that rely on distance to opportunities, as in Hanson and Schwab's essay) are inadequate in today's world because they ignore the distance-distorting presence of ICTs. These authors point out that whereas conventional access measures have emphasized the friction of distance as it impedes interaction, they see the role of distance in accessibility as declining in importance with the spread of ICTs. Exactly how ICTs affect access is a matter of considerable interest and debate (see, for example, Janelle and Hodge, 2000), but Kwan's and Weber's essay illustrates the importance of the historical–geographical context within which authors are working and the need for scholars now to conceptualize and measure ICT-sensitive access. These authors also note that time of day, not considered in traditional measures of accessibility, is likely to affect access significantly, especially via congestion. Measures including time of day are now within the realm of feasibility, thanks to GIS. In general, the availability of sophisticated GIS now enables measures of access that are far more detailed than was the case 30 years ago.

But is access to opportunity what transportation is all about? Is the demand for travel truly dependent on, or derived from, the individual's desire to participate in some activity (for example, work, shopping or visiting a doctor) at some location that is distant from home or from wherever that person happens to be? Or might people travel simply because they want to experience mobility? Challenging the received wisdom that travel is a derived demand, Patricia Mokhtarian and Ilan Salomon ask these questions in their essay 'How Derived is the Demand for Travel?' (Chapter 4). They show, however, that travel is not always merely a by-product of people wanting to accomplish activities at distant destinations; in fact, mobility has its own positive utility, indicating that transportation is not just about access and the mobility that access induces, but also about mobility pure and simple. This finding raises interesting and complicated questions for those who advocate dense settlements as a way of reducing VMT, energy consumption and greenhouse gas emissions. Note how Mokhtarian and Salomon go about trying to find out whether travel has a positive utility independent of its connection with satisfying a demand to conduct an activity at a place away from home.

Access, Networks and Development

The four essays in this section mobilize the concept of accessibility (thought of as the cost of overcoming the friction of distance) to consider questions about land-use change and development. In other words, the authors of these essays examine how access is related to the fate of places. Why do some places thrive and others wither? What role does accessibility (in terms of transportation costs) play in determining the answer to this question? Because travel is always easier on a network than off one, networks are part and parcel of access. The first two essays in this section deal with the United States and the second two with developing countries. Two are from the 1960s, and two are more recent.

Note that in each of the essays published in the 1960s the authors first present a general model and then test it with empirical data; note also that both of these earlier essays use areas as the units of analysis. In these respects, these essays are representative of the time they were written.

In both 'Spatial Reorganization: A Model and Concept' by Donald Janelle and 'New Directions for Understanding Transportation and Land Use' by Genevieve Giuliano, the authors review the theory surrounding the role of transport costs in location decisions and note that the theoretical basis for expectations that transportation improvements (that is, improved accessibility) should lead to increased land values and therefore altered land uses. Both discussions reflect the influence of economic theory on framing conventional thinking about the relationship between transportation and place. Measuring access (transportation costs) in terms of travel time and using places (nodes in a network) and network links as units of analysis, Janelle shows how change in accessibility on the network is related to change in economic activity at the nodes. By contrast, Giuliano stresses the complexity of the relationship between transport costs and land-use change and the striking lack of correspondence between theoretical expectations and empirical findings regarding the nature of this relationship. In examining some possible reasons for this disjuncture, she highlights factors that are rooted in the specifics of contemporary American cities.

In 'Transport Expansion in Underdeveloped Countries: A Comparative Analysis' (Chapter 7) Edward Taaffe, Richard Morrill and Peter Gould relate the development of the transport network in an area to the area's population and commercialization (urbanization or commercial agriculture). Although the setting is Ghana and Nigeria – not Southern Michigan – the essay is similar in purpose, conceptualization and conclusions to Janelle's. In concert with Giuliano's argument, Taaffe's, Morrill's and Gould's analysis underlines the significance of an *initial* increase in accessibility through the provision of new roads or railroads; such large improvements to access usually do lead to changes in the nature of places, as such places become nodes in new networks of mobility.

Thomas Leinbach's 'Mobility in Development Context: Changing Perspectives, New Interpretations, and the Real Issues' (Chapter 8) updates understandings of the relationship between transport and development and, like Giuliano's essay, emphasizes complexity. Leinbach advocates using a disaggregate (individual-level) approach in examining this relationship so that analysts can discover who really profits from transport improvements in poor areas – is it the NGOs and the wealthy landowners or do women and children also benefit? In this sense Leinbach's analysis differs from that of Taaffe, Morrill and Gould by demanding that we look *within* places to see the impacts of transportation improvements.

Both Leinbach and Giuliano highlight the importance of local (distinctive) factors, not just those general factors posited by theory, and both call for new conceptual development and new theory on the role of transport costs in land-use development. The essays by Giuliano and Leinbach are based on – and made possible by – the dozens of empirical studies on transportation and development that have been completed in the decades since the essays by Janelle and by Taaffe, Morrill, and Gould first appeared. Note the many ideas for needed future research that Leinbach and Giuliano each identify.

Equity

Each of the three essays in this section provides a framework for thinking about the equity, or fairness, of transportation systems, which are both costly to build and designed to serve the public good. In addition, as the essays in the previous section have described, because of their networked nature, transport systems provide improved access to only a portion of the population in a region. Questions of equity ask who pays (in terms of local tax structures, as well as in terms of noise and air pollution, car–pedestrian conflicts and the like) and who benefits from transportation investments. Concern over (in)equities in transportation arrangements has a long history, dating at least to John Kain's (1968) classic paper on the spatial mismatch between residential and job locations for African-Americans in US cities.

Melvin Webber's 1976 essay, 'The BART Experience: What Have We Learned?' (Chapter 9), was published only four years after the Bay Area Rapid Transit (BART) began service in San Francisco.[1] Webber, who was involved in the planning of BART, takes us back to the thinking of the 1950s and 1960s about transit and cities. He describes what rapid rail transit was expected to do: first, get people to abandon their cars for public transit, thereby reducing road congestion; and, second, direct urban development to the city centre and to land around transit stations, thereby transforming the landscape by deterring suburban sprawl and central city decline. In assessing BART's success in meeting these goals, Webber engages the concepts of accessibility and mobility, land values and land uses that we have encountered in the previous sections of this volume. But he also highlights the power of politics in public transportation and reminds us that all transportation investment decisions are intensely political and that they are so precisely because every such decision yields winners and losers.

In this context, Webber, as well as David Hodge in his essay 'Geography and the Political Economy of Urban Transportation' (Chapter 10) and David Levinson in 'Identifying Winners and Losers in Transportation' (Chapter 11) stress the importance of identifying what groups of people, where, are gaining or losing from a particular system. Depending on their place of residence and places of work, certain groups have good access to a system, whereas others have virtually no access. In other cases, as Webber and Hodge demonstrate, certain groups pay a disproportionate share of their income to support a system they rarely use. In the most recently published equity essay included here, Levinson discusses different kinds of equity and the different population subgroups to consider when asking questions about the equity of a system or a policy. To ensure that equity issues are identified and addressed, he proposes that transportation investments undergo an equity impact statement as well as an environmental impact statement.

---

1 For a more recent assessment of BART, see Cervero and Landis (1997).

Costs Associated with Transport

The final fundamental concept we explore has to do with the costs of reliance on the car, or the costs of automobility, in our society. It is now well known that drivers do not bear the full costs of their car use, but scholars voiced little concern for the externalities associated with the car until 15–20 years ago. John Whitelegg's short but pithy 'Time Pollution' (Chapter 12) is thought-provoking and asks us to reflect on how we value and use our time (see also Whitelegg, 1993, ch. 5 for more of Whitelegg's thoughts on this subject). Whitelegg is right in saying that the entire urban transportation planning edifice is built on the idea of 'the value of time' (in which monetary values are given to travel time) and that the amounts of time 'saved' by road improvements, while relatively small (see also Giuliano, Chapter 6, this volume), are nevertheless used to justify huge expenditures for creating more road capacity. Whitelegg is clearly an advocate for non-motorized transport, but how, then, will your food reach the supermarket?

In 'A Review of the Literature on the Social Cost of Motor Vehicle Use in the United States' (Chapter 13) James Murphy and Mark Delucchi systematically examine the costs of automobility to US society. Their analysis is notable for its attempt to identify all the major ways in which the motor vehicle incurs costs for which users do not pay directly (for example, accidents, pollution, congestion, urban sprawl, road maintenance) and for the care with which they assess the studies in their review. Note the variety of methods used in the studies Murphy and Delucchi review and the consequent difficulty they have in arriving at the monetary costs of automobility. Although both essays in this section address costs in the context of industrialized nations, the frameworks they provide are probably useful for thinking through costs in other contexts as well.

Individual Behaviour in Urban Spatial Context

The travel-activity behaviour of people in specific geographical contexts has been an important theme in transportation studies in geography since the early 1970s. Initially growing out of behavioural geography, research in this area has expanded to encompass the impacts of a variety of individual, household and geographical factors. In addition to spatial factors (or constraints), gender, race/ethnicity, and employment status have been identified as the most important 'dimensions of difference' shaping individuals' travel-activity patterns. Time geography, which emphasizes the role of spatial and temporal constraints, has informed some of this work (for example, Kwan, 1999). All of the five essays in Part II look at how geographical context and individual/household factors affect people's travel-activity behaviour.

Susan Hanson's 'The Determinants of Daily Travel-Activity Patterns: Relative Location and Sociodemographic Factors' (Chapter 14) is one of the earliest studies to explore the role of the built environment in shaping people's travel behaviour.[2] The study examines the effects of sociodemographic variables (attributes of the individual and household) and spatial constraints (the location of the individuals relative to potential destinations) on different aspects of individuals' travel-activity patterns. Using 35-day travel diaries collected in Uppsala, Hanson not only finds that both sets of factors are important to particular travel variables

2 For a recent review of this topic see Ewing and Cervero (2001).

(for example, frequency of travel), but also identifies gender as an important determinant of people's activity-travel patterns. The study demonstrates the role of geographical context in addition to individual and household factors in shaping individual activity-travel patterns. Because these patterns are the outcomes of the combined effect of contextual/spatial factors and individual/household factors, this study raises a question that still remains challenging: how can we disentangle the role of geographical context or spatial constraints from that of individual or household factors?

In addition to the spatial constraints that Hanson examines, several studies have also looked at the role of temporal constraints (the need for individuals to undertake activities at particular times of the day) in people's travel-activity behaviour. Informed by Hägerstrand's time geography, Pip Forer's and Helen Kivell's 'Space–Time Budgets, Public Transport, and Spatial Choice' (Chapter 15) pays explicit attention to the role of space–time constraints on people's activity patterns. The authors emphasize that all activities have particular requirements which can be understood in terms of the spatial and temporal restrictions that they impose on a person's activity choice (for example, the location and opening hours of a day-care centre and the need to be at home at meal times for primary schoolchildren). Using estimates of the free time available in a typical day, the authors found that both public transport and the temporal constraints of activities are important in shaping people's access to, and choice of, potential destinations. They suggest that the interaction between transport and the geography of local opportunities can lead to considerable variation in the quality of life for people located in different areas of a city. The study also identifies life-stage variables (child-rearing, for example) to be important for understanding individual constraints and access to urban opportunities. As in other studies based on the time-geographic perspective (for example, Palm, 1981; Tivers, 1985), the authors show that the interaction of time and space is crucial in determining the activity choices of individuals, especially those with gender or social roles that impose stringent restrictions.

In 'Gender and individual access to urban opportunities: A study using space-time measures' (Chapter 17) Mei-Po Kwan continues this emphasis on the importance of gender and space–time constraints in shaping people's activity and travel choice. This study, however, challenges the notion that individual accessibility should be understood and measured in terms of locational proximity between a person's home (or workplace) and the surrounding opportunities because this approach ignores the role of complex travel behaviour and space–time constraints. Drawing upon time geography, Kwan conceptualizes individual accessibility as space–time feasibility and provides formulations of accessibility measures based on the space–time prism construct. Using a network-based GIS method, she finds that women have lower levels of individual access to urban opportunities than do men. A surprising observation is that individual accessibility as evaluated by space–time measures has no relationship with the length of the commuting journey. Through adopting a time-geographic perspective the study reveals the complex interaction among gender, employment status and people's spatial and temporal constraints, as well as the effects of these relationships on people's access to urban opportunities.

While the previous three essays call our attention to the importance of many individual and household variables (such as gender, age and life-stage) in shaping people's activity patterns, Sara McLafferty's and Valerie Preston's 'Gender, Race, and Commuting among Service Sector Workers' (Chapter 18) highlights the importance of race and its complex interaction with gender. They revisit the observation that women tend to work closer to home and have shorter commuting times than men. Using a large sample of service sector workers in the New York metropolitan area, these authors find that gender differences in commuting vary significantly among racial groups. McLafferty and Preston observe that black and Hispanic women commute as far as black and Hispanic men and that their commuting times far exceed those of white males and females. The study reveals the complex ways in which gender and race interact to affect urban commuting times and highlights the problems of looking at individual travel behaviour without considering the combined effects of gender, race and ethnicity.

A long-standing theme in behavioural and transport geography has centred on the recognition that the objective built environment might not be as important as the perceived environment in shaping people's travel-activity patterns. That is, in addition to the spatial constraints, as measured, for example, in Chapters 14 and 17 by Hanson and Kwan, a person's activity and travel choices are also influenced by her spatial knowledge of the opportunities and routes that can be used to reach these opportunities. Understanding how people acquire spatial knowledge and the relationship of this knowledge to route-learning can help shed light on the behavioural foundations of travel behaviour. In 'Spatial Knowledge Acquisition by Children' (Chapter 16) Reginald Golledge, Nathan Gale, James Pellegrino and Sally Doherty examine how children recognize, recall and use the spatial properties of routes in route-related spatial tasks. They found that route-learning does not by itself produce the type of spatial knowledge that is required to understand the layout and spatial configuration of a child's neighbourhood. Children can learn routes without having clearly defined procedures for operating on a declarative knowledge base or for integrating that knowledge into a configurational understanding. This finding suggests that spatial learning involves a highly selective process, and much information can be ignored in specific task situations such as those of navigating over preset routes

Interregional Transport

The three essays in Part III examine interregional transport issues. They focus on specific network structures (for example, hub-and-spoke networks), the use of these systems for the movement of passengers and/or goods, and the evolution of their spatial structure over time. At the interregional level, flows of passengers and goods are often channelled through networks with prominent hubs that enhance connectivity between places. In 'A Geographer's Analysis of Hub-and-Spoke Networks' (Chapter 19) Morton O'Kelly examines hub-and-spoke networks by considering the linkages, hinterlands and hierarchies formed by such networks by means of an analysis of different hub-and-spoke outcomes between the air passenger and air freight sectors. Through a simulation exercise, he finds that a single hub allocation model minimizes the number of links but leads to inconvenient routing between city-pairs, whereas the structure generated by the multiple assignment model is more convenient to air passengers. O'Kelly suggests that the single hub model is more suitable for regional freight or communications systems than for passenger travel. The study concludes that the hub-and-spoke structure of an interregional transport network can have different impacts on passenger travel and freight transport in terms of level of service and convenience.

There are, however, important differences between air and rail transport. In his network study 'Intermodal Transportation in North America and the Development of Inland Load Centers' (Chapter 20) Brian Slack analyses the radical changes that transformed intermodal transportation in the United States and Canada in the late 1980s. Examining the emergence of railnets and hubs and the underlying forces behind their development, Slack finds that traffic becomes concentrated at inland centres that serve as regional truck distribution points and are linked by high-volume train service. One consequence revealed by his exploration of the spatial impacts of the restructuring of intermodal transportation is the concentration of economic advantage, which Slack predicts will negatively affect the cities (particularly port cities) and regions that lose access to rail, rendering them less accessible and less attractive to industries with high transport needs. Slack's study calls attention to the practical and theoretical questions raised by the spatial shifts in service activity that hub development may bring about. The process of network rationalization raises questions about relative accessibility and regional development. Geographical analysis of spatial shifts and their differential consequences for the prosperity of places can help to alert the public, as well as government, to the process of network change.

In contrast to O'Kelly and Slack, Thomas Leinbach and John Bowen focus on commodity flows and the organization and decision characteristics of firms. In their essay 'Air Cargo Services and the Electronics Industry in Southeast Asia' (Chapter 21) they highlight the importance of air cargo services for manufacturers with international production networks. Such services include the air transport of goods as well as related ground-based air logistics services. Examining the demand for air cargo services in the electronics industry in Southeast Asia, Leinbach and Bowen highlight the importance of drawing together research on advanced producer services, the global scale of production networks and the spatial configuration of those networks. This study is distinctive because it examines the demand for a producer service across several economies at different levels of industrialization and situates advanced producer services within global production networks.

Policy Issues

Many contemporary cities, whether in developed or less developed countries, are facing serious transport-related problems, such as congestion and environmental degradation. Although transport planners worldwide have devised a variety of policies for dealing with these problems, few measures seem to have been effective in reducing the negative social and environmental impacts of passenger travel by private vehicle. The seven essays in Part IV focus on a number of policy issues, including social equity, sustainability and the insatiable demand for automobility. Recognizing the importance of policy decisions in shaping transportation patterns, all the authors present lessons from various contexts, including cities in the United States, Europe and the developing world.

How effective has US public transit policy been in meeting the mobility needs of the transportation disadvantaged? In 'Reconsidering Social Equity in Public Transit' (Chapter 22) Mark Garrett and Brian Taylor reveal that transit policy in the United States has focused on expanding suburban public transit routes and capturing new affluent customers rather than on improving well-patronized transit service in low-income, central-city areas. As a result, the quality of public transit diverges between inner city and suburbs, with the former experiencing higher demand and poorer levels of service. The authors find this situation to be both economically inefficient and socially inequitable, and they raise a number of questions regarding the value of public transit. Among the key questions are: should public transit policy strive for greater geographic mobility for riders with options or for accessibility for people who lack other mobility options? Are current policies increasing or decreasing social equity?

Is land-use policy (for example, promoting denser rather than less dense urban settlements) an effective means of reducing vehicle miles travelled (VMT)? In 'Land Use and Transportation: Why We Won't Get There From Here' (Chapter 23) Genevieve Giuliano considers the effectiveness of land-use policy as an instrument for reducing environmental and other external costs associated with the private car. Giuliano argues that land-use patterns, specifically decentralization and low-density development, are the result of complex factors over which policy has little control. Thus policies aimed at altering such patterns are, therefore, unlikely to reduce individual travel by private vehicle. Giuliano acknowledges, however, the necessity of advocating change in land-use policy to promote greater social equity, as, for instance, in the US case of poor and minority households who are spatially segregated owing to exclusionary practices such as zoning.

Although the notion that information and communication technologies (ICTs) will soon be substituted for travel and thereby lead to reductions in VMT, Kwan and Weber (Chapter 3) have argued that the interaction between people's activity-travel behaviour and the use of ICTs is highly complicated. New concepts that allow for the effective analysis of such interaction are sorely needed. In 'Telecommunications and Travel: The Case for Complementarity' (Chapter 24) Patricia Mokhtarian examines the conceptual, theoretical and empirical arguments in support of the claim that telecommunications will increase travel rather than reduce it. Focusing on the movement of people and goods, she finds that substitution, complementarity, modification and neutrality within and across communication modes are occurring simultaneously. Thus if current trends continue, both telecommunications and travel will grow. Yet a paramount question remains: how much of the simultaneous increase observed in telecommunications and travel reflects a true causal complementarity, and how much is it due to spurious correlation with other variables?

How can transportation, which is so essential to the fabric of contemporary life, become more sustainable? In 'Sustainable Transportation: U.S. Dilemmas and European Experiences' (Chapter 25) Elizabeth Deakin considers this question in light of the experiences of European countries and presents specific strategies to support sustainable transport. Though initially concerned with the reduction of greenhouse gases, sustainable transport initiatives have come to emphasize more broadly social, economic, and environmental concerns. Despite barriers to implementing sustainable transportation plans at the national level in the United States, efforts are underway at the local level. In the European context such plans are supported by strong policy commitments, government incentives and new planning processes emphasizing collaboration and performance measurement.

Why have policies designed to discourage car use been spectacularly unsuccessful just about everywhere? In 'The Demand for Cars in Developing Countries' (Chapter 26) Eduardo Vasconcellos analyses the misunderstandings surrounding private versus public transportation issues in developing countries. Understanding the car as a transport technology embedded in social reproduction, Vasconcellos argues that automobile demand has been created by policies that have targeted the affluent, who now see automobility as being essential to their identity and social reproduction. The emphasis on pro-car policies to win the political support of the middle class in developing countries has detracted from the development of alternative forms of urban transport that might serve the needs of the less fortunate.

In contrast to Vasconcellos, Ralph Gakenheimer in 'Urban Mobility in the Developing World' (Chapter 27) focuses on how increased congestion, resulting from the rapid increase in automobility, has reduced mobility and accessibility for both private and public transit users. Contributing to this decline in mobility have been inadequate road maintenance and disagreement among officials on planning/policy approaches. Gakenheimer outlines strategies to deal with rising motorization and falling mobility and concludes that policies to meet mobility needs in developing cities are likely to include building more highways, improving the management of public transport, pricing travel to pay for infrastructure and managing traffic more efficiently. Gakenheimer's essay calls attention to the serious congestion problems (and the accompanying policy concerns) facing the numerous large cities in the developing world where car ownership is increasing apace.

As Gakenheimer's essay suggests, congestion is perhaps the most widely recognized transportation problem around the world. Although charging more to use facilities during peak-demand periods has been widely used to ration limited infrastructure supply in telecommunications for example, the use of congestion pricing as a transport policy to ration limited road space has faced strong opposition. In 'An Assessment of the Political Acceptability of Congestion Pricing' (Chapter 28) Genevieve Giuliano asks whether planners and policy-makers finally accepted the rationale for congestion pricing. Are they willing to take the political risk of actually testing the idea? Will the proponents of congestion pricing be defeated once again by public opposition? In Giuliano's view, despite the apparent need for the policy, it is unlikely that congestion pricing will be implemented to any significant extent in the United States. In reviewing the reasons for her skepticism, Giuliano reminds us of the intensely political context within which any policy decision is taken (or not).

Conclusion

Many of the policy-related essays in this volume return to the key concepts we outlined at the outset of this chapter: accessibility, mobility, equity, costs to society and the relationship of access and mobility to the fate of places. Taken together, the policy essays also pose serious questions about what kinds of places we earthly inhabitants are creating and what kinds of places we might prefer to dwell in. Access and some degree of mobility are essential to life in any place, but the type and level of mobility – like space itself – are not pre-given; they are the outcome of human choices. Moreover, as a number of essays in this volume demonstrate, the type and level of access/mobility connecting places at various scales are hugely important in shaping the quality of economic and social life in those places. The choices involved are complex and difficult; we hope that the essays in this volume will be the beginning of your becoming curious and informed citizens on questions of transportation, space, and place.

Acknowledgement

Mei-Po Kwan thanks Suzanna Klaf for her capable assistance in this project.

References

Berry, Brian J. L. (1967), *The Geography of Market Centers and Retail Distribution.* Englewood Cliffs, NJ: Prentice Hall.

Cervero, Robert and Landis, J. (1997), 'Twenty Years of BART: Land Use and Development Impacts', *Transportation Research A*, **31**(4), pp. 309–33.

Cox, Kevin and Golledge, Reginald (1969), *Behavioral Problems in Geography*, Evanston, IL: Northwestern University Press.

Dacey, Michael (1966), 'A County-Seat Model for the Areal Pattern of an Urban System', *Geographical Review*, 61(4), pp. 527–42.

Ewing, Reid and Cervero, Robert (2001), 'Travel and the Built Environment: A Synthesis', *Transportation Research Record 1780*, pp. 87–114.

Garrison, W.L., Berry B.J.L., Marble, D.F., Nystuen, J.D. and Morrill, R.L. (1959), *Studies in Highway Development and Geographic Change*, Seattle: University of Washington Press.

Giddens, Anthony (1986), *The Constitution of Society: Outline of the Theory of Structuration*, London: Polity Press.

Hanson, Susan (ed.) (1986). 'Getting There', in *The Geography of Urban Transportation* (2nd edn), New York: Guilford Press, pp. 3–25.

Janelle, Donald and Hodge, David (eds) (2000), *Information, Place, and Cyberspace: Issues in Accessibility*, Berlin, Heidelberg, New York: Springer-Verlag.

Kain, John (1968), 'Housing Segregation, Negro Employment, and Metropolitan Decentralization', *Quarterly Journal of Economics*, **82**, pp. 32–59.

Kwan, Mei-Po (1999), 'Gender, the Home-Work Link, and Space–Time Patterns of Non-employment Activities', *Economic Geography*, **75**(4), pp. 370–94.

Marble, Duane (1959), 'Transport Inputs at Residential Sites', *Essays and Proceedings of the Regional Science Association*, **5**, pp. 253–66.

Palm, Risa (1981), 'Women in Non-metropolitan Areas: A Time–Budget Survey', *Environment and Planning A*, **13**, pp. 373–78.

Rushton, G. (1969), 'The Scaling of Locational Preferences', in K.R. Cox and R.G. Golledge (eds), *Behavioral Problems in Geography: A Symposium*, Evanston, Ill: Department of Geography, Northwestern University, Studies in Geography, 17, pp. 197–227.

Tivers, Jacqueline (1985), *Women Attached: The Daily Lives of Women with Young Children*, London: Croom Helm.

Whitelegg, John (1993), *Transport for a Sustainable Future: The Case of Europe.* London: Belhaven Press.

Part I
Fundamental Concepts

Access and Mobility

[1]

WHAT ABOUT PEOPLE IN REGIONAL SCIENCE?

by Torsten Hägerstrand*

Since this occasion is the first time in the annals of the Regional Science Association that the presidential address is being delivered at a congress in Europe, it seems appropriate to explore the past to see whether there has been a difference in emphasis or tone between the European and the North American meetings. I think there has been a difference although I am not prepared to show statistical evidence. When looking over the proceedings of the sixties, one gets the impression that participants in this part of the world have preferred to remain closer to issues of application rather than to issues of pure theory. We in Europe seem to have been looking at Regional Science primarily as one of the possible instruments with which to guide policy and planning. I have chosen to proceed along this line by suggesting that regional scientists take a closer look at a problem which is coming more and more to the forefront in discussions among planners, politicians, and street demonstrators, namely, the fate of the individual human being in an increasingly complicated environment or, if one prefers, questions as to the quality of life. The problem is a practical one and, therefore, for the builder of theoretical models, a 'hard nut to crack.'

Now, first of all, does the problem fall within the scope of Regional Science? I think it does. A forest economist remarked some time ago that, "forestry is people, not trees." How much more accurate it would be to say that Regional Science is about people and not just about locations. And this ought to be so, not only for reasons of application. Regional Science defines itself as a social science, thus its assumptions about people are also of scientific relevance. Regional scientists differ in their attitudes toward concepts related to the quality of life. In his presidential address of 1962 Ullman [10] concluded that the, "problem remains to design cities to take advantage of scale economies and the other advantages of concentration, and at the same time to provide optimum livability." This formulation indicates a belief in 'livability' as a worthwhile problem for research and a goal for planning. In 1967 Lowry [6] sounded more skeptical, at least as far as the idea of an optimum in physical planning was concerned. "People seem able to extract apparently equivalent values from diverse environments, so long as the mechanics of the environment are comprehensible, and so long as its responses to individual initiatives are predictable." However, the next sentence takes back some of the conviction of that statement by arguing that, "when our cities become too dismal for comfort, we retire to the suburbs and substitute the amenities of

* The author is associated with the Department of Social and Economic Geography, University of Lund, Sweden.

gardening for those of museums and bright lights."

One frequently notices that economists are very quick to suggest that we solve our problems by moving somewhere else. It is convenient in theory and often in reality, but the idea implies two things: first, that there is a worthwhile place to go to; and second, that it is of no relevance that some have to be left behind. To earn money and to find desirable things to spend it on is a basic part of livability and Regional Science has a lot to say about that. But it is also important that there be easy access to schools, other educational facilities, universities, libraries, theatres and concert-halls, doctors and hospitals, security services, playgrounds, parks, even silence and clean air. One does not find much written in Regional Science publications on the location and dimensioning of such items in relation to the spatial distribution of needs. Perhaps the problems involved fit better into the more restricted framework of specialism or operations research. I do not feel that this is good research policy. The sum total of these items is regionally too important to make it reasonable to leave them entirely in the hands of people who view them predominantly from the inside.

I am not going to carry this point further and it is not my intention to remain on quite so practical a level. Let me instead raise a question as to what regional scientists assume about people at the level of first principles. Have the efforts to give spatial realism and generality to economic matters also brought human realism and generality to matters of spatial organization? It is hard to find an answer to this, since, as Isard and Reiner [4] have pointed out, "models of human behavior over space have been almost entirely oriented to mass probabilistic behavior." These models of large aggregates are often presented without explicit statements about the assumed social organization and technology that exist at the micro-level from which the individual tries to handle his situation.

It may well be that when a region is given a certain areal size, which is well above the daily range of the majority of its population, it does not matter very much (as far as the aggregate spatial outcome is concerned) what forms micro-arrangements have happened to take. Such possible insensitivity would, in itself, be a problem for analysis. Nothing truly general can be said about aggregate regularities until it has been made clear how far they remain invariant with organizational differences at the micro-level. As an illustration, let me refer to the great number of studies of consumer and commuter behavior. In one case only I found the straightforward statement to the effect that in, "the modal case, the male traverses the habitat to exchange labor for money, and the female traverses it to exchange this money for food and other objects of value." See Fox and Kumar [2]. It could be argued that a modal case of that kind is a particular solution, typical of a given culture area and of a given period of time. What about a modal case where both man and wife exchange labor for money? Or what about doing away entirely with many of the retail trade establishments by equipping dwellings with refrigerators and storerooms beside the mailbox and having them filled from cruising delivery vehicles without the presence of the customer? Since we know that social roles can be re-defined and that experts on physical distribution are working on new technical approaches, it would be rather interesting to determine to what extent variations in basic assumptions at the household level would affect the principles of Central Place Theory or those of traffic generation models.

In a different problem area it is unquestionable that there are fundamental direct links to be explored between the micro-situation of the individual and the large scale aggregate outcome. I am referring to migration. In spite of the intuitive feeling among all workers in the field that micro-environmental factors are important in the decision to move, nearly all models involve only the extrapolation of current aggregate behavior. These observations are sufficient to illustrate that there is a purely theoretical case for taking a closer look at the individual human being in his situational setting. To do so would improve our ability to relate the behavior of small scale elements and large scale aggregates. Failure in this respect is a common, fundamental weakness of all social sciences. To focus "on the locational dimension of human activities," as Isard and Reiner [4] have argued the regional scientist is obliged to do, should be a point of departure as promising as most others, or perhaps more so, for tackling the problem of establishing coherence between the two ends of the scale.

The initial task is, I think, to eliminate imprecise thought processes which conceptually deceive us into handling people as we handle money or goods once we commence the process of aggregation. In order to illustrate this I would like to relate an experience which can hardly be unique. When I was three or four years old my father tried to teach me the principles of banking and we trotted along to the local establishment to deposit what I had accumulated in my savings box, including a very shiny silver crown. The next day I insisted on walking back to the bank to make sure that the people had really guarded my money. The clerk was very understanding and showed me the correct mix of coins. But the shiny crown was not among them and it could not be produced. I decided that savings banks did not really save money.

It was primitive economics to assume that banks should worry about the identity of coins. Is it advanced or primitive social science to disregard the identity of people over time in the same fashion? This is what we do in most cases when we treat a population as a mass of particles, almost freely interchangeable and divisible. It is common to study all sorts of segments in the population mass, such as the labor force, commuters, migrants, shoppers, tourists, viewers of television, members of organizations, etc., each segment being analyzed very much in isolation from the others. As one of my students put it, "we regard the population as made up of 'dividuals' instead of individuals." Of course, we cannot focus on every single individual in the aggregate. We have to leave it to the historian to concern himself with biographies of sample individuals. But on the continuum between biography and aggregate statistics, there is a twilight zone to be explored, an area where the fundamental notion is that people retain their identity over time, where the life of an individual is his foremost project, and where aggregate behavior cannot escape these facts.

With a concern for the individual, it follows that we need to understand better what it means for a location to have not only space coordinates but also time coordinates. It might be quite reasonable to eliminate time by concealing it in costs of transportation and storage, as long as the handling of material is the main concern of locational analysis. But it is hardly reasonable to do so when the problems of people are brought in. When, for example, in a general equilibrium model, it is assumed that every individual performs a multitude of roles, it is also implicitly admitted that location in space cannot effectively be separated from the flow of time. Sometimes, of course, an individual plays several roles at the same moment. But more often the roles exclude each other. They have to be carried out within a given duration, at given times and places, and in conjunction with given groups of other individuals and pieces of equipment. They may have to be lined up in non-permutable sequences.

Of equal importance is the fact that time does not admit escape for the individual. He cannot be stored away for later use without complications for himself or society. As long as he is alive at all, he has to pass every point on the time-scale. Every point in space does not demand the same of him; he need only be somewhere in an environment which grants at least minimum conditions for survival. But this 'somewhere' is always critically tied to the 'somewhere' of a moment earlier. Jumps of non-existence are not permitted. To argue that time has to be taken into account along with space does not necessarily mean that studies of change and of development trends should take precedence over examinations of equilibria and steady states. See Stewart [9]. It means primarily that time has a critical importance when it comes to fitting people and things together for functioning in socio-economic systems, whether these undergo long-term changes, or rest in something which could be defined as a steady state. What I have in mind is the introduction of a time-space concept which could help us to develop a kind of socio-economic web model. The model would be asked what sorts of web patterns are attainable if the threads in the web (i.e., the individuals) may not be stretched beyond agreed levels of 'livability.' And when I speak of a web model, this is not just a metaphoric expression but a way of indicating what kind of mathematics one would need in order to handle it. Let me try to illustrate these ideas in an informal and surely very 'half-baked' fashion. I will not be concerned with a research technique. I am stressing a point of view by indicating the outlines of a model presently under study. As you will see, the various traditional concepts will be tied up in only a few packets with new labels.

In time-space the individual describes a *path*, starting at the point of birth and ending at the point of death. (Inanimate things also follow time-space paths but the characteristics of these are excluded here although they are needed in the complete web model). The concept of a life path (or parts of it such as the day path, week path, etc.) can easily be shown graphically if we agree to collapse three-dimensional space into a two-dimensional plain or even a one-dimensional island, and use perpendicular direction to represent time. In a Garden of Eden in which life was so entertaining that we did not even feel the need for regular rest, with a continually pleasant climate, ubiquitous self-replacing fruits to consume, and no social responsibilities, the path could be a true time-space random walk. In a more earthly environment it cannot be so, even if some drop-outs would have us believe otherwise. Assuming that continued survival is the first choice of those who have already set out on their life path, then some sort of counter-randomness programming has to occur.

When Robinson Crusoe found himself alone on his island, he could make up his program without regard to a pre-existing socio-economic system. The natural resources were all his to develop under his specific set of biological and technical constraints. An individual who migrates into an established society, either by being born into it or by moving into it from outside, is in a very different position. He will at once find that the set of potentially possible actions is severely restricted by the presence of other people and by a maze of cultural and legal rules. In this way, the life paths become captured within a net of constraints, some of which are imposed by physiological and physical necessities and some imposed by private and common decisions. Constraints can become imposed by society and interact against the will of the individual. See Vining [11]. An individual can never free himself from such constraints. To migrate during a pressing situation involves substituting a known pattern of constraints for one which is largely unknown. And, being a forward-looking animal, the individual probably tries to compare not just the prevailing situation but the anticipated situation in the life-perspective of himself and members of his family.

Several different ways of investigating the socio-economic web come to mind. One is to sample life paths. Biologists found this to be useful a long time ago when they invented the world-wide system of bird-banding. In countries with a continuously up-dated population register, it would be feasible (after computerization) to sample paths between dwellings on a very broad scale. Some experiments in that direction have already been made. See Jakobsson [5]. But it would be difficult to dig deeply enough to unveil the really critical events. Similarly, the short-term paths, days and weeks, can be sampled by observation or by some diary method. In either case, one risks becoming lost in a description of how aggregate behavior develops as a sum total of actual individual behavior, without arriving at essential clues toward an understanding of how the system works as a whole. It seems to be more promising to try to define the time-space mechanics of constraints which determine how the paths are channeled or dammed up. Some authors believe that the study of negative determinants might be the safest kind of social science. In the following pages, I am going to look at the matter entirely from the point of view of constraints.

Even if many constraints are formulated as general and abstract rules of behavior we can give them a 'physical' shape in terms of location in space, areal extension, and duration in time. Even a universal rule such as, "thou shalt not kill," means that a set of configurations of paths are not permitted, except in war and in traffic. It would be impossible to offer a comprehensive taxonomy of constraints seen as time-space phenomena. But three large aggregations of constraints immediately present themselves. The first of these could be tentatively described as 'capability constraints,' the second as 'coupling constraints,' and the third as 'authority constraints.'

'Capability constraints' are those which limit the activities of the individual because of his biological construction and/or the tools he can command. Some have a predominant time orientation, and two circumstances are of overwhelming importance in this connection: the necessity of sleeping a minimum number of hours at regular intervals and the necessity of eating, also with a rather high degree of regularity. Both needs determine the bounds of other activities as continuous operations. Other constraints are predominantly distance oriented, and as a consequence, enable the time-space surrounding of an individual to be divided up into a series of 'concentric' tubes or rings of accessibility, the radii of which depend on his ability to move or communicate *and* on the conditions under which he is tied to a rest-place. The inner tube or ring covers the small volume which the individual can reach with his arms from a fixed place such as a position at a machine or desk. It follows him as a shadow when he moves. Two such tubes can never be brought to coincide completely but they have to be close to coincidence for procreation, nursing, and some kinds of playing and fighting. Handtools can enlarge this tube but normally not by very much. Food has somehow to be brought inside the tube at regular intervals.

The second tube is defined by the range of the voice and the eye as combined instruments of communication. The boundary is by no means sharp but it is clear that the convenient spatial size of this entity varies between the normal living room and the assembly hall or its outdoor counterpart, the agora of the Greek city, for example. Historically, this uninstrumented tube has had a tremendous significance for the chosen forms of social, political, military and industrial organization. It was only after the introduction of the loudspeaker that really big outdoor political rallies became practicable. I am sure that we are still far from understanding the locational implications of the next enlargement of the range of this tube (i.e., telecommunications), which have entirely broken up this once so narrow spatial boundary. One hears the most divergent opinions about future possibilities of having television screens substitute for face-to-face meetings around a table. The amount of travelling undertaken by functionaries these days indicates that a breakthrough in terms of new behavior patterns is still on the waiting list. The two kinds of time-space compartments just mentioned have, to a small extent, been systematically studied by biologists, psychologists, and sociologists. But mostly they have remained the practical concern of architects, engineers, and time-and-motion experts.

The next tube in the hierarchy bring us directly into the business of regional science. People need to have some kind of home base, if only temporary, at which they can rest at regular intervals, keep personal belongings and be reached for receiving messages. And once a place of this sort has been introduced, one can no longer avoid considering more closely how time mixes with space in a non-divisible time-space. Assume that each person needs a regular minimum number of hours a day for sleep and for attending to business at his home base. When he moves away from it, there exists a definite boundary line beyond which he cannot go if he has to return before a deadline. Thus, in his daily life everybody has to exist spatially on an island. Of course, the actual size of the island depends on the available means of transportation, but this does not alter the principle.

Improvements in transport technology have enlarged the size of the island considerably over the centuries. The difference in range between the walker and the motorist is tremendous. For the flyer the entity has been broken up into an archipelago of smaller islands around the airports which are within reach. While in the air, he is imprisoned in a narrow time-space tube without openings and he does not therefore effectively exist in the geographic locations over which he is flying. During the era of more primitive transport technology, the population was nearly homogenous with respect to daily range. Today differences between groups within the same area and differences between areas can be very great. On most days the effective size of an individual's island is much smaller than the potential size which is delineated by his ability to move. The purposes of movement from the home base

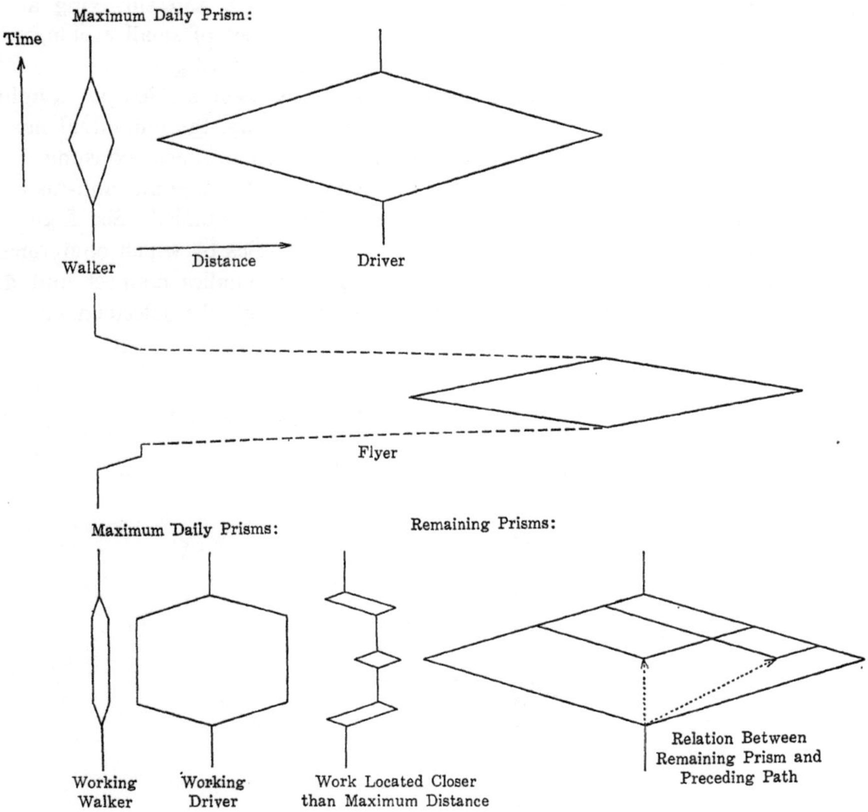

FIGURE 1. Daily Prisms

include going to work, collecting goods, meeting other people, etc. If we look closer at the time-space volume within reach, it turns out to be not a cylinder but a prism. It not only has a geographical boundary; it has time-space walls on all sides. See Figure 1. Depending on where the stops are located and how long they last, the walls of the prism might change from day to day. However, it is impossible for the individual to appear outside the walls. Every stay at some station means that the remaining prism is shrinking in a certain proportion to the length of the stay. A stay at a work place for eight hours might cause the remaining prism to disappear entirely if the stopping point lies at a maximum distance from the home base. A more normal situation for a week day in a Western society would be one in which the remaining prism breaks up into three portions, one in the morning before work, one at the lunch hour, and one in the evening after work.

Wherever the location and duration of stops inside the daily prism, the path of the individual will always form an unbroken line inside the prism without backward loops. He cannot pass a certain point in time-space more than once but he always has to be at some point. Over a lifetime he steers his path through a string of daily prisms, growing in radius during earlier years of his life and shrinking at an advanced age. Life becomes an astronomically large series of small events, most of which are routine and some of which represent very critical gates.

The path inside the daily prism is to a pronounced degree ruled by 'coupling constraints.' These define where, when, and for how long, the individual has to join other individuals, tools, and materials in order to produce, consume, and transact. Here, of course, the clock and the calendar are the supreme anti-disorder devices. We may refer to a grouping of several paths as a 'bundle.' See Figure 2. In the factory, men, machines, and materials form bundles by which components are connected and disconnected. In the office, similar bundles connect and disconnect information and channel messages. In the shop, the salesmen and the

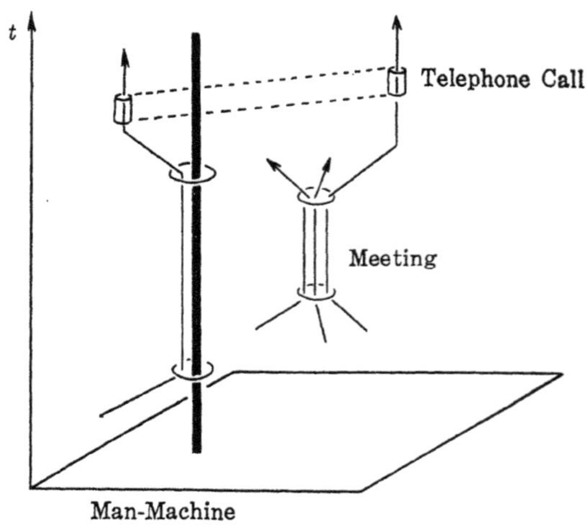

FIGURE 2. Grouping of Several Paths

customer form a bundle to transfer articles and in the classroom, students and teachers form a bundle to transfer information and ideas. Bundles are formed according to various principles. Many follow predetermined time-tables, often the same, weekday after weekday. This principle, which exists in the factory and the school, generally operates over the head of the participating individual. His freedom lies in his choice of work or place of work. After that, he has to obey the choreography of his superior, as long as he wants to maintain this contractual arrangement. The schoolchild, however, does not have the freedom to select in most cases. And always, families have to adjust to compulsory timetables.

Shops, banks, doctors and barbers permit random access between given hours. In many functions, particularly the managerial ones in firms and organizations, bundles have to be formed and located some time in advance in a kind of trial-and-error fashion. Today, hoards of administrators and secretaries spend their regular working hours trying to get other people together for future meetings. The more principles of participation come into fashion, the more this business will expand. Appointments seem to be moving more and more into the future, indicating a growing congestion. A person, who wants great freedom to maneuver, now has to extend his programming ahead by a year to eighteen months. The bundles formed with family members and friends are subject to private administration during the time remaining after outside demands and associated transport requirements. Private administration does not mean that the bundles are entirely outside the general social and legal control.

The bundles tend to be closely interdependent because individuals, materials, and bits of information have to move from one to the other in an orderly way. (The principles of maximum packing would be an interesting area of research related to critical path analysis.) An individual, bound to his home base, can participate only in bundles which have both ends inside his daily prism and which are so located in space that he has time to move from the end of one to the beginning of the following one. This means, for example, that if a doctor holds his clinic during the working hours of his patient, the latter cannot see the doctor except by obtaining permission to be absent from work. It is also clear that the car-owner, because of his random access to transport, has a much greater freedom to combine distant bundles than the person who has to walk or travel by public transportation. The difference is not so much a matter of speed as one of loss of time at terminals and junctions. See Figure 3.

A further kind of bundle deserves some passing comment. Telecommunication allows people to form bundles without (or nearly without) loss of time in transportation. Radio and television are of interest in this connection mostly because they take time from alternative activities. Everyone can jump on and off the bundle as he likes. But the telephone has a great significance from the point of view of social organization. It is true that a call may save much time, especially when it concerns the arrangement of future meetings. But at the same time, it is an outstanding instrument for breaking other activities. So one may sometimes wonder about the net outcome. In this regard, a world-wide dialing system seems to be a mixed blessing, since all too often people may forget differences in local time around the globe.

The third family of constraints, which I would like to discuss, relates to the time-space aspects of authority. The world is filled with a device which we may call the 'control area' or 'domain.' These words are essentially spatial. However, I would suggest that the concept of a domain be redefined to refer to a time-space entity within which things and events are under the control of a given individual or a given group. The purpose of domains (they are almost natural phenomena and many animals have them) seems to be to protect resources, natural as well as artificial, to hold down population density, and to form containers which protect an efficient arrangement of bundles, seen from the inside point of view of the principal. In time-space, domains appear as cylinders the insides of which are either not accessible at all or are accessible only upon invitation or after some kind of payment, ceremony, or fight. Some smaller domains are protected only through immediate power or custom, e.g., a favorite chair, a sand cave on the beach, or a place in a queue. Others, of varying size, have a very strong legal status: the home, land property, the premises of a firm or institute, the township, county, state, and nation. Many of these have a long, almost permanent duration, such as nations, British universities, and Japanese companies. Others are only temporary such as a seat in the theater or a telephone booth at the roadside.

Thus, there exists a hierarchy of domains (see Figure 4) and certain kinds are beyond escape. Those who have access to power in a superior domain frequently use this to restrict the set of possible actions which are permitted inside subordinate domains. Sometimes they can also oblige the subordinate domains to remove constraints or to arrange for certain activities against their will. Decision-makers in domains on equal or nearly equal levels cannot command each other. They have to influence each other by trading, by negotiation, or (in primitive cases) by invasion and warfare. Gaining access to power within a domain is a problem that may be solved in a variety of ways, of which only some are economic in the ordinary sense.

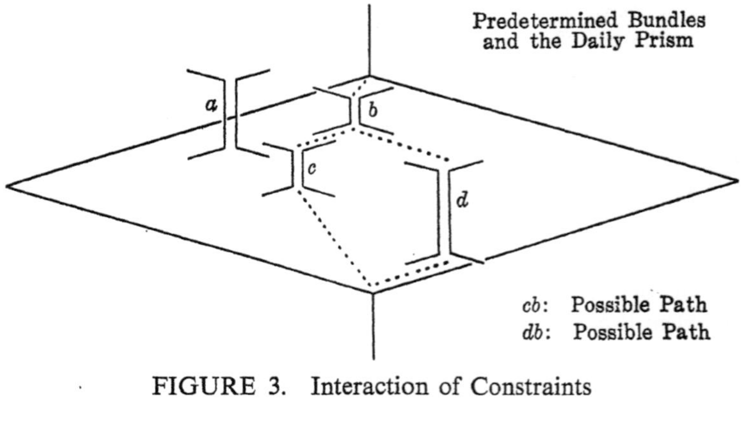

FIGURE 3. Interaction of Constraints

The three aggregations of constraints (i.e., capability, coupling, authority) interact in many ways; in direct obvious ways, and in indirect ways which are less easily detectable. See Figure 4. A few cases are discussed as illustrations. It is obvious that a low-income job, compared to a higher income job, gives access to fewer or inferior domains. Inability to rent a dwelling close to a place of work may, in the first instance, directly lead to long commuting times but may also lead to more concealed repercussions such as incursions on the time available for other activities. It may well be that the low rate of participation in cultural activities by large groups of people has less to do with the lack of interest than the prohibitive time-space locations of dwelling, work, and cultural activities. Even in nations where medical care is free, considerable numbers of people do not get their intended share. The reasons for this could be similar.

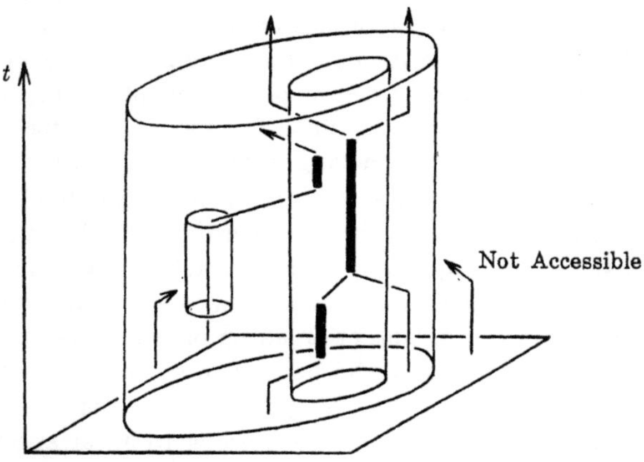

FIGURE 4. Hierarchy of Domains

Of special interest in terms of complicated interactions are the dependent members of families. The child has a small daily prism, unless a parent can spend a good deal of the day taking him from place to place. This means that the coverage and quality of the local training establishments and types of social contacts in the neighborhood will have very long-term effects on the life paths, since training as well as friendship ties provide the keys which open or close the gates to domains later in life. In all probability, the manner in which things are arranged for the child also affects the spatial structure of both the population composition and the labor market. Anderson [1], writing on location of residential neighborhoods, points out that, "if more satisfactory arrangements were devised for providing for the children of working mothers, then many larger families now suburban, might move into more central locations."

Important for interregional relations is the birth and death of jobs in relation to the effect of the life-path system on prism distance. First, as Self [8] has observed, "It is false to think of regional balance simply in terms of numbers of jobs, when a better index is the *range* of jobs . . . ." Thus, if training and the spectrum of jobs do not match in a given time perspective then the need for migration inevitably arises. Migration, of course, is not necessarily a bad thing, unless it detracts from living standards in the exporting and/or importing areas. The timing of complementary events is also of importance here. One can find cases where there is a balance over the year in the number of new jobs available on one hand, and demand for jobs on the other, and still much outmigration occurs. Job opportunities may arise at times of the year other than exactly when a group of impatient young people are finishing school and are searching for jobs. Similar observations could be made with respect to dwellings. The influence of migration is not confined to the mover and his dependents: there is also an effect on the situation around him. An out-migrant partially breaks up an established network and removes some skills, information and purchasing power. This does not mean that the situation is always made worse by migration. It might improve the situation as would be the case when the remaining population has more elbow room in a formerly crowded rural area. Furthermore, the in-migrant may cause positive or negative external effects. There are surely cases where migration might assume forms which disorganize communities. We know very little, I suppose, about the convenient proportion between the stable and the moving parts of a population.

A society is not made up of a group of people which decides in common what to do a week ahead of time. It consists primarily of highly institutionalized power and activity systems. A majority of domains and bundles within them have a location in space, a duration over time, and a composition according to consciously or habitually preestablished programs of organization which are made up with no particular regard to the individuals who happen to enter these systems and play the needed roles for portions of their life-paths. A company, a university, and a government department are structured according to an arrangement which exists as a time-space pattern, even if the people are not there. The same is true of the multitude of barriers and channels formed by legislation, administration (e.g., taxation), entries to professions, maximum speeds on roads or building codes. In total, seen from the point of view of the individual, this is an enormous maze about which he personally can do very little. Of course, there is a slow response in the system to people's reactions and this means that the set of domains and bundles changes over long periods of time. One might perhaps say that technology, which changes the capability constraints, is the prime mover. Domains and bundles thus change position in time-space. New units are born, existing ones grow in size, dwindle, or die. However, because so many domains have a very strong legal status and consequently, a long life (as for example, units of land ownership or municipal boundaries) and because so much is constrained within buildings having typically long lives, reactions often seem (from a system-wide point of view) local and not very purposeful. It is sufficient to note that despite the new ranges created by improved transportation, local government units have tended to remain medieval in size. A farmstead attitude to the domain problem survives in political life, in strong contrast to the sophisticated conception of space which industry tends to possess.

Viewed in a time-space perspective, then, we have two diverse systems in interaction. One is the predominantly time-directed warp of individual life-paths, which make up the population of an area and the concomitant capability constraints. The other is the more space-oriented set of imposed constraints of domains and bundles to which the individual may or may not have access according to his needs and wants. The population forms a kind of traffic flow in a road net with generally rusty gates. Loose ends of life-paths either have to discover for themselves how to open new roads and domains, locally or after migration into more permissive regions, or they disappear. I think it is true to say that the system of domains is much better understood with respect to flows of goods and money than with respect to flows of people. Social scientists know very little about interactions of constraints, as seen from the point of the life-path of the individual. In the main, people are viewed as parts of activities to be perfomed within each domain in isolation, and not as entities who need to make sense out of their paths between and through domains. It may well be that the more we, as optimizers, estimate efficiency within domains in the use of bundles of people, machines, materials and information, the more loose ends, which do not know how to move on, will appear in the population flow. Extrapolated to the limits, the question of life-paths between domains has certain strange, even repulsive, aspects. If the transplantation of hearts becomes a standard surgical procedure, then a continued high rate of accidents in traffic and in industry will be necessary to maintain the balance.

When asked about the concept of livability, people would express very different opinions. Nevertheless, I do not think it would be an entirely impossible task to make up a widely acceptable list of items which are fundamental for survival, comfort, and satisfaction. The individual who saw his life-path as an eighty year scheme, would need these items to be distributed along the time axis in characteristic ways. Thus, to consider the simplest items, it would be necessary or desirable, as the case may be, to have access: to air and a dwelling continuously, to food several times a day, to some daily and weekly recreation, to play and training during early years, to security of work and continuity of education at irregular intervals during a career, to assistance at an advanced age, and at all times random access to means of transportation, relevant information, and medical care. However, access involves much more than the simple juxtaposition of supplies in regions of arbitrary size. It involves a time-space location which really allows the life-path to make the required detours. It further involves the construction of barriers, physical, legal, economic and political, which serve to give to everyone his full share of the fundamental requirement listed. The study of livability would need a large political science component, but one which does not hesitate to look into the micro-manifestations of power. In this latter area exist the direct links between the macro and micro realms, links which have been largely unexplored by regional scientists. As mentioned before, those who have access to power in domains use much of their energy within their area of competence to superimpose (or sometimes remove) constraints on activities in lower level domains. At least such upper echelons as national, regional, and municipal governments and sometimes big organizations tend to do this in a formalistic way without much understanding in terms of time-space interactions for the population involved. Therefore, even with the best of intentions, outcomes are often questionable.

Given the list of needs and their statistical biography, it would be the task of the analyst to try to find out how much of various kinds of livability items would be simultaneously attainable under various assumptions of technical, economic and social organization. And since, once born, everybody has to be somewhere, everybody should be included in the picture, the child as well as the entrepreneur. This means, for example, that calculations of the demand for medical care must be seen as a function of the total population's state of health and not as a function of revealed demand. A time-space web model in the sense of a flow of life-paths, controlled by given capabilities and moving through a system of outside constraints which together yield certain probability distributions of situations for individuals, should, in principle, be applicable to all aspects of biology, from plants to animals to men. However, although some animals make buildings, defend domains, and believe in social rank, it is only man who can to a large extent choose between different constraints and, by restricting the number of offspring, even control the size of the population flow. The choice of constraints has always been a very piecemeal affair, more like a natural process than conscious planning. History and cultural anthropology show that it is possible to live under a tremendous variety of constraint systems, even if all have specific drawbacks, as seen from the viewpoint of individual survival and welfare.

The striking drawback of the so-called developed industrial society has been, and perhaps still is, the poverty problem, i.e., the fact that large groups of people have continued to live at the margin of famine or at least below what to people with a sense of fairness seems to be an acceptable standard. Systematic studies of poverty, started in Britain toward the end of the nineteenth century, eventually led to the concept of the Welfare State. Perhaps because of the initial limitations of their goals, even the best conceived versions of the Welfare State are not well prepared to cope with the new forms of poverty problems which are tending to affect everybody, e.g., ugly landscapes, simultaneous overcrowding and loneliness, alienation from crucial decision-making in work and society, etc. It seems that the main focus of our practical problems are moving away from the allocation of money towards the physical allocation of the uses of space and time.

Neurath [7] suggested some decades ago that we should be looking at, "markets and finance and at the whole reckoning in money as an institution like any other, such as funeral rites, golf, rowing, and hunting. To regard money as a historically given institution does not involve any objection to its use—though there may be such objections—but an objection to the application of arguments, valid in the field of higher bookeeping, to the analysis of social problems and human happiness in general." Now, when looking at the other methodological extreme, we also do not get very far by running around and questioning people about their likes and dislikes. First of all, we need some way of finding out the workings of large socio-environmental mechanisms. To me, a physical approach involving the study of how events occur in a time-space framework is bound to yield results in this regard. In order to be realistic, our models would have to recognize the fact that the individual is indivisible and that his time is limited. Further, we would have to note that the individual in dealing with space not only considers distance, but also has a strong (and perhaps logically necessary) drive towards organizing space in sharply bounded territories.

It was said before that the choice of constraints has always been a piecemeal affair. Even in theoretical studies, social scientists have tended to take most of them for granted according to available experience. With a suitable technique for grouping constraints in time-space terms, one could perhaps hope to be able to boil down their seemingly tremendous variety into a tractable number. Simulation comes to mind as a way of analysis until more general mathematical tools become available. Reasonably good simulations should improve our ability to survey whole systems and help to reduce the considerable trial and error component in applications. A purely theoretical, even artistic, satisfaction for the regional scientist would then be the ability to invent entirely fictitious societies which were still founded on realistic first principles. The technological forecasts which edify us these days and often seem so promising, at least superficially, cry out for instruments which could help us to judge the impacts on social organization and thereby the impact on the ordinary day of the ordinary person.

REFERENCES

[1] Anderson, T. R. "Social and Economic Factors Affecting the Location of Residential Neighborhoods," *Papers and Proceedings of the Regional Science Association*, Vol. 9 (1962), pp. 161–170.

[2] Fox, K. A. and T. K. Kumar. "The Functional Economic Area: Delineation and Implications for Economic Analysis and Policy," *Papers of the Regional Science Association*, Vol. 15 (1965), pp. 57–85.

[3] Harris, B. "The City of the Future: the Problem of Optimal Design," *Papers of the Regional Science Association*, Vol. 19 (1967), pp. 185–195.

[4] Isard, W. and T. A. Reiner. "Regional Science: Retrospect and Prospect," *Papers of the Regional Science Association*, Vol. 16 (1966), pp. 1–16.

[5] Jakobsson, A. "Omflyttingen i Sverige 1950–1960," *Meddelanden från Lunds Universitets Geografiska Institution*, Avhandlingar 59 (1969).

[6] Lowry, I. S. "Comments on Britton Harris" [3], *Papers of the Regional Science Association*, Vol. 19 (1967), pp. 197–198.

[7] Neurath, O. "Foundations of the Social Sciences," *International Encyclopedia of Unified Science*, Vol. 2, No. 1. Chicago: University of Chicago Press, 1944.

[8] Self, P. "Regions: the Missing Link," *Town and Country Planning*, Vol. 36 (June 1968), pp. 282–283.

[9] Stewart, J. Q. "Discussion: Population Projection by Means of Income Potential Models," *Papers and Proceedings of the Regional Science Association*, Vol. 4 (1958), pp. 153–154.

[10] Ullman, E. L. "The Nature of Cities Reconsidered," *Papers and Proceedings of the Regional Science Association*, Vol. 9 (1962), pp. 7–23.

[11] Vining, R. "An Outline of a Stochastic Model for the Study of the Spatial Structure and Development of a Human Population System," *Papers of the Regional Science Association*, Vol. 13 (1964), pp. 15–40.

[2]

Accessibility and intraurban travel

S Hanson, M Schwab
Graduate School of Geography, Clark University, Worcester, MA 01610, USA
Received 8 August 1985; in revised form 22 July 1986

Abstract. This paper contains an examination of the fundamental assumption underlying the use of accessibility indicators: that an individual's travel behavior is related to his or her location vis-à-vis the distribution of potential activity sites. First, the conceptual and measurement issues surrounding accessibility and its relationship to travel are reviewed; then, an access measure for individuals is formulated. Using data from the Uppsala (Sweden) Household Travel Survey and controlling for sex, automobile availability, and employment status, the authors explore the relationship between both home- and work-based accessibility and five aspects of an individual's travel: mode use, trip frequencies and travel distances for discretionary purposes, trip complexity, travel in conjunction with the journey to work, and size of the activity space. From the results it can be seen that although all of these travel characteristics are related to accessibility to some degree, the travel-accessibility relationship is not as strong as deductive formulations have implied. High accessibility levels are associated with higher proportions of travel by nonmotorized means, lower levels of automobile use, reduced travel distances for certain discretionary trip purposes, and smaller individual activity spaces. Furthermore, the density of activity sites around the workplace affects the distances travelled by employed people for discretionary purposes. Overall, accessibility level has a greater impact on mode use and travel distance than it does on discretionary trip frequency. This result was unexpected in light of the strong trip frequency-accessibility relationship posited frequently in the literature.

That there exists an intimate and reciprocal relationship between transportation and land-use patterns has become part of the conventional wisdom quickly absorbed by every novice transportation student. The individual's location vis-à-vis the distribution of potential activity sites has long been considered an important determinant of travel behavior. Altshuler (1979, page 374), for example, states confidently that "trip-making patterns, volumes, and modal distributions are largely a function of the distribution of activities".

One indication of the pervasiveness of this idea is the widespread use of accessibility measures in transportation planning, highway routing, and the siting of public facilities (for example, Morris et al, 1979; Bach, 1980; Koenig, 1980). Implicit in the calculation and use of accessibility indicators is the conviction that individuals with different levels of accessibility will exhibit distinctly different travel patterns, especially different trip frequencies. Theoretical formulations of trip generation reflect this same assumption; they have either assumed or deduced a close connection between travel and accessibility. Sheppard (1980), for example, has constructed an elaborate deductive model showing that the demand for travel is a function of accessibility to opportunities. Similarly, Ghosh and McLafferty's (1984) model of multipurpose shopping trip frequency yields the conclusion that "the rate of multipurpose shopping depends on the consumer's location (transport cost) vis-à-vis shopping opportunities" (page 249). The reasoning behind these deductively reached conclusions (and indeed behind the use of accessibility measures) is the following: because individuals with high levels of accessibility can reach many places at relatively low cost, they will make more trips than will people with similar sociodemographic characteristics but lower levels of access.

Although there is a large body of literature on methods of measuring accessibility, few researchers have questioned the fundamental assumption that underlies the construction and use of such measures. One reason for the dearth of studies relating accessibility to individual travel patterns might be the prevailing tendency (especially in the past) to compute access indicators for areas rather than for individuals, a practice that would hamper efforts to relate individual travel characteristics to accessibility levels. In fact, very little is known about the impact of relative accessibility on trip frequencies, trip-purpose mixes, or destination choices at the aggregate level, and "even less is known about [the] implications [of accessibility] at the level of individual or household behavior" (Wachs and Koenig, 1979, page 706).

In this paper we first review the conceptual and measurement issues surrounding accessibility and its relationship to travel. We then present the results of an empirical test of the notion that accessibility (that is, the individual's location vis-à-vis opportunities within an urban area) is related to certain aspects of an individual's intraurban travel pattern. Such an empirical investigation should have both theoretical and applied value. With regard to the theoretical value, there is a growing literature on the determinants of travel behavior, and any theory of urban travel must consider the role of the spatial environment in the individual's travel decisions. On an applied level, an improved understanding of the relationship between accessibility and travel would be useful in formulating policies in such areas as energy conservation and public transportation. For example, in many studies of energy consumption and urban form [recently reviewed by Talarchek (1984)] it has been assumed that lower density environments are associated with longer travel distances and, therefore, with higher per capita energy consumption. These studies have, however, lacked the data on nonmotorized travel necessary to examine the trade-offs people make between motorized and nonmotorized travel at different densities (for example, see Cheslow and Neels, 1980).

The travel-accessibility relationship

For many years the primary goal of transportation planning was to increase the population's vehicular mobility, and in pursuit of this goal planners focused on evaluating highway network and speed alternatives (Dalvi, 1979). Dalvi has traced the interest in accessibility indicators to a concern for equity issues in making such evaluations, because accessibility indicators have been seen as a way to measure "what is possibly the major function of cities: i.e. [the provision of] opportunities for easy interaction or exchange" (Koenig, 1980, page 169).

In a number of recent studies it has been demonstrated how access measures can be used to evaluate the distribution and quality of opportunities provided to urban residents. Accessibility indicators have been used, for example, to gauge the equity of potential locations for public and private facilities (for example, Knox, 1978; Bach, 1980; Jones and Kirby, 1982), to assess the net utility of alternative road networks (Koenig, 1980), and to evaluate the availability of employment opportunities to different groups, for example, men and women (Wachs and Kumagai, 1973; Black and Conroy, 1977; Black et al, 1982).

Implicit in such studies is the notion that there exists a fundamental and significant relationship between accessibility (which can be construed as a measure of 'potential for travel') and actual travel behavior. At the *inter*urban scale there is some evidence in support of the idea that spatial structure variables (for example, the density of activity sites) affect travel patterns (for example, Daniels and Warnes, 1980, chapter 3; Fotheringham, 1981). But the use of accessibility indicators at the *intra*urban level seems to assume that this relationship is transferable across scales.

Most studies of access at the intraurban scale have concentrated on developing appropriate measures of accessibility but have not gone on to relate these to observed travel patterns. Those that have investigated this relationship have concentrated primarily on trip generation rates as the dimension of travel that would most be affected by accessibility, and these studies have obtained mixed results.

Trip frequency

If distance of the residence from the central business district (CBD) can be taken as a valid surrogate for the density of opportunities around the residence (and therefore used as a surrogate for accessibility level), then there is some evidence that low accessibility levels are related to lower trip frequencies (for example, Von Rosenbladt, 1972; Wiseman, 1975). Robinson and Vickerman (1976) also report a positive association between trip frequencies for shopping and accessibility levels. Perhaps the strongest support for an accessibility-trip generation relationship comes from Koenig's (1980) study of five French cities. Stratifying his sample by mode availability and stage in the life cycle, Koenig shows not only that different population subgroups have different levels of accessibility, but also that access is an important determinant of trip generation for nonworkers.

Contrary evidence comes from Fawcett and Downes (1977), Wermuth (1982), Brunso and Hartgen (1983), and Herz (1983, page 393)—all of whom report no relationship between variables measuring density around the residence and trip generation rates. Furthermore, in a study in which tripmaking was explicitly examined as a function of accessibility, Dalvi and Martin (1976) found no strong relationship between accessibility (measured as a function of interzonal distance and zonal retail employment) and the automobile trip generation rates of nonworkers. As these and other authors note, however, accessibility measures are strongly influenced by the measure of attractiveness used. Because the location of *employment* opportunities might not be a good surrogate for the locations of other (nonwork) activity sites, it is perhaps not especially surprising that Dalvi and Martin found only a weak relationship between overall trip generation for nonworkers and access measures based on employment locations.

The existence of a strong positive relationship between trip generation and accessibility measures is still, therefore, uncertain. The frequency of discretionary tripmaking is, however, only one (though certainly the most studied) way in which we might expect accessibility levels to affect individuals' travel activity patterns. Although the literature offers very little guidance as to how access might affect other aspects of travel, we hypothesize that accessibility levels are also related to mode use, discretionary travel distances, trip complexity, discretionary travel carried out in conjunction with the journey to work, and the overall size of the individual's activity space. We briefly explain our reasoning for each as follows.

Mode use

It seems reasonable to propose that people living in areas with many activity sites will conduct more of their travel on foot or by bicycle, whereas those living in low-density areas will make greater use of the automobile. There is some support for this idea in the literature, though most studies have lacked data on nonmotorized travel and therefore have been unable to address the mode use-accessibility relationship directly. In Cheslow and Neels's (1980) study of the relationship between travel and urban form, for example, neighborhood density was found to affect vehicle trip frequency (with a lower proportion of automobile trips being made by people living in high-density neighborhoods); this suggests that walking trips are probably substituted for vehicular trips in higher density areas, but Cheslow and Neels did not have the data on nonmotorized trips necessary to test this idea.

Also, in an appendix to their paper, Brunso and Hartgen (1983) provide modal split data which show that in the period studied a higher proportion of trips on foot were made in downstate New York (that is, greater New York City) than in upstate New York (18.2% versus 7.0%). Thus, even if trip rates were found to be stable across accessibility levels, we would expect to find mode-use patterns varying according to an individual's accessibility.

Travel distance
If in fact closer destinations are preferred over farther ones, then there are grounds for expecting a relationship between travel distance and accessibility. People living (or working) in high-density environments will select destinations that are closer to home (or work) than would be the case were they living in low-density settings. In support of this idea, Cheslow and Neels (1980) report lower trip lengths in higher density neighborhoods. An individual's overall travel distance has, however, been shown to be highly correlated with the home-to-work distance (McLynn, 1976; as cited by Cheslow and Neels, 1980); the effect of accessibility on travel distances is therefore likely to be most evident in travel undertaken for discretionary purposes such as shopping and socializing. We expect people living in high-density areas to travel shorter distances for discretionary purposes than do those living in low-density areas.

Trip complexity
Trip complexity (the proportion of the individual's trips that are single stop versus multistop) has been shown to be an important dimension of travel (Hanson and Hanson, 1981). Because individuals with high levels of accessibility should be able to reach activity sites with less effort than those living in low-density areas, we expect that they will make a higher proportion of their journeys as single-stop trips. This is so because people living in high-density settings, with, for example, shops close at hand, can satisfy needs as they arise, rather than delaying a trip until a long list has accumulated. They can acquire the proverbial quart of milk or loaf of bread with little effort. By the same line of reasoning, individuals who must travel farther (from either home or work) to engage in nonwork activities should make fewer trips, a higher proportion of which should be multistop in order to minimize the total effort expended. [The same line of thinking is outlined by Westelius (1973).]

Travel in conjunction with the journey to work
Individual-level accessibility measures are invariably calculated around the person's place of residence; yet for people employed outside the home, the distribution of shops, banks, restaurants, and so on around the workplace is also likely to have an impact on the individual's travel pattern. We therefore expect that the density of activity sites around the workplace will affect the amount of nonwork travel an individual does in conjunction with the work trip. A higher proportion of an employed person's discretionary stops should be made on the journey to or from work for those working in high-density locations. It follows that the relative density of opportunities about home and work might also be important, with those who work in high-density areas making many of their discretionary trips on working days (weekdays) rather than on weekends[1].

Size of the activity space
If accessibility level does affect travel distances (at least for discretionary purposes), then it should also have an impact on the overall size of a person's activity space.

[1] This would especially be the case for areas (such as the one studied here) where shops are not open weekday nights.

Fotheringham (1981) has established that aggregate spatial interaction models are context-dependent; that is, the density and spatial arrangement of destinations around an origin affect the size of the distance-decay parameter. This suggests that at the disaggregate level the spatial extent of an individual's activity pattern should be related to the distribution of potential destinations, but to our knowledge this notion has never been tested with disaggregate data at the intraurban scale. We expect that people with high accessibility levels will have spatially more restricted activity spaces than will those with low levels of access.

To sum up, although there is reason to expect a relationship between accessibility and actual travel, there is no coherent body of evidence specifying the nature of that relationship. Most of the studies that have sought to establish that accessibility affects travel have used trip frequency as the only measure of travel, but illuminating the relationship between accessibility and other aspects of travel should have important implications for theoretical and applied work. Any study of accessibility raises numerous methodological issues, and it is to these that we now turn.

Data

In this study we use data collected in the Uppsala (Sweden) Household Travel Survey to examine the relationship between access and travel behavior. The adult members of 278 households (constituting a stratified random sample, stratified by life-cycle stage) kept travel diaries of all out-of-home movements (including those made on foot) for thirty-five consecutive days [see Marble et al (1972) or Hanson and Hanson (1981) for details of the survey design]. In addition to the diary data, the Uppsala survey contained information on individual and household characteristics and on the location and type of all retail and service establishments in the Uppsala urbanized area. As the establishments have been geocoded to street addresses, these data can be used to calculate the desired accessibility measures for each individual in the sample.

Deriving an accessibility measure

Most accessibility measures involve calculating an access index for each zone or group of individuals. The formulation of such an indicator is by no means straightforward, as Bach (1981, page 957) points out:

> "Even though operational definitions for accessibility and access opportunity in the form of mathematical formulas might, to the superficial observer, suggest scientific objectivity, they in fact do no more than hide basic issues of spatial politics."

Important methodological issues include determining the appropriate level of aggregation, measuring the separation between destinations and origins, measuring the attraction of different destinations, and formulating the actual accessibility indicator. We describe here the approach we have taken on each of these.

Although accessibility measures are most commonly calculated for zones, it is clear that our problem formulation requires computing accessibility indices for individuals because we seek to relate each individual's relative location to his or her travel pattern. As Pirie (1979, page 303) has noted, zonal accessibility measures not only neglect the distribution of activity sites within the zone, but also essentially assume that all individuals within a zone face the same set of opportunities. Zonal measures also overlook the fact that individuals within any zone differ in their access to transport modes and therefore in their ability to reach activity sites (Pirie, 1979). Last, an individual-level access measure enables the investigator to generate measures of the traveller's accessibility to opportunities from the home and from the workplace, two important organizational nodes in the individual's travel pattern.

Another consideration is the measure of spatial separation to employ. There are many to choose from (for example, time or monetary cost; Euclidean, network, or perceived distance). Morris et al (1979) believe that systematic errors are associated with all approaches; they further contend that although a measure of perceived separation may be desirable on behavioral grounds, an objective measure is to be preferred on operational ones (page 94). Because Euclidean distance is the easiest to measure and is a good surrogate for other (nontransit) travel times and distances (see Bach, 1981), it is usually the preferred indicator and is the one used here.

The appropriate way to measure destination attractiveness has also been the subject of debate. When the focus is on access to nonwork sites, as it is here, attractiveness has usually been measured by the number of facilities per zone, where the type of facility depends on the focus of the study. Some, however, have used residential density as a surrogate for the density of potential sites for diverse kinds of activities (including shopping, working, and recreation) (for example, Allaman et al, 1982). Because we hypothesize that the effect of accessibility on the individual's overall travel pattern is likely to be particularly evident in discretionary travel, our attractiveness measure focuses on the density of sites for nonwork activities. And, because we seek to use a disaggregate accessibility index, it is necessary to know the precise location of these opportunities. We make use of an inventory of establishments, together with the Cartesian coordinates for the street address of each, to measure the density of opportunities round the individual's home and workplace.

Last, there is the issue of how these measures are to be combined in a model. Pirie (1979) reviews the strengths and limitations of different formulations of accessibility, including topological, gravity, and cumulative-opportunity measures. Here we employ an individual-level cumulative-opportunity index which provides the number of opportunities that can be reached within a given distance from a person's home or work. The attractiveness measure is the number of retail and service establishments, summed over one-half kilometer increments, from the origin. As Pirie (1979) has noted, the size of isochrone increments used in this sort of measure is invariably arbitrary; we chose 0.5 km annuli as being fine enough to capture spatial-structure-induced behavioral variations, yet large enough to measure significant increments in the number of opportunities[2]. The index measures accessibility to opportunities within 5 km of the individual's home (or work):

$$A_i = \sum_{n=1}^{10} \frac{R_n}{0.5n},$$

where

A_i is the accessibility index for the ith individual, and

R_n is the number of establishments within the nth annulus, that is, between $0.5n$ km and $0.5(n-1)$ km from the ith individual's home or work.

As the 0.5 km (0.31 mile) annulus is a relatively small increment, this cumulative-opportunity index is sensitive to small-scale changes in the opportunity structure. Because home and work locations are both important organizational nodes in the individual's travel activity pattern, the accessibility index is calculated separately for

[2] The increment of 0.5 km is significantly less than the 1.0 km distance noted by Jones and Kirby (1982, page 306) as the distance beyond which spatial interaction begins to diminish. Also, we prefer the use of a fine annulus to the use of zonal increments (employed in most other studies, for example, Black and Conroy, 1977; Doubleday, 1977; Koenig, 1980; Black et al, 1982) because it permits a more fine-grained spatial analysis.

each individual relative to home and, where appropriate, to work, providing one or two accessibility measures per person. This is in line with Wachs and Koenig's (1979) call for accessibility measures that include more than the widely used home-based ones.

Method

Because accessibility is not the only factor that affects travel behavior, we examine the travel-accessibility relationship while controlling for a number of other variables known to affect travel. Specifically, following Pirie (1979) and Koenig (1980), we identify subgroups of individuals on the basis of the sociodemographic and role factors that have been shown in previous studies to be important determinants of travel behavior. This enables the formulation of a separate model for each group, in recognition of the fact that all individuals do not have equivalent means of or opportunities for travel. It has been pointed out that many aspects of travel are strongly related to role differences, especially those measured by *sex* and *employment status* (Hanson and Hanson, 1981; Pas, 1984). Notable travel differences between workers and nonworkers, especially with regard to trip frequency and trip length, have been documented (Doubleday, 1977; Hanson and Hanson, 1981). The significance of *automobile availability* in explaining nonwork travel frequency is also well recognized (Paaswell et al, 1973; Doubleday, 1977; Hanson and Hanson, 1981). We therefore investigate the effect of accessibility on travel while controlling

Table 1. Relationships[a] between the home-based accessibility measure and travel variables[b].

Variable[c]	Men (222)				Women (266)			
	work (179)		no work (43)		work (145)		no work (121)	
	car (142)	no car (37)	car (21)	no car (22)	car (61)	no car (84)	car (44)	no car (77)
Mode use								
% nonmotorized	0.37[d]	0.41[d]	0.64[d]	0.32	0.43[d]	0.45[d]	0.29[e]	0.36[d]
% automobile	−0.30[d]	−0.15	−0.59[d]	−0.22	−0.36[d]	−0.12	−0.20	−0.23[e]
% car driver	−0.28[d]		−0.55[d]		−0.33[d]		−0.09	
Discretionary travel								
nonwork stops	0.15[f]	0.31[f]	0.36[f]	−0.27	0.16	0.07	0.23	0.06
shop/pb stops	0.06	0.28[f]	0.24	−0.27	0.11	0.04	0.11	0.04
social stops	0.03	0.15	0.54[d]	0.53[d]	0.25[e]	−0.05	0.27[f]	−0.09
km nonwork	0.02	−0.06	0.01	−0.33	0.13	−0.19[f]	0.05	−0.18
km shop/pb	−0.40[d]	−0.47[d]	−0.51[d]	−0.66[d]	−0.43[d]	−0.50[d]	−0.57[d]	−0.48[d]
km social	−0.14	−0.40	0.13	−0.72[d]	−0.01	−0.14	0.13	−0.30[d]
Trip complexity								
% short trips	0.12	0.16	0.30	0.21	0.08	0.31[d]	0.09	0.20[f]
% long trips	−0.18	−0.03	0.29	−0.32	−0.13	−0.32[d]	0.04	−0.32[f]
Travel in conjunction with work trip								
% weekdays	0.01	0.22	−0.16	−0.30	−0.31[d]	−0.03	0.38[d]	0.02
% work trip	0.07	0.08			0.11	0.18		
Size of activity space								
points to home	−0.31[d]	−0.58[d]	−0.32	−0.65[d]	−0.61[d]	−0.50[d]	−0.58[d]	−0.65[d]
points to centroid	0.00	0.11	−0.04	−0.20	0.06	−0.16	−0.40[d]	−0.47[d]

[a] Entries are correlation coefficients (r).
[b] Figures in parentheses are sample sizes.
[c] See table 2.
[d] Significant at the $p \leq 0.01$ level.
[e] Significant at the $p \leq 0.05$ level.
[f] Significant at the $p \leq 0.10$ level.

for sex, employment status[3], and automobile availability[4]. The eight groups into which the sample is divided (and the size of each) are shown along with the results in table 1.

The travel variables used in the analysis are listed in table 2. To determine how closely each of these is related to accessibility, we computed zero-order correlations[5] between accessibility and each travel variable, for each of the sample subgroups[6].

Table 2. Travel variables[a].

Variable[b]	Abbreviation
Mode use	
proportion of an individual's stops made by nonmotorized means (foot and bike)	% nonmotorized
proportion of stops made by automobile (car driver and car passenger)	% automobile
proportion of stops made as car driver	% car driver
Discretionary travel	
number of stops for discretionary (nonwork) purposes	nonwork stops
number of stops made for shopping and personal business	shop/pb stops
number of stops made for social functions	social stops
distance travelled for discretionary purposes	km nonwork
distance travelled for shopping and personal business	km shop/pb
distance travelled for social functions	km social
Trip complexity	
proportion of individual's trips that had one or two stops	% short trips
proportion of trips with more than four stops	% long trips
Travel in conjunction with work trip	
proportion of individual's stops that were made on weekdays	% weekdays
proportion of nonwork stops that were made on the work trip	% work trip
Size of activity space	
average distance from destinations (weighted by contact frequency) to home	points to home
average distance from destinations (weighted by contact frequency) to centroid of activity space	points to centroid

[a] All travel variables measure an individual's behavior over a thirty-five-day period.
[b] A trip was defined as a home-to-home circuit, consisting of one or more intermediate stops.

(3) Workers were defined as those employed outside the home for at least 10 hours per week.

(4) Neither the number of driver's licenses nor household automobile ownership is alone a sufficient indicator. The family vehicle may be tied up on the work trip for most weekdays or multiple vehicles may be available but there is only one driver. The measure of automobile availability used here is the number of automobiles owned by the household divided by the number of persons in the household with a valid driver's license. All individuals with both a license and either full-time or part-time access to one or more vehicles are classified in the 'car' category. Those without either a car or a license are classified as 'no car'.

(5) Inspection of the scattergrams revealed no distinctly nonlinear relationships.

(6) In an effort to make our results as comparable as possible with those of Koenig (1980), we also investigated the relationships using an accessibility measure similar to the one he describes in his equation (3) (page 148). Ours is similar to his except that our measure of separation is distance, not travel time:

$$A_i = \ln \sum_{n=1}^{10} (R_n - 0.5n) ,$$

where R_n is the number of establishments in the nth annulus. The results of these analyses were so close to the results using the simpler, more easily interpreted, previously defined access measure that we report here only the results using our original access measure.

Results

Before investigating the travel-accessibility relationship, we examined the frequency distribution and the spatial distribution of the accessibility index. A look at the frequency distribution of the accessibility measure within each group indicates that the measure is normally distributed and that there is a good deal of variation in the measure from person to person; as others (for example, Wachs and Kumagai, 1973; Forer and Kivell, 1981) have shown for larger cities, considerable interpersonal variability in accessibility levels exists even within the relatively smaller city of Uppsala (population 120000). When the home-based accessibility index is mapped for the individuals in each of the subgroups, it is clear that distance of the residence from the CBD is a reasonably good surrogate for a person's accessibility to retail and service establishments (figure 1). This pattern is to be expected because at the time of the study Uppsala was close to being an archetypical single-center city.

The most remarkable finding overall was that the relationship between an individual's accessibility level and his or her travel behavior was relatively weak. Of the five aspects of travel examined [(1) mode use; (2) discretionary travel; (3) trip complexity; (4) travel in conjunction with the work trip; and (5) size of activity space], all were related to accessibility to some extent, but the relationship was only modest in most cases. The relationships that did emerge as being statistically significant weave an interesting pattern, however (see tables 1 and 3), and it is to these that we now turn.

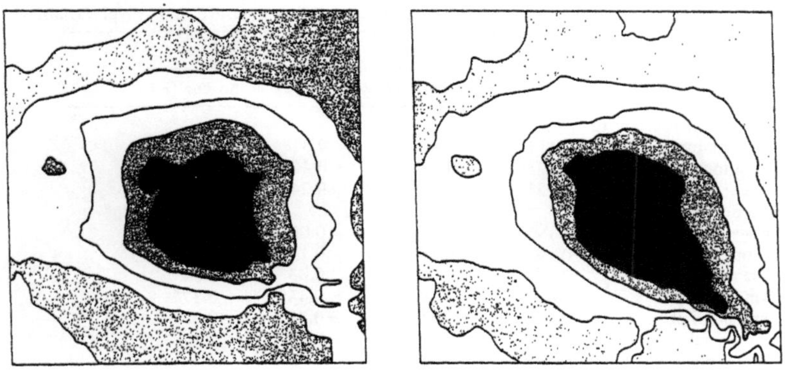

(a) (b)

Figure 1. Accessibility levels (A_i) for (a) working women with cars and (b) working women without cars. Contour interval is 200. The darker the shading, the higher the accessibility level.

Mode use

Among the strongest relationships identified were those between the access measure and the mode use variables. In particular, people with many opportunities close to home make a higher proportion of their stops with nonmotorized modes (on foot or by bicycle) (see table 1). Nonmotorized modes are also more likely to be used on the journey to work by individuals having higher work-based accessibility (table 3). (Note that this is for all trips made on the home-to-home circuit that include a stop at work.) Especially for those with cars, higher accessibility levels are also associated with lower levels of automobile use. This relationship does not hold, however, for nonworking women with automobiles available. These women might be discouraged from substituting foot or bike for automobile because of their larger average family size[7] (and therefore larger number of parcels to carry) and because of the presence of young children on trips.

Discretionary travel

Accessibility level was found to have relatively little impact on overall discretionary *trip frequency* (see nonwork stops in tables 1 and 3). The relationship between home-based access and the number of discretionary stops is statistically significant for three groups of men; and, for working men with cars, work-based accessibility affects discretionary trip frequency. It is interesting that these relationships are significant for men but not for women; the indication is that women undertake discretionary travel regardless of their access levels, whereas for men the discretionary trip frequency is, to a limited degree, sensitive to accessibility level. The correlation coefficients are unimpressive, though, especially in light of the strong positive trip frequency-accessibility relationship that has been assumed to exist. Similarly, the r for the relationship between both home- and work-based accessibility and the number of stops for shopping and personal business purposes are uniformly low. High home-based access is associated with higher social trip frequencies for a few groups (especially unemployed men), but the relationship is insignificant for half of the groups. In sum, the trip frequency results lend support to those (for example, Fawcett and Downes, 1977; Wermuth, 1982; and Herz, 1983) who found little empirical basis for the claim that accessibility level and trip frequency are closely (and positively) related.

Table 3. Relationships[a] between the work-based accessibility measure and travel variables[b].

Variable[c]	Men		Women	
	car (142)	no car (37)	car (61)	no car (84)
Mode use on journey to work				
% nonmotorized	0.41[d]	0.44[d]	0.31[e]	0.21[f]
% automobile	−0.29[d]	−0.38[e]	−0.42[d]	−0.27
% car driver	−0.25[d]		−0.24[e]	
Discretionary travel				
nonwork stops	0.19[e]	0.15	0.02	0.05
shop/pb stops	0.16	0.16	0.14	0.15
social stops	0.07	0.01	0.03	−0.04
km nonwork	−0.49[d]	−0.50[d]	−0.44[d]	−0.37[d]
km shop/pb	−0.07	−0.15	−0.08	0.22[e]
km social	−0.06	−0.22	−0.09	0.11
Trip complexity				
% short trips	−0.14	−0.14	−0.16	−0.23[e]
% long trips	0.12	0.22	0.18	0.20[f]
Travel in conjunction with work trip				
% weekdays	0.15	0.22	0.13	0.20[f]
% work trip	0.03	−0.07	−0.20	−0.09
Size of activity space				
points to centroid	−0.42[d]	−0.38[e]	−0.34[d]	−0.15

[a] Entries are correlation coefficients (r).
[b] Figures in parentheses are sample sizes.
[c] See table 2.
[d] Significant at the $p \leqslant 0.01$ level.
[e] Significant at the $p \leqslant 0.05$ level.
[f] Significant at the $p \leqslant 0.10$ level.

(7) The average family size for nonworking women with cars was 3.3, compared with an average for the other groups of 2.4.

The effect of accessibility on *travel distances* for certain discretionary purposes is, however, a different story. Although the relationship between travel distance and home-based accessibility is not strong when all discretionary travel is considered, it *is* so for travel distances associated specifically with shopping and personal business: for all the groups, higher home-based access levels mean lower travel distances for these purposes (see km shop/pb in table 1), and for nonworking men and women without cars high home-based accessibility is associated with lower travel distances for social purposes (see km social in table 1). Moreover, workers with a high density of opportunities around the workplace have lower overall discretionary travel distances than do workers with lower work-based accessibility levels (see km nonwork in table 3), although work-based access affects distance travelled for shopping and personal business only for women without access to cars. As expected, social travel is unrelated to the density of opportunities around the workplace.

There is some evidence then that whereas higher density settlement patterns do not have a dramatic impact on trip frequency, they do yield reduced travel distances. Even for the travel distance-accessibility relationship, however, the *r*, though statistically significant, are not as large as might be expected from the amount of attention that intraurban accessibility measures have commanded.

Trip complexity

Although the directions of the relationships between accessibility and the travel complexity variables are as hypothesized, insignificant associations are revealed for most of the groups. Higher levels of home-based accessibility are related to higher proportions of short trips only for the two groups of women without access to automobiles; this may reflect (1) women's greater responsibility for travel related to household maintenance (for example, shopping and personal business), which would explain why the relationship is not found among men without cars, and (2) the effect of a carrying constraint (which would explain why the relationship is insignificant for women with cars). For women without cars there is also a significant (negative) correlation between home-based accessibility and the proportion of very long trips (those with five or more stops). Also, this is the only group of workers whose trip complexity is at all related to the density of opportunities around work. Accessibility level does then appear to affect the trip complexity for women without access to automobiles.

Travel in conjunction with the journey to work

There is some evidence that accessibility around home and work affects the timing and location of tripmaking, although the correlation coefficients are again not large despite being statistically significant. *Working* women with cars and with a high density of opportunities around home make a lower percentage of their trips on weekdays, a higher proportion on the weekends; this implies that these women make discretionary trips in the vicinity of home (on the weekends) rather than making them in the vicinity of work. For *nonworking* women with cars, however, a higher home-based accessibility level means that a higher proportion of their travel occurs during the week (see % weekdays in table 1). Home-based accessibility is unrelated to the proportion of discretionary travel that is undertaken in conjunction with travel to work (see % work trip in table 1).

More surprising, though, is the fact that the proportion of an individual's discretionary stops that are made on the work trip is not related to the density of opportunities around the workplace (table 3). As mentioned above, however, high work-based accessibility is associated with lower overall discretionary travel distances (see km nonwork in table 3); the indication is, then, that although work-based accessibility does not affect discretionary trip frequency on the work trip it does affect discretionary travel distances.

Size of the activity space

The centrographic measures used to assess the size of each individual's activity space were computed on the point set of destinations visited during the thirty-five-day study period, each destination being weighted by frequency of contact. The centroid of this two-dimensional weighted-point distribution is akin to the mean of a one-dimensional frequency distribution. For every group but one, the average distance from weighted points to home is significantly (and inversely) related to the home-based accessibility index; a higher density of opportunities around home means, essentially, a smaller activity space (see points to home in table 1). It is interesting that the average distance from weighted points to centroid is unrelated to the home-based accessibility measure for individuals in all of the groups *except* the two groups of nonworking women. In other words, the activity space centroid of women who are not in the labor force is close to home; for people in the other groups it is not (see points to centroid in table 1).

For three groups of *workers* (all except for women without cars), however, the average distance from points to centroid *is* significantly related to the *work*-based accessibility measure, indicating that individuals working in high-density areas make so many of their trips to or near to the workplace that the centroid of the activity space is in fact close to work (and not to home) (see table 3). These results are in line with the findings for discretionary travel distance reported above: people working in high-density areas make their discretionary stops closer to the workplace than do those working in lower density areas. Moreover, the finding that the activity space centroid is closer to the workplace for people working in high-density settings emphasizes the importance of considering accessibility vis-à-vis the workplace as well as vis-à-vis the home.

Conclusions

The question posed at the outset was whether an individual's accessibility to activity sites had any discernable impact on the person's travel pattern. Few have posed this question before at the intraurban scale, and none to our knowledge has examined it with diverse aspects of travel behavior, has looked at the impact of accessibility on travel over some extended period of time, or has considered the impact of both home- and work-based accessibility on travel. Many have, however, assumed a strong travel-accessibility link; among these are studies of travel and energy consumption (particularly those with a predictive bent), which have been long on scenario building and short on data that would permit specification of the relationship on which much of the scenario rests. For these reasons, then, the results of this study should hold considerable interest.

The results are mixed. On the one hand, the evidence here suggests that the travel-accessibility relationship is not as strong as the existing literature would lead us to believe. Through our analysis of the relationship between accessibility indices and measures of travel over a five-week period, it was shown that, even where the relationship is statistically significant, it is in most cases surprisingly weak. On the other hand, the pattern of findings in these weak relationships is fascinating; high densities are related to more nonmotorized travel, less travel by automobile, reduced travel distances for certain discretionary purposes, and a more spatially restricted activity sphere. Here is some evidence, then, that higher density settlement patterns could yield energy savings.

Our finding of only a relatively weak travel-accessibility relationship is essentially in line with the results of other studies that have enabled an assessment of the relative importance of spatial/environmental versus sociodemographic variables in affecting travel patterns: the shape of the spatial environment is generally less influential on travel than are the personal and household characteristics of travellers (for example, Hanson, 1982; Recker and Schuler, 1982; Wermuth, 1982; Pas, 1984). It is not surprising then that the results reported here suggest that the accessibility-travel relationship is mediated through the three sociodemographic characteristics that define the subgroups, with accessibility affecting only certain aspects of travel for each of the subgroups. For example, accessibility level affects the individual's propensity to make short (versus long) trips only for women without cars and affects the proportion of travel undertaken on weekdays (as opposed to weekends) only for women with cars. Similarly, accessibility is related to distance travelled for socializing only for nonworkers without cars.

The use of both a home-based and a work-based accessibility measure enables us to assess the relative importance of the two accessibility indices to observed movement patterns. Overall, the individual's travel is more affected by home-based accessibility; among employed people, for example, the home-based measure was related to distance travelled for shopping and personal business, whereas the work-based measure was not (except for women without cars). Work-based access was, however, linked to several aspects of travel, including mode use on the journey to work and distance travelled for discretionary purposes. Moreover, for three groups of workers the average distance between places visited (over the thirty-five days) and the activity space centroid was related to the work-based but not to the home-based access measure. It would seem productive, then, for future studies to include measures of work-based and perhaps other nonhome-based accessibility levels in addition to the widely used home-based ones.

Finally, our results suggest that the past focus on the relationship between accessibility and travel frequency might have been misguided, at least at the intraurban scale, for when all out-of-home movements are examined (including those made on foot) discretionary trip frequency is only weakly linked to accessibility level. Other aspects of travel behavior (particularly mode use and travel distances) are more closely related to accessibility than is trip frequency. This suggests that, at the intraurban scale, accessibility level does not need to be included in models of trip generation [as some (for example, Wachs and Koenig, 1979; Sheppard, 1980) have suggested it should be] but that accessibility should be considered for inclusion in models of mode choice and trip distribution.

References

Allaman P M, Tardiff T, Dunbar F C, 1982, "New approaches to understanding travel behavior" National Research Council, 2101 Constitution Ave, Washington, DC 20418

Altshuler A, 1979 *The Urban Transportation System: Politics and Policy Innovation* (MIT Press, Cambridge, MA)

Bach L, 1980, "Locational models for systems of private and public facilities based on concepts of accessibility and access opportunity" *Environment and Planning A* **12** 301-320

Bach L, 1981, "The problem of aggregation and distance for analyses of accessibility and access opportunity in location-allocation models" *Environment and Planning A* **13** 955-978

Black J, Conroy M, 1977, "Accessibility measures and the social evaluation of urban structure" *Environment and Planning A* **9** 1013-1031

Black J A, Kuranami C, Rimmer P J, 1982, "Macroaccessibility and mesoaccessibility: a case study of Sapporo, Japan" *Environment and Planning A* **14** 1355-1376

Brunso J M, Hartgen D T, 1983, "An update on household-reported trip generation rates" Transportation Analysis Report number 31, New York State Department of Transportation, Planning Division, Albany, New York 12232

Cheslow M D, Neels J K, 1980, "The effect of urban development patterns on transportation use" presented at the 59th Annual Meeting of the Transportation Research Board, January 1980; copy available from the first author, Evaluation Research Corporation, Vienna, VA 22180

Dalvi M Q, 1979, "Behavioral modelling, accessibility, mobility and need: concepts and measurement" in *Behavioral Travel Modelling* Eds D A Hensher, P R Stopher (Croom Helm, Beckenham, Kent) pp 639-653

Dalvi M Q, Martin K M, 1976, "The measurement of accessibility: some preliminary results" *Transportation* **5** 17-42

Daniels P W, Warnes A M, 1980 *Movement in Cities: Spatial Perspectives on Urban Transport and Travel* (Methuen, Andover, Hants)

Doubleday C, 1977, "Some studies of the temporal stability of person trip generation models" *Transportation Research* **11** 255-263

Fawcett F, Downes J D, 1977, "The spatial uniformity of trip rates in the Reading area, 1971" report 797, Transport and Road Research Laboratory, Crowthorne, Berks

Forer P C, Kivell H, 1981, "Space-time budgets, public transport, and spatial choice" *Environment and Planning A* **13** 497-509

Fotheringham A S, 1981, "Spatial structure and distance decay parameters" *Annals of the Association of American Geographers* **71** 425-436

Ghosh A, McLafferty S, 1984, "A model of consumer propensity for multipurpose shopping" *Geographical Analysis* **16** 244-249

Hanson S, 1982, "The determinants of daily travel-activity patterns: relative location and sociodemographic factors" *Urban Geography* **3** 179-202

Hanson S, Hanson P, 1981, "Travel-activity patterns of urban residents: dimensions and relationships to sociodemographic characteristics" *Economic Geography* **57** 332-347

Herz R, 1983, "Stability, variability, and flexibility in everyday behavior" in *Recent Advances in Travel Demand Analysis* Eds S Carpenter, P Jones (Gower, Aldershot, Hants) pp 385-400

Jones K, Kirby A, 1982, "Provision and wellbeing: an agenda for public resources research" *Environment and Planning A* **14** 297-310

Knox P L, 1978, "The intraurban ecology of primary medical care: patterns of accessibility and their policy implications" *Environment and Planning A* **10** 415-435

Koenig J G, 1980, "Indicators of urban accessibility: theory and application" *Transportation* **9** 145-172

Marble D F, Hanson P O, Hanson S E, 1972, "Household travel behavior study: field operations and questionnaires" The Transportation Center Research Report, Northwestern University, Evanston, IL 60201

Morris J M, Dumble P L, Wigan M R, 1979, "Accessibility indicators for transport planning" *Transportation Research* **13A** 91-109

Paaswell R E, Recker W W, Milione V, 1973, "A profile of a carless population" Department of Civil Engineering, State University of New York at Buffalo, Amherst, NY 14260

Pas E I, 1984, "The effect of selected sociodemographic characteristics on daily travel-activity behavior" *Environment and Planning A* **16** 571-581

Pirie G H, 1979, "Measuring accessibility: a review and proposal" *Environment and Planning A* **11** 299-312

Recker W W, Schuler H J, 1982, "An integrated analysis of complex travel behavior and urban form indicators" *Urban Geography* **3** 110-120

Robinson R V F, Vickerman R W, 1976, "The demand for shopping travel: a theoretical and empirical study" *Applied Economics* **8** 267-281

Sheppard E, 1980, "Location and the demand travel" *Geographical Analysis* **12** 111-128

Talarchek G M, 1984, "Energy and urban spatial structure: a review of forecasting research" *Urban Geography* **5** 71-86

Von Rosenbladt B, 1972, "The outdoor activity system in an urban environment" in *The Use of Time* Ed. A Szalai (Mouton, The Hague) pp 335-355

Wachs M, Koenig J G, 1979, "Behavioral modelling, accessibility and travel need" in *Behavioral Travel Modelling* Eds D A Hensher, P R Stopher (Croom Helm, Beckenham, Kent) pp 698-710

Wachs M, Kumagai T G, 1973, "Physical accessibility as a social indicator" *Socio-economic Planning Science* **7** 437-456

Wermuth M J, 1982, "Hierarchical effects of personal, household, and residential location characteristics on individual activity demand" *Environment and Planning A* **14** 1251-1264

Westelius O, 1973, "The individual's way of choosing between alternative outlets" National Swedish Building Research, Gävle, Sweden

Wiseman R F, 1975, "Location in the city as a factor in trip making patterns" *Tijdschrift voor Economische en Sociale Geographie* **66** 167-177

[3]

Individual Accessibility Revisited: Implications for Geographical Analysis in the Twenty-first Century

Mei-Po Kwan
Joe Weber

Analytical methods for evaluating accessibility have been based on a spatial logic through which the impedance of distance shapes mobility and urban form through processes of locational and travel decision making. These methods are not suitable for understanding individual experiences because of recent changes in the processes underlying contemporary urbanism and the increasing importance of information and communications technologies (ICTs) in people's daily lives. In this paper we argue that analysis of individual accessibility can no longer ignore the complexities and opportunities brought forth by these changes. Further, we argue that the effect of distance on the spatial structure of contemporary cities and human spatial behavior has become much more complicated than what has been conceived in conventional urban models and concepts of accessibility. We suggest that the methods and measures formulated around the mid-twentieth century are becoming increasingly inadequate for grappling with the complex relationships among urban form, mobility, and individual accessibility. We consider some new possibilities for modeling individual accessibility and their implications for geographical analysis in the twenty-first century.

1. INTRODUCTION

Accessibility has traditionally been conceptualized as the proximity of one location (whether zone or point) to other specified locations. Analytical methods for evaluating accessibility have been based on a spatial logic through which the impedance of distance shapes mobility and urban form through processes of locational and travel decision making. As a result, traditional models of urban form and accessibility are based upon a similar conceptual foundation and spatial logic, and the relationships

The authors thank the three anonymous reviewers for helpful comments that significantly improved this paper. Mei-Po Kwan would like to acknowledge the support of a grant from the Information Technology Research (ITR) Program of the U.S. National Science Foundation (NSF) and an Ameritech Faculty Research Grant. The assistance of the Geography and Regional Science Program of NSF is also gratefully acknowledged.

Mei-Po Kwan is an associate professor of geography in the Department of Geography, The Ohio State University. E-mail: kwan.8@osu.edu. *Joe Weber is an assistant professor in the Department of Geography, University of Alabama. Email:* jweber2@bama.ua.edu between models of urban form and conceptualizations of accessibility are inextricably intertwined.

Conventional concepts and measures of accessibility are useful for studying a variety of phenomena, especially the aggregate analysis of social groups within an area-based spatial framework. Gravity-based and cumulative-opportunity measures, for example, are helpful for identifying changes in the accessibility of different locations (place accessibility) and the effect of competition on access to urban opportunities (e.g., Shen 1998; van Wee, Hagoort, and Annema 2001; Wachs and Kumagai 1973). Nodal measures are also useful for addressing issues of accessibility within transportation or information networks (e.g., Lee and Lee 1998). These conventional accessibility measures have the general form $A_i = \Sigma W_j f(d_{ij})$, where A_i is the accessibility at location i, W_j is the weight representing the attractiveness of location j, d_{ij} is a measure of physical separation between i and j (in terms of travel time or distance), and $f(d_{ij})$ is the impedance function. The most commonly used impedance functions in gravity measures are the inverse power function and the negative exponential function, while an indicator function is used in cumulative-opportunity measures to exclude opportunities beyond a given distance limit. In nodal measures, the impedance function takes the form of an indicator function that reflects the presence or absence of a network link between two nodes.

These conventional measures are, however, less suitable for understanding individual experiences because of recent changes in four broad areas: (a) the processes that shape urban form and contemporary urbanism; (b) the complexities of and individual difference in human spatial behavior; (c) the availability of new technologies, especially GIS, and data for modeling individual accessibility; and (d) the increasing importance of information and communications technologies (ICTs) in people's everyday lives.

In light of these important changes, there is an urgent need to re-examine the concepts and methods in accessibility research, especially in the context of the lived experience of individuals. We suggest that analysis of individual accessibility can no longer ignore the complexities and opportunities brought forth by these changes. Further, we argue that the effect of distance on the spatial structure of contemporary cities and human spatial behavior has become much more complicated than what has been conceived in conventional urban models and concepts of accessibility. We suggest that the methods and measures formulated around the mid-twentieth century are becoming increasingly inadequate for grappling with the complex relationships among urban form, mobility, and individual accessibility. We consider some new possibilities for modeling individual accessibility and their implications for geographical analysis in the twenty-first century. We use the analysis of accessibility as an example to argue for the need to question the usefulness of many accepted notions and methods, and to indicate ways in which limitations of past studies may be overcome.

The next four sections of the paper discuss the changes outlined above and their implications for accessibility research and geographical analysis. They focus on the limitations of traditional urban models and concepts of accessibility. We then discuss an alternative approach to evaluating individual accessibility, using our recent studies on space-time accessibility measures as examples. These studies use individual-level, georeferenced activity-travel diary data and involve the development of dedicated GIS-based geocomputational algorithms and the establishment of a comprehensive geographic database of the study areas (including a digital street network and all land parcels). We examine the strengths of this approach and the ways in which it overcomes some of the limitations of traditional methods. The concluding section discusses the implications of this kind of research for geographical analysis in the twenty-first century.

2. URBAN FORM AND CONTEMPORARY URBANISM

For much of the twentieth century, most conceptualizations and formal models of urban structure have been based on a spatial logic that shapes individual mobility and urban form through the impedance of distance. In these models, distance between locations (e.g., home and workplace), expressed in terms of travel or transport cost and modeled with an impedance function, has been the central mechanism of spatial choice of individuals and firms. For example, proximity to shops, facilities, and services is a critical factor that determines individuals' residential and travel choice. This is the central mechanism underlying the idea of monocentric cities (Burgess 1925; Hoyt 1939; Harris and Ullman 1945; Alonso 1964; Muth 1969) and the related polycentric model (Garreau 1991; Vance 1964). In such models, commuting refers to movement from peripheral home locations to the CBD or suburban employment centers (Holzer 1991), and accessibility is understood largely as an attribute of places (defined mainly as distance from the CBD or other centers), not people.

Traditional models of urban form and accessibility are therefore based upon a similar conceptual foundation and spatial logic. Such representation of urban form continues to influence the description of urban patterns and travel behavior (Davis 1998; Marshall 2000). Although straightforward, such a conception does not represent the complexities either of real urban environments or of human spatial behavior at the beginning of the twenty-first century (Dear and Flusty 1998; Giuliano and Small 1991; Waddell and Shukla 1993). This is largely the result of the belated observation of new urban forms that cannot be easily explained by such an organizing principle. It is also observed that these new urban forms cannot be described adequately in the analytical terms of traditional urban models or accessibility concepts (e.g., in terms of distinct land use types in measurable economic or social activities; see the helpful discussion in Sui 1998). A variety of research has provided new insights into these topics, and has significant implications for the study and understanding of individual accessibility.

The most significant view is perhaps that distance as conventionally understood is of declining importance as an organizing principle of urban form and accessibility. For example, distance to the CBD or suburban centers appears to provide little explanation for variations in housing value or the level of office employment within cities (Archer and Smith 1993; Heikkila et al. 1989; Hoch and Waddell 1993; Waddell, Berry, and Hoch 1993). Researchers have also shown that both distance to employment centers and the geographical distribution of urban opportunities do not have a consistent relationship with individual accessibility (Weber 2003; Weber and Kwan 2002).

This decline in the importance of distance has been attributed to a variety of forces, such as the construction of freeway networks that have increased mobility and reduced the locational advantage of central locations (Giuliano 1989). The influence of distance has therefore been reduced by the creation of a more uniform and efficient transport system. Recent trends of globalization and the operations of multinational corporations also constitute a powerful force that shapes urban processes and form but transcends the geographical boundaries of contemporary cities. Some therefore argue that the urban changes observed in recent decades represent an entirely new pattern of urbanism that requires new perspectives. This view holds that contemporary cities should be conceptualized as having instead been shaped by the various locational logics of industries, firms, and individuals (e.g., Dear and Flusty 1998; Fishman 1990). Land use patterns and individual accessibility appear to be determined by much more complicated processes in contemporary cities, and the relationships between the spatial logic of traditional urban models and notions of accessibility are becoming increasingly unclear.

3. COMPLEXITIES OF HUMAN SPATIAL BEHAVIOR

In addition to a number of studies that have found mixed relationships between distance and land uses, property values, rents, or employment levels (e.g., Waddell, Berry, and Hoch 1993), it is increasingly difficult to understand human spatial behavior within contemporary cities in terms of distance as conceived in traditional urban models. Actual commuting distances, for example, strongly suggest that very few people are acting to minimize their journey to work by relocating either their home or workplace in the intraurban context (Hamilton 1982, Small and Song 1992). The relatively minor influence of distance as conceived in traditional urban models on residential location decision making can be largely explained by the fact that homes and workplaces are not simply points in space, but instead are tightly connected within a web of economic and social relations (England 1991; Gilbert 1998; Wachs et al. 1993; Waddell 1993). As a result, residential (and workplace) mobility is far from easy or automatic.

The emphasis of traditional urban models and concepts of accessibility on either home or workplace location is therefore incomplete. Relocating a home or job will likely incur significant social and economic costs, such as the difficulty in selling a home and/or obtaining new housing, losing contact with friends and neighbors, or switching schools, potentially resulting in a high degree of inertia or "spatial fixity" (England 1993). Home locations for many individuals may therefore tend to be fixed regardless of the presence of suitable nearby employment (especially for women) or changes in workplace location (Hanson and Pratt 1995). As a result, "at any given time, a large number of (rational) household and employment locations may in fact be 'suboptimal' with respect to transport cost" (Giuliano 1989, 152). This spatial fixity and a lack of mobility can be important to accessibility, not just because of local variations in mobility or opportunities, but because of the way they influence individuals' knowledge of cities, employment prospects, and attitudes.

The role of distance as conceived in conventional accessibility measures is of declining importance in shaping human spatial behavior in another sense. These measures have conceptualized accessibility largely in terms of the proximity of one's home and/or workplace (whether represented as a zone or a point) to other specified locations. Measuring access to employment or services from these locations, however, assumes that they are the center of an individual's daily activities and the origin of each individual's daily travel. This is not always appropriate as it denies the existence of considerable amounts of multi-stop trips over the course of the day, and a person may spend considerable time away from home or the workplace during the day (Kwan 1998, 1999a). The length of the commute is not simply the distance between home and the workplace because of these multi-purpose trip chains (Kitamura, Nishii, and Goulias 1990; Michelson 1985). By not taking into account the ways in which individuals combine various activities and destinations into a single trip (trip chaining), these conventional accessibility measures may be underestimating an individual's accessibility.

4. THE SPATIAL AND TEMPORAL FRAMEWORKS OF CONVENTIONAL ACCESSIBILITY MEASURES

Traditional urban models and measures of accessibility were formulated at times when modern GIS technology and digital geographic data of the urban environment and human spatial behavior were not available. Such limitations in tools and data had an inevitable influence on the type of models and methods formulated, and on the kind of real-world complexities represented by these models and methods—as how we conceptualize the world and represent places, distances, and time depends heavily on the available conceptual and operational tools. As a result, both the spatial and temporal frameworks of conventional proximity-based accessibility measures reflect many simplifications to which contemporary analysts no longer need to be restricted (because of the new capabilities of modern GIS technology and digital geographic data; see discussion in Kwan 2000a and Kwan et al. 2003).

4.1. The Spatial Framework of Conventional Measures

Conventional accessibility measures, for example, are often based on a zonal spatial framework as most data used in past studies were area-based (e.g., census tracts or traffic analysis zones). The notion of accessibility operationalized by these measures is therefore more a property of geographical entities such as zones than the experience of households or individuals. Households and individuals possess accessibility only by virtue of living at a particular location, which creates several problems for the use of these measures when evaluating individual accessibility (Pirie 1979). First, the zonal scheme used in a particular study may not correspond to individual perception of urban opportunities in contemporary cities, and accessibility measures based on such a spatial framework cannot capture the complexities arising from the effect of personal idiosyncrasies and individual perception of the geographical and temporal availability of urban opportunities (e.g., Fishman 1990; Kwan and Hong 1998).

Second, conventional zone-based accessibility measures cannot easily take into account certain types of individual differences even when disaggregate versions of these measures are used (although the effect of age, income, or access to different transport modes may be incorporated into disaggregate versions of these measures using individual-level data [e.g., Guy 1983]). This problem was shown in recent studies on the differential impact of space-time constraints on men's and women's spatial mobility and accessibility (e.g., Kwan 1999a, 2000b). These studies revealed that the failure to take into account the effect of individual space-time constraints on women's accessibility has rendered significant gender differences invisible.

Third, any use of zonal data is also subject to the modifiable areal unit problem (MAUP), in which zones of varying sizes and configurations will yield different results or relationships between accessibility and other characteristics. A related issue of the MAUP is to identify the proper scale of analysis, for which several methods exist (Green and Flowerdew 1996; Jones and Duncan 1996). However, even when accessibility has been measured at multiple scales, for example by measuring local access to neighborhood retail and grocery stores separately from regional access to major employment and retail centers (Handy 1992; Handy and Niemeier 1997), there is also no easy way to combine local and regional access into a single measure, or to determine how one may substitute for the other. Lastly, the issue of how to handle self-potential (the accessibility of a zone to itself) is also crucial when implementing gravity-based accessibility measures within a zone-based framework (Pooler 1987).

Fourth, another problem is the issue of representing and specifying the highly complex influence of distance on individual accessibility. A number of choices are possible, such as whether to use linear distance, travel times, or travel costs as a measure of distance, and whether to measure this distance as a straight line or through a transport network. While the representation of distance using Euclidean distance is straightforward, it is also not very accurate for most intraurban applications in which movement is confined to street networks. Further, few studies have considered the multi-modal characteristics of actual travel and trip making (e.g., one needs to walk to the car or to a transit station); while variations of travel speed among various parts of a city, road segments and times of the day are only beginning to be addressed in recent research (e.g., Weber and Kwan 2002).

4.2. The Temporal Framework of Conventional Measures

Time is an integral element of individual accessibility. This refers not only to the amount of time available to individuals for carrying out travel and activities but also to the scheduling of activities throughout the day (Forer and Huisman 2000). Given that men and women tend to have different time constraints on their activities, as well as a different temporal scheduling of activities throughout the course of the day, the absence of time from conventional measures ignores an important source of accessibility variation (Kwan 1998, 1999a, 1999b).

These measures also ignore the importance of time to mobility due to various types of delays while traveling, traffic congestion, or changes in transit schedules at different hours of the day, which has been significant in recent research (Weber and Kwan 2002, 2003). The varying availability of opportunities throughout the day associated with different patterns of business hours is also not represented. This potentially overestimates mobility and accessibility while ignoring that many opportunities will not be available in the evening. Further, not all opportunities within reach are relevant unless the time one can spend at the activity site exceeds the threshold required for meaningful participation in that activity (Kim and Kwan 2003). Conventional accessibility measures therefore take a static, timeless view of mobility and accessibility, which denies the ways in which behavior, activity patterns, and even population composition varies by time of the day (Goodchild and Janelle 1984; Harvey and Macnab 2000).

5. ICT AND INDIVIDUAL ACCESSIBILITY

The spatial and temporal frameworks of conventional accessibility measures encounter serious limitations when the impact of information and communications technologies (ICTs) on spatial behavior is recognized (Janelle and Hodge 2000). An important transformation that may result from the increasing use of ICT is the relaxation of many space-time constraints that limit human spatial mobility and activity space. For example, as many activities no longer need to be performed at certain places or times (e.g., through e-shopping), more time may become available for undertaking other activities, and more flexible spatial and temporal arrangements of human activities become possible.

Given the importance of space-time constraints in determining individual accessibility, this increase in the flexibility in activity scheduling—for example, conducting e-banking on weekends or in the evening rather than during a very limited span of time during weekdays—is likely to have a significant influence on individual accessibility. But this does not mean that space-time constraints no longer exist. Access to ICTs may be restricted to certain times of the day or week such as while at work or when public libraries are open. Further, the existence of time zones and human sleep patterns means that instantaneous communications can reach only a portion of the world's population at any given time (Harvey and Macnab 2000).

Associated with the relaxation of space-time constraints in individuals' everyday lives are changes in individual activity patterns and hence accessibility. The time people spend using ICTs takes time away from other activities, and there may be distinctive geographical consequences associated with such time displacement, which may be described as the space-time displacement effect of ICTs (Kwan 2002; Lee and Whitley 2002; Robinson et al. 2000). For example, if people spend more time using the Internet, they may spend less time on social activities; and as Internet users purchase goods online, they may spend less time shopping in and making trips to stores in the physical world (as found by Nie and Erbring 2000; UCLA-CCP 2003). Recent studies have also observed a considerable reduction in work-related travel and a contraction of activity space by telecommuters (Henderson and Mokhtarian 1996; Pendyala, Goulias, and Kitamura 1991; Saxena and Mokhtarian 1997).

It seems quite clear, however, that the potential impacts of ICTs on human behavior and individual accessibility are highly complex, with substitution for physical travel being only one of many possibilities (Gillespie and Richardson 2000; Graham 1998; Kitchin 1998; Mokhtarian and Meenakshisundaram 1999; Warf 2001). It has been asserted that the most important impact of ICT is that "it permits much more flexibility in whether, when, where, and how to travel, and thus loosening the constraint of having to be at a certain place at a certain time" (Mokhtarian 1990, 240). But it is far from clear how traditional urban models and accessibility measures can take the effect of constraint relaxation, space-time displacement, and telesubstitution into account.

Finally, the spatial and temporal frameworks of conventional accessibility measures cannot take into account the additional opportunities accessible to individuals through the use of ICTs. This issue arises because ICTs will give people greater ability to make use of space-adjusting technologies to overcome distance and therefore will increase their extensibility (Black 2001; Janelle 1973, 1995; Kwan 2000c). This means that ICTs will allow them to access information resources and participate in activities at the global scale although their physical activities are largely confined at the local scale. There are many issues, however, concerning how concepts of accessibility in the physical world can be extended to incorporate individual access to the opportunities in cyberspace (or cyberspatial accessibility). For example, what are the meanings of concepts like distance, impedance, attractiveness, or feasible opportunity set for cyberspace, and what space-time framework can be used to represent them (Kwan 2001)? As conventional measures are based on measures of distance in the physical world, they are not able to accommodate the increased accessibility that result from the use of ICTs. These issues are only beginning to be addressed by recent studies (e.g., Dodge 2000; Shen 1999). Further, the traditional temporal scale (hour/minute) is not adequate for studying individual accessibility in cyberspace, as cyberspatial activities may be accomplished within a few seconds.

6. ALTERNATIVE FRAMEWORKS FOR ANALYZING INDIVIDUAL ACCESSIBILITY

We have argued in previous sections that traditional urban models and accessibility measures are inadequate for understanding individual experiences in contemporary cities for various reasons. As the effect of distance is much more complex than before, and as current GIS technologies and digital geographic data allow us to incorporate and model real-world complexities in ways inconceivable before, we are no longer restricted to use traditional models or methods whose conceptual foundation was laid around the mid-twentieth century. In this section, we consider some new possibilities for modeling individual accessibility and their implications for geographical analysis in the twenty-first century. The discussion draws upon our recent research on space-time accessibility measures. While we believe that these measures are viable alternatives that may be used to overcome some limitations of conventional measures, we do not argue that they are the only good alternatives. Other new approaches for dealing with the complexities brought forth by the four recent changes discussed in the paper are also highly desirable.

6.1. Space-Time Accessibility Measures

Space-time accessibility measures are based on the time-geographic framework of Hägerstrand (1970), which conceives individuals' activities and travel as continuous trajectories or paths in three-dimensional space-time. These paths do not exist randomly in space-time but are subject to a range of personal and social constraints, including the limits on mobility resulting from the available transport technology and the biological need for resting time. For example, individuals must be in certain places for certain lengths of time for work or must arrive at or depart businesses by certain times. These times and places constitute "fixed" activities that cannot be ignored or rescheduled and so provide the framework of an individual's daily activities (Burns 1979). Activities that allow for more flexible scheduling, duration, or location must be fit into the time available between successive fixed activities

The mobility allowed by the fixity constraints that an individual faces can be described in the form of a space-time prism—though because of the difficulty of working with a three-dimensional prism, these have commonly been simplified into a two-dimensional projection in geographic space, known as a potential path area, or PPA (Dijst and Vidakovic 2000; Dijst, de Jong, and van Eck 2002; Lenntorp 1976, 1978). A potential path area is the geographic area that can be reached within the space-time constraints established by an individual's fixed activities. It is the area that an individual can physically reach after one fixed activity ends while still arriving in time for the next fixed activity. Each individual will create their own PPAs through their daily activities.

The approach offered by space-time measures—coupled with the geocomputational ability of GIS, the technology for massively parallel computing and individual-level activity-travel data—has potential for overcoming many limitations of conventional analytical frameworks. This is not just because the use of GIS may greatly enhance our ability to represent real-world complexities or allow for ever larger data sets. Rather, with GIS and increasingly available disaggregate data, highly refined space-time measures of individual accessibility can be operationalized, and the conceptualization and modeling framework are no longer conditioned by a priori schema of areal units or spatial frames. Concepts of space-time accessibility based on individual-level data have been widely implemented within vector and raster GIS (Forer and Huisman 2000; Kim and Kwan 2003; Kwan 1998, 1999a; Miller 1999; O'-Sullivan, Morrison, and Shearer 2000; Weber 2001, 2003; Weber and Kwan 2002, 2003), creating a resurgence of interest in these concepts. Several examples are provided below to illustrate the advantages of using space-time measures for evaluating individual accessibility in contemporary cities.

6.2. *Sensitivity to Individual Differences*

Perhaps the clearest example of the issues faced by conventional concepts and measures of accessibility can be seen by directly comparing the results produced by conventional and space-time measures within a contemporary U.S. city and examining their ability to identify important factors that influence individual accessibility. One such study examined access to commercial, educational, and recreational land uses for a sample of 39 men and 48 women in Columbus, Ohio (Kwan 1998). In the study, 20 conventional measures of the gravity and cumulative opportunity variants were evaluated using the home locations of the 87 individuals as origins and 10,727 property parcels representing destinations. Distances were computed using point-to-point travel times through a digital street network with 47,194 arcs and 36,343 nodes.

Three space-time measures were also computed for each individual using a geocomputational algorithm implemented in ARC/INFO GIS. Instead of providing different representations of the importance of distance for accessibility, these three measures evaluate the size of the space that can be reached, the number of opportunities that can be reached, and the size or attractiveness of those opportunities. The results of the study reveal the contrast between conventional and space-time measures. While the values produced by most gravity and cumulative opportunity measures were highly correlated and produced similar spatial patterns, space-time measures were very different.

Gravity measures tended to replicate the geographical patterns of urban opportunities in the study area by favoring areas near major freeway interchanges and commercial developments, while cumulative opportunity measures emphasized centrality within the city by showing the downtown area to be the most accessible place. In contrast, space-time measures produced different spatial patterns, and the patterns for men resembled the spatial distribution of opportunities in the study area while the women's patterns were considerably different. These gender differences, however, were invisible when using conventional accessibility measures, showing the ability of space-time measures to capture certain types of differences among individuals (e.g., the effect of space-time constraints on individual accessibility).

6.3. *Temporal Representation of the Urban Environment*

Recent research on space-time measures has also shown that facility opening hours and variable travel speeds at different times of the day and parts of the city are important to individual accessibility. Weber and Kwan (2002) examine the effect of travel time variations and facility opening hours on individual accessibility using a range of data. These include data of 101 men and 99 women from an activity-travel diary data set of Portland, Oregon; a digital street network with estimates of free flow and congested travel times (with 130,141 arcs and 104,048 nodes); and a comprehensive geographic database of the study area. This digital geographic database, containing 27,749 commercial and industrial land parcels, was used to represent potential activity opportunities in the study area.

The analytical procedures involved creating a realistic representation of the temporal attributes of the transport network and urban opportunities in the study area, as well as developing a geocomputational algorithm for implementing space-time accessibility measures within a GIS environment. The algorithm was developed and implemented using Avenue, the object-oriented scripting language in the ArcView 3.x GIS environment. Five space-time accessibility measures were computed. The first is the length of the road segments contained within the daily potential path area (DPPA). The second is the number of opportunities within the DPPA. The total area and total weighted area (to reflect the importance of major employment centers) of the land parcels within the DPPA were the third and fourth space-time accessibility measures computed. Finally, to incorporate the effect of business hours on accessibility measures, opportunity parcels were assumed to be available (and could therefore be accessible to an individual) only from 9:00 A.M. to 6:00 P.M. This creates a fifth accessibility measure.

The results show that link-specific travel times produce very uneven accessibility patterns, with access to services and employment varying considerably within the study area. The time of day that activities were carried out also has had an effect on accessibility, as evening congestion sharply reduced individual's access throughout the city. The effect of this congestion on mobility is highly spatially uneven. Further, the use of business hours to limit access to opportunities at certain times of the day shows that non-temporally restricted accessibility measures produce inflated values by treating these opportunities as being available at all times of the day. It is not just that incorporating time reduces accessibility, but that it also produces a very different, and perhaps unexpected, geography of accessibility (Weber and Kwan 2002). This geography depends much on individual behavior and so cannot be discerned from the location of opportunities or congestion alone. The study observed that the role of distance in predicting accessibility variations within cities is quite limited.

6.4. *Scale and Frame Independence*

While aggregate measures must contend with the MAUP due to their reliance on zones represented at particular scales, recent work with space-time measures suggests that accessibility is actually frameless; variations have little or no relationship with common spatial zones or distinct geographic scales. Evidence for the scale independence of space-time measures was discussed in Weber (2001) and Weber and Kwan (2003), in which multilevel modeling with individual-level data was used to examine the effect of geographical scale and zonal scheme on accessibility variations over the study area (Portland, Oregon). Three zonal schemes were used. In the first scheme, individuals were grouped according to residence within the cities of the study area. The second grouped individuals according to location within the school districts of the study area. In the last zonal scheme, individuals were grouped within the commutershed of the closest regional center or the CBD. The study shows that there are no substantial accessibility variations among discrete zones or at different scales within the study areas. It shows that the spatial configuration of zones would likely play a small part, if any, in MAUP issues. Selecting different neighborhood boundaries would therefore be unlikely to affect the parameter estimates or model fit.

In summary, space-time measures of individual accessibility have a number of advantages when compared to conventional measures. Accessibility is an attribute of individuals, who create it through their daily activities and movements. Time, space, and individual activity patterns are integral elements of these measures. They can also incorporate certain interpersonal differences that cannot be captured by conventional measures, even among those living in the same household. Space-time measures are built upon a conceptual foundation that corresponds more closely with theoretical expectations about urban form and human spatial behavior in contemporary cities. Rather than being proximity-based, they can be thought of as context-based measures that incorporate both the individual's own activities and constraints as well as characteristics of the individual's urban environment. Further, space-time measures may be adapted to take into account of individual access to opportunities in cyberspace (Kwan 2001). They also appear to be independent of the spatial scale or 'frames' used, and this framelessness suggests that the MAUP need not be a problem for accessibility analysis (Weber and Kwan 2003).

7. CONCLUSIONS

In this paper we suggest that the methods and measures formulated around the mid-twentieth century are becoming increasingly inadequate for grappling with the complex relationships among urban form, mobility, and individual accessibility. Distance, as the foundational construct of spatial analysis, is playing a declining but much more complex role in shaping contemporary cities and human behavior. We therefore argue for the need to go beyond conventional spatial and temporal frameworks and to rethink how we approach conceptualizing and evaluating individual accessibility. We call for approaches that are sensitive to the complexities of urban form and differences among individuals across multiple axes. Space-time accessibility measures are discussed as alternative methods that, to a considerable extent, overcome many limitations of the conceptual foundation and spatial and temporal frameworks of traditional models.

This examination of past studies on individual accessibility represents a challenge to many accepted notions and methods in geographical analysis. It is apparent that there is an urgent need not only to recognize the new opportunities offered by modern GIS technologies and digital geographic data, but also to recognize that we are no longer restricted to traditional models or methods whose conceptual foundation was established around the mid-twentieth century. GIS not only allows us to incorporate and model real-world complexities in ways inconceivable before, it also allows us to go beyond the simplifications necessitated by the use of conventional urban models and proximity-based accessibility measures. Consider the capabilities of GIS to model accessibility using the temporal attributes of more than 130,000 links of the transport network of a city (Weber and Kwan 2002), to take into account the internal structure of a building when considering human mobility (Church and Marston 2003), to visualize over 400,000 land parcels (Kwan 2000d) or GPS data with 800,000 space-time points (Kwan and Lee 2003) in a three-dimensional vector GIS environment. Recent research is just beginning to explore new representational possibilities like these. But how to materialize this potential and establish a new conceptual foundation for geographical analysis remains one of the most serious challenges for geographical analysis in the twenty-first century.

LITERATURE CITED

Alonso, W. (1964). *Location and Land Use: Toward a General Theory of Land Rent.* Cambridge, Mass.: Harvard University Press.

Archer, W. R., and M. T. Smith (1993). Why Do Suburban Offices Cluster? *Geographical Analysis* 25, 53–64.

Black, W. R. (2001). An Unpopular Essay on Transportation. *Journal of Transport Geography* 9, 1–11.

Burgess, E. W. (1925). The Growth of the City: An Introduction to a Research Project. In *The City*, edited by R. E. Park, E. W. Burgess, and R. D. McKenzie, pp. 47–62. Chicago: University of Chicago Press.

Burns, L. D. (1979). *Transportation, Temporal, and Spatial Components of Accessibility*. Lexington. Mass.: Lexington Books.

Church, R. L., and J. Marston (2003). Measuring Accessibility for People with a Disability. *Geographical Analysis* 35, 83–96.

Davis, M. (1998). *Ecology of Fear: Los Angeles and the Imagination of Disaster*. New York: Henry Holt and Company.

Dear, M., and S. Flusty (1998). Postmodern Urbanism. *Annals of the Association of American Geographers* 88, 50–72.

Dijst, M., and V. Vidakovic (2000). Travel Time Ratio: The Key Factor of Spatial Reach. *Transportation* 27, 179–99.

Dijst, M., T. de Jong, and J. R. van Eck (2002). Opportunities for Transport Mode Change: An Exploration of a Disaggregated Approach. *Environment and Planning B* 29, 413–30.

Dodge, M. (2000). Accessibility to Information within the Internet: How Can It Be Measured and Mapped? In *Information, Place, and Cyberspace: Issues in Accessibility*, edited by D. G. Janelle and D. C. Hodge, pp. 187–204. Berlin: Springer.

England, K. V. L. (1991). Gender Relations and the Spatial Structure of the City. *Geoforum* 22, 135–47.

———. (1993). Suburban Pink Collar Ghettos: The Spatial Entrapment of Women? *Annals of the Association of American Geographers* 83, 225–42.

Fishman, R. (1990). Megalopolis Unbound. *The Wilson Quarterly* 14, 25–45.

Forer, P., and O. Huisman (2000). Space, Time and Sequencing: Substitution at the Physical/Virtual Interface. In *Information, Place, and Cyberspace: Issues in Accessibility*, edited by D. G. Janelle and D. C. Hodge, pp. 73–90. Berlin: Springer.

Garreau, J. (1991). *Edge City: Life on the New Frontier*. New York: Doubleday.

Gilbert, M. (1998). "Race," Space, and Power: The Survival Strategies of Working Poor Women. *Annals of the American Association of Geographers* 88, 595–621.

Gillespie, A. and R. Richardson (2000). Teleworking and the City: Myths of Workplace Transcendence and Travel Reduction. In *Cities in the Telecommunications Age: The Fracturing of Geographies*, edited by J. O. Wheeler, Y. Aoyama, and B. Warf, pp. 229–45. New York: Routledge.

Giuliano, G. (1989). New Directions for Understanding Transportation and Land Use. *Environment and Planning A* 21, 145–59.

Giuliano, G., and K. A. Small (1991). Subcenters in the Los Angeles Region. *Regional Science and Urban Economics* 21, 163–82.

Goodchild, M. F., and D. G. Janelle (1984). The City around the Clock: Space-Time Patterns of Urban Ecological Structure. *Environment and Planning A* 16, 807–20.

Graham, S. (1998). The End of Geography or the Explosion of Place?: Conceptualizing Space, Place, and Information Technology. *Progress in Human Geography* 22, 165–85.

Green, M., and R. Flowerdew (1996). New Evidence on the Modifiable Unit Problem. In *Spatial Analysis: Modelling in a GIS Environment*, edited by P. Longley and M. Batty, 41–54. New York: John Wiley and Sons.

Guy, C. M. (1983). The Assessment of Access to Local Shopping Opportunities: A Comparison of Accessibility Measures. *Environment and Planning B* 10, 219–38.

Hägerstrand, T. (1970). What about People in Regional Science? *Papers of the Regional Science Association* 24, 7–21.

Hamilton, B. W. (1982). Wasteful Commuting. *Journal of Political Economy* 90, 1035–53.

Handy, S. (1992). Regional versus Local Accessibility: Neo-Traditional Development and Its Implications for Non-Work Travel. *Built Environment* 18, 253–67.

Handy, S., and D. A. Niemeier (1997). Measuring Accessibility: An Exploration of Issues and Alternatives. *Environment and Planning A* 29, 1175–94.

Hanson, S., and G. Pratt (1995). *Gender, Work, and Space*. London: Routledge.

Harris, C. D., and E. L. Ullman (1945). The Nature of Cities. *The Annals of the American Academy of Political and Social Sciences* 242, 7–17.

Harvey, A., and P. A. Macnab (2000). Who's Up?: Global Interpersonal Temporal Accessibility. In *Information, Place, and Cyberspace: Issues in Accessibility*, edited by D. G. Janelle and D. C. Hodge, pp. 147–70. Berlin: Springer.

Heikkila, E., P. Gordon, J. I. Kim, R. B. Peiser, and H. W. Richardson (1989). What Happened to the CBD-Distance Gradient?: Land Values in a Policentric City. *Environment and Planning* A 21, 221–32.

Henderson, D. K. and P. L. Mokhtarian (1996). Impacts of Center-Based Telecommuting on Travel and Emissions: Analysis of the Puget Sound Demonstration Project. *Transportation Research* D 1, 29–45.

Hoch, I., and P. Waddell (1993). Apartment Rents: Another Challenge to the Monocentric Model. *Geographical Analysis* 25, 20–31.

Holzer, H. J. (1991). The Spatial Mismatch Hypothesis: What Has the Evidence Shown? *Urban Studies* 28, 105–22.

Hoyt, H. (1939). *The Structure and Growth of Residential Neighborhoods in American Cities*. Washington, D.C.: Federal Housing Administration.

Janelle, D. G. (1973). Measuring Human Extensibility in a Shrinking World. *Journal of Geography* 72, 8–15.

———. (1995). Metropolitan Expansion, Telecommuting, and Transportation. In *The Geography of Urban Transportation, Second Edition,* edited by S. Hanson, pp. 407–34. New York: Guilford.

Janelle, D. G., and D. C. Hodge, eds. (2000). *Information, Place, and Cyberspace: Issues in Accessibility*. Berlin: Springer.

Jones, K., and C. Duncan (1996). People and Places: The Multilevel Model as a General Framework for the Quantitative Analysis of Geographical Data. In *Spatial Analysis: Modelling in a GIS Environment*, edited by P. Longley and M. Batty, pp. 79–104. Cambridge: Geoinformation International.

Kim, H.-M., and M.-P. Kwan (2003). Space-Time Accessibility Measures: A Geocomputational Algorithm with a Focus on the Feasible Opportunity Set and Possible Activity Duration. *Journal of Geographical Systems* 5, 71–91.

Kitamura, R., K. Nishii, and K. Goulias (1990). Trip Chaining Behavior by Central City Commuters: A Causal Analysis of Time-Space Constraints. In *Developments in Dynamic and Activity-Based Approaches to Travel Analysis*, edited by P. Jones. 145–70. Aldershot: Avebury.

Kitchin, R. M. (1998). Towards Geographies of Cyberspace. *Progress in Human Geography* 22, 385–406.

Kwan, M.-P. (1998). Space-Time and Integral Measures of Individual Accessibility: A Comparative Analysis Using a Point-Based Framework. *Geographical Analysis*: 30, 191–217.

Kwan, M.-P. (1999a). Gender and Individual Access to Urban Opportunities: A Study Using Space-Time Measures. *The Professional Geographer* 51, 210–27.

Kwan, M.-P. (1999b). Gender, the Home-Work Link, and Space-Time Patterns of Nonemployment Activities. *Economic Geography* 75, 370–94.

Kwan, M.-P. (2000a). Analysis of Human Spatial Behavior in a GIS Environment: Recent Developments and Future Prospects. *Journal of Geographical Systems* 2, 85–90.

Kwan, M.-P. (2000b). Gender Differences in Space-Time Constraints. *Area* 32, 145–56.

Kwan, M.-P. (2000c). Human Extensibility and Individual Hybrid-Accessibility in Space-Time: A Multi-Scale Representation Using GIS. In *Information, Place, and Cyberspace: Issues in Accessibility*, edited by D. G. Janelle and D. C. Hodge, pp. 241–56. Berlin: Springer.

Kwan, M.-P. (2000d). Interactive Geovisualization of Activity-Travel Patterns Using Three-Dimensional Geographical Information Systems: A Methodological Exploration with a Large Data Set. *Transportation Research* C 8, 185–203.

Kwan, M.-P. (2001). Cyberspatial Cognition and Individual Access to Information: The Behavioral Foundation of Cybergeography. *Environment and Planning B* 28, 21–37.

Kwan, M.-P. (2002) Time, Information Technologies and the Geographies of Everyday Life. *Urban Geography* 23(5), 471–82.

Kwan, M.-P., and X.-P. Hong (1998). Network-Based Constraints-Oriented Choice Set Formation Using GIS. *Geographical Systems* 5, 139–62.

Kwan, M.-P., and J. Lee (2003). Geovisualization of Human Activity Patterns Using 3D GIS-A Time-Geographic Approach. In *Spatially Integrated Social Science*, edited by M. F. Goodchild and D. G. Janelle (forthcoming) New York: Oxford University Press.

Kwan, M.-P., A. T. Murray, M. E. O'Kelly, and M. Tiefelsforf (2003). Recent Advances in Accessibility Research: Representation, Methodology and Applications. *Journal of Georaphical Systems* 5, 129–38.

Lee, H., and E. A. Whitley (2002). Time and Information Technology: Temporal Impacts on Individuals, Organizations, and Society. *The Information Society* 18, 235–40.

Lee, K., and H. Lee (1998). A New Algorithm for Graph-Theoretic Nodal Accessibility Measurement. *Geographical Analysis* 30, 1–14.

Lenntorp, B. (1976). *Paths in Space-Time Environments: A Time-Geographic Study of the Movement Possibilities of Individuals*. Lund Studies in Geography B: Human Geography. Lund: Gleerup.

———. (1978). A Time-Geographic Simulation Model of Individual Activity Programmes. In *Human Activity and Time Geography*, edited by T. Carlstein, D. Parkes, and N. Thrift, pp. 162–80. London: Edward Arnold.

Marshall, A. (2000). *How Cities Work: Suburbs, Sprawl, and the Roads Not Taken*. Austin: University of Texas Press.

Michelson, W. (1985). *From Sun to Sun: Daily Obligations and Community Structure in the Lives of Employed Women and Their Families*. Totowa, N.J.: Rowman and Allanheld.

Miller, H. (1999). Measuring Space-Time Accessibility Benefits within Transportation Networks: Basic Theory and Computational Procedures. *Geographical Analysis* 31, 187–212.

Mokhtarian, P. L. (1990). A Typology of Relationships between Telecommunications and Transportation. *Transportation Research A* 24, 231–42.

Mokhtarian, P. L., and R. Meenakshisundaram (1999). Beyond Tele-Substitution: Disaggregate Longitudinal Structural Equations Modeling of Communication Impacts. *Transportation Research C* 7, 33–52.

Muth, R. F. (1969). *Cities and Housing: The Spatial Pattern of Urban Residential Land Use*. Chicago: University of Chicago Press.

Nie, N., and L. Erbring (2000). *Internet and Society: A Preliminary Report*. Stanford Institute for the Quantitative Study of Society (SIQSS), Stanford University, California.

O'Sullivan, D., A. Morrison, and J. Shearer (2000). Using Desktop GIS for the Investigation of Accessibility by Public Transport: An Isochrone Approach. *International Journal of Geographical Information Science* 14, 85–104.

Pendyala, R. M., K. Goulias, and R. Kitamura (1991). Impact of Telecommuting on Spatial and Temporal Patterns of Household Travel. *Transportation* 18, 411–32.

Pirie, G. H. (1979). Measuring Accessibility: A Review and a Proposal. *Environment and Planning A* 11, 299–312.

Pooler, J. (1987). Measuring Geographical Accessibility: A Review of Current Approaches and Problems in the Use of Population Potentials. *Geoforum* 18, 269–89.

Robinson, J. P., M. Kestnbaum, A. Neustandtl, and A. Alvarez (2000). Mass Media Use and Social Life among Internet Users. *Social Science Computer Review* 18, 490–501.

Saxena, S. and P. L. Mokhtarian (1997). The Impact of Telecommuting on the Activity Spaces of Participants. *Geographical Analysis* 29, 124–44.

Shen, Q. (1998). Location Characteristics of Inner-City Neighborhood and Employment Accessibility of Low-Wage Workers. *Environment and Planning B* 25, 345–65.

———. (1999). Transportation, Telecommunications, and the Changing Geography of Opportunity. *Urban Geography* 20, 334–55.

Small, K. A., and S. Song (1992). Wasteful Commuting: A Resolution. *Journal of Political Economy* 100, 888–98.

Sui, D. (1998). GIS-Based Urban Modeling: Practices, Problems, and Prospects. *International Journal of Geographical Information Science* 12(7), 651–71.

UCLA-CCP. (2003). *The UCLA Internet Report: Surveying the Digital Future*. UCLA Center for Communication Policy. University of California, Los Angeles.

Vance, J. E., Jr. (1964). *Geography and Urban Evolution in the San Francisco Bay Area*. San Francisco: Institute of Governmental Studies.

van Wee, B., M. Hagoort, and J. A. Annema (2001). Accessibility Measures with Competition. *Journal of Transport Geography* 9, 199–208.

Wachs, M., and T. G. Kumagai (1973). Physical Accessibility as a Social Indicator. *Socio-economic Planning Science* 7, 437–56.

Wachs, M., B. D. Taylor, N. Levine, and P. Ong (1993). The Changing Commute: A Case-Study of the Jobs-Housing Relationship over Time. *Urban Studies* 30, 1711–29.

Waddell, P. (1993). Exogenous Workplace Choice in Residential Location Models: Is the Assumption Valid? *Geographical Analysis* 25, 65–82.

Waddell, P., B. J. L. Berry, and I. Hoch (1993). Housing Price Gradients: The Intersection of Space and Built Form. *Geographical Analysis* 25, 5–19.

Waddell, P., and V. Shukla (1993). Manufacturing Location in a Polycentric Urban Area: A Study in the Composition and Attractiveness of Employment Subcenters. *Urban Geography* 14, 277–96.

Warf, B. (2001). Segueways into Cyberspace: Multiple Geographies of the Digital Divide. *Environment and Planning B* 28, 3–19.

Weber, J. (2001). *Evaluating the Effects of Context and Scale on Individual Accessibility: A Multilevel Approach*, Ph.D. diss., Department of Geography, Ohio State University.

———. (2003). Individual Accessibility and Distance from Major Employment Centers: An Examination Using Space-Time Measures. *Journal of Geographical Systems* 5, 51–70.

Weber, J., and M.-P. Kwan (2002). Bringing Time Back in: A Study on the Influence of Travel Time Variations and Facility Opening Hours on Individual Accessibility. *The Professional Geographer* 54, 226–40.

Weber J., and M.-P. Kwan (2003). Evaluating the Effects of Geographic Contexts on Individual Accessibility: A Multilevel Approach. *Urban Geography* (forthcoming).

[4]

How derived is the demand for travel? Some conceptual and measurement considerations

Patricia L. Mokhtarian [a,*], Ilan Salomon [b,1]

[a] *Department of Civil and Environmental Engineering, Institute of Transportation Studies, University of California, One Shields Avenue, Davis, CA 95616, USA*

[b] *Department of Geography, Hebrew University, Jerusalem 91905, Israel*

Abstract

This paper contests the conventional wisdom that travel is a derived demand, at least as an absolute. Rather, we suggest that under some circumstances, travel is desired for its own sake. We discuss the phenomenon of undirected travel – cases in which travel is not a byproduct of the activity but itself constitutes the activity. The same reasons why people enjoy undirected travel (a sense of speed, motion, control, enjoyment of beauty) may motivate them to undertake excess travel even in the context of mandatory or maintenance trips. One characteristic of undirected travel is that the destination is ancillary to the travel rather than the converse which is usually assumed. We argue that the destination may be to some degree ancillary more often than is realized. Measuring a positive affinity for travel is complex: in self-reports of attitudes toward travel, respondents are likely to confound their utility for the activities conducted at the destination, and for activities conducted while traveling, with their utility for traveling itself. Despite this measurement challenge, preliminary empirical results from a study of more than 1900 residents of the San Francisco Bay Area provide suggestive evidence for a positive utility for travel, and for a desired travel time budget (TTB). The issues raised here have clear policy implications: the way people will react to policies intended to reduce vehicle travel will depend in part on the relative weights they assign to the three components of a utility for travel. Improving our forecasts of travel behavior may require viewing travel literally as a "good" as well as a "bad" (disutility).

Keywords: Travel attitudes; Travel time budget; Travel behavior; Excess travel

* Corresponding author. Tel.: +1-530-752-7062; fax: +1-530-752-7872; http://www.engr.ucdavis.edu/~its/telecom/.
E-mail addresses: plmokhtarian@ucdavis.edu (P.L. Mokhtarian), msilans@mscc.huji.ac.il (I. Salomon).

[1] Tel.: +972-2-5883345; fax: +972-2-5820549.

1. Introduction

Since the origin of transportation as a field of scientific inquiry, the tenet that "travel is a derived demand" has been accepted with little question. This view pervades modern transportation planning approaches. For example, in demand models travel is assumed to involve a disutility to be endured for the sake of achieving a desired destination, but one that is minimized. This disutility is modeled as a function primarily of time and cost, and is assumed to increase with each. In project evaluation, the assumed monetary value of travel time savings typically constitutes the largest share of the quantified benefits of a proposed improvement (e.g., Welch and Williams, 1997). Policies directed at the problem of urban congestion often attempt to reduce travel by increasing its cost (disutility) or by bringing destinations closer to origins (through denser and more mixed land use patterns or through information/communications technology (ICT) substitutes). And current efforts to improve regional transportation models take an "activity-based" approach whose premise is that to understand travel we need to understand the demand for the activities that generate the travel.

In a previous paper (Salomon and Mokhtarian, 1998), we reviewed some conceptual and empirical evidence challenging the derived demand paradigm as a behavioral absolute. This paper continues to reassess the assumption that the demand for travel is completely derived from the demand for spatially separated activities. It expands on and extends some of the concepts presented previously, and discusses some important measurement issues that need to be addressed if an intrinsic desire for travel is to be properly identified. As in that previous work, our discussion in this paper refers to personal travel rather than goods movement, but we place no restrictions on mode, purpose, or distance.

The organization of this paper is as follows. Section 2 discusses the phenomenon of undirected travel, and what it can tell us about more destination-oriented travel. Section 3 explores the role of the activity/destination in the demand for travel. Section 4 describes the tripartite nature of an affinity for travel and why it presents a measurement challenge, with survey respondents likely to confound their feelings about travel as an end in itself, with the benefits provided by travel as a means to an end. Section 5 illustrates those measurement difficulties while presenting some specific results, in the context of an ongoing empirical study of the desire for mobility. These results offer partial support for the claim of the existence of a desire to travel for its own sake, and point to productive directions for improving our ability to identify and understand that desire. Section 6 discusses non-travel alternatives for potentially achieving similar levels of utility, together with implications for the theory of a constant travel time budget (TTB). Section 7 summarizes the key points of the paper and makes some concluding observations.

2. The phenomenon of undirected travel

Clearly, the desire to engage in activities at different locations underlies a great deal of the demand for travel. But sometimes, can travel itself not be the activity that is demanded?

There are a variety of activities consisting of what might be called "undirected travel". Joyriding (simply "taking the car out for a spin") is one such activity, [2] but there are many others. Examples include traveling in an off-road vehicle, recreational boating or flying, taking a recreational vehicle cross-country, recreational walking/jogging/cycling/skating/skateboarding, horseback riding, hiking, skiing, hang-gliding, scuba diving, spelunking, taking amusement park rides, and others. These differ widely in terms of distance traveled, typical location, mode used, and impacts on the environment and energy consumption, but they are fundamentally similar in one respect – that travel *is* the activity, movement *is* the object, and a destination, if there is one (or more) in the usual sense of the word, is to varying degrees incidental. [3]

Even sports such as auto racing (or horse racing, 10 K or marathon runs, the Tour de France, or any other form of racing involving a human being driving, riding, or providing the motive power) qualify as undirected travel. Although in those cases the destination (finish line) is arguably of crucial importance, it is an arbitrarily chosen point that is meaningless as a destination in its own right (in that people would not travel to the finish line independently of the race). It is *traveling* to that arbitrarily selected destination in the context of the race (whether faster than others, faster than one's own record, or simply at all) that is the main point of the activity.

Vigorous physical effort is neither a necessary nor a sufficient condition for an activity to constitute undirected travel, although physical exercise may be one motivation for engaging in such an activity. To see that it is not necessary, note that in the list above, the activities involving operating or riding in a vehicle do not require much physical human energy. To see that it is not sufficient, note that there are essentially stationary alternatives to a number of the above activities (e.g., working out in a gym), which can involve considerable human energy expenditure. Movement through space, on the other hand, is a necessary but not sufficient condition for undirected travel. It is necessary, of course, because travel by definition involves movement through space. It is not sufficient because most travel, as has been repeatedly noted, is largely ancillary to reaching a desired destination and engaging in a desired activity. Thus, just "going for a walk" in the neighborhood after dinner is undirected travel; walking through the grocery store to purchase food is directed travel.

---

[2] Automobile advertisements still play to this phenomenon, sometimes in a nostalgic appeal. Consider the recent campaign for the Chevrolet Impala, appearing, for example, in a four-page foldout on the inside front cover of the 7 June 1999 issue of *Newsweek* magazine: "Remember how great it was just to get in your car and drive? We do. It didn't matter where you were going. All you needed was an open road and a full tank of gas. The world streaming by your window, wind in your hair, sun through the trees, tires humming and the radio on. Hot summer days, dusty dirt roads. Not a care in the world. Whatever happened to that? The pure joy of a long drive, a great car, and no particular place to go. Isn't it time somebody brought that back? The New Chevy Impala. Let's go for a drive". A similar theme is portrayed in a current television advertisement, with the 1964 Chuck Berry song "No Particular Place to Go" playing in the background: "Ridin' along in my automobile/My baby beside me at the wheel/I stole a kiss at the turn of a mile/My curiosity runnin' wild/Cruisin' and playin' the radio/With no particular place to go".

[3] In this context, it is important to distinguish between the general location of an activity and the micro-scale destination, or lack thereof, of the activity itself. For example, the general location at which scuba diving occurs is obviously not incidental: the Great Barrier Reef is preferable to the community swimming pool. But the actual activity of scuba diving may involve a more or less random path within a general area, with no particular spot being the target of the activity. Thus, travel *to* the Reef is directed, but the scuba diving activity itself represents undirected travel.

What characterizes undirected travel, then, is movement through space for which the *destination* rather than the travel is ancillary. Whereas the strict view of travel as a derived demand would hold that the destination is always 100% primary, we suggest that the set of all travel for which destination is primary is a fuzzy one (see, e.g., Smithson, 1987; Zimmermann, 1985). Stated another way, the relative proportions of "primariness" of the travel and the destination constitute a continuum, as shown in Fig. 1. These proportions can vary by person and situation, even for the same type of activity. For example, strolling through the shopping district of a foreign city may in one case be largely undirected (mainly to absorb the novel ambience), in another case largely directed (mainly to buy souvenirs), and in yet another case nearly equal parts of both. One message of this paper is that the relative degrees to which travel and the destination are ancillary are often difficult to measure, especially when the traveler/survey respondent herself may not have consciously articulated the distinction. As will be argued in Section 3, our predisposition to view travel as a derived demand may cause us to overestimate the degree to which travel is ancillary to the destination instead of a situation more toward the middle or even the opposite end of the spectrum.

Almost by definition, undirected travel is for the most part a leisure activity (except for the relatively few professional practitioners of each type). This is not however to dismiss its importance as an indicator of the positive utility of travel in general, for three reasons. In the first place, rather than diminishing that importance, it strengthens it to realize that so many people, for so much of their limited discretionary time, choose to spend it not just traveling to activities, but on traveling *as* an activity. Just how many people, and how much time, is difficult to determine from current data collection instruments that do not distinguish travel as a (leisure) activity from either travel to an activity or other leisure activities. It would be valuable to make that distinction in the future.

Second, contrary to popular complaint, leisure time in developed countries does not seem to be declining. For the US, Robinson and Godbey (1997) report that the average weekly hours of free time (which would include stationary free-time activities, undirected travel as an activity, and travel to free-time activities) rose from 35 in 1965 to 40 by 1985, remaining approximately stable since then. In Germany, Chlond and Zumkeller (1997) note that increases in paid vacation time and decreases in weekly work hours have resulted in greater leisure time. Further, total travel is growing and travel for leisure purposes appears to be a growing share of total travel. Anable (1999), Lanzendorf (1999) and Tillberg (1999) indicate that leisure activities currently account for

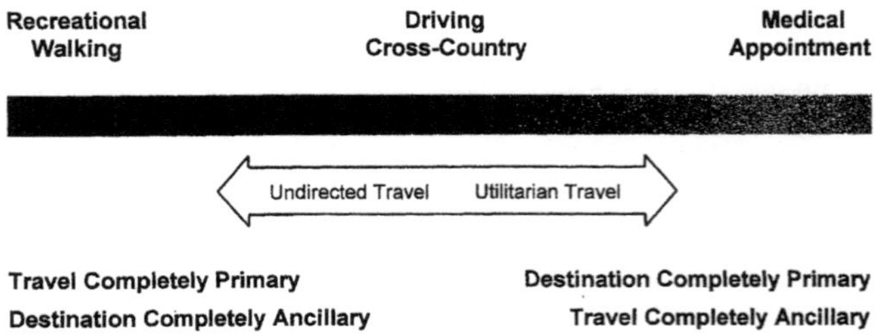

Fig. 1. Relative degrees to which destination and travel are primary.

half of total distance traveled in the UK, Germany, and Sweden, respectively. Robinson and Godbey find that only 3 h a week are spent on all free-time travel, out of 10 h a week total travel time, but as indicated above, much undirected travel is likely to be classified as an activity rather than as travel. Even so, the amount of time spent on undirected travel is doubtless small now, and apt to remain a small proportion of the total. However, it is also likely to increase in the aggregate over time, as rising incomes continue to result in rising amounts of leisure time and leisure travel (Schafer and Victor, 1997; Tanner, 1981).

Third, the fundamental nature of undirected travel may hold to some degree for more directed travel. The examples of undirected travel given above serve to illustrate some of the aspects intrinsic to travel that contribute to its positive utility: the sensation of speed, the exposure to the environment and movement through that environment, the ability to control movement in a demanding and skillful way, the enjoyment of scenic beauty or other attractions of a *route*, not just a *destination* (Hupkes, 1982). It is likely that those same positive aspects of travel apply, to some extent, to ancillary or directed travel as well. Many authors (Berger, 1992; Flink, 1975; Marsh and Collett, 1986; Sachs, 1992; Wachs and Crawford, 1992) have commented on the sense of independence, control, expression of status or identity, and mastery of a skill afforded by driving a personal automobile. Individuals who place a high value on those attributes may, for example, choose to drive to work in a congested central business district even when public transportation is actually both faster and cheaper. A desire for exposure to and movement through the environment is doubtless partially responsible for some people choosing not to telecommute even when they are able to do so (Mokhtarian and Salomon, 1997). The beauty or novelty or some other characteristic of a particular route may motivate an individual to travel that route even when it is not the shortest way to a desired destination. These outcomes are examples of excess travel in the sense that lower cost, time, and/or vehicle-kilometers-traveled alternatives are available but not chosen because of an intrinsic desire (or a positive utility) for travel. (We define excess travel more formally in Section 4.4. Here, we make the following semantic distinction: undirected travel is a subset of excess travel, but excess travel can also constitute or, more often, augment a trip that is basically directed or utilitarian, as in the examples above.)

3. Which came first, the activity or the trip?

In the previous section we pointed out that the destination of a trip may in some cases play a more ancillary role to the trip itself. In this section we discuss further the role of the destination in the demand for travel.

Conventional trip distribution (Papacostas and Prevedouros, 1993) or destination choice models (Barnard, 1987; Jones, 1978) consider the utility of a given destination to be inversely related to the generalized cost of reaching it and directly related to some measure of the attractiveness of the destination. Hence, a tradeoff between the disutility of travel and the utility of the activity at the destination is explicit, and the choice of a more distant destination is completely consonant with the concept of travel as a derived demand when the increased attractiveness of that more distant destination outweighs the increased disutility of travel required to reach it. Thus, for example, a more distant shopping center may be chosen if it has more variety or better prices or a particular hard-to-find item. A more distant restaurant may be chosen when the decision-maker is in the mood for the kind of food it serves, or the atmosphere it possesses.

However, we suggest that there are situations in which a more distant destination is chosen, not entirely because the utility of its inherent attractiveness exceeds the disutility of travel, but because a positive component to the utility for travel contributes to making the net utility of that destination–trip combination the highest among the alternatives. Consider the situation in which, in a dense urban environment, there are a number of franchises of the same "favorite" restaurant or coffee house. Only one is "nearest". Yet an individual may habitually visit more distant ones as well as the closest, not because of an intrinsic greater attractiveness of the more distant franchises (in fact they may look and "feel" virtually identical to the nearest one), nor even particularly because of a greater attractiveness of the neighborhoods in which the more distant facilities are located, nor because of trip chaining economies, but purely out of a variety-seeking impulse.

In this example, a variety-seeking orientation leads to excess travel. It should be understood that the attribute "contributes variety" is not entirely intrinsic to the destination itself nor to the vicinity of the destination – to the extent that it is, variety can be considered part of the attractiveness measure of the destination. Instead, at one extreme, variety is a property of the route rather than the destination, and hence (apparently) excess travel is an inevitable accompaniment to the achievement of variety.

What about another prevalent human characteristic, curiosity? Curiosity (often, to be sure, mixed with more directed goals such as the search for physical resources or commercial opportunities) may be the trait that launched a thousand ships, and pedestrian forays, and horses, and covered wagons, and airplanes, and rockets. One could argue that novelty or uncertainty should be part of the attractiveness measure of the (often unknown) destination, but it seems at least equally useful to view curiosity (in its particular manifestation as an exploration impulse) as a generator of what must surely be considered excess travel. Today, curiosity still impels us to travel "out of our way", whether to see a new development on the other side of town or to visit an intriguing location on the other side of the planet, and stimulates us to dream of traveling to the other side of the solar system and beyond.

Thus we see that there are several related traits such as variety- (or adventure- or novelty-) seeking and curiosity that have the result of increasing the utility of more distant destinations and/or inevitably generating travel in order to satisfy those traits. We suggest that in many of these situations, the demand for travel is not so much derived from the demand for a specific activity at a specific location, but that both the travel and the activity/location are derived from the demand to satisfy the impulse in question.

This in turn suggests that viewing the desire for a particular activity as antecedent to and causative of the demand for traveling to that activity may not always be accurate – although it is presumably the most common situation. But in some cases, as just indicated, the demand for both travel and activity may be caused by a third set of factors. And in other cases, the complete reverse of the usual situation may occur: the demand for an activity may arise *as a consequence* of the desire to travel.

Consider, for example, the choice to eat out instead of eating at home, even though ample food is available at home. In some cases, eating out may be preferred because a certain type of food or a certain neighborhood or a certain ambience is actively desired. In these situations, the decision to eat out and the destination may be chosen simultaneously, and the utility of a *particular* destination–travel combination (the net of the positive attractiveness of the destination and the putatively negative utility of travel) exceeds the utility of the home alternative (the net of a lower attractiveness plus zero travel). In other cases, the disutility of cooking and cleaning up is the primary motivation for going out to eat, and the destination may be a secondary choice. In the present context, it is a third type of situation that is of interest. In these cases, also involving a sequential rather than simultaneous choice, the desire just to get out and go *somewhere* (another form of variety-seeking) manifests itself in deciding to eat out instead of staying at home. The destination/activity becomes an excuse or justification for the desired travel. Many other such examples are possible, in which the (perhaps subconscious) decision to travel is made first, and then a destination/activity is invented to support that decision and yes, increase its utility. The "Sunday drive", which was so common during the early popularization of the automobile, probably often fit this situation, although a desire to see the scenic countryside was often a destination-specific motivation as well (Muller, 1986).

Such cases may arise more often than we realize, because we have not tried to measure them as such. We see that a destination is reached and an activity is carried out, and we assume that activity to have generated the trip. Instead it may be the trip that generated the activity!

4. The tripartite nature of the affinity for travel

If a positive utility for travel exists at all, it is important to understand it better than we do now. How does such a positive regard for travel differ by personality type and other individual characteristics, by travel purpose, by mode and trip length? Can we identify the impact a positive utility for travel has on the objective amount an individual travels – that is, its contribution to excess travel?

Measuring an individual's affinity or liking for travel is a fundamental first step in this process. If travel affinity can be appraised in some generic way, it becomes possible to explore causes and effects of that affinity. Obtaining a reliable measurement of travel liking, however, is a non-trivial matter. This is because an individual's expressed affinity for travel is likely to be a composite of positive utilities for three different elements, in unknown and varying proportions. These three elements are conceptually distinguishable but empirically apt to be confounded. They are:

1. the activities conducted at the destination;
2. activities that can be conducted while traveling;
3. the activity of traveling itself.

We briefly discuss each of these elements in turn, and then use them to define excess travel.

4.1. Activities conducted at the destination

When a respondent reports that she "loves" vacation travel, it is unlikely that she is referring to the 15 h in one or more crowded and noisy airplanes, the 6 h waiting in uncomfortable airports eating overpriced and unpalatable food, and the 3 h of ground access travel in peak-period urban traffic. It is more likely that a halo effect (Sommer and Sommer, 1997) is at work, so that she is confounding the positive appeal of the destination with the travel required to reach it (the halo effect is a type of response bias identified by survey researchers, in which the respondent bases the answer to a specific question on a general impression about the subject). The implication, to which we return in Section 6, is that if she could forgo the travel to the destination, she would. However, it is also possible to be cognizant of the unpleasant aspects of travel itself but for those to be outweighed by the positive aspects *of travel* (not just the destination), such as those discussed in Sections 4.2 and 4.3.

4.2. Activities that can be conducted while traveling

In reporting an affinity for traveling, individuals may in part be considering the utility of activities they can conduct while traveling. In some cases, it is in fact the "anti-activity" (or the absence of other activities) that is important – that is, the ability to use the time for relaxing or thinking, including "shifting gears" mentally between the origin and destination activities and roles. As one analyst put it, "Thanks to the construction of interstate highways, the entry of women into the work force, and several other social revolutions, driving has become America's most important source of quiet time" (Edmonson, 1998, p. 46). In other cases, the concomitant activity is external: making and receiving mobile phone calls (including shopping and checking stock quotes on or off the Internet, as well as engaging in conversation); reading; listening to music, talk shows, or books on CD, radio, or cassette; watching television or videos (not only in airplanes but now in some personal vehicles such as the Oldsmobile Silhouette Premier and the Ford Econoline Conversion Van). The phenomenon of "carcooning" is one manifestation of this aspect, in which the personal vehicle is customized for the traveler's comfort, almost as a sanctuary-escape from the world (Crawford, 1992; Larson, 1998). [4] But as the preceding list indicates, this aspect of a liking for travel is not restricted to the automobile; in fact some people prefer public transportation to the private auto precisely because not having to operate a vehicle offers the opportunity to engage in other activities while traveling. Cycling and walking as modes of directed travel also offer opportunities for quiet time, listening to music, and the additional benefit of physical exercise while traveling.

Several researchers have noted that for some people the commute to work fulfills various positive roles (Richter, 1990; Salomon, 1985; Shamir, 1991; Mokhtarian and Salomon, 1997). Some of these roles relate to the utility of the commute as a desired transition between work and home, which allows for the types of activities and anti-activities described above. Work-related travel for mobile professionals often fulfills similar functions. Anecdotally, a number of such professionals have remarked that long trips represent "the only time for thinking" they have, or "the chance to catch up" on reading or other neglected but important tasks.

[4] A recent advertisement for the Toyota 4Runner sport utility vehicle (appearing, for example, on the inside back cover of the 7 June 1999 issue of *Newsweek*) plays to this component of utility: "Escape. Serenity. Relaxation. The 1999 Toyota 4Runner Limited puts them all well within your reach. With features like a leather-trimmed interior, a CD sound system as well as more than a dozen new refinements, you might actually find the journey to be as rewarding as the destination".

4.3. The activity of traveling itself

The third element of a liking for travel is a consequence of intrinsic aspects of travel itself. These include the characteristics discussed in Section 2: the sensation of speed, movement through and exposure to the environment, the scenic beauty or other attraction of a route. Arguably, only this element represents a true affinity for travel itself. Whereas in the other two categories travel is valued as a means to an end (either performing activities at a fixed destination or performing activities in transit), in this case travel is (at least in part) the end in itself (Reichman, 1976). [5] For instance, an individual may in fact actively choose 24 h of traveling in an automobile or recreational vehicle over a much shorter time of travel in an airplane to the same farthest point, for the opportunity to see many sights on the way to a "final" destination. In cases where there is not so much a single major destination as many linked ones, the airplane may not even be a realistic alternative. This situation is discussed further in Section 6.1.

Traveling in response to a variety-seeking or curiosity impulse may represent a somewhat more indirect relationship, since those personality traits may be less specific to travel than attitudes directly related to characteristics of travel such as movement and speed. However, these personality and attitudinal impulses are similar in that (a) both have alternate, non-travel ways of potentially satisfying them (as discussed in Section 6.3), but (b) in both cases, travel for its own sake is likely to be an often-preferred way of satisfying them. Thus the personality traits of variety-seeking or curiosity are possible causal variables generating a liking of travel for its own sake, just as the attitudes of "loving speed" or "loving scenic beauty" are other possible causal variables for travel affinity.

4.4. A definition of excess travel

If it is considered desirable to try to quantify the impact that a positive utility for travel has on an individual's objective amount of travel, then it is important to be clear about what constitutes such "excess travel". Simply equating "excess" to "unnecessary" is problematic. Leisure activities are discretionary, and hence in some sense unnecessary. Is all travel for leisure activities excess? We do not adopt that extreme a view: we would not classify as excess a shortest-path trip generated by the pre-existing demand for a leisure activity, although it may be unnecessary.

Some examples of excess travel were offered in Section 2: cases in which lower-VKT (or lower-time/cost) alternatives were available but not chosen because of a positive utility for travel. This implicit definition can now be made more explicit based on the foregoing discussion of the components of an affinity for travel. Namely, we specify excess travel to be that portion of travel that is prompted by the second and third elements of an affinity for travel, that is, any travel not derived from the utility of the destination itself. Thus, excess travel would include the subset of leisure activities identified as undirected travel in Section 2, which are a manifestation of the third element – a positive utility for travel itself. But it would not include the travel *to* those activities,

[5] Again, automobile advertisements frequently play to this concept. A 16-page Chrysler ad in the center of the 18 October 1999 issue of *Newsweek* included tag lines such as: "Because driving should be a destination in itself" and "it does something no other minivan can: make you wish the journey were a bit longer."

which is derived from the demand to be in a location where the undirected travel can be performed.

This definition may make sense theoretically, but it is not easy to operationalize it. For one thing, without perfect knowledge of a person's choice set (the alternatives that are truly feasible under the circumstances), it is impossible to know whether the chosen alternative is in fact lowest-VKT or not – it may only appear to involve excess travel (a longer route, a non-optimal mode) to the analyst. Second, as indicated in Section 3, despite the utility of the destination itself, it may not always be the most important generator of the trip. This is especially true for leisure activities, but as previous examples have shown, it can also be true for mandatory activities (commuting even when telecommuting is feasible) and maintenance activities (eating out as a solution to "cabin fever"). Third, in assessing and reporting the attractiveness of a destination, a respondent may be partly influenced by his utility for the second and third elements of an affinity for travel. Thus, even some travel which appears to fit the derived demand paradigm may be "excess" in ways that are difficult to disentangle.

5. Empirical indications

5.1. Background

The preceding discussion has made clear some of the difficulties associated with empirically measuring the existence and impact of a positive utility for travel. Nevertheless, the importance of the issues raised here makes the measurement challenge worth undertaking. One goal may be to quantify the amount of excess travel that occurs, and under what circumstances. Such insight could inform the design of policies more responsive to natural inclinations (including policies that attempt to influence or channel inclinations in socially beneficial ways), and improve our predictions of the reaction to various policies.

Independently of attempting to calculate kilometers of excess travel, however, it is useful to explore further the general concept of travel affinity and its distribution in society. We have developed and administered a survey with that second goal in mind. As is often the case, our thinking has continued to evolve after the completion of the survey, so that many of the ideas presented here were not fully articulated at the time of data collection. Further, as is also often the case, in designing the survey we consciously traded off depth against breadth, in this context favoring breadth. That is, we chose to obtain somewhat general data on a large number of concepts of interest, rather than situation-specific data in a more narrowly-defined context. The latter approach is probably essential to a goal of quantifying the amount of excess travel, but the former approach is consistent with our goal of increasing understanding of general concepts (although with limitations even in that respect). As a consequence of these factors, we can (and do) suggest a number of ways in which future related studies can build on and refine the data collected in this one. Nevertheless, we believe that the preliminary empirical results reported here are still strongly suggestive, even if not definitive.

Our 14-page questionnaire collects data on general attitudes toward travel and related issues, affinity for travel, objective and perceived amounts of travel, satisfaction with one's amount of travel, personality traits, lifestyle orientation, and demographic characteristics. The questions relating to affinity for travel and amounts of travel distinguish between short distance and long distance (more than 100 miles one way, consistent with the definition of long distance leisure travel in the American Travel Survey), and within each of those categories obtains an overall measure and separate measures for several different purposes and modes.

Some 8000 surveys were sent to residents of three communities in the San Francisco Bay Area, representing a variety of land use patterns. With an overall response rate of more than 25%, after discarding responses with too much missing data we retained about 1900 cases for further study. Due to sampling biases (in the selection of particular neighborhoods, although sampling within neighborhoods was entirely random) and self-selection in responding, the sample (and hence the distributions of variables discussed here) cannot be assumed to be perfectly representative of the general population. Nevertheless, the findings serve to support the existence of a positive utility for travel, even if the precise distribution of that utility across the population is uncertain.

Detailed analysis of the data is underway, and future papers will present results from a variety of empirical explorations. Here, we focus on a few summary results that illustrate some of the issues we have presented in this paper. As background to interpreting the results, it should be noted that in the cover letter to the survey, travel was defined as "moving any distance by any means of transportation – from walking around the block to flying around the world." In questions relating to the amount of travel conducted or desired by respondents, they were asked (borrowing wording from the American Travel Survey) to exclude "travel you do as an operator or crew member on a train, airplane, truck, bus, or ship".

5.2. Travel affinity

To directly measure the affinity for travel, the question was asked, "How do you feel about *traveling* in each of the following categories? We are *not* asking about the activity at the destination, but about the travel required to get there". Respondents were then asked to rate short and long distance travel, overall and by purpose and mode, on a five-point scale from strongly dislike to strongly like. Despite our attempt to alert respondents to distinguish the destination activity from the travel, it is likely that even many of those who actually read the instructions (and more of those who did not) were unsuccessful at doing so.

Clear differences between overall ratings for short and long distance travel emerge, as shown in Fig. 2(a) (where for economy of presentation the five-point scale has been collapsed to the three points dislike, neutral, like). Levels of dislike are similar for both short distance (13%) and long distance (11%) travel. But a majority (55%) of respondents are neutral about short distance travel, whereas an even larger majority (63%) are positive about long distance travel. Thus, there is clearly a stronger affinity for long distance travel, but even short distance travel is not viewed negatively.

Differences are also apparent by purpose and mode, in the expected directions. Figs. 2(b) and (c) show that, for short-distance travel, respondents have greater affinity for entertainment/recreation/social activity-related travel than for travel related to other kinds of activities, and greater affinity for travel by personal vehicle and walking/jogging/bicycling than for travel by public transportation. For long distance travel (Fig. 2(d)), respondents like travel for entertainment/recreation/social activities far more than travel for work, and travel by plane somewhat more than travel by car. Again, it is probable that respondents are partly confounding the utility

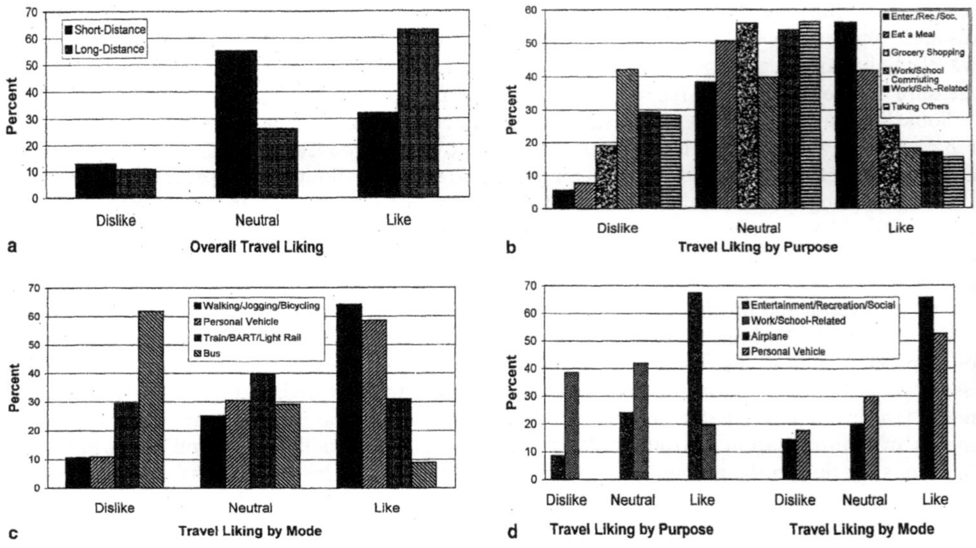

Fig. 2. (a) Overall travel liking by distance category ($N = 1904$). (b) Liking for short-distance travel by purpose ($N = 1904$). (c) Liking for short-distance travel by mode ($N = 1904$). (d) Liking for long-distance travel by purpose and mode ($N = 1904$).

for the activity at the destination with the utility for the travel required to reach the destination, as well as potentially including the second element, utility for activities conducted while traveling. For example, vacation travel may be better liked than work travel if one brings work to do on the work trip but novels to read or knitting to do on the vacation trip. Being with family members on the vacation may also increase its utility for many.

It is tempting to argue that the mode-specific ratings are more indicative of a true travel affinity (the third element of utility) than are the purpose-specific ones. Theoretically, "travel is travel" – if differences in destination activity and in activities conducted while traveling are factored out, a 10-h flight is a similar physical experience whether it is for work or for leisure. Differences in liking for travel by auto and plane, on the other hand, may reflect genuine differences in comfort, convenience, control, and other attributes intrinsic to those modes. However, the situation is not that simple. First, even the mode-specific ratings are not immune to confounding with utility for activities conducted while traveling: for example, an airplane flight may be more conducive to relaxing or multitasking and hence have higher utility than a trip of comparable length as a solo automobile driver. Second, likely interaction effects between purpose and mode complicate making the appropriate inference. A higher expressed affinity for plane than for auto may be partly based on the fact that for the respondent in question, plane is more often associated with leisure travel and auto is more often associated with work travel. Conversely, a higher rating of leisure travel compared to work travel may reflect a higher content in leisure travel of those undirected travel activities (often by unusual modes) described in Section 2, in which attributes intrinsic to travel contribute heavily to utility. But the desirability of obtaining ratings for each mode–purpose combination must be traded off against the added burden on the survey respondent.

Nevertheless, the extent of the affinity for travel across most of the categories is striking. We have already discussed some reasons why commuting may have positive utility, but even trips for activities that most people would consider chores (chauffeuring, grocery shopping) are liked by 15–25% of the sample. The most disliked type of travel is that which takes place in a bus, but even that is rated neutrally or positively by more than a third (38%) of the sample. (However, some proportion of the neutral responses may simply reflect a lack of experience of the respondents with that mode, rather than a considered opinion.)

5.3. *Indicators of excess travel*

Individuals with a liking of travel should manifest that predisposition in their travel behavior. To help assess that behavioral outcome, the survey asked respondents about their participation in 13 different indicators of excess travel. The question was kept as mode- and context-neutral as possible. Specifically, respondents were asked, "Keeping in mind that travel is going any distance by any means, how often do you travel..." in each of the 13 ways shown in Fig. 3, with possible responses of never/seldom, sometimes, and often.

For two of the indicators shown in Fig. 3 – to explore new places and to see beautiful scenery – it could be argued that the utility of the destination is prompting the travel behavior. It may therefore not be particularly surprising that those were the two most popular choices based on combining the "sometimes" and "often" responses (only 13% of the sample "never" did each of those two indicators). Some other indicators (to relax, when time is needed to think, to clear one's head, and mainly to be alone) are based on the utility of what can be accomplished while traveling. These represent the 6th, 8th, 10th, and 11th most common choices, respectively. The remaining seven indicators are intended to reflect a desire to travel for its own sake – and appear to be quite common although perhaps not universal. For example, traveling "just for the fun of it" was ranked 4th, done sometimes or often by three-quarters of the sample. Traveling "by a longer route to experience more of your surroundings" ranked 5th, done by nearly two-thirds of the sample. Going "to a more distant destination than necessary, partly for the fun of traveling there" was 7th, done by more than half of the sample.

Overall, more than half of the sample sometimes or often engaged in seven or more of these 13 indicators. More than one-fifth did 10 or more. Only 2% of the sample never did any of them. Focussing only on the seven measures most purely indicative of a desire to travel for its own sake, half of the sample engaged in four or more of those seven, and only 6% never did any of them.

5.4. *Personality traits*

The survey asked respondents to rate 17 personality characteristics in terms of how well each one described them. Again collapsing a five-point scale into a three-point scale, Fig. 4 presents the

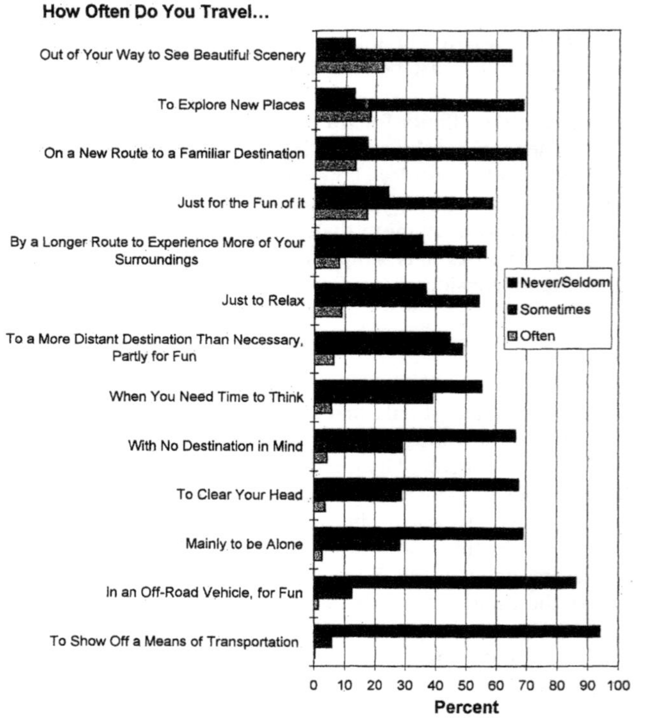

Fig. 3. Engagement in excess travel ($N = 1904$).

responses to five of the traits most relevant to the discussion in this paper. More than half the sample felt that "variety-seeking" or "adventurous" described them very well or almost completely. A third described themselves as liking to move at high speeds, and nearly a fifth considered themselves restless. Only 18% "liked to stay close to home", while more than a third of the sample felt that phrase described them not very well or hardly at all.

Apparently, the raw ingredients for an impulse to travel for its own sake are present in a sizable portion of the sample. The extent to which this is the case is probably overestimated due to a social desirability bias (Dillman, 1978) toward traits perceived to be positive. However, such a bias is unlikely to account for all the responses of that type, especially since sizable portions of the sample were willing to describe themselves in the opposite way (indicating the absence of such a bias for at least those respondents, and presumably others). In future research, a specialized survey could be designed to measure these traits more indirectly, through responses to a variety of questions or statements related to each trait. Such an approach would minimize response bias compared to the direct self-classifications elicited here.

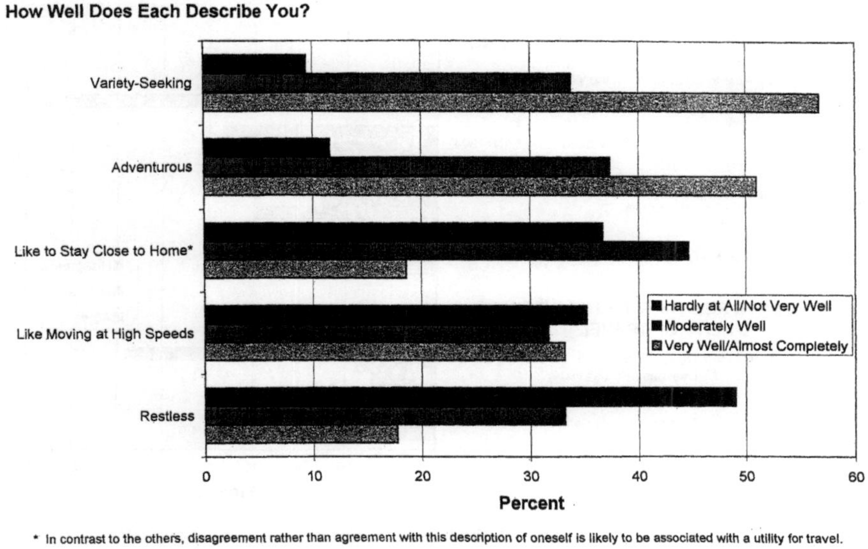

Fig. 4. Distribution of key personality traits ($N = 1904$).

5.5. *Attitudes toward travel*

Additional evidence for both the second and third elements of a positive utility for travel is found in a series of attitudinal statements to which responses were measured on a five-point Likert-type scale. Fig. 5 presents the responses to the statements from this section that relate to those aspects of the utility for travel, collapsed to a three-point scale of disagree, neutral, agree. With respect to the ability to conduct other activities while traveling, more than four-fifths of the sample sees value in linking errands to the commute trip. Nearly half disagree that travel time is generally wasted time. More than a third see their commute trip as a useful transition, and use that time productively. With respect to traveling itself, more than two-thirds of the respondents disagree that "the only good thing about traveling is arriving at your destination", and nearly half agree that "getting there is half the fun".

These latter responses in particular suggest the existence of a large group of people seeing at least some intrinsic utility for travel. Clearly, however, responses to these and many of the other attitudinal statements of this section (primarily relating to disadvantages of travel) are likely to vary by mode and purpose, whereas we only obtained a "generic" response. Future research may need to narrow the focus of these statements in order to obtain data that may relate more strongly to specific travel outcomes.

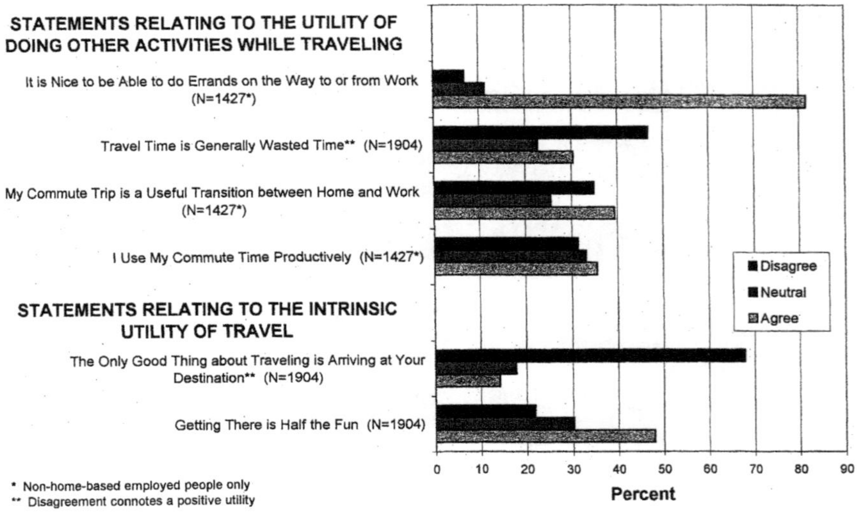

Fig. 5. Attitudes toward travel.

5.6. Ideal commute time

Another measure that is likely to be a complex function of all three elements related to a positive utility for travel is a respondent's ideal commute time. An individual's ideal commute time may be different from 0 because (as seen in Section 5.5) she values the transition between home and work and the ability to use the time productively (the second element), or because she values the opportunity to drive a status-oriented automobile or the chance to experience the environment by traveling (the third element), or because she values a non-home destination for work due to the social/professional interaction opportunities, the scenic location, or the shopping and other locational amenities it offers (the first element).

The ideal commute time question in our survey was placed immediately following the series of attitudinal questions relating to both positive and negative aspects of travel, so that respondents would be more likely to have a range of attributes in mind when they answered this question. The wording of the question itself also attempted to project a balanced perspective: "Some people may value their commute time as a transition between work and home, while others may feel it is stressful or a waste of time. For you, what would be the ideal one-way commute time?"

For the 1384 current workers who responded to this question, the average reported ideal one-way commuting time was just over 16 min. Only 3% desired a 0–2 min commute, suggesting that entirely eliminating the commute does not resonate with most people as a desirable aspect of telecommuting – for any or all of the three reasons described above. Almost one-half of the respondents preferred a commute of twenty minutes or more. Further analysis of this variable is found in Redmond and Mokhtarian (1999). In future studies, it may be valuable to obtain data on satisfaction with a range of commuting times (i.e., to obtain an entire satisfaction curve), similar to the approach of Young and Morris (1981), rather than obtaining just the single maximum point of the curve.

6. What are the alternatives?

6.1. Applying the teleportation test

It is important to try to distinguish the three elements of an affinity for travel discussed in Section 4, both in our conceptual formulation of the subject, and in our measurement of individuals' possession of that affinity. A stated preference approach could prove helpful in the latter regard, allowing the analyst to systematically vary one of the three elements while keeping the other two fixed. Another, whimsical but potentially useful, way to help make the distinction – in either the conceptual or the empirical context – may be to apply the "teleportation test". The question is, "if you could snap your fingers or blink your eyes and instantaneously teleport yourself to the desired destination, would you do so?"

For the three elements described in Section 4, the answers seem to be yes, maybe, and no, respectively. That is, in circumstances where an expressed utility for travel actually derives completely from the activity at the destination (the first element), a person should not hesitate to eliminate the undesired travel while still achieving the desired spatial separation from the origin. In circumstances where utility is derived from multitasking while traveling (the second element), the answer might depend on the perceived ability to accomplish the same tasks without the travel, as discussed in Section 6.3.

In circumstances where there is a utility to traveling itself (the third element), a person would choose to travel even if a teleportation alternative were available. For example, the 48% of Section 5.5 who agreed that "getting there is half the fun", or the 68% who disagreed that "the only good thing about traveling is arriving at your destination", may not be receptive to teleportation. The individual who wants to tour the US by car, or Europe by rail, may selectively teleport himself between some desired destinations, but complete teleportation from spot to spot is unlikely to appeal to those who want a sense of connectivity between locations, linkage to the surrounding geographical and cultural context, and/or enjoyment of a route as well as a destination. This orientation contributes to a preference, for some people in some circumstances, of ground-based alternatives over air travel (which begins to approach teleportation in its disconnection between origin and destination).

Of course, neither the utility nor the response to the teleportation test must fall into only a single one of the three categories we have identified – another case of fuzzy membership. An individual's utility for her activity set (and even for any single activity) may have all three elements to varying degrees, and hence with a teleportation alternative, travel would be eliminated or retained in similarly varying degrees.

Applying the teleportation test is more than just an exercise in futile fantasy; it offers insight into the likely reaction to real changes in travel that move us closer to the teleportation extreme. As lower-cost, higher-speed travel alternatives become available, will individuals take advantage of these improvements to travel the same distances but at less time and lower cost? To the contrary, some aggregate studies indicate that people increase the amount they travel under such circumstances (Bieber et al., 1994; Chlond and Zumkeller, 1997), although at least one disaggregate study (Kitamura et al., 1997) reported that only 3.6% of a 10-min savings in travel time would be spent on additional travel.

6.2. TTB, or not TTB? That is the question

The idea that, when travel becomes faster, people will travel longer distances in the same amount of time rather than equal distances in less time is consistent with the theory of a constant TTB (Hupkes, 1982; Marchetti, 1994; Zahavi and Talvitie, 1980). Some researchers (Goodwin, 1981; Gunn, 1981) have questioned this theory, citing the relative variability in daily travel times at the disaggregate level. We believe that the empirical evidence on both sides of this debate may fit a modified version of the TTB theory, grounded in the considerations presented here.

Specifically, we hypothesize the existence of an unobserved desired level of mobility, that varies both across individuals (as a function not only of the demographic characteristics usually used to model TTBs at the disaggregate level, but also of the travel-related attitudes and personality traits described in this paper), and within the same individual across time. Rather than uniformly trying to minimize travel, people seek to decrease their travel if it exceeds the desired optimum, but seek to increase travel if it falls short of their ideal amount.

Previous work (Mokhtarian et al., 1997) has classified potential individual responses to congestion as travel-maintaining, travel-reducing, or long-term lifestyle/location changes (assumed to reduce travel). But even areawide congestion will not impact every individual to the same degree, and we further suggest that even an individual who faces congestion may still wish to increase travel (although not necessarily at peak periods). Thus, it would be fruitful to consider responses to dissatisfaction with travel time in *either* direction, and extend the types of responses analyzed to travel-increasing strategies as well, including relocations that result in longer trips for work and non-work purposes than before. Again, the challenge is to distinguish the extent to which such strategies are adopted for reasons other than an intrinsic utility for travel (e.g., because the greater attractiveness of the new location outweighs the negative utility of the increased travel).

From the perspective of this paper, the TTB becomes an unobserved, variable ideal toward which people strive rather than an observed, stable quantity. The instinct that the demand for travel is not purely derived may account for the persistent appeal within the profession of the concept of the TTB. The empirical regularities that have been found may be due to the fact that, at the aggregate level, fluctuations on either side of the ideal will tend to cancel. On the other hand, the possibility that desired mobility both varies as a function of seldom-measured internal characteristics, and is not always achieved due to constraints, can account for the lack of stability that is often found at more micro-scales of analysis.

If there is a "desired TTB", it is useful to measure what that budget is. It is also difficult to do so: a direct question is unlikely to elicit a reliable response, at least for travel time in total. One specific context in which the direct question may obtain useful answers is for the repetitive and familiar case of commuting: the ideal commute time discussed in Section 5.6 is in fact a subset of an individual's desired TTB. That question on our survey appears to have produced meaningful and interesting results, notably that the desired commute time is greater than 0 for nearly the entire sample (see also Redmond and Mokhtarian, 1999).

We did not otherwise attempt to measure a desired TTB in minutes per day. Instead, following Ramon (1981), we obtained measures that she referred to as travel satisfaction, and that we now refer to as relative desired mobility. Specifically, our survey asked how much respondents would like to travel compared to what they do now, with a five-point response scale ranging from "much less" to "much more". As with the travel liking question, the relative desired mobility question was asked separately for short distance and long distance, and within those two categories it was asked "overall", and by purpose and mode. The results are shown in Fig. 6.

The numbers following each label in the legend of the figure are the correlations of each variable with its travel liking counterpart in Fig. 2. As might be expected, the correlations are positive and strongly significant, ranging from 0.3 to 0.6. One implication is that the analysis of travel affinity in Section 5.2 roughly applies to relative desired mobility as well. Another implication is that the more a person likes to travel in a particular category, the more he wants to increase his travel in that category – even without controlling for how much he already is traveling.

Further study of these data will undertake more sophisticated analysis of the relationships among travel liking, objective, perceived, and relative desired mobility, together with the attitudinal, personality, lifestyle, and demographic variables measured by the survey. The key point at this stage is that the results do lend some support to the concept of a desired TTB, with respondents indicating a number of circumstances under which they want to maintain or increase the travel they are doing now. If travel were purely a derived demand, we would expect a much

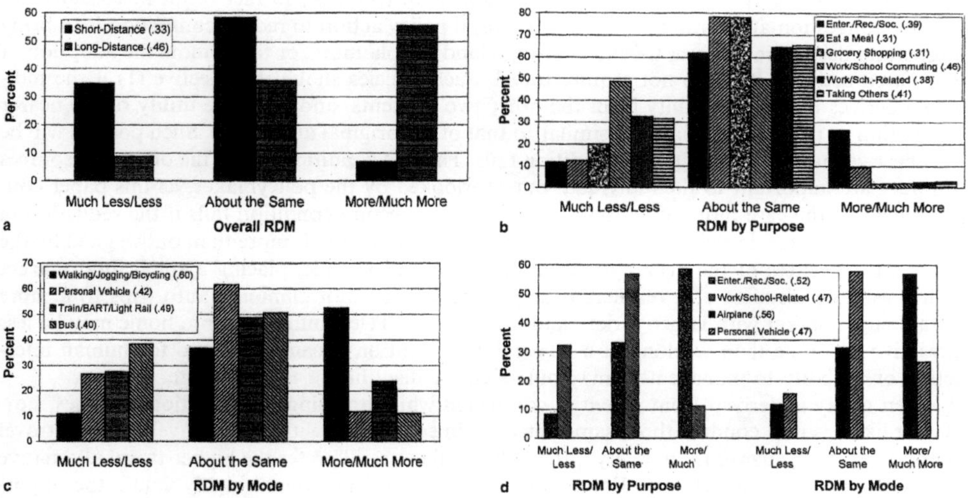

Fig. 6. (a) Overall relative desired mobility by distance category ($N = 1904$). (b) Relative desired mobility for short-distance travel by purpose ($N = 1904$). (c) Relative desired mobility for short-distance travel by mode ($N = 1904$). (d) Relative desired mobility for long-distance travel by purpose and mode ($N = 1904$). The numbers following each label in the legends are the correlations of each relative desired mobility variable with its travel liking counterpart in Fig. 2.

more universal desire to decrease it. Even if respondents are partially confounding the travel itself with the benefits of being at the destination, these results at a minimum point to substantial latent demand for those benefits, and hence for the travel required to achieve them. However, future studies should more carefully attempt to distinguish the respondent's desire to travel in order to engage in certain kinds of activities from his desire to travel for its own sake, perhaps through application of some form of the teleportation test.

6.3. Alternatives to a travel–activity combination

Applying the teleportation test leads naturally to an examination of alternatives within an individual's choice set. Understanding the choice set is crucial to modelling or predicting a behavioral response (Genç, 1994; Thill, 1992). That is, to be able to predict a choice, we need some knowledge of the competing alternatives. In particular, for policy reasons we are interested in identifying non-travel or lower-travel alternatives that may appeal to an individual.

A person's logical alternatives differ depending on the source of her utility. Consider each of the three elements of Section 4 in turn. Again, for simplicity we discuss each case as if it were "pure", but in reality the utility for a particular travel–activity combination may be composed of all three elements to varying degrees. Hence, all three elements should be considered in identifying alternatives in a specific choice context, and in evaluating the individual's utility for each alternative.

When utility is based entirely on the activities conducted at the destination, the scenario resolves to a conventional activity/destination choice. In the absence of a teleportation alternative that would potentially render network congestion moot (but may in fact result in considerable "point" congestion at desirable locations), a typical policy action to reduce congestion is to bring origins and destinations closer together through land use planning, or to promote the adoption of ICT-based activities that do not require travel. Such policies should be effective (1) if the individual derives little or no utility from the other two elements, and (2) if the utility of the nearer destination or the ICT alternative is similar to that of the original alternative. Such policies will be less effective than expected if either condition fails. The first condition fails if the other two sources of utility are important to the individual but overlooked by the policymaker, as this paper contends may be the case more often than we realize. The second condition fails if the reduction in travel for the nearer destination (or the no-travel ICT alternative) is more than outweighed by the lower attractiveness of the nearer destination. Thus, for example, placing a small food market within walking distance of a residential neighborhood may not eliminate auto trips to a more distant supermarket with larger variety and lower prices. Telecommuting from home may be less attractive for some than working at a conventional location because the need for human interaction or visibility to management may outweigh the disutility of the commute.

When utility is derived from activities conducted while traveling, the question becomes, how easy or likely is it to conduct those same activities in connection with a non-travel or lower-travel alternative? If the answer is "not very", the individual may still prefer the higher-travel alternative for its multitasking opportunity. For example, suppose a "carcooner" highly values the opportunity of listening to his favorite music at elevated decibel levels during the daily commute. Telecommuting may not offer an equivalent utility if the presence of others in the household prohibits exercising that particular proclivity at home. The traveler who values the opportunity to think or read while traveling may recognize that without the enforced physical idleness and isolation of traveling, those activities would be crowded out by competing demands for attention. It can be noted in passing that the Intelligent Transportation Systems goal of vehicle automation will have the effect of conferring some of the "category 2 appeal" currently characteristic of public means of transportation (that is, the opportunity to conduct other activities while traveling without having to operate a vehicle) onto the personal automobile, thus further increasing the attractiveness of auto relative to transit for some. For others, however, surrendering control of their vehicle might materially diminish the utility of driving.

When utility is drawn from the activity of traveling itself, it is possible to identify non-travel alternatives that may offer a similar experience. The characteristics described earlier – the sensation of speed, the experience of and movement through the environment, operational skill, scenic beauty – can all be simulated through various applications of virtual reality. Driving or flight simulators can provide an increasingly realistic operational experience. Multimedia tours of museums and scenic attractions are increasingly available on the Internet. The question is, will these ersatz alternatives be as satisfactory as the real thing? Clearly, there may be situations in which constraints such as cost, time, or physical ability have the result that the utility of the virtual alternative exceeds that of the real one. But it seems probable that for many people under many circumstances, the real thing will continue to have an edge no matter how realistic the virtual alternatives become (Kenner, 1998). Indeed, the possibility cannot be overlooked that the growing reality and pervasive availability of virtual alternatives may increasingly stimulate the appetite of society in the aggregate for the real thing. If that is true, then the closer we get to teleportation (that is, the faster and cheaper travel becomes), the more local congestion is likely to escalate in especially desirable locations and along especially desirable/scenic routes. Already, our ability to offer ecologically sustainable tourism opportunities, in an era of rising global prosperity and hence rising demands on fragile environments, is a matter of serious concern (see, e.g., Williams, 1998).

7. Summary and conclusions

This paper contests the conventional wisdom that travel is a derived demand, at least as an absolute. We do not dispute the principle that most travel is derivative, but we also argue that humans possess an intrinsic desire to travel, a point previously made by a number of researchers in a variety of disciplines and geographic locations. We discuss the phenomenon of undirected travel – cases in which travel is not a byproduct of the activity but itself constitutes the activity. We note that while undirected travel is predominantly a leisure activity, it is relevant to our understanding of directed travel as well. For example, the same reasons why people enjoy undirected travel (a sense of speed, motion, control, enjoyment of beauty) may motivate them to undertake excess travel even in the context of mandatory or maintenance trips.

One characteristic of undirected travel is that the destination is ancillary to the travel rather than the converse which is usually assumed. We argue that the destination may be more or less ancillary more often than is realized. The paradigm of travel as a derived demand requires that we view the destination/activity as generating the trip, when on some occasions it may be the desire to travel that prompts the invention of a spatially-removed activity to satisfy that desire. It is impossible to distinguish those opposite directions of causality without further insight into attitudes toward traveling.

The share of total travel that is completely undirected is presumably relatively small, although we do not know how small because travel/activity/time use diaries are likely to classify such travel (sailing, skiing, etc.) as a leisure activity rather than as undirected travel. However, it is theoretically straightforward to modify standard data collection instruments to make that distinction, if it is deemed worth doing. It is not at all straightforward to identify the extent of excess travel undertaken in connection with destination-oriented activities, including mandatory and maintenance ones. The total is probably still small, but it may be large enough to matter at some level, and it may also be growing (as a function of improving economic conditions worldwide).

Even if the amount of excess travel is small, it is important to clarify our understanding of individuals' basic motivation to travel. If travel has a positive utility to some extent, it is desirable to develop ways to measure the extent and circumstances for which that is true. But in self-reports of attitudes toward travel, respondents are likely to confound their utility for traveling itself with their utility for the activities conducted at the destination and for activities conducted while traveling. Despite this measurement challenge, preliminary empirical results from a study of more than 1900 residents of the San Francisco Bay Area provide suggestive evidence of a positive utility for travel. For example, more than three-quarters of the sample reported sometimes or often traveling "just for the fun of it". More than two-thirds disagreed that "the only good thing about traveling is arriving at your destination".

Instead of the constant (observed) TTB favored by some transportation analysts, we hypothesize the existence of an unobserved desired TTB, which varies by attitudes, personality traits, demographic variables, mode, and purpose. The individual's ideal commute time is a subset of this desired TTB; the average ideal one-way commute time in our sample was 16 min. Measures of relative desired mobility (the amount an individual wants to travel compared to the present) identified a variety of circumstances under which our respondents wanted to maintain or increase their travel.

With the current data collection instrument, we are only imperfectly able to distinguish the three components of a utility for travel mentioned above. For example, our measurements of travel affinity, ideal commute time, and relative desired mobility probably confound all three elements. However, future analyses of the data will focus on modelling these variables as functions of other indicators of the presence of each of those three elements. Doing so will help identify the extent to which each element influences the dependent variable. It would also be desirable to search for segments within the sample who weight the three elements differently. These initial results and further analyses of these data will provide valuable insight into the desire to travel, and will point the way to refinement of future data collection instruments having similar goals.

The concepts presented here may raise some questions about the current practice of transportation planning. For example, one might ask whether, if people are not time-or cost-minimizers when it comes to travel, using travel time savings as a basis for valuing capacity enhancements is valid. Our results do not necessarily point to discarding that approach (although other legitimate concerns have been expressed in this regard, especially about the practice of monetizing time savings; see, e.g., Atkins, 1984; Calfee and Winston, 1998; Welch and Williams, 1997). Suppose people do have a non-zero desired TTB. Then, first, many people will be exceeding that desired budget, and even though they would not want to eliminate all time spent traveling, they still want to reduce the time they are currently spending. Second, even those people who are currently traveling their desired amount probably want to minimize their exposure to congested travel conditions. Hence, they would value a capacity enhancement that reduced the time they spend in congestion – even if they then allocated that time savings to additional travel in order to fill up their desired budget. All things considered, then, travel time savings probably still serve as a useful, measurable proxy for the theoretically superior but nebulous and impracticable "increase in utility" that is the real measure of traveler benefit.

On the other hand, it may be productive to pursue a more nuanced approach toward the treatment of travel time in regional modelling. The literature has already identified considerable variation in the monetary value of travel time by characteristics of the trip and the traveler. Mode choice models typically differentiate between in-vehicle travel time and out-of-vehicle travel time, with the latter considered more onerous (and therefore having a more negative coefficient in the utility function). Hupkes (1982) takes this a step farther by distinguishing the "derived" component of travel utility from its "intrinsic" component, and suggesting that both of them initially rise with travel time (at different rates), and then fall after the point that is optimal for each component. Under this conceptualization, travel time would not always be assumed to have a negative impact on utility (as is currently the case even for those studies that identify different coefficients of travel time under different circumstances); rather a non-linear function is obtained, containing a segment over which the net impact of travel time on utility is positive.

In sum, the issues raised here have important implications for transportation planning in general and travel demand analysis in particular. The way people will react to policies intended to reduce vehicle travel will depend in part on the relative weights they assign to these three components of a positive utility for travel, and on whether they desire more or less mobility than they currently experience. Only in cases in which the positive utility derives completely from the activities conducted at the destination will a lower-travel alternative be preferred, *ceteris paribus*. Although non-travel alternatives are available that may partially satisfy each of the three components of utility, those alternatives will often not be as desirable as traveling. Ultimately, improving our forecasts of travel behavior may require viewing travel literally as a "good" as well as a "bad" (a disutility), and modelling the demand for that good as we do for other goods. As we have seen, the demand for travel is a function of fundamental human characteristics as well as the external variables typically measured, and those relationships need to be understood much better than we do at present.

Acknowledgements

This work is funded by the University of California Transportation Center and the Daimler-Chrysler Corporation. Michael Bagley provided early input to the survey design. Lothlorien Redmond has done an outstanding job of managing the survey design and production, and has carefully overseen the data collection, entry, cleaning, and analysis to date. She has been recently assisted by Richard Curry, who also prepared the graphics for this paper. Naomi Otsuka has ably and intelligently assisted with data entry and cleaning, and literature searches. Some of the ideas presented in this paper have been sharpened by discussion with participants at the European Science Foundation/National Science Foundation conference on Social Change and Sustainable Transport, Berkeley, CA, 10-13 March 1999, and by discussions with Qiuzi (Cynthia) Chen and Hani Mahmassani. Insightful comments from Rick, Lorien, Naomi, two referees, and Associate Editor Yossi Berechman were most helpful in improving earlier drafts.

References

Anable, J., 1999. Picnics, pets and pleasant places: The distinguishing characteristics of leisure travel demand. Paper presented at the European Science Foundation/National Science Foundation Conference on Social Change and Sustainable Transport (SCAST), Berkeley, CA, 10–13 March.

Atkins, S.T., 1984. Why value travel time? The case against. Highways and Transportation 31 (7).

Barnard, P.O., 1987. Modelling shopping destination choice behaviour using the basic multinomial logit model and some of its extensions. Transport Reviews 7 (1), 17–51.

Berger, M.L., 1992. The car's impact on the American family. In: Wachs, M., Crawford, M. (Eds.), The Car and the City: The Automobile, The Built Environment and Daily Urban Life. University of Michigan Press, Ann Arbor, MI, pp. 57–74.

Bieber, A., Massot, M.-H., Orfeuil, J.-P., 1994. Prospects for daily urban mobility. Transport Reviews 14 (4), 321–339.

Calfee, J., Winston, C., 1998. The value of automobile travel time: Implications for congestion policy. Journal of Public Economics 69, 83–102.

Chlond, B., Zumkeller, D., 1997. Future time use and travel time budget changes – estimating transportation volumes in the case of increasing leisure time. Paper presented at the Eighth Conference of the International Association for Travel Behaviour Research, Austin, TX, 21–25 September.

Crawford, M., 1992. The fifth ecology: Fantasy, the automobile, and Los Angeles. In: Wachs, M., Crawford, M. (Eds.), The Car and the City: The Automobile, the Built Environment and Daily Urban Life. University of Michigan Press, Ann Arbor, MI, pp. 222–233.

Dillman, D.A., 1978. Mail and Telephone Surveys: The Total Design Method. Wiley, New York.

Edmonson, B., 1998. In the driver's seat. American Demographics (March), 46–52.

Flink, J.J., 1975. The Car Culture. MIT Press, Cambridge, MA.

Genç, M., 1994. Aggregation and heterogeneity of choice sets in discrete choice models. Transportation Research B 28 (1), 11–22.

Goodwin, P.B., 1981. The usefulness of travel budgets. Transportation Research A 15, 97–106.

Gunn, H.F., 1981. Travel budgets – a review of evidence and modelling implications. Transportation Research A 15, 7–23.

Hupkes, G., 1982. The law of constant travel time and trip-rates. Futures, 38–46.

Jones, P.M., 1978. Destination choice and travel attributes. In: Hensher, D., Dalvi, Q. (Eds.), Determinants of Travel Choice. Praeger, New York, pp. 266–311.

Kenner, Hugh, 1998. The Elsewhere Community. House of Anansi Press, Concord, Ont.

Kitamura, R., Fujii, S., Pas, E.I., 1997. Time-use data, analysis and modeling: Toward the next generation of transportation planning methodologies. Transport Policy 4 (4), 225–235.

Lanzendorf, M., 1999. Social change and leisure mobility. Paper presented at the European Science Foundation/ National Science Foundation Conference on Social Change and Sustainable Transport (SCAST), Berkeley, CA, 10–13 March.

Larson, J., 1998. Surviving commuting. American Demographics (July).

Marchetti, C., 1994. Anthropological invariants in travel behavior. Technological Forecasting and Social Change 47, 75–88.

Marsh, P., Collett, P., 1986. Driving Passion: The Psychology of the Car. Faber and Faber, Boston, MA.

Mokhtarian, P.L., Raney, E.A., Salomon, I., 1997. Behavioral response to congestion: Identifying patterns and socio-economic differences in adoption. Transport Policy 4 (3), 147–160.

Mokhtarian, P.L., Salomon, I., 1997. Modeling the desire to telecommute: The importance of attitudinal factors in behavioral models. Transportation Research A 31 (1), 35–50.

Muller, P.O., 1986. Transportation and urban form: Stages in the spatial evolution of the American metropolis. In: Hanson, S. (Ed.), The Geography of Urban Transportation. Guilford Press, New York, pp. 24–48.

Papacostas, C.S., Prevedouros, P.D., 1993. Transportation Engineering and Planning, second ed. Prentice-Hall, Englewood Cliffs, NJ.

Ramon (Perl), C., 1981. Sociological aspects in the analysis of travel behavior in an urban area: Jerusalem as a model. Ph.D. Dissertation, The Hebrew University, Jerusalem (in Hebrew).

Redmond, L.S., Mokhtarian, P.L., 1999. The positive utility of the commute: Modeling ideal commute time and relative desired commute amount. Transportation, submitted.

Reichman, S., 1976. Travel adjustments and life styles – a behavioral approach. In: Stopher, P.R., Meyburg, A.H. (Eds.), Behavioral Travel-Demand Models. D.C. Heath and Company, Lexington, MA, pp. 143–152 (Chapter 8).

Richter, J., 1990. Crossing boundaries between professional and private life. In: Grossman, H., Chester, L. (Eds.), The Experience and Meaning of Work in Women's Lives. Lawrence Erlbaum, Hillsdale, NJ, pp. 143–163.

Robinson, J.P., Godbey, G., 1999. Time for Life: The Surprising Ways Americans Use Their Time. Pennsylvania State University Press, University Park, Penn.

Sachs, W., 1992. For Love of the Automobile: Looking Back into the History of Our Desires. Translated from German by Don Reneau, University of California Press, Berkeley, CA (originally published as Die Liebe zum Automobil: ein Rückblick in die Geschichte unserer Wünsche, 1984).

Salomon, I., 1985. Telecommunications and travel: Substitution or modified mobility? Journal of Transport Economics and Policy 19, 219–235.

Salomon, I., Mokhtarian, P.L., 1998. What happens when mobility-inclined market segments face accessibility-enhancing policies? Transportation Research D 3 (3), 129–140.

Schafer, A., Victor, D., 1997. The past and future of global mobility. Scientific American (October), 58–61.

Shamir, B., 1991. Home: The perfect workplace? In: Zedeck, S. (Ed.), Work and Family. Jossey-Bass, San Francisco, CA, pp. 273–311.

Smithson, Michael, 1987. Fuzzy Set Analysis for Behavioral and Social Sciences. Springer-Verlag, New York.

Sommer, B., Sommer, R., 1997. A Practical Guide to Behavioral Research: Tools and Techniques, fourth ed. Oxford University Press, New York.

Tanner, J.C., 1981. Expenditure of time and money on travel. Transportation Research A 15, 25–38.

Thill, J.-C., 1992. Choice set formation for destination choice modelling. Progress in Human Geography 16 (3), 361–382.

Tillberg, K., 1999. The relations between residential location and daily mobility patterns: A Swedish case study of households with children. Paper presented at the European Science Foundation/National Science Foundation Conference on Social Change and Sustainable Transport (SCAST), Berkeley, CA, 10–13 March.

Wachs, M., Crawford, M. (Eds.), 1992. The Car and the City: The Automobile, the Built Environment and Daily Urban Life. University of Michigan Press, Ann Arbor, MI.

Welch, M., Williams, H., 1997. The sensitivity of transport investment benefits to the evaluation of small travel-time savings. Journal of Transport Economics and Policy 31 (3), 231–254.

Williams, S., 1998. Tourism Geography. Routledge, New York.

Young, W., Morris, J., 1981. Evaluation by individuals of their travel time to work. Transportation Research Record 794, 51–59.

Zahavi, Y., Talvitie, A., 1980. Regularities in travel time and money expenditures. Transportation Research Record 750, 13–19.

Zimmermann, H.J., 1985. Fuzzy Set Theory – and its Applications. Kluwer-Nijhoff, Boston, MA.

Access, Networks and Development

[5]

SPATIAL REORGANIZATION: A MODEL AND CONCEPT[1]

DONALD G. JANELLE
USAF Academy, Colorado

ABSTRACT. Travel-time connectivity is a key factor in defining a process of the spatial reorganization of man's functional establishments. A case study relating highway development with the growth in wholesale activity for selected cities in the upper midwest of the United States indicates that, aside from being a good surrogate of transport efficiency, travel-time connectivity is also a good measure of the relative advantage of a given place in attracting to itself the centralization and specialization of human activity.

A functional framework which includes a measure of the friction of distance, such as time or cost of travel, seems essential in a study of central place development. Furthermore, as Blaut noted, structure (the areal arrangement of earth-space phenomena) and process (the rearrangement of these phenomena over time) are one and the same thing—that is: " . . . structures of the real world are simply slow processes of long duration."[2] Inherent in Blaut's view is the implicit existence of a temporal pattern in each and every spatial pattern.[3] Thus, these two factors, the friction of distance (measured in travel-time) and historical development, have been incorporated into the following statement of a model of spatial reorganization.[4]

A MODEL OF SPATIAL REORGANIZATION

In this study, the concept of spatial reorganization identifies a process by which places adapt both the locational structure and the characteristics of their social, economic, and political activities to changes in time-space connectivity (the time required to travel between desired origins and destinations). As an example of such areal reorganization, Fox noted how, for the food retailing industry, spatial adaptations to advances in transportation have tended towards fewer, larger, and more distantly spaced establishments—an abandonment of the corner grocery store in favor of the supermarket.[5]

A model has been designed to depict a normative process of such areal development. Later this model (the basic model) will be

---

Accepted for publication January 15, 1968.

---

[1] This study represents a revision of portions of the author's doctoral dissertation, *Spatial Reorganization and Time-space Convergence,* completed at Michigan State University in 1966. For their kind assistance, the author extends a special thanks to Dr. Julian Wolpert, Dr. Allen K. Philbrick, Dr. Robert C. Brown (Simon Fraser University), and to Dr. Donald A. Blome. This investigation was supported (in part) by a Public Health Service fellowship (number 1-F1-CH-31, 220) from the Division of Community Health Services. The views expressed herein are those of the author and do not necessarily reflect the views of the U. S. Air Force or the Department of Defense.

[2] J. M. Blaut, "Space and Process," *The Professional Geographer,* Vol. 13 (July, 1961), p. 4.

[3] This notion, first recognized by the physicists and physical philosophers, has been acknowledged by many geographers and other social scientists, including W. J. Cahnman, "Outline of a Theory of Area Studies," *Annals,* Association of American Geographers, Vol. 38 (1948), pp. 233–43; A. H. Hawley, *Human Ecology: A Theory of Community Structure* (New York: The Ronald Press, 1950), p. 288; W. Isard, *Location and Space-Economy* (Cambridge, Mass.: The Massachusetts Institute of Technology Press, 1956), p. 11; F. Lukermann, "The Role of Theory in Geographical Inquiry," *The Professional Geographer,* Vol. 13 (March, 1961), p. 1; and R. L. Morrill, "The Development of Spatial Distributions of Towns in Sweden: A Historical-Predictive Approach," *Annals,* Association of American Geographers, Vol. 53 (1963), pp. 2–3.

[4] The term "spatial reorganization" is not new. It has been used by W. L. Garrison. See "Notes on Benefits of Highway Improvements," in W. L. Garrison, B. J. L. Berry, D. F. Marble, J. D. Nystuen, R. L. Morrill, *Studies of Highway Development and Geographic Change* (Seattle: University of Washington Press, 1959), p. 23. This and other impact studies of the post-1956 period suggests that although the term has not seen wide use, the concept is one of immediate concern.

[5] K. A. Fox, "The Study of Interactions Between Agriculture and the Nonfarm Economy: Local, Regional and National," *Journal of Farm Economics,* Vol. 44 (February, 1962), pp. 1–34.

expanded so as to present a more comprehensive view. Although these models are intended to be applicable to urban-exchange economies typical of the United States and Western Europe, the writer believes that they may have some predictive value in forecasting the areal development of areas which have only recently begun progressing through the industrial-commercial revolution. Before describing the models, a concept which is central to the overall process of reorganization needs to be considered—this is the notion of *locational utility.*

Locational Utility

Very simply, utility is a measure of value. However, the term locational utility used in this study should be distinguished from *place utility* as defined by Wolpert.[6] Wolpert recognized in his discussion on the decision to migrate that utility is inherently individualistic. Thus, place utility is an individual's subjective measure of the degree to which the opportunities at a particular place permit his perceived or actual achievement level to be in as close as possible accordance with his aspiration level. By integrating this individualistic concept with information on the life cycles, life styles, and life spaces of specific socioeconomic groups, Wolpert developed an aggregate measure of the utility of specific places relative to the mover-stayer decision.

In contrast to place utility, locational utility is defined in a context which, in part, overlooks the individualistic and subjective connotation of value. It is a measure of the utility of specific places or areas, which in this case is defined by the aggregate time-expenditure (cost or effort) in transport required for that place or area to satisfy its operational needs.[7] Operational need refers to those natural and human resource requirements which permit the place or area to fulfill its functional roles in the larger spatial system of places and areas. The alternative possibilities of a place, either to decrease, maintain, or increase its existing competitive

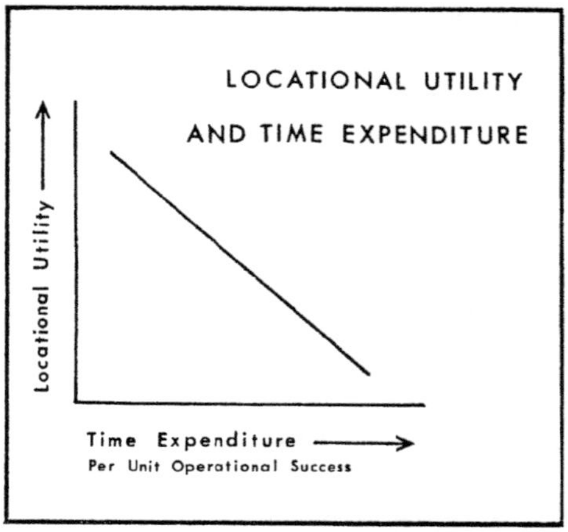

FIG. 1. Locational utility and time expenditure per unit operational success. Some curvilinear function might express the reality of this relationship better than the linear function represented in this graph.

status within the bounds of either its present spatial system of socioeconomic activities or in an expanded sphere of influence, are here considered to be functions of its locational utility. Thus, as shown in Figure 1, the locational utility or value of places and areas increases as travel-time expenditure per unit of operational success (profit or some other form of amenity benefit) decreases. Whereas

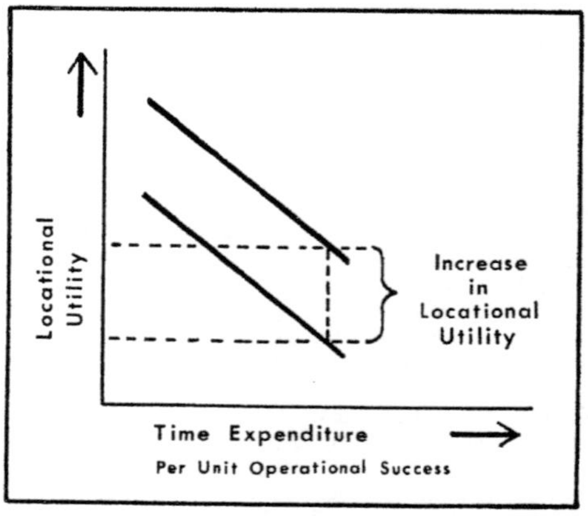

FIG. 2. The increase in locational utility for time period t_1 through t_2 resulting from the introduction of a transport innovation in time t_2.

[6] J. Wolpert, "Behaviorial Aspects of the Decision to Migrate," *Papers, Regional Science Association,* Vol. 15 (1965), pp. 159–69.

[7] The terms place and area are used as designators of areal scale. In this study they are used interchangeably.

a first degree linear function is used to express this relationship, it is likely that a second, third, or higher degree function would be more appropriate.

In reality, the spatial variance in the locational utility for a system of places may be characterized by surfaces of utility. In a given spatial system there exists one surface for each of the many possible functional roles to be performed. Theoretically, with possible loss of much information, it might be feasible to treat these surfaces in an additive sense and to arrive at a surface of composite locational utility for the system.

Once the surface of utility has been described, one can then focus attention on a more significant problem—the dynamics of surface change. For example, the depletion or the discovery of a resource which is an operational need for the success of a given economic activity would alter the utility surface for that activity, and could necessitate the selection of a new production site.

In that locational utility is defined as a function of time-expenditure, it is evident that innovations which speed transportation will also lead to changes in the utility surface. Thus, for a given place, the increase in locational utility from time t_1 to time t_2 that is derived from a transport innovation at time t_2 is indicated in Figure 2. Such changes pose many questions of practical relevance. For example, are these innovations and certain distributive forces leading towards greater equilibrium in the utility surface and, thus, possibly towards a more homogeneous distribution of man's socioeconomic activities? Or, do transport improvements and certain agglomerative forces lead to increasing spatial variance in locational utility and, thus, to-

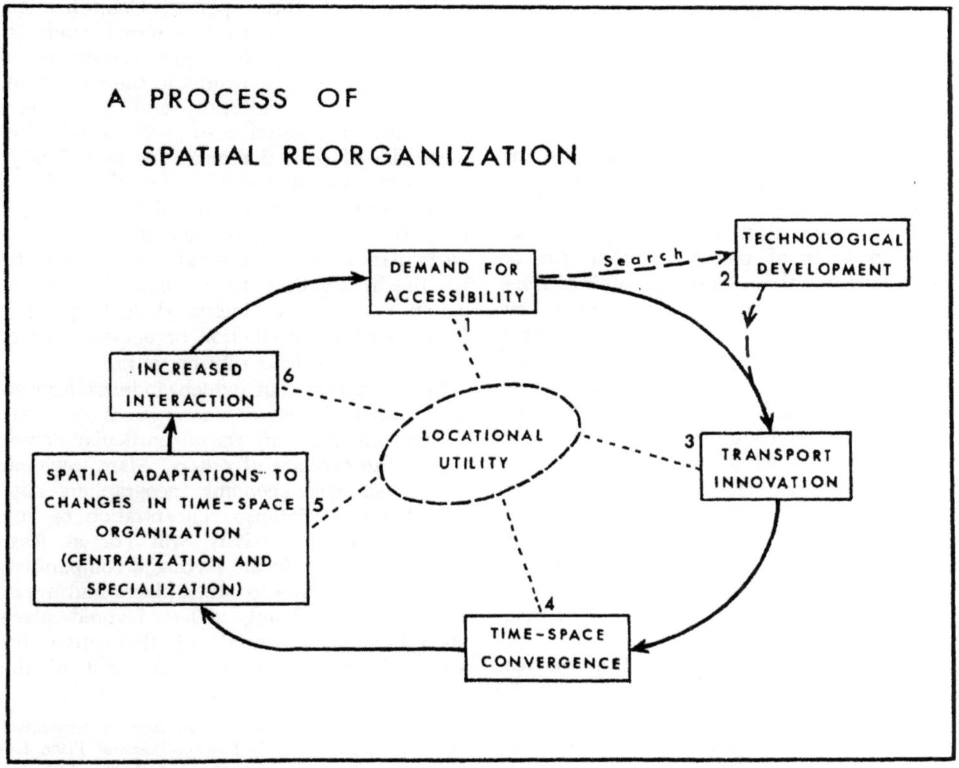

FIG. 3.

wards greater place-concentration of human enterprise?

These questions, along with the process of spatial reorganization will be clarified as the concepts integrated into the model in Figure 3 are defined. These concepts include:

1) Demand for accessibility;
2) transport innovations;
3) time-space convergence;
4) spatial adaptations—centralization and specialization; and
5) spatial interaction.

Demand for Accessibility

Accessibility is a measure of the ease (time, cost, or effort) in which transfer occurs between the places and areas of a system. The demand for accessibility, then, is really a quest to decrease the transport effort expended per unit of operational success or, very simply, to augment locational utility. A useful and more objective measure of accessibility (not used in this study) is provided by the graph theoretic approaches employed by Garrison, Kansky, and others.[8]

Transport Innovations

In this study, transport innovations are any technologies or methods which serve to increase accessibility between places or which permit an increase in the quantity of goods or the number of passengers that can be moved between these places per unit of time. Thus, a transport innovation may be a new and faster type of carrier, improved traffic routing procedures, better gasoline, improved lighting for night travel, the straightening of angular routes, and so forth. All such introductions are likely to result in what the author describes as time-space convergence (step 4 of the model).

Time-space Convergence

By time-space convergence, the writer is implying that, as a result of transport innovations, places approach each other in time-space; that is, the travel-time required between places decreases and distance declines in significance.[9] An example of this phenomenon is illustrated in Figure 4 for travel between Detroit and Lansing, Michigan. As a consequence of such convergence, man has found that it is possible and practical to adapt the spatial organization of his activities to their evolving time-space framework (step 5 of the model).

Spatial Adaptations to Changes in Time-space Organization

In the basic model under consideration, the spatial adaptations of man's activities to their changing time-space framework will lead to the centralization and specialization of secondary and tertiary economic activities in specific places and, as is frequently the case, to the specialization of primary economic activities in the resource-oriented hinterlands of these places. Centralization (of which urbanization is a form) refers to the increasing focus of human activity upon a particular place; it results in the growth of an economically, culturally and, sometimes, politically integrated area over which this particular place is dominant (its hinterland). The economies that result when the scale of an economic, political, or cultural endeavor is increased at a particular place or in a particular area are generally considered to be the motivating forces behind centralization. As a rule, increased scale permits lower per-unit production or operation costs —unless diminishing returns set in.

Specialization (of which industrialization is a form) develops when places or areas concentrate their efforts on particular activities at the expense of others. Many regional economists and economic geographers note that the most intense concentration of any given economic activity will (or at least should) be in a locale having a comparative advantage relative to other places and areas. On the other hand, a less favored place should choose to specialize in that activity for which it has (relative to the rest of the

[8] W. L. Garrison, "Connectivity of the Interstate Highway System," *Papers and Proceedings of the Regional Science Association,* Vol. 6 (1960), pp. 121–37; K. J. Kansky, *Structure of Transportation Networks* (Chicago: Department of Geography Research Paper No. 84, University of Chicago Press, 1963).

[9] For a more thorough discussion of time-space convergence, see D. G. Janelle, "Central Place Development in a Time-space Framework," *The Professional Geographer,* Vol. 20 (1968), pp. 5–10.

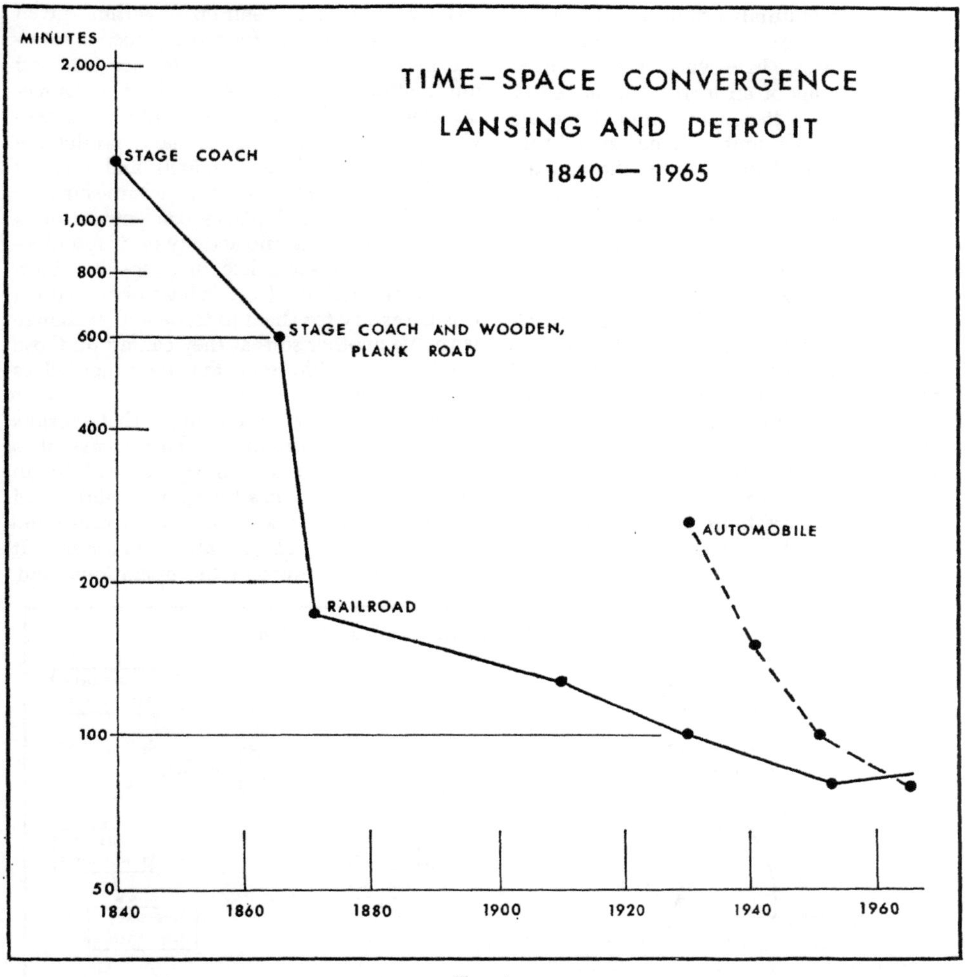

FIG. 4.

system) a least comparative disadvantage.[10] For the mechanistic model in question, the surfaces of locational utility dictate the specialties of places and areas.

The greater the centralization and specialization of man's activities, the greater is the need for efficient transport and increased locational utility (steps 1–4 of the basic model). As man speeds up his means of movement, it becomes possible for him to travel further in a given time, to increase his access to a larger surrounding area and, possibly, to more and better resources. This idea is in line with Ullman's concept of transferability.[11] Likewise, secondary and tertiary functions can serve more people; and the

[10] For a discussion on the law of comparative advantage, see P. A. Samuelson, "The Gains from International Trade," *Canadian Journal of Economics and Political Science,* Vol. 5 (May 1939).

[11] E. L. Ullman, "The Role of Transportation and the Bases of Interaction," in W. L. Thomas, Jr. (Ed.), *Man's Role in Changing the Face of the Earth* (Chicago: University of Chicago Press, 1956), pp. 862–80.

perishable agricultural products and other primary products can be profitably marketed over a larger area. In essence these changes are manifestations of an increasing degree of locational utility (greater operational success can be derived per unit of time-expenditure from a given place) that permits the increasing centralization and specialization of human endeavors. Thus, these scale economies are, in part, both forms of spatial adaptation to an evolving time-space framework.

Unlike centralization and specialization, suburbanization (a form of spatial decentralization) represents an alternative response to time-space convergence which is not treated in this basic model. Improvements in individual mobility have made it possible for some families and for some firms to trade off central accessibility for the amenities associated with suburban life and industrial parks. These adaptations are considered in the expanded model of spatial reorganization (Fig. 5).

tertiary activities centralize within given places, it is necessary for those places to interact in the forms of products, service, and information exchange with their resource-oriented hinterlands. These hinterlands provide the necessities of primary production and they demand the products and services of secondary and tertiary establishments. Similarly, if central places concentrate on a given type of economic activity or, if resource-oriented areas specialize in a specific form of primary activity (*e.g.*, wheat or iron ore), it is necessary for them to trade and exchange with one another so that they can attain those needs or desirable items that they themselves do not produce.

The increasing intercourse that results from the concentration of human activities at particular places is likely to lead to an overextension of man's transport facilities and result in their deterioration from overuse and in the development of traffic congestion. It is, therefore, likely that the operational success of these places can only be continued through increased costs. Consequently, the increasing interaction that results from centralization and specialization leads to further demands for increased accessibility, greater degrees of locational utility, and transport innovations (steps 6, 1–3 in basic model). Thus, the spatial reorganization of human activities is perpetuated in what, theoretically

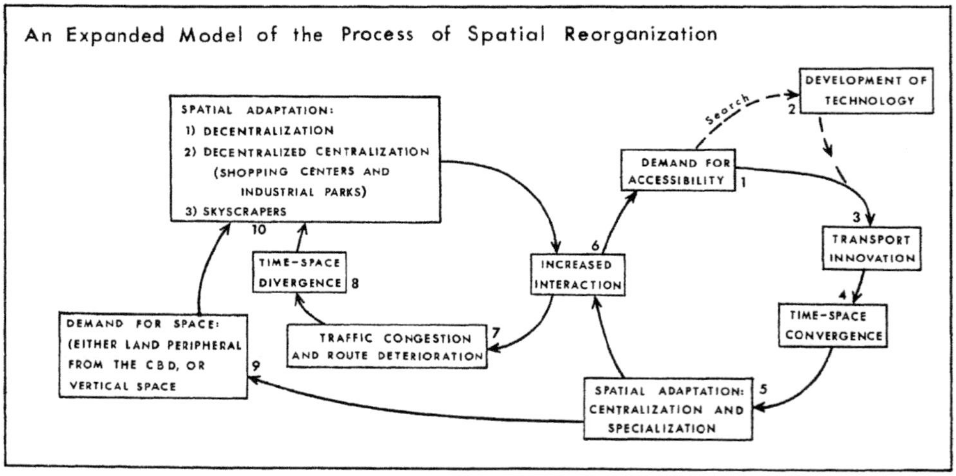

FIG. 5.

Interaction

Step 6 of the basic model indicates that an increase in interaction results between places and areas that experience increasing centralization and specialization.[12] As secondary and

[12] For a more complete treatment of this notion, see Ullman, *op. cit.*, footnote 11.

is a never ending and accelerating cycle. This notion of a multiplier effect or positive feedback[13] implies that the state of a system (that is, the degree of convergence between interacting settlements, their demands for accessibility, and so forth) at a given time is determined completely from within the system and by the previous state of the system.[14] Thus, the positive feedback system, as indicated by the completed circuit in the basic model, is self-perpetuating.

Support for the notion that a transport improvement, in itself, encourages increased interaction is also available. Studies by Coverdale and Colpitts show that improvements in highway facilities result in traffic volumes greater than the number accounted for by the diverted traffic.[15] That is, many new facilities (*i.e.*, bridges and freeways) will attract considerably more traffic than would be expected had the previous facilities continued to operate alone. This increase is frequently termed induced traffic. The volumes of induced traffic encouraged by bridges replacing ferries have ranged, in many instances, from sixty-five percent to seventy-five percent of that before the improvement. For the Philadelphia-Camden Bridge it was seventy-eight percent, and for the San Francisco-Oakland Bay Bridge it was about sixty-four percent. This finding lends additional support for the inclusion of a positive feedback system in the basic model of spatial reorganization.

AN EVALUATION OF THE BASIC MODEL

Changes in the time-spatial and spatial organization of human endeavors present places and areas with possibilities for greater scale economies and with problems of developing more efficient means of transport. It is man's awareness or perception of these possibilities and problems that enables him to take advantage of the changes in the time-space structure of his activities. In reality, however, the process that has been described is not so simple—not all men will perceive the changes described in the model nor will they see the implications of these changes in the same way. Furthermore, some of the assumptions of the model lack complete accord with reality.

Varying Conceptions of Utility and Time

Whereas the basic model is based on an objective measure of locational utility (time-expenditure), it was indicated earlier that utility is inherently individualistic; that is, it is perceived according to one's values, goals, and technical and institutional means of living. At the level of places and areas it is likely that the criteria for utility are based on factors other than just the expenditure of time.

It is also apparent that man's perceived value of a given unit of time has increased as the tempo of his activities has increased. A component to represent this change is not included in the model. Yet, by sole reason of the tremendously greater commodity, passenger, and information flows today as compared with past periods, man is motivated to seek greater utility for his expenditure of time. Imagine the magnitude of storage that would be necessary if New York City had to store food for its population to meet their needs over the winter months. With faster transport, the city can rely on more distant sources. Food can be moved to the city when it is needed, thus reducing its storage costs and increasing its operational success.

There is the additional likelihood that a person's perception of the utility of time will differ for various travel purposes. For example, an individual may be willing to spend an hour in travel to receive the medicinal services of an eye specialist; but, he may only grudgingly give up ten minutes to purchase a loaf of bread. No model of the process of spatial reorganization could account for all of the multitudinous goals and criteria of all persons, places, and areas, and the

---

[13] The application of the feedback concept in this context was suggested to the author by James H. Stine, Department of Geology and Geography, Oklahoma State University.

[14] An excellent discussion on both positive and negative feedback systems is provided by M. Maruyama, "The Second Cybernetics: Deviation-Amplifying Mutual Causal Processes," *American Scientist*, Vol. 51 (1963), pp. 164–79; also, see L. von Bertalanffy, "General System Theory," *General Systems*, Vol. 1 (1956), pp. 1–10.

[15] Coverdale and Colpitts, Consultant Engineers, *Report on Traffic and Revenues, Proposed Mackinac Straits Bridge* (New York: Coverdale and Colpitts, January 22, 1952), p. 18.

changing value of time for each. Therefore, in the development of the model, a standard pattern of human place-behavior has been assumed.

A Basic Assumption: Rationality in Human Place-Behavior

The principal assumption upon which the spatial process model is based is that man is rational. This concept of a rational man or the economic man has been well developed elsewhere and only the implications relevant to this discussion are presented.[16] These include the following:

1) Man has perfect knowledge. Thus, in an aggregate sense, places and areas show complete awareness of all factors operative in the areal reformation of their activities; they are aware of all their operational needs and of all the possibilities for fulfilling these needs.
2) Man has no uncertainty—he has perfect predictability. Thus, the rational place foresees the time-space convergence that will result from any transport innovation; it foresees the degree of increased interaction that will be derived from greater centralization and specialization of its activities.
3) Man is interested solely in maximizing the utility of time at a given place. This permits the necessary spatial reorganization to augment the operational success of places. Net social benefits, inclusive of all possible benefits—whether economic, political, or cultural—could be substituted for operational success.

Limitations and Omissions of the Model

Although the inclusion of the rational place concept may limit the correspondence of the basic model with reality, it does permit one to consider the process of reorganization under controlled circumstances with a minimum of conflicting factors (*i.e.*, changing criteria and varying degrees of rationality that have to be accounted for). Other factors which are, in part, attributable to a lack of perfect rationality may be summed up as perceptive, responsive, and technological lags. Spatial change is not necessarily characterized by a smooth flow through the six-step process identified in the basic model.

There may be lags or delays in the process resulting from man's inefficient behavior—his slowness in adapting the spatial organization of human activity to its changing time-space framework, or his slowness in introducing more efficient forms of transportation. It is also possible that improvements in transfer technology will lag behind the need for such development. It will be noted from Figure 3 that the development of technology, although intimately related, is considered exogenous to the system depicting the process of spatial change. Such development may take place independently of any need present within the system—innovations developed for an entirely different purpose may be readily applicable to transportation.

AN EXPANDED MODEL OF SPATIAL REORGANIZATION

If the restraints of rationality, as defined above, are relaxed, and, if another factor, the demand for land, is introduced, then the mechanism of the basic model breaks down. In reality, places and areas do not always seek to maximize their degree of locational utility and, in many cases, they find it impossible to do so. Thus, if there is no demand for increased accessibility in response to increases in interaction or if there is no technology available for meeting demands for greater accessibility, then it is likely that either traffic congestion, route deterioration, or both will occur. This, in turn, would lead to time-space divergence (places getting further apart in time-space). This is indicated by steps 7 and 8 of the expanded model depicted in Figure 5.

The demand for land or space (step 9) is a form of decentralization which is a direct consequence of the centralization and specialization associated with time-space convergence. Factories, warehouses and so forth, which seek to augment scale economies, find

[16] R. M. Cyert and J. G. March, *A Behavioral Theory of the Firm* (Englewood Cliffs, N. J.: Prentice-Hall, Inc., 1963); J. H. Henderson and R. E. Quandt, *Micro-Economic Theory, A Mathematical Approach* (New York: McGraw-Hill Book Co., Inc., 1958); H. A. Simon, "Some Strategic Considerations in the Construction of Social Science Models," in P. F. Lazarsfeld (Ed.), *Mathematical Thinking in the Social Sciences*, (Glencoe, Ill.: The Free Press, 1954), pp. 388–415.

land scarce and expensive in the central areas of cities and, thus, move to the peripheries of the built-up areas where it is available and comparatively cheap. Jobs created by this expansion may increase the population attraction power of places and lead to further demands for land. Additional factors accounting for a demand for land peripheral to the built-up areas include the population holding power of the urban area itself and the amenity goal to gain more elbow room—to get out of the noisy, crowded city. This demand to leave the central city results as interaction accelerates beyond a tolerable threshold. It seems likely that this demand for land coupled with time-space divergence will lead to a completely different form of areal adaptation than was the case with convergence.

Spatial Adaptation: Decentralized Centralization and the Expansion of the City

Because the land available for expansion is generally peripheral to that portion of the city area which is already developed, the new and relocated establishments (residents, retail and service firms, and so forth) find themselves at a time-disadvantage in attaining goods and services that are only offered in the central core of the city. To obviate this problem these families and firms can either demand greater transport access (steps 1–3), or they can encourage the location of new establishments in the city's peripheral area to serve and to employ them (step 10). Frequently, the demand for new commercial, industrial, and cultural establishments is met prior to any substantial improvements in transport access. The pattern of such development is typified by shopping centers carrying on many retail and service functions and by the nucleation of secondary activities in planned industrial parks.

The decentralized nucleation of man's activities in planned shopping centers and in industrial parks may owe to the desire to reduce the number of trips or the distance of movement needed to attain a given quantity of goods and services.[17] This is made possible by grouping many functions at one center. Such nucleation of activities within given subregions of the urban area may lead to increased interaction within the subregion and, eventually, to an even greater demand for accessibility (step 1). Thus, in this manner, the subregion finds itself in a new stage of areal rearrangement—it is operative within the basic model of spatial reorganization and will develop greater centralization and specialization (steps 1–5).

With the continuance of this process, it is easily seen how subnucleated secondary and tertiary activities can eventually become a part of the very core of the urban area—the increasing concentration of activities within the urban core and within the subnucleated secondary and tertiary centers leads to further demands for land (step 9). It is possible that they will engulf each other in their expansion and become fused into one highly integrated unit. Without some form of control or planning, this process could lead to one vast urban-society—a megalopolis.[18]

In the absence of planning, it is evident that decentralization is merely an intermediate or lag-stage in the general process (described by the basic model) leading towards an expanded area of centralization and specialization. This model highlights only the basic components of spatial reorganization and clearly expresses the cyclical tendency towards the increasing centralization and specialization of human activity.

An Evaluation of the Expanded Model

Unlike the basic model of spatial reorganization, the expanded model accounts for what happens when the degree of rationality, as defined for the basic model, is lessened or when the criteria of rationality change. This model permits consideration of the spatial consequences to the alternative demands of either accessibility or space (land and air).

In concluding the discussion on the de-

[17] For information on the multiple nuclei concept of urban growth, see C. D. Harris and E. L. Ullman, "The Nature of Cities," *Annals of the American Academy of Political and Social Science,* Vol. 242 (November, 1945), pp. 7–17; E. L. Ullman, "The Nature of Cities Reconsidered," *Papers and Proceedings of the Regional Science Association,* Vol. 9 (1962), pp. 7–23.

[18] J. Gottman, "Megalopolis, Or the Urbanization of the Northeastern Seaboard," *Economic Geography,* Vol. 33 (1957), pp. 189–200.

velopment and evaluation of the model, one further observation is necessary. Spatial reorganization is not operative everywhere to the same degree and it does not occur simultaneously at all points in earth space. Therefore, it is essential to determine why this process is so selective and why some places undergo a more rapid areal reorganization than others.[19]

THE PROCESS OF SPATIAL REORGANIZATION AND THE CONCEPT OF RELATIVE ADVANTAGE

The concept of relative advantage states that the process of spatial reorganization in the form of centralization and specialization will accelerate most rapidly at those places which stand to benefit most from increasing accessibility.[20] In other words, transport innovations are most likely between those places which will benefit most from a lessening in the expenditure of time (cost or effort) to attain needed and desirable goods and services. Relative advantage is defined in terms of the benefits of operational success (inclusive of all economic, political, and cultural benefits) that can be derived from a particular place with a given expenditure of time. The concept is based on the same assumptions of rationality as were the process models.

Since locational decentralization, as defined in the expanded model, is simply an intermediate or lag-stage in the overall trend towards centralization (given the continuance of the process and the assumption that a point of diminishing returns does not set in), it is possible to confine the evaluation of the relative advantage concept to the basic model of spatial reorganization. The question is, where will this process be likely to accelerate most rapidly? Or, where is man most likely to introduce a transport improvement? In seeking answers to these questions, the concepts of relative advantage and spatial reorganization will be applied to a selected set of cities in the northern, midwest of the United States.

Relative Advantage and Spatial Reorganization in the Upper Midwest

Because of their significance as times of automobile and highway innovation, the periods of 1900 to 1925, and 1940 to 1965, were selected for evaluating the real world applicability of the concepts proposed in this study. In the early twentieth century, prior to about 1930, railways and electric interurban lines not only dominated intercity travel in the United States, they also had a definite speed advantage over the automobile. For example, although in 1930 a typical forty-five mile auto trip from Dexter, Michigan, to Detroit took three hours, interurban lines averaging anywhere from forty to sixty miles per hour connected most of the nation's major cities.[21] Nonetheless, people increasingly sought the personal convenience and versatility of the automobile and demanded better roads.[22] The tangible results of this demand are illustrated in Figure 6 by a series of five highway status maps for southern Michigan.

Relative Advantage for Transport Improvement in Southern Michigan

In Figure 7A a closed system of seven major Michigan cities and eleven highway links has been selected to evaluate the concept of relative advantage. The immediate objective is to predict highway status for 1925 on the basis of information for 1900 and, similarly, to project the status of highways in

[19] The skyscraper is an alternative choice in attaining more space while still retaining central access. The skyscraper, however, is a form of centralization which fosters greater interaction and additional demand for accessibility. Thus, this spatial adaptation also helps to perpetuate the trend towards greater centralization.

[20] Interesting statements on a concept of relative advantage similar to that proposed here are provided by Z. Griliches, "Hybrid Corn and the Economics of Innovation," *Science*, Vol. 132 (July 29, 1960), pp. 275–80; and "Hybrid Corn: An Exploration in the Economics of Technological Change," *Econometrica*, Vol. 25 (1957), pp. 501–22.

[21] G. W. Hilton and J. F. Due, *The Electric Interurban Railways in America* (Stanford, Calif.: Stanford University Press, 1960).

[22] The number of automobiles in Michigan increased from 2,700 in 1905 to more than 60,000 by 1913. See Michigan State Highway Department, *History of Michigan Highways and the Michigan State Highway Department* (Lansing, Mich.: Michigan State Highway Department, 1965), pp. 6–7.

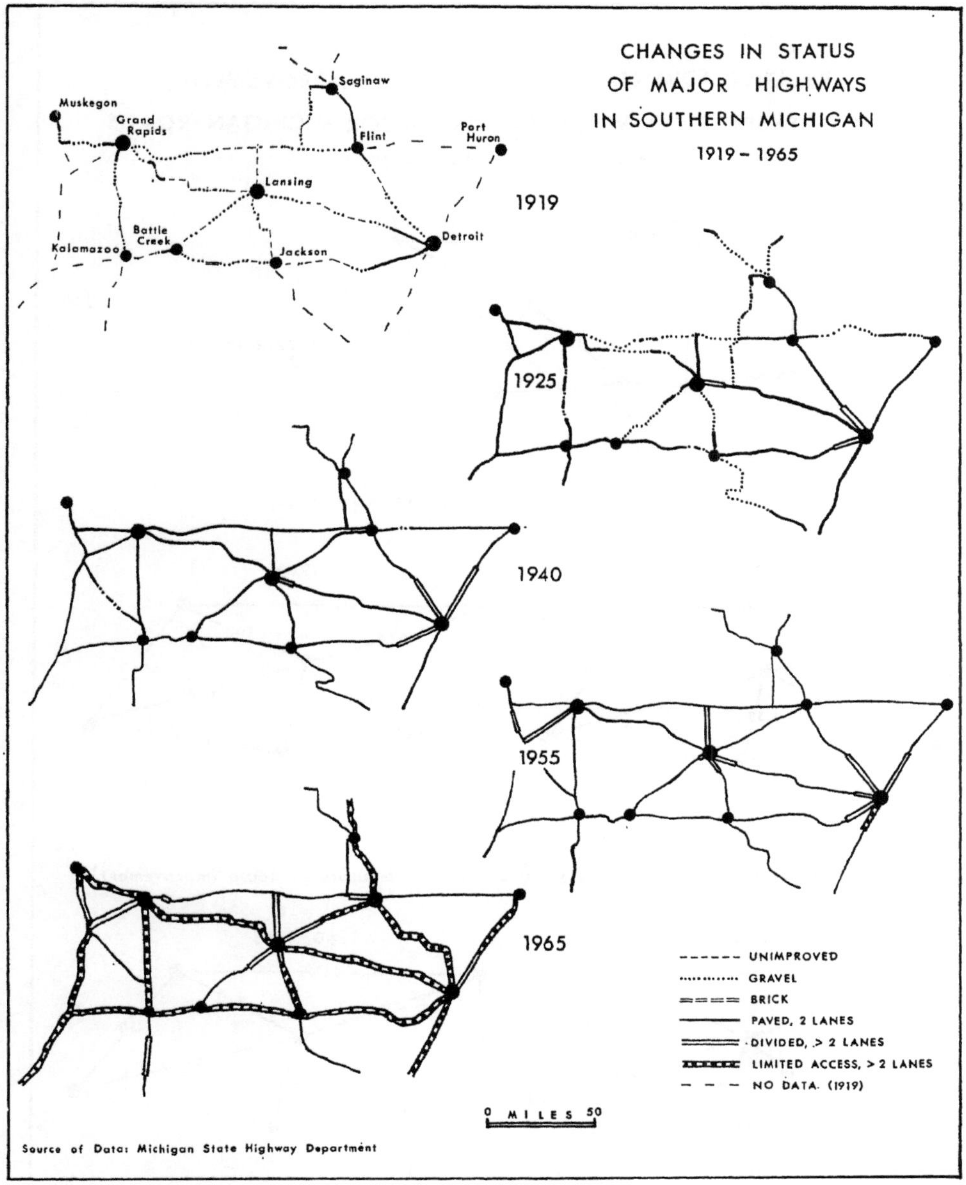
CHANGES IN STATUS
OF MAJOR HIGHWAYS
IN SOUTHERN MICHIGAN
1919 – 1965
Saginaw
Muskegon
Grand Rapids
Flint
Port Huron
Lansing
Battle Creek
Kalamazoo
Jackson
Detroit
1919
1925
1940
1955
1965
UNIMPROVED
GRAVEL
BRICK
PAVED, 2 LANES
DIVIDED, > 2 LANES
LIMITED ACCESS, > 2 LANES
NO DATA. (1919)
0 MILES 50
Source of Data: Michigan State Highway Department

FIG. 6.

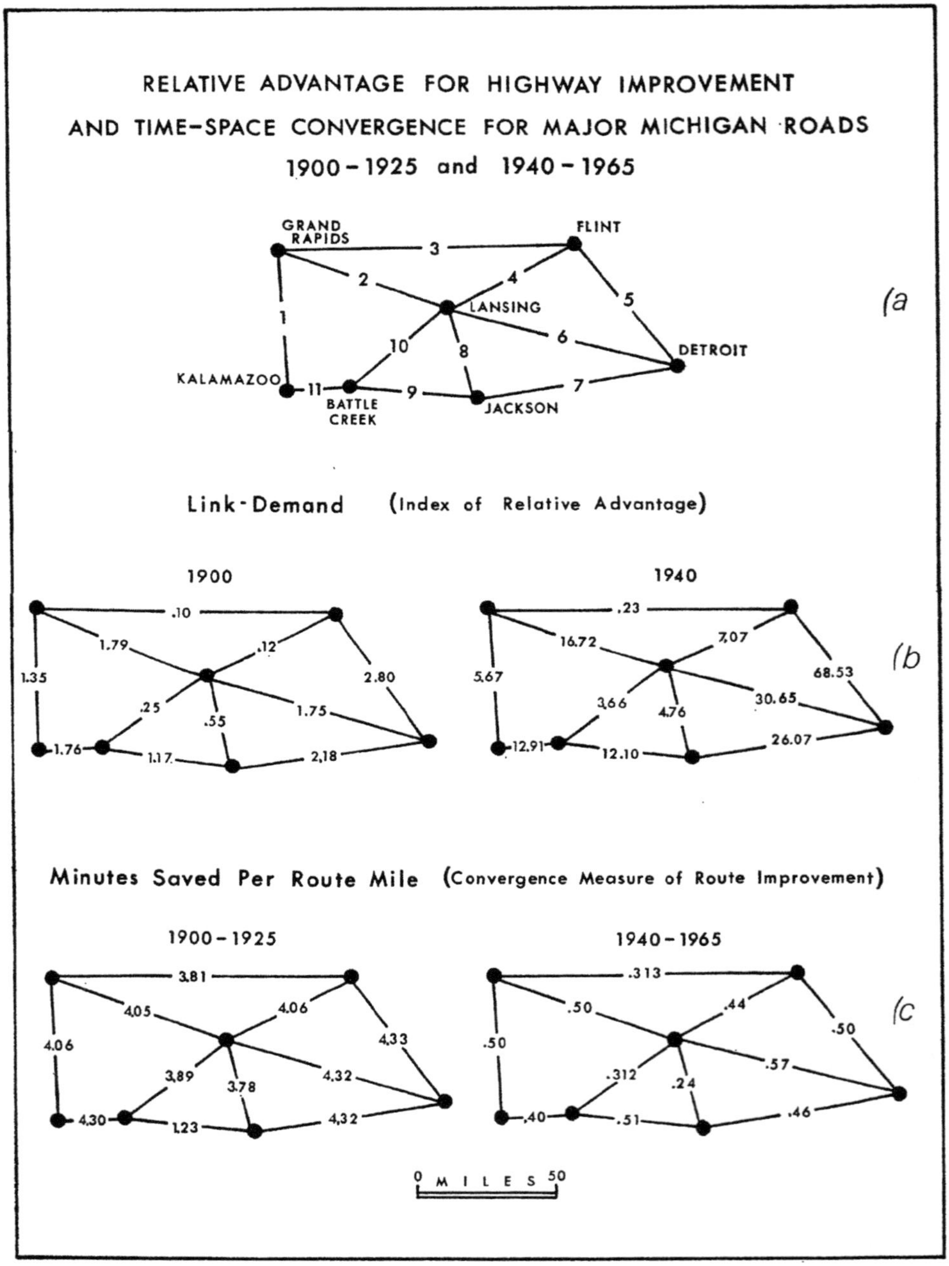
RELATIVE ADVANTAGE FOR HIGHWAY IMPROVEMENT
AND TIME-SPACE CONVERGENCE FOR MAJOR MICHIGAN ROADS
1900 - 1925 and 1940 - 1965
GRAND RAPIDS
FLINT
LANSING
DETROIT
KALAMAZOO
BATTLE CREEK
JACKSON
1
2
3
4
5
6
7
8
9
10
11
(a
Link-Demand (Index of Relative Advantage)
1900
.10
1.79
.12
1.35
2.80
.25
.55
1.75
1.76
1.17
2.18
1940
.23
16.72
7.07
5.67
68.53
3.66
4.76
30.65
12.91
12.10
26.07
(b
Minutes Saved Per Route Mile (Convergence Measure of Route Improvement)
1900 - 1925
3.81
4.05
4.06
4.06
4.33
3.89
3.78
4.32
4.30
1.23
4.32
1940 - 1965
.313
.50
.44
.50
.50
.312
.24
.57
.40
.51
.46
(c
0 M I L E S 50

FIG. 7.

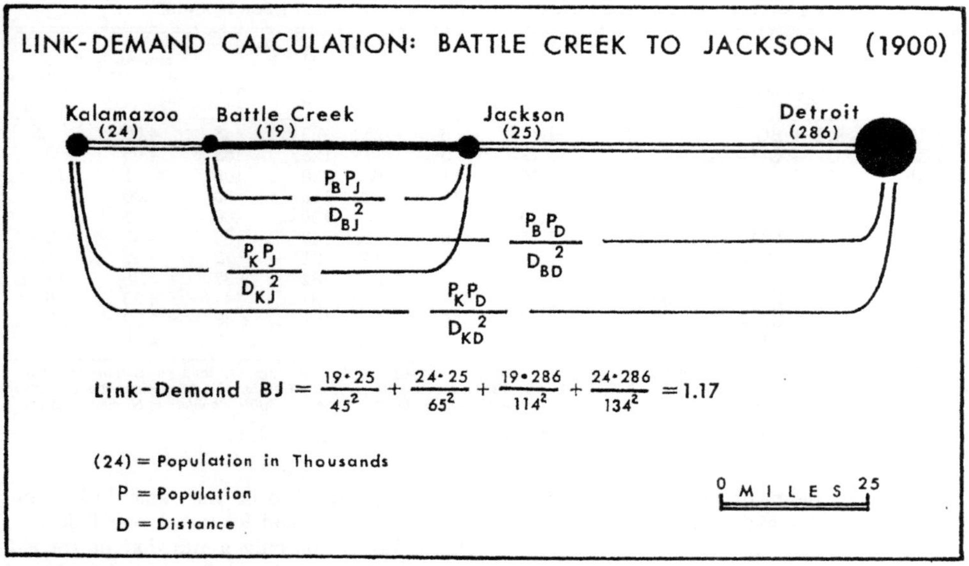

FIG. 8.

1965 from information known in 1940. For the initial years of each period, 1900 and 1940, the principal highway trunklines were nearly homogeneous in quality—mostly unimproved clay and sand roads in 1900 and, as shown in Figure 6, mostly two-lane paved roads in 1940.[23] Thus, the calculation of travel-times between cities for these two years assumes standard speeds of ten miles per hour for 1900 and forty miles per hour for 1940. For the years 1925 and 1965, travel-times are based on the following criteria:

1925—unimproved roads (10 miles per hour),
—gravel roads (25 miles per hour),
—brick roads (35 miles per hour),
—paved two-lane roads (40 miles per hour).

1965—where possible, actual travel-time data from the Michigan State Highway Department are used.[24]

Otherwise:

—paved two-lane roads (45 miles per hour),
—divided highways (55 miles per hour),
—limited access roads (60 miles per hour).

Through application of the above criteria in calculating travel-times, a convergence measure of actual route improvement—minutes saved per route mile—is derived for the two periods in question. This convergence measure will be used to evaluate the success of the predictive variable—relative advantage. The hypothesis under investigation is as follows: the degree of innovation will increase as relative advantage increases. The surrogate used to represent relative ad-

[23] Of the 68,000 miles of roads in Michigan in 1905, only 7,700 miles were graveled and only 245 miles were stone or macadam. See Michigan State Highway Department, *op. cit.*, footnote 22, p. 6.

[24] Michigan State Highway Department, *Highways Connecting Pertinent Cities with O'Hare Field (Chicago) or Metropolitan Airport (Detroit)* (Lansing, Mich.: Michigan State Highway Department, 1963).

TABLE 1.—LINK-DEMANDS FOR 1900 AND TIME-SPACE CONVERGENCE FOR 1900–1925

Linkages (see Fig. 7)	Calculations of link-demand*	Link-demand value (LDV)	Rank (LDV)	Travel-time over link** (minutes) 1900	1925	Route miles	Minutes saved per route mile 1900–1925	Rank (Minutes saved)
1 (KG)	KG + BG	1.35	6	288	93	48	4.06	6.5
2 (GL)	GL + GD + GJ	1.79	3	378	123	63	4.05	8
3 (GF)	GF	.10	11	630	230	105	3.81	10
4 (LF)	LF + BF + KF + JF	.12	10	360	116	60	4.06	6.5
5 (FD)	FD	2.80	1	360	100	60	4.33	1
6 (LD)	LD + GD	1.75	5	510	143	85	4.32	2.5
7 (JD)	JD + BD + KD	2.18	2	432	121	72	4.32	2.5
8 (JL)	JL + FJ + GJ	.55	8	222	82	37	3.78	11
9 (BJ)	BJ + KJ + KD + BD	1.17	7	270	84	44	4.23	5
10 (BL)	BL + KL + BF + FK	.25	9	288	101	48	3.89	9
11 (KB)	KB + KJ + KD + KL + KF	1.76	4	120	34	20	4.30	4

* Link-demands for 1900 were calculated as indicated in text and are based on the following city limit populations (in thousands): Battle Creek (B), 19; Detroit (D), 286; Flint (F), 13; Grand Rapids (G), 88; Jackson (J), 25; Kalamazoo (K), 24; and Lansing (L), 16. Source: U.S. Bureau of Census, *Twelfth Census of the United States: 1900, Population, Number and Distribution of Inhabitants*, Vol. I.

** Based on criteria established by author (see text). Source: Compiled and calculated by author.

vantage is an index of link-demand derived from the simple gravity model

$$\frac{p_i\, p_j}{d_{ij}^2}$$

where $p_i\, p_j$ is the product of the populations of the two places joined by the link, and d_{ij}^2 is the square of the route mileage between them.[25]

The above procedure is complicated somewhat when a system has several places demanding travel over the same link. For example, the demand for travel over link 9 in Figure 7B is not only a function of travel-demand between Battle Creek and Jackson, but it is also a function of the demands for travel between Detroit and Battle Creek, Detroit and Kalamazoo, and Kalamazoo and Jackson. Thus, as illustrated in Figure 8, the link-demand for a highway improvement between Battle Creek and Jackson represents the sum of the gravity model indices for each pair of places whose interconnection requires use of link 9. The demand values for the other ten links were determined in similar fashion and are shown in Figure 7B for the

[25] For a good review and appraisal of the gravity model, see G. Olson, *Distance and Human Interaction* (Philadelphia: Regional Science Research Institute, 1965).

TABLE 2.—LINK-DEMANDS FOR 1940 AND TIME-SPACE CONVERGENCE FOR 1940–1965

Linkages (see Fig. 7)	Calculations of link-demand*	Link-demand value (LDV)	Rank (LDV)	Travel-time over link** (minutes) 1940	1965	Route miles	Minutes saved per route mile 1900–1925	Rank (minutes saved)
1 (KG)	KG + BG	5.67	8	72	48	48	.50	4
2 (GL)	GL + GD + GJ	16.72	4	93	62	62	.50	4
3 (GF)	GF	.23	11	158	125	105	.313	9
4 (LF)	LF + BF + KF + JF	7.07	7	75	53	50	.44	7
5 (FD)	FD	68.53	1	90	60	60	.50	4
6 (LD)	LD + GD	30.65	2	126	78	84	.57	1
7 (JD)	JD + BD + KD	26.07	3	108	70	72	.46	6
8 (JL)	JL + JF + JG	4.76	9	56	40	37	.24	11
9 (BJ)	BJ + KJ +KD + BD	12.10	6	68	45	45	.51	2
10 (BL)	BL + KL + BF + KF	3.66	10	72	57	48	.312	10
11 (KB)	KB + KJ + KD + KL + KF	12.91	5	30	22	20	.41	8

* Link-demands for 1940 were calculated as indicated in text and are based on the following city limit populations (in thousands): Battle Creek (B), 43; Detroit (D), 1,623; Flint (F), 152; Grand Rapids (G), 164; Jackson (J), 50; Kalamazoo (K), 54; and Lansing (L), 79. Source: U.S. Bureau of Census, *Sixteenth Census of the United States: 1940. Population, Number of Inhabitants by States*, Vol. I.

** Based on criteria established by author (see text). Source: Compiled and calculated by author.

TABLE 3.—TRAVEL-TIME AND TIME-SPACE CONVERGENCE (1940–65) BETWEEN SELECTED CITIES IN THE NORTHERN MIDWEST

	D	F	GR	J	K	L	M	PH	S	SB	T	Chicago
BATTLE CREEK	170 105 .58	147 114 .34	93 60 .53	68 45 .51	30 22 .40	72 57 .31	146 101 .46	248 230 .11	179 150 .24	131 91 .46	168 135 .29	234 157 .49
DETROIT		90 60 .50	219 140 .54	108 70 .53	200 135 .49	126 78 .57	276 182 .51	86 64 .39	137 91 .51	272 188 .46	86 53 .58	402 268 .50
FLINT			158 125 .31	131 98 .38	177 135 .36	75 53 .44	215 165 .35	100 80 .30	47 31 .52	278 225 .29	162 115 .44	381 265 .46
GRAND RAPIDS				146 100 .47	72 48 .50	93 62 .50	57 40 .44	258 200 .34	173 145 .24	168 118 .44	246 195 .31	254 180 .44
JACKSON					98 67 .46	56 40 .43	203 150 .39	193 149 .34	162 125 .34	174 135 .34	100 85 .18	291 200 .46
KALAMAZOO						102 81 .32	123 82 .50	278 208 .34	209 165 .32	100 70 .45	198 150 .36	204 136 .50
LANSING							150 100 .50	176 140 .31	107 80 .38	203 150 .39	156 130 .25	306 215 .45
MUSKEGON								315 252 .30	192 180 .09	177 158 .16	303 240 .31	263 180 .47
PORT HURON									147 120 .28	357 260 .41	171 120 .45	488 333 .48
SAGINAW										309 234 .36	209 140 .50	413 300 .41
SOUTH BEND											219 135 .58	131 80 .59
TOLEDO												349 215 .58

Key: 170 (1940), 105 (1965) — Travel-time in minutes; .58 — Minutes saved per route mile (1940–65)

Source: Based on criteria established by author (see text).

years 1900 and 1940. Shown in Figure 7C are the convergence values of actual route improvement for the periods 1900 to 1925 and 1940 to 1965. Data pertinent to these calculations are included in Tables 1 and 2.

Spearman's rank correlation technique was used to measure the statistical association of the rankings of the demand and improvement variables. This technique yielded *R* values (significant at the ninety-five percent level) of .74 for the 1900–1925 period and .69 for the 1940–1965 period.[26] The results, though not conclusive, are encouraging. It is evident that the inclusion of places outside the chosen system of seven cities such as Chicago, Toledo, Saginaw, and others might have greatly altered both the rankings of the link-demand and the results. Furthermore, the gravity model used here may be a comparatively crude measure of the relative advantage for transport improvement.

Although a comparison of the rankings of the link-demands for 1900 and for 1940 show a high degree of stability (*R* equals .87 at the ninety-nine percent level of significance), a similar comparison of the convergence rankings for the two periods reveals some signs of significant change in the time-space connectivity of Michigan's transport network. For example, link 2 from Grand Rapids to Lansing moved from eighth in the convergence ranking for 1900–1925 to fourth during the 1940–1965 period. On the whole, however, the changes for 1940–1965 were consistent with those of the 1900–1925 period of route improvement (*R* equals .67 at the ninety-five percent level of significance).

In general, this evaluation suggests that highway development in Michigan has varied with the changes in relative advantage. And, owing to the pronounced stability in the rankings of the link-demands, it is evident that transport innovations helped to confirm and to augment the existing advantages in time-space connectivity for dominant places. For example, during both periods, Detroit ranked first among the seven cities in the average number of minutes saved per route mile along each of its radiating links. In essense, Detroit has been favored by a greater increase in locational utility than any of the other six places in the system. Thus, in accordance with the norm of spatial reorganization as outlined in the basic model (Figure 3), Detroit should also be favored by the greatest increase among the seven cities in the centralization and specialization of human activity. This concept will be tested against the background of wholesale enterprise in thirteen upper midwestern United States cities.

Spatial Reorganization: Wholesale Activity in the Upper Midwest

Wholesale activity is a form of economic specialization which, according to Philbrick, shows dominant centralization in places of

26 The test of the significance of *R* values was based on the use of Student's *t* as suggested by M. J. Moroney, *Facts From Figures* (Baltimore: Penguin Books, 1956), pp. 334–36. In this study, findings based on Spearman's rank correlation technique are to be regarded as tentative rather than conclusive.

TABLE 4.—TIME-SPACE CONVERGENCE (1940–1965) AND WHOLESALE GROWTH (1939–1963) FOR SELECTED METROPOLITAN AREAS IN THE NORTHERN MIDWEST

		Indices of wholesale growth (1939–1963)**									Rankings of Change			
	Avg. TSC* 1940–1965 (minutes saved per route mile)	Wholesale sales (millions of dollars)			No. establishments			No. paid employees (thousands)						
Place		1939	1963	Change	1939	1963	Change	1939	1963	Change	TSC	Sales	Estab.	Employ.
Battle Creek	.393	18	117	99	107	180	73	.6	1.5	.9	9	12	11	10
Detroit	.513	1,392	9,952	8,560	3,100	5,640	2,540	29.7	65.0	35.3	1	2	2	2
Flint	.391	76	386	310	181	420	239	2.1	7.6	5.5	10	7	6	5
Grand Rapids	.422	91	1,008	917	500	1,000	500	3.6	10.3	6.7	3	4	3	3
Jackson	.403	16	147	131	109	200	91	.9	1.6	.7	7.5	10	10	12
Kalamazoo	.416	23	205	182	134	290	156	1.1	2.6	1.5	4	9	8	8
Lansing	.404	45	411	366	260	440	180	1.7	4.0	2.3	6	6	7	7
Muskegon	.373	16	123	107	113	180	67	.6	1.6	1.0	11	11	12	9
Port Huron	.338	18	62	44	76	130	54	1.3	.7	−.6	13	13	13	13
Saginaw	.349	42	240	198	190	290	100	1.9	2.7	.8	12	8	9	11
South Bend	.411	50	512	462	190	460	270	1.7	4.8	3.1	5	5	5	6
Toledo	.403	137	1,096	959	530	1,000	470	5.7	11.5	5.8	7.5	3	4	4
Chicago	.486	4,150	23,682	19,532	8,200	12,210	4,010	98.3	171.0	72.7	2	1	1	1

* This is the average convergence of each place to the other twelve places in the system. Calculated by author from data in Table 3.

** The values of wholesale indices are for the SMSA's as defined for the 1960 U.S. Census. County data are used for Battle Creek and Port Huron.

Source: U.S. Bureau of Census, *Sixteenth Census of the United States: 1940. Census of Business 1939, Wholesale Trade*, Vol. II; and *1963 Census of Business, Wholesale Trade Area Statistics*, Vol. V.

third order and above.[27] It seems reasonable to assume that wholesale firms would fare best if they located at those places which are most accessible to their customers. If this is so, then places offering high degrees of locational utility relative to other places should be dominant wholesale centers. However, as indicated for the highway network of southern Michigan, the time-space surface of locational utility is in a state of sporadic flux —differential transportation development induces variations in the relative rates with which places improve their time-space connectivity with one another. Thus, this factor of non-homogeneous transport change is incorporated in the general hypothesis that the wholesale activity in a place will increase as the time-expenditure per unit of operational success decreases. In other words, that place which experiences the greatest degree of time-space convergence, compared to all other places in the system, will be expected to show the greatest absolute growth in wholesale activity.

A system of thirteen metropolitan areas in the Upper Midwest has been selected to test this hypothesis. Indicated in Table 3 are the travel-times for 1940 and 1965 and the time saved per route mile between each city and the other twelve places in the system. In Table 4 the average convergence of each place to all other places in the system between 1940 and 1965 is included along with various indicators of wholesale growth during roughly the same period—1939 to 1963. These cities were ranked from one to thirteen on the convergence and wholesale variables, and Spearman's rank correlation was used to measure the statistical association of the rankings of time-space convergence with each of the indicators of wholesale growth. The three wholesale measures showed close association with the convergence factor; R values, significant at the ninety-nine percent level, equaled .76 for the increase in dollar sales, .81 for the increase in the number of wholesale establishments, and .77 for the increase in the number of paid wholesale employees. These findings lend cautious support for the notion that, at least for the wholesale function, time-space convergence is a useful surrogate for estimating the centralization and specialization possible at given places.

In this example, convergence was least

[27] A. K. Philbrick, "Principles of Areal Functional Organization in Regional Human Geography," *Economic Geography*, Vol. 33 (1957), pp. 299–336.

effective in suggesting the rank changes resulting from the growth in wholesale activities for Kalamazoo, Flint, and Toledo. Among the thirteen cities, Kalamazoo stood four positions lower in wholesale growth than it did in convergence. In contrast, Flint and Toledo each ranked four positions higher in change of wholesale activity than they did in the ranking of time-space convergence. Thus, relative to the locational utility of other places in the system, it is possible that Kalamazoo has increased, and Flint and Toledo have declined, in status as potential sites for wholesale activity. Interestingly, dollar sales by wholesale firms in the Kalamazoo metropolitan area increased 8.9 times between 1939 and 1963 in contrast to increases of 5.1 and 8.0 for the Flint and Toledo metropolitan areas. The average growth factor for the thirteen areas was 7.5. It appears, therefore, that even in those cases where convergence showed comparatively little rank association with wholesale growth, wholesale activity did gradually shift in association with changes in the surface of locational utility. As indicated in the evaluation of the basic model, there is an inherent inefficiency in human place-behavior and, therefore, such lags in spatial reorganization are to be expected.

CONCLUSION

The objective of this study has been threefold:

1) To conceptualize the role of transport technology as a key factor in the spatial reorganization of man's activities;
2) to outline the "steps" in the process of spatial reorganization; and
3) to propose a conceptual framework that accounts for the differential operation of this process at various places.

The premise upon which the study was based was that man adapts the areal structure of his activities in response to changes in transport technology which enable him to travel faster and to have access to larger areas and to more resources.

Given the assumptions of rationality as developed earlier in this study, the proposed theses seem tenable: 1) that time-space convergence is a significant factor leading towards spatial adaptation, and that 2) spatial reorganization will accelerate most rapidly at those places which stand to benefit most from increasing accessibility. However, in light of the limitations posed by the assumption of rationality, it is evident that the concept of relative advantage, as applied to the process of spatial reorganization, is not in complete accord with reality. The relative advantage concept seeks to explain man's decisions on the basis of motivation (*i.e.*, desire for net social benefits, locational utility, operational success, and the like). The decisions of man, however, are often conditioned by his lack of information, or non-maxima goals.

In that much of man's areal development is presently directed by various government agencies (at the city, region, state, and nation levels) rather than by the demand for locational utility, it seems that complete understanding of the process of spatial reorganization may rest upon one's knowledge of the decision-making processes of these agencies.[28] Upon what information are their decisions made? By what goals are they motivated? Answers to these questions were beyond the scope of this study. Nonetheless, until answers to such questions are found, the understanding of spatial reorganization will be limited to mechanistic concepts similar to those outlined herein. It is believed that a more refined understanding of spatial reorganization must await efforts to account for lags in the process and to evaluate the significance of both spatially and temporally variant goals. Dynamic programming techniques may hold some promise for the solution of these problems.[29]

---

[28] For an excellent review of studies on decision theory and their applicability in planning, see J. W. Dyckman, "Planning and Decision Theory," *Journal of the American Institute of Planners,* Vol. 27 (1961), pp. 335–45.

[29] A brief review of dynamic programming is provided by R. Bellman, "Dynamic Programming," *Science,* Vol. 153 (1 July, 1966), pp. 34–37.

[6]

Research policy and review 27. New directions for understanding transportation and land use

G Giuliano¶
Institute of Transportation Studies, University of California, Irvine, CA 92717, USA
Received 19 May 1988

Abstract. Theories of relationships between land use and transportation, and the empirical research conducted to test these relationships are reviewed. Recent empirical research seldom supports theoretical expectations. These results are explained by the changes in urban structure that have occurred over the past three decades. The paper concludes with some suggestions for revising the theories to represent conditions in contemporary urban areas better.

1 Introduction

The relationship between transportation and land use has been an enduring subject of both theoretical and empirical study (Fujita, 1984; Lerman et al, 1978; Meyer and Gomez-Ibanez, 1981; Meyer and Miller, 1984; Putman, 1975). Land-use-transportation relationships are not only key factors to understanding the nature and evolution of urban form, but also have important policy implications. Transportation investments are among the largest of public investments, and thus their anticipated impacts play a critical role in the public cost-benefit calculus.

Despite the rich and varied literature on this issue, the theoretical basis of land-use-transportation relationships has been subject to a variety of criticisms, and recent empirical research seldom supports theoretical expectations. In this paper I argue that these results suggest a reevaluation of existing theory and of the role of transport costs in location choice. In part 2 of the paper I present a brief overview of existing theory and outline its major weaknesses. Part 3 is a review of empirical research that illustrates these changing conditions, and part 4 provides some explanations for these findings. The paper concludes with some suggestions for rethinking the relationship between land use and transportation.

2 Overview of existing theory

Land-use-transportation relationships have been examined in the context of different paradigms. Two major streams may be distinguished: economic-behavioral and mathematical programming or network equilibrium. Economic-behavioral approaches are based on economic principles of utility and profit maximization (for example, Alonso, 1964; Mills, 1972; Muth, 1969). Land is allocated to different uses on the basis of factor prices, including transport costs. Accessibility is implicit in the theory; the value of location relative to opportunities is reflected in land prices. Mathematical programming models optimize flows for a given allocation of activities (for example, Boyce, 1980; Harris and Wilson, 1978), or allocate activities for a given fixed set of transport costs (for example, Herbert and Stevens, 1960), or for given transport characteristics (for example, Kim, 1983; Los, 1979). The economic-behavioral and mathematical programming approaches both predict a unique solution in which total costs are minimized. Since the programming models have only recently

¶ Present address: School of Urban and Regional Planning, University of Southern California, Los Angeles, CA 90089-0042, USA.

been empirically tested (Boyce, 1986; Boyce et al, 1987), the emphasis of this discussion is on the behavioral models.

2.1 *Residential location*
Residential location theory expresses household location choice as utility-maximization problem in which choice depends on land rent, transport costs, and the cost of all other goods. The basic model (for example, Muth, 1969) employs important simplifying assumptions: (1) the total amount of employment is fixed and located at the center of the city, (2) each household has one worker and only work-trip travel is considered, (3) housing is homogeneous (that is, location is the distinguishing factor), and (4) unit transportation cost is constant and uniform in all directions. The utility-maximizing location for a given household is the point at which the marginal savings of housing cost is equal to the marginal cost of transportation. Location theory gives rise to the familiar declining land-rent and density gradients as the value of access to the center is capitalized in land rent. The theory predicts decentralization in response to a reduction in transportation cost, as households consume more housing at greater distances. Land values reflect these shifts; the land-rent gradient flattens as the relative price of land closer to the center declines.

Most criticisms of residential location theory deal with the simplifying assumptions employed in the basic monocentric model, rather than with its substantive basis (for example, Muth, 1985). Criticisms of the theory can be categorized into three types: (1) those which add complexity but do not affect basic validity of the model, (2) those which affect the predictive ability of the model, and (3) those which affect the underlying assumptions. These are illustrated with the following examples.

2.1.1 *Monocentricity* Criticisms of the monocentricity assumption is an example of the first type. The assumption of a single employment center has been widely challenged, and several empirical studies have demonstrated the multicentered structure of metropolitan areas (Gordon et al, 1986; Greene, 1980; Odland, 1978). Relaxing the assumption of monocentricity adds to the complexity of the model, but does not affect its behavioral validity. Several polycentric models have been developed (for example, Hartwick and Hartwick, 1974; Ogawa and Fujita, 1980; White, 1976). Although they differ in many ways, these models are based on the trade-off between land rent and transport cost with respect to a given (unique) workplace location as in the monocentric model. The spatial outcome (where determinant) is similar to the monocentric case, with declining density gradients emanating from each center.

2.1.2 *Multiple centers and multiple-worker households* If the assumptions of monocentricity and single-worker households are both relaxed, a more serious problem emerges. Residential location theory can no longer define a unique optimal location solution, and the predictive ability of the model is challenged. For example, combined commute costs for a two-worker household are constant at any point along the ray connecting two workplaces. Therefore, all other things being equal, any location on the ray is equally optimal. Given that the most recent available census data suggest that the ratio of US workers to households with members of working age is about 1.5[1], the appropriateness of applying this theory to contemporary household location behavior is questionable.

[1] Ratio calculated from population and employment data in *Statistical Abstract of the United States, 1986* US Department of Commerce and Bureau of the Census, 1985.

2.1.3 *Capital stock durability* Residential location theory contains an implicit assumption of nondurable capital stock. With durable capital stock, responses to changes in parameters (for example, a change in transport cost) are difficult to predict and can lead to counterintuitive outcomes (Wheaton, 1982). Durability cannot be effectively incorporated in location theory because it implies a dynamic process, whereas the theory is deterministic and static. The inability of the theory to incorporate dynamic (historical) processes is a criticism of the third type; it is a fundamental shortcoming which brings into question the substantive basis of the theory itself (Batten and Johansson, 1987).

Criticisms of the traditional monocentric model are serious only when the utility of the model comes into question. The strengths of the model are its elegance and simplicity, and these have made it an ideal tool for urban policy analysis. The first category of criticisms does not materially reduce the utility of the model for analyzing urban phenomena. The second category limits applicability of the model, and the third category questions whether it is appropriate at all. Thus any assessment of the relevance of this model for urban anlaysis depends on the extent of the second and third types of criticisms.

2.2 *Employment location*

Three different employment location theories have been developed. Business location theory is an extension of residential theory described previously (Solow, 1973; Strotz, 1965). Employment location is a function of land rents, commuter costs, and all other costs. Since a reduction in transport cost results in lower land values at the center (that is, a flattening land-value gradient), the effect will be a centralization of employment.

Business location theory does not, of course, apply to market-sensitive employment (activities which require access to consumers). The classic theory of market-sensitive activities is central place theory, formulated by Christaller (1966) and Lösch (1954). It predicts the distribution and size of markets as a function of population and market-threshold requirements. The resulting distribution is that which minimizes total consumer transport cost and fulfills the minimum threshold requirements. Central place theory has enjoyed renewed interest in recent years because it predicts a multinucleated land-use pattern. For example, recent theoretical work has shown that profit-maximizing behavior by firms can give rise to a hierarchial multicentered city structure given certain assumptions (von Boventer, 1976; Carruthers, 1981).

Industrial location theory is based on minimization of total transport costs for production inputs and outputs for a given optimal level of production. This is the 'classical model' of location, first formulated by Weber (1928) and later developed by Hoover (1948), Isard (1956), and Moses (1958). This model applies primarily to manufacturing activities. Location choice depends on the relative costs of shipping inputs and outputs as well as on economies of scale in production.

These theories are quite different from one another. Each applies only to certain activities; there is no integrated theory that applies to all types of employment. In every case, however, transport costs play a key role. All of these theories have been subject to a variety of criticisms. Business location theory is subject to the same criticisms as residential location theory. Also, recent theoretical work has shown that location can be indeterminant in the industrial location model under some production conditions, even when transport costs are taken into account (Karlson, 1985).

These theories may also conflict with one another. For example, location theory predicts that a transportation improvement has a decentralizing incentive for households, but a centralizing incentive for business. Centralizing business, however, creates an incentive for households to locate closer to the center. At the same time, market-dependent firms must follow the population. A transportation improvement enlarges the market area for such firms, allowing greater decentralization. Meyer and Gomez-Ibanez (1981) point out that these conflicts obviously make it impossible to predict the impact of transportation changes on urban form on the basis of location theory alone.

3 The empirical evidence

Evidence from empirical tests of location theory hypotheses demonstrates that the relationship between location and transport expressed in the conventional theories is generally not supported in contemporary metropolitan settings. Two types of empirical studies are presented here in illustration: studies of employment and household location behavior, and studies of land-use impacts of major transportation investments.

3.1 *Employment location*

Recent studies of employment location provide the greatest support for the conventional theories. Labor-force access is identified as an important or significant locational consideration at the *regional* level in several different studies.

A study of firms that had relocated from Milwaukee City to its suburbs between 1964 and 1974 showed that agglomeration economies and labor-force availability were the most significant factors explaining location choices for all industries. Estimates of the share of firms locating in a given municipality were generated via separate regressions for six different industrial sectors. Labor-force availability was measured by the number of workers in the sector located within a given radius of the municipality; agglomeration was measured as the sector's share of total employment within each municipality. Other access-related variables were significant only for construction and wholesale trade: distance from the Milwaukee CBD (central business district) was positively related to location choice, presumably reflecting demand for the cheaper land available in more distant suburbs (Erickson and Wasylenko, 1980).

The importance of access to the labor force was examined in a different way in a study of high-technology industry. Herzog et al (1986) hypothesized that, if access to a highly skilled work force is important and if highly skilled workers have special locational preferences, high-technology firms should locate in areas that reflect these locational preferences. Results showed that high-technology workers' locational preferences were not substantially different from those of other workers, but that these workers demonstrate a higher degree of geographic mobility. The authors conclude that considerations of human capital play an important role in location choice among high-technology workers, and that consequently metropolitan officials have little control over retention of specialized labor resources because of the relative unimportance of the specific attributes of local areas.

It could be argued that labor-force access at the metropolitan or regional choice level is an obvious requirement for any firm, and it does not adequately reflect transport (commute) cost considerations. Site-related access may be more relevant. If firms locate to minimize total costs, then, ceteris paribus, shorter commutes for workers are preferable. Herzog et al list the following factors that influence site choice (in addition to site availability and cost): labor-force characteristics, local taxes, transportation for workers, quality of schools, and proximity to recreational and cultural opportunities.

Similar results are presented in a British survey of relocating firms (Patterson and May, 1979). The surveyed firms varied greatly in size and age. Among manufacturing firms the following criteria for choice of location site were identified, in order of importance: site availability, site cost, labor-market accessibility, and site access. Other factors identified were the local political environment and personal preferences of location decisionmakers. Site cost and availability were considered the key factors that ultimately determine where the firm relocates.

These results suggest that labor force access is a significant factor in firm-location choice, at least at the metropolitan level. However, in addition to the trade-off between transport and land cost in the traditional theory, these studies suggest that transport cost may be traded for more favorable local development policies, more esthetically attractive locations, better schools and amenities, and a host of other factors.

3.2 *Residential location*

Recent studies of residential location also suggest a much more complex relationship between transport cost and location choice, but provide less support for the traditional theory. A comparison of actual commuting behavior in a sample of US and Japanese cities with that prediced by a monocentric location model showed actual commuting to be about eight times greater than the predicted value based on commute-cost minimization. Correcting for possible sources of bias (for example, employment decentralization, two-worker households) reduced the actual versus predicted disparity to a factor of about five. In contrast, a random location model overpredicted total commuting by only 25% for the same sample. These results question whether the trade-off between work trip and housing cost has any significant role in residential location decisions (Hamilton, 1982).

Other studies provie additional evidence that residential location is not adequately explained as a function of (fixed) job location. Simpson (1987) proposed a simultaneous model of residential and workplace location. Workplace location is expressed as a spatial job-search model, in which location choice is a function both of residential location (for example, the distance of potential job opportunities from home) and level of job skill. Residential location is expressed in the traditional form, with independent variables including workplace location and household characteristics. Empirical results with data from Toronto showed that the workplace-location equation had far greater explanatory power, whether estimated separately or simultaneously. Moreover, homeowners and those who were not heads of households (that is, those with less mobility) were found to be more sensitive to local employment conditions than were renters and household heads. These results suggest that residential location has a greater effect on job choice than job location has on residential choice.

Another study which used Toronto data focused on occupational status as a key location factor. Findings showed that different occupational groups have varying sensitivities to travel time, with income and other relevant factors held constant, and different preferences with respect to suburban residential location. The study also documented the existence of locational segregation between occupational groups, again with household income and demographic characteristics held constant (Miller and Cubukgil, 1981). These studies suggest that important location factors are not captured in traditional residential location models, and these factors may overwhelm considerations of transport cost.

Another way of testing the hypothesis that residential location depends on job location is by examining household relocation behavior. It has been argued that the high cost of relocation may be an important reason for suboptimal location

(Hamilton, 1982). Thus, when households move, the move should be closer to work, all other things being equal. An empirical analysis of household relocation in the Milwaukee area with data from 1962 to 1963 revealed a different pattern: the probability of moving closer to the workplace increases at a constant, then decreasing rate with distance from the workplace. Beyond some critical distance (approximately nine miles in this case) the probability approaches a constant value of about 0.8. *Below* some commute distance (about three miles) there appears to be no relationship between housing and job location (Clark and Burt, 1980). The implication here is that households may be indifferent to commute costs until these costs become significantly large, and that household preferences with respect to commuting are highly variable.

3.3 *Land-use impacts of transportation investments*

If transport costs play an important role in location choice, then transportation improvements should influence location choices. The basic concept underlying the relationship between land use and transportation is accessibility. Any significant improvement in the transportation system (for example, new highway or mass transit link) increases accessibility and reduces transport costs. Location theory thus predicts that the improvement in accessibility will be capitalized in land values and reflected in land-use changes responding to the shift in land value.

Numerous studies of the land-use impacts of new highways and mass transit lines have been conducted. (For a review, see Knight and Trygg, 1977; Lerman et al, 1978; Meyer and Miller, 1984.) Most of the work on the impact of rail took place in the 1970s in response to construction of the 'new generation' rail systems in San Francisco, Washington, DC, and Atlanta. Results of these studies showed that rail transit has had little impact on land values. Various explanations have been advanced for these findings. Local zoning practices and political attitudes that constrain intensification of development have been identified in some cases (Boyce et al, 1972; Knight and Trygg, 1977). It is also argued that the durability of capital stock implies long time lags in land-market response to changes in the transportation system, and that the methodological complexities involved in isolating the effect of any one factor on land values over several years make it unlikely that impacts can be measured, even if they do exist (Giuliano, 1986). Last, some would claim that rail systems do not have a significant impact on accessibility—they serve few origins and destinations, and they carry a very small share of the total number of trips in an area—and therefore should not be expected to affect land use (Hamer, 1976; Meyer and Gomez-Ibanez, 1981).

Highway investments are a different matter. Given that urban highways carry over 90% of all person-trips and a large proportion of all goods movement in the USA, it seems reasonable to expect that highway investments would generate significant land-use impacts if transport-cost considerations are important in location decisions. Two generations of highway studies have been conducted in the era after World War 2. The first studies were performed in conjunction with construction of the interstate highway system in the late 1950s through the 1960s, and the second series have been conducted during the present decade. The first-generation studies very consistently showed significant, positive land-value impacts of new highways, despite sometimes significant methodological weaknesses (Adkins, 1959; Czamanski, 1966; Mohring, 1961). In every case, these were studies of the first freeway constructed within the metropolitan area.

In contrast, the second-generation studies show no consistent relationship between highway improvements and land-use change. A national study of beltway (circumferential highways) impacts concluded that there was no consistent relationship between the presence of beltways and land use. Rather, land-use impacts were dependent upon (1) overall local economic conditions, (2) access to medium-income or high-income residential areas, (3) availability of land to develop, and (4) favorable local zoning policies (Payne-Maxie Consultants, 1980). A recent study of highway impact which used Minnesota data reported comparable results: a positive long-term impact of highway expenditures (defined as an increase in local employment beyond the regional trend) was identified only for the Minneapolis – St Paul Metropolitan Area, and negative impacts were identified in the adjacent counties. These findings were attributed to the relatively greater capacity to absorb economic growth within the regional center (Stephanedes and Eagle, 1987).

Despite the lack of firm quantitative evidence, the conviction that highways affect land use remains. Several descriptive or historical studies have documented land-use development along freeways, noting the tendency for clustering around major interchanges and for linear development along freeway frontages (Baerwald, 1982; Erickson and Gentry, 1985; Muller, 1981). In these studies land availability, local public policy, and the durability of infrastructure were identified as factors that affect land-use impacts.

The empirical evidence presented above may be summarized briefly. Studies of the impacts of transportation investments document the expected theoretical relationships only in the case of the early freeways. Studies of rail impact, as well as more recent freeway studies, show no consistent relationship. Similarly, employment and household location studies both document the presence of other more important factors that affect locational decisionmaking. There are two possible interpretations of this recent evidence. The first is that the concept of trade-offs between transport cost and location choice continues to be valid, but the observed pattern is affected by other (random) factors. The second is that transport cost is no longer a key factor in locational decisionmaking.

Taken as a whole, the empirical record suggests that the second interpretation is more accurate. Transport cost is a much less important factor than location theory predicts. There are several possible explanations for these results, and they can provide guidance for the development of more appropriate theories.

4 Some explanations for the evidence

Contemporary metropolitan areas differ in many ways from the cities on which location theory is based. Some major differences are described here.

4.1 *Accessibility in contemporary metropolitan areas*

With development of a cheap ubiquitous transportation system and the decentralization both of residences and of businesses over the past thirty years, accessibility has been greatly increased in US metropolitan areas. The highway system is well developed, with linkages to the national interstate system as well as to the local system. Consequently, new facilities, even if large scale, have little *relative* impact when viewed from a regional perspective. Use of the transport system is also cheap in relative terms. Although turn-of-the-century streetcar commuters spent about 20% of their daily wages on the journey to work, for example (Hershberg et al, 1981), urban auto commuters today spend about 7%. The overwhelming majority of metropolitan residents enjoy a very high level of mobility for relatively little cost.

Assessibility has also been affected by decentralization of economic activities. With the commercial, manufacturing, and service activities dispersed throughout the metropolitan area, relative differences in accessibility have declined.

Contemporary metropolitan areas are perhaps better characterized by a homogenous activity and transportation surface than by the traditional negative density gradient.[2] The approximation of a homogeneous accessibility surface is not intended to preclude the existence of nodes or subcenters. Rather, in multicentered metropolitan areas, alternative locations have approximately equal access to the set of activity concentrations—which may themselves display a high level of homogeneity. Given these conditions, then, any number of locations are equally accessible, because locational differences have declined. The relative unimportance of access in locational decisionmaking is a logical outcome, as most location changes in contemporary metropolitan areas will lead to negligible changes in accessibility. That is, the changes in accessibility resulting from ubiquitous transportation systems and decentralization have led to new criteria for locational differentiation, for example, neighborhood characteristics, local public services, political attitudes, etc. Under these conditions, the empirical findings discussed above appear reasonable: once some basic level of accessibility has been fulfilled, it is no longer a primary consideration for either workers or firms.

4.2 *The costs of relocation*

Directly related to changes in accessibility is the issue of relocation. The long life and immobility of fixed capital makes relocation costs significant both for households and for firms. Relocation costs also include less easily quantifiable considerations: search costs for a better location, possible loss of key employees, and the information costs associated with reestablishing household activities in a new location. Relocation costs are thus a significant factor in any location-choice decision. Given these costs, it follows that the expected benefits of a new location must be at least as great as the cost of moving before a relocation will take place. If accessibility differences between alternative locations are small relative to relocation costs, accessibility considerations will not be sufficient to cause a move to take place. Therefore, at any given time, a large number of (rational) household and employment locations may in fact be 'suboptimal' with respect to transport cost. The importance of relocation costs has also been theoretically demonstrated in recent research. In his work on endogenous development of subcenters, Clapp (1984) has shown that no unique spatial equilibrium exists given positive relocation costs. Rather, relocation of a firm does not occur until the benefits of the new location are greater than the cost of moving.

4.3 *The scale of residential and employment development*

The scale of development has also increased dramatically in the era after World War 2. In the 1950s, the residential development industry consisted primarily of many builders developing small tracts as land parcels became available. In the 1980s, the industry is characterized by many fewer builders, and the 'typical' development is the planned community, which involves very large land parcels and is usually developed over several years. Similar changes have taken place in the development of industrial and commercial projects, evolving from individual buildings to vast industrial parks and mixed-use office centers. The consequences of these changes are twofold. First, the availability of large tracts of land becomes a critical factor. Such projects cannot be realized unless sufficient land is available, and large parcels are most likely to be found at the periphery of metropolitan areas. Second, land-use change is determined by fewer decisionmakers—and fewer decisions. Under these conditions, unique local characteristics, as well as the

(2) Studies of density gradients document the decreases in the slope over time (for example, Guest, 1975), and zero or positive gradients have been reported (Jackson, 1979).

unique preferences of decisionmakers, should play a major role in determining land use, once the necessary condition of land availability has been fulfilled.

4.4 *Changing structure of economic activity*
Much has been written on the changes within the economy over the past few decades (for example, Harris, 1985). The share of manufacturing activity has declined, and service activities have significantly increased. Integration of the economy has resulted in growth of national and international markets. How do these changes affect transportation and land-use relationships? First, the relative importance of transport cost in economic activity has declined as the transportation of information has been substituted for transportation of goods. Second, the market orientation of firms is shifting more heavily to national and international networks. Thus for many firms, access to the interstate highway system and to major airports may be far more important than relative access within the metropolitan area. The exceptions here are market-dependent activities which require access to an adequate consumer base. However, given an approximately homogenous population distribution (and homogeneous accessibility), it follows again that traditional access considerations may be relatively unimportant.

Agglomeration economies are also related to the structure of economic activity. Although both theoretical and empirical understanding of agglomeration economics continues to be limited, it does appear that agglomeration continues to be a significant force in location (Erickson and Wasylenko, 1980; Stevens, 1985). Given that the service sector is the fastest growing sector in the US economy and that agglomeration economies are associated with service activities, one might expect agglomeration forces to be very important in contemporary metropolitan areas. Several empirical studies document the concentration of economic activities in suburban areas, particularly (but not exclusively) around key highway intersections or airports (Baerwald, 1982; Erickson, 1983; Greene, 1980). A recent study of land values in the Dallas area showed proximity to subcenters to be a significant explanatory factor (Peiser, 1987). And downtowns in major US metropolitan areas continue to grow, though at a less dramatic rate than suburban areas.

In the presence of agglomeration economies, the role of transport costs depends on trade-offs between the benefits of agglomeration and the associated congestion costs. The historical record suggests that congestion is much less important: development intensification continues in the downtowns of some of the most congested cities (for example, Los Angeles, New York, San Francisco, Chicago), despite the absence of new transportation facilities, and traffic problems associated with the rapid growth of suburban subcenters have become major public policy issues (Cervero, 1986a; 1986b). It may be argued, however, that this trade-off is inefficient, because congestion costs are largely external, whereas agglomeration benefits accrue directly to the firm. In the absence of efficient pricing, then, excessive levels of agglomeration might be expected.

4.5 *Local preferences and public policy*
Local governments have the potential to control and influence land-use change. If strong community preferences are present, local jurisdictions can exercise zoning power to prevent or downgrade development despite favorable market conditions. They can also promote development through tax breaks, provision of infrastructure, financial assistance, and other incentives. Recent studies of highway and transit projects have demonstrated the critical role of community preferences (Knight and Trygg, 1977; Payne-Maxie Consultants, 1980). For example, specific instances of down-zoning around transit stations in response to local opposition have been documented in Washington, DC, and in the San Francisco area (Dvett et al, 1979;

Lerman et al, 1978). A variety of development constraints (for example, floor–area ratio maximums, building permit limits) have more recently been imposed in suburban centers in response to traffic congestion generated by rapid growth (Cervero, 1986a).

Contemporary metropolitan areas are characterized by multiple local jurisdictions, each representing different community goals and preferences. These jurisdictions may compete for development (or for no development) through exercise of regulatory power, and thus may have a significant impact on land-use decisions and consequently on urban structure. Given the scale of most contemporary developments, local government approvals are a key factor in the development process. If it is accepted that accessibility considerations are of declining importance in locational decision-making, it is only logical to expect local conditions to play a more significant role.

The explanations presented here suggest that existing theory does not capture major explanatory factors of land-use change in today's metropolitan areas. The available evidence shows that transport cost has decreased in importance as a locational consideration both for households and for firms. Briefly summarized, this change is primarily the result of the decentralization and a well-developed transportation system that have reduced differences in accessibility between locations. Local characteristics have correspondingly increased in importance because of scale economies in development, agglomeration economics, and regulatory influence of local governments.

5 New directions for revising the theory

In light of the discussion in this paper, it is necessary to identify concepts that both are more appropriate for these changed conditions and could lead to the development of a better theory. Some possibilities for residential and firm location are presented here.

5.1 *Residential location*

There are several ways to expand the traditional model to provide a more adequate representation of the household location-decision process. First, a temporal element might be added. In this case, transport cost would be measured as total household-commute costs over the expected tenure at a given location. The idea here is that households have some expectations about changes in job location (and job locations are distributed throughout the region), and the residential location decision is an attempt to minimize travel to these possible locations. That is, households may maximize access to possible employment opportunities over the expected housing tenure. Thus households with stable job histories should live closer to work than households with a high degree of job turnover, all other things being equal. Also, households with shorter housing-tenure (renters) should live closer to work than those with longer housing-tenure (owners).

Second, the 'all other goods' term in the traditional model could be treated more explicitly. Existing evidence suggests that neighborhood characteristics and access to activities other than work play an important role in location choice. National travel-survey data are also supportive. Total vehicle trips per US household increased by 6.4% from 1969 to 1983. During the same period, the share of work-related trips dropped by 15%, while the share of family and personal business trips increased by 27% (Klinger and Kuzmyak, 1986). Thus a more accurate model would incorporate access to services, recreational opportunities, etc in the calculation of trade-offs between transport cost and housing cost.

Third, a more flexible form of household utility with respect to transport costs may be considered. For example, households may be quite indifferent to increases in travel cost when total cost is low, but very sensitive to increases when total cost is high. Given the relatively low unit cost of travel (in time and money) in metropolitan areas today, it is not unreasonable to assume that households would willingly trade off additional travel cost for considerations of residential choice, as long as the work commute is not made intolerably long.

There are many reasons why further research in these directions makes sense. First, locational characteristics are emphasized and the spatial variation in residential areas is taken into account. Second, access to activities outside work, which may be relevant to decisions about location choice are incorporated. Third, in light of the high mobility rate of US households, and given that housing is the single largest household investment, location choice may depend on expectations about rate of return. If so, more general location characteristics (for example, access to significant amenities, 'good' schools, etc) would be more important than access to a particular job site. Fourth, job tenure has also shortened in recent years; most workers change jobs several times over their working careers. Thus distance from a specific work site may be much less important than access to other potential employment opportunities. A recent analysis of residential land values in Los Angeles lends support to the concept of multiple-access considerations. Access to several different employment centers, as well as to the ocean, were found to be significant explanatory variables (Heikkila et al, 1989) for residential property values. Last, large-scale residential development might have reduced spatial variation in the housing market. That is, the variety of housing available in a given area must decline as housing tracts become larger, and thus the housing search area may of necessity increase. All of these considerations point to the declining relevance of commute distance as a key explanatory factor in residential location.

5.2 *Firm location*

The research reviewed here suggests that labor-force access continues to be an important factor in firm location. However, within contemporary urban areas, labor is ubiquitous, given the relatively even distribution of the population. Under such conditions any number of possible locations may be equally likely, and thus random events have a great influence. Combining this idea with the presence of agglomeration economies implies that initial events play a critical role in urban spatial structure, because once a location has been established, growth and concentration will follow. Simply stated, history matters in contemporary metropolitan development.

The importance of initial conditions or events has been recognized in a number of different ways. In his simulation study of central place theory, Carruthers (1981) demonstrated that the hierarchical structure of a linear central place system is dependent upon the initial location of the first firm and its characteristics, since subsequent firms locate in response to the locations both of other firms and of employees. A second example is an extension of the new urban economics model that explains the development of endogenous suburban subcenters. Subcenter development is based on the linkages between firms. As growth occurs and (dominant) firms relocate in search of more and cheaper land, other dependent firms will also relocate, thus generating a subcenter (Clapp, 1984). In this context, the location choice of the initial relocating firm determines the location of the subcenter.

A related theory of endogenous regional growth is based on the concept of local potential (Coffey and Polèse, 1984). In this model, the entrepreneurial ability of the local population (that is, the human capital) is the key development factor. Growth of local firms depends on the ability of local managers, and, to the extent that these firms expand while remaining under local control, continued local growth is assured. This model was used to explain regional economic differences; however, it is applicable to intrametropolitan growth as well. One might identify major firms or institutions which dominate the local economy, and thus determine the subsequent location of additional economic activity. Similarly, aggressive 'growth-oriented' local governments may determine activity concentrations by attracting key firms to locate within their jurisdiction. The important point here is the endogenous nature of the process, that is, the extent to which local characteristics can direct and influence the urban development process.

If local factors play an important role, how does transportation affect development or firm location decisions? Obviously, transportation must have at least some indirect effect which is reflected in the need for firms to have adequate access to the labor market. This is nothing more than a restatement of the geographic concept of commute fields which have been used to define functional boundaries of urban areas (Berry, 1973). The urban field, or daily urban system, is defined by the maximum distance workers are willing to commute. In this case, however, the functional area may define the development potential of a given center. Once again, central place concepts seem relevant. All types of firms are in some way market dependent, as seekers either of consumers or of workers. The historical process of employment decentralization following population decentralization also supports this concept.

The role of transportation may be conceptualized in a couple of ways to reflect these conditions. Firms may treat transportation (or access) as a boundary condition or constraint. Some minimum must be fulfilled, but additional levels of accessibility have little value. This constraint-based concept implies that there is no unique spatial equilibrium with respect to land-use and transportation considerations—any number of locations may be equally acceptable, and the choice of location depends on other factors. Alternatively (to preserve the basis of the urban economic model), transportation may be considered a minor but still significant factor. Although agglomeration tendencies, land availability, and local institutions may be the major determinants for firms, transport or access considerations continue to have some relevance. In fact, these considerations may become more important as congestion increases.

6 Conclusion

The concepts presented here fit existing empirical evidence better than the traditional theory of land use and transportation. Research is necessary to develop and test these ideas more fully. The theoretical implications of treating transport in other ways should be explored. In the case of residential location, additional studies of household relocation patterns and household commute patterns could determine whether any of the alternative concepts presented here are appropriate. Simulation studies both of households and of firms may be useful as well. Stochastic processes, for example Monte Carlo simulations, could be used to test the influence of random factors on location. Last, detailed historical case studies should be conducted to identify key factors in the development process, with particular attention paid to changes in the transportation system. Case studies can isolate events or decisions which have materially affected urban structure, and thus provide guidance for generalizing land-use change and the role of transportation in the process.

The purpose of this paper has been to present some alternative interpretations of the relationship between transportation and land use. These interpretations are based on previous empirical research that calls into question some of the basic tenets of traditional location theory. The ideas presented here should provide direction for a better understanding of this complex relationship.

Acknowledgements. This research was supported by the Institute of Transportation Studies, University of California, Irvine, as part of a project on Activity and Transportation Systems Development. I am grateful to Peter Gordon, Kenneth Small, and Martin Wachs for comments on an earlier draft.

References

Adkins W G, 1959, "Land value impacts of expressways in Dallas, Houston, and San Antonio, Texas", Bulletin 227, Highway Research Board, Washington, DC, pp 50-65

Alonso W, 1964 *Location and Land Use* (Harvard University Press, Cambridge, MA)

Baerwald T J, 1982, "Land use change in suburban clusters and corridors" *Transportation Research Record* number 891, pp 7-12

Batten D, Johansson B, 1987, "The dynamics of metropolitan change" *Geographical Analysis* **19** 189-199

Berry B J L, 1973 *The Human Consequences of Urbanization* (Macmillan, London)

Boventer E von, 1976, "Transportation costs, accessibility, and agglomeration economics: centers, subcenters, and metropolitan structures" *Papers of the Regional Science Association* **37** 167-183

Boyce D E, 1980, "A framework for constructing network equilibrium models of urban location" *Transportation Science* **14** 77-96

Boyce D E, 1986, "Using transportation network equilibrium models in strategic planning", unpublished paper, Department of Civil Engineering, University of Illinois, Urbana, IL

Boyce D, Allen W B, Mudge R, Slater P, Isserman A, 1972, "Impact of rapid transit on suburban residential property values and land development: analysis of the Philadelphia-Lindenwold high-speed line", final report to the US Department of Transportation, Department of Regional Science, University of Pennsylvania, Philadelphia, PA

Boyce D E, Lee C-K, Lundqvist L, 1987, "Network equilibrium models of urban residential location and travel choice", paper presented at the TIMS/ORSA Joint National Meeting, New Orleans; copy available from Department of Civil Engineering, University of Illinois, Urbana, IL 61801

Carruthers N, 1981, "Central place theory and the problem of aggregating individual location choices" *Journal of Regional Science* **21** 243-261

Cervero R, 1986a *Suburban Gridlock* (Center for Urban Policy Research, New Brunswick, NJ)

Cervero R, 1986b, "Unlocking suburban gridlock" *Journal of the American Planning Association* **52** 389-406

Christaller W, 1966 *Central Places in Southern Germany* translated by C W Bushin (Prentice-Hall, Englewood Cliffs, NJ)

Clapp J M, 1984, "Endogenous centers: a simple departure from the NUE model" *Papers of the Regional Science Association* **54** 13-24

Clark W A V, Burt J, 1980, "The impact of workplace on residential relocation" *Annals of the Association of American Geographers* **70** 59-67

Coffey W J, Polèse M, 1984, "The concept of local development: a stages model of endogenous regional growth" *Papers of the Regional Science Association* **55** 1-12

Czamanski S, 1966, "Effects of public investments on urban land values" *Journal of the American Institute of Planners* **2** 204-217

Dyett M, Dornbusch D, Fajans M, Falcke C, Gussman V, Merchant J, 1979, "Land use and urban development impacts of BART", final report, DOT-P-30-79-09, US Department of Transportation and Department of Housing and Urban Development; John Blayney Associates—David M Dornbusch and Company Inc. A Joint Venture, 177 Post Street, Suite 750, San Francisco, CA 94108

Erickson R A, 1983, "The evolution of the suburban space economy" *Urban Geography* **4** 95-121

Erickson R A, Gentry M, 1985, "Suburban nucleations" *Geographic Review* **75** 19-31

Erickson R A, Wasylenko M, 1980, "Firm relocation and site selection in suburban municipalities" *Journal of Urban Economics* **8** 69-85

Fujita M, 1984, "Urban land-use theory", working papers in Regional Service and Transportation, Department of Regional Science, University of Pennsylvania, Philadelphia, PA

Giuliano G, 1986, "Land-use impacts of transportation investments: highway and transit", in *The Geography of Urban Transportation* Ed. S Hanson (Guildford Press, New York)

Gordon P, Richardson H W, Wong H L, 1986, "The distribution of population and employment in a polycentric city: the case of Los Angeles" *Environment and Planning A* **18** 161-173

Greene D L, 1980, "Recent trends in urban spatial structure" *Growth and Change* **11** 29-40

Guest A M, 1975, "Population suburbanization in American metropolitan areas" *Geographical Analysis* **7** 267-283
Hamer A, 1976 *The Selling of Rail Rapid Transit: A Critical Look at Urban Transportation Planning* (Lexington Books, Lexington, MA)
Hamilton B, 1982, "Wasteful commuting" *Journal of Political Economy* **90** 1035-1053
Harris B, 1985, "Urban simulation models in regional science" *Journal of Regional Science* **25** 545-567
Harris B, Wilson A G, 1978, "Equilibrium values and dynamics of attractiveness terms in production-constrained spatial-interaction models" *Environment and Planning A* **10** 371-388
Hartwick P G, Hartwick J M, 1974, "Efficient resource allocation in a multinucleated city with intermediate goods" *Quarterly Journal of Economics* **88** 340-352
Heikkila E, Dale-Johnson D, Gordon P, Kim J I, Peiser R B, Richardson H W, 1989, "What happened to the CBD-distance gradient?: land values in a polycentric city" *Environment and Planning A* **21** 221-232
Herbert J S, Stevens B H, 1960, "A model of the distribution of residential activity in urban areas" *Journal of Regional Science* **2** 21-36
Hershberg T, Light D Jr, Cox H, Greenfield R, 1981, "The journey to work: an empirical investigation of work, residence and transportation, Philadelphia, 1850 and 1880", in *Philadelphia: Work, Space and Group Experience in the Nineteenth Century* Ed. T Herschman (Oxford University Press, New York) pp 128-173
Herzog H W, Schlottmann A M, Johnson D L, 1986, "High-technology jobs and worker mobility" *Journal of Regional Science* **26** 445-460
Hoover E, 1948 *The Location of Economic Activity* (McGraw-Hill, New York)
Isard W I, 1956 *Location and Space-economy* (MIT Press, Cambridge, MA)
Jackson J R, 1979, "Intraurban variation in the price of housing" *Journal of Urban Economics* **6** 464-479
Karlson S H, 1985, "Spatial competition with location-dependent costs" *Journal of Regional Science* **25** 201-214
Kim T J, 1983, "A combined land use-transportation model when zonal travel demand is endogenously determined" *Transportation Research B* **17** 449-462
Klinger D, Kuzmyak R, 1986 *Personal Travel in the United States, Volume 1, 1983-1984 National Personal Transportation Study* final report, Department of Transportation, Federal Highway Administration, Washington, DC
Knight R, Trygg L, 1977, "Evidence of land use impacts of rapid transit systems" *Transportation* **6** 231-248
Lerman S, Damm D, Lam E L, Young J, 1978 *The Effect of the Washington Metro on Urban Property Values* final report UMTA-MA-11-0004-79-1, Urban Mass Transportation Administration (MIT Press, Cambridge, MA)
Los M, 1979, "Combined residential-location and transportation models" *Environment and Planning A* **11** 1241-1265
Lösch A, 1954 *The Economics of Location* translated by W H Woglom, W F Stolper (Yale University Press, New Haven, CT)
Meyer J R, Gomez-Ibanez J A, 1981 *Autos, Transit and Cities* (Harvard University Press, Cambridge, MA)
Meyer M D, Miller E J, 1984 *Urban Transportation Planning: A Decision-oriented Approach* (McGraw-Hill, New York)
Miller E J, Cubukgil A, 1981, "Land use trends and transportation demand forecasting: an empirical and theoretical investigation", RR-75, University of Toronto/York University Joint Program in Transportation, Toronto
Mills E S, 1972 *Studies in the Structure of the Urban Economy* (Johns Hopkins University Press, Baltimore, MD)
Mohring H, 1961, "Land values and the measurement of highway benefits" *Journal of Political Economy* **79** 236-249
Moses L N, 1958, "Location and the theory of production" *Quarterly Journal of Economics* **72** 259-272
Muller P, 1981 *Contemporary Suburban America* (Prentice-Hall, Englewood Cliffs, NJ)
Muth R, 1969 *Cities and Housing* (University of Chicago Press, Chicago, IL)
Muth R M, 1985, "Models of land-use, housing, and rent: an evaluation" *Journal of Regional Science* **25** 593-606
Odland J, 1978, "The conditions for multi-centered cities" *Econmic Geography* **54** 234-245

Ogawa H, Fujita M, 1980, "Equilibrium land use patterns in a non-monocentric city" *Journal of Regional Science* **30** 455-475

Patterson N S, May A D, 1979, "The impact of transport problems on inner city firms: a review" WP-112, Institute for Transport Studies, University of Leeds, Leeds

Payne-Maxie Consultants, 1980, "The land use and urban development impacts of beltways", final report DOT-OS-90079, US Department of Transportation and Department of Housing and Urban Development; Blayney-Dyett Urban and Regional Planner, 177 Post Street, Suite 750, San Francisco, CA 94108, subcontractors to Payne-Maxie Consultants, prime contractor

Peiser R, 1987, "The determinants of non-residential land values" *Journal of Urban Economics* **22** 340-360

Putman S H, 1975, "Urban land use and transportation models: a state of the art summary" *Transportation Research* **9** 187-202

Simpson W, 1987, "Workplace location, residential location, and urban commuting" *Urban Studies* **24** 119-128

Solow R M, 1973, "On equilibrium models of urban location", in *Essays in Modern Economics* Ed. J M Parkin (Longman, Harlow, Essex) pp 2-16

Stephanedes Y J, Eagle D, 1987, "Highway impacts on regional employment" *Journal of Advanced Transportation* **21** 68-79

Stevens B H, 1985, "Location of economic activities: the JRS contribution to the research literature" *Journal of Regional Science* **25** 663-685

Strotz R H, 1965, "Urban transportation parables", in *The Public Economy of Urban Communities* Ed. J Margolis (Resources for the Future, Washington, DC) pp 127-169

Weber A, 1928 *Theory of the Location of Industries* translated by C J Friedrich (University of Chicago Press, Chicago, IL)

Wheaton W, 1982, "Urban spatial development with durable but replaceable capital" *Journal of Urban Economics* **12** 53-67

White M J, 1976, "Firm suburbanization and urban subcenters" *Journal of Urban Economics* **3** 323-343

[7]

TRANSPORT EXPANSION IN UNDERDEVELOPED COUNTRIES: A COMPARATIVE ANALYSIS*

EDWARD J. TAAFFE, RICHARD L. MORRILL, AND PETER R. GOULD

IN THE economic growth of underdeveloped countries a critical factor has been the improvement of internal accessibility through the expansion of a transportation network. This expansion is from its beginning at once a continuous process of spatial diffusion and an irregular or sporadic process influenced by many specific economic, social, or political forces. In the present paper both processes are examined as they have been evident in the growth of modern transportation facilities in several underdeveloped areas. Certain broad regularities underlying the spatial diffusion process are brought to light, which permit a descriptive generalization of an ideal-typical sequence of transportation development. The relationship between transportation and population is discussed and is used as the basis for examination of such additional factors as the physical environment, rail competition, intermediate location, and commercialization. Throughout the study, Ghana and Nigeria are used as examples.[1]

Sequence of Transportation Development

Figure 1 presents the authors' interpretation of an ideal-typical sequence of transport development. The first phase (A) consists of a scattering of small ports and trading posts along the seacoast. There is little lateral interconnection except for small indigenous fishing craft and irregularly scheduled trading vessels, and each port has an extremely limited hinterland. With the emergence of major lines of penetration (B), hinterland transportation costs are reduced for certain ports. Markets expand both at the port and at the interior center. Port concentration then begins, as illustrated by the

* This study is based on a combination of the findings in Edward J. Taaffe and Richard L. Morrill: Transportation Geography Research: Part 2, Investigation of the Internal Spatial Distribution of Transportation Facilities (unpublished report to the U. S. Army Transportation Research Command under the auspices of the Transportation Center at Northwestern University, July 1, 1960); and Peter R. Gould: The Development of the Transportation Pattern in Ghana, *Northwestern Univ. Studies in Geogr. No. 5*, Evanston, Ill., 1960.

[1] Other areas examined in some detail, though from secondary sources, are Brazil, Kenya, Tanganyika, and Malaya.

➤ Dr. Taaffe is professor of geography at The Ohio State University, Columbus. Dr. Morrill is assistant professor of geography at the University of Washington, Seattle. Dr. Gould is assistant professor of geography at The Pennsylvania State University, University Park.

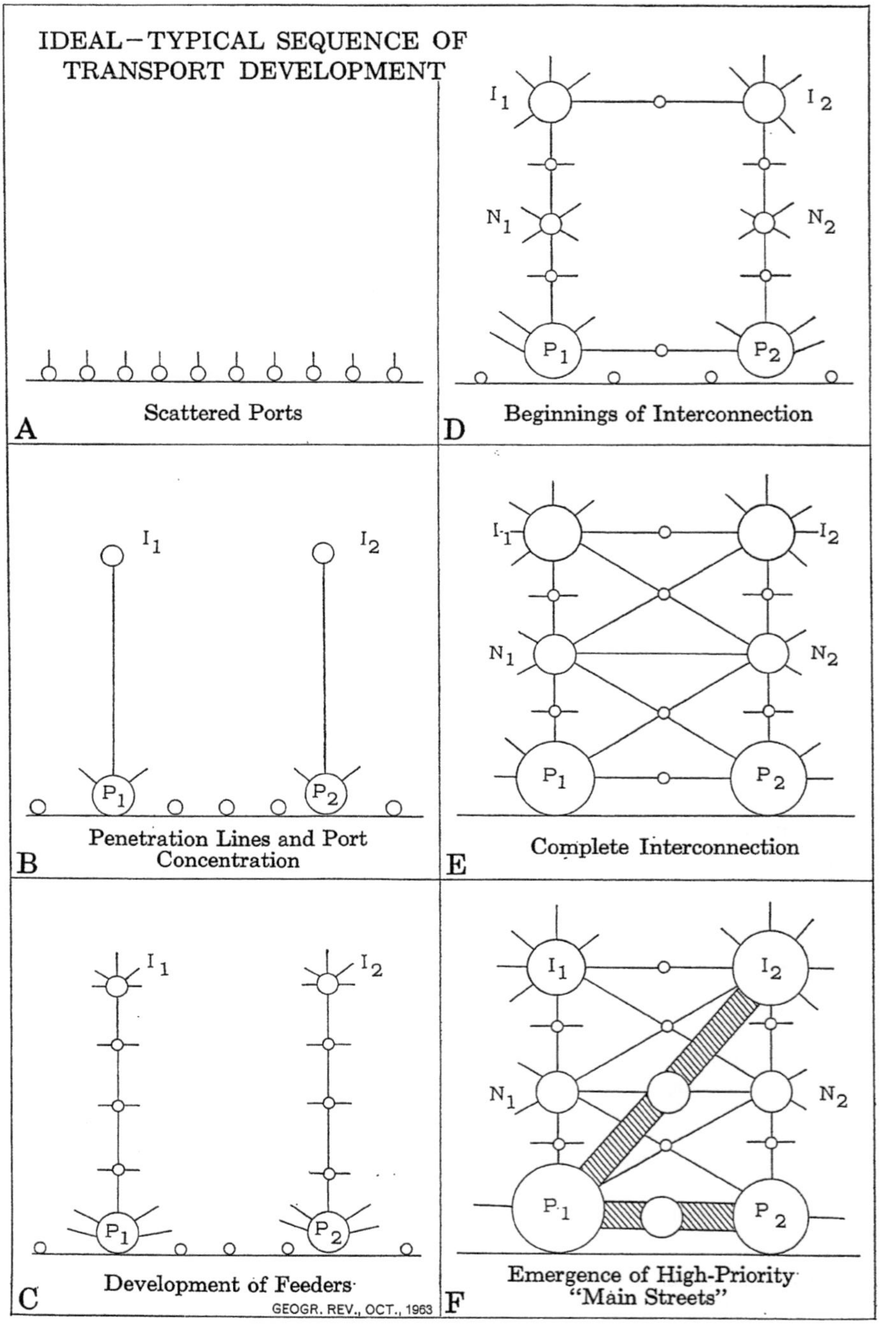
IDEAL–TYPICAL SEQUENCE OF TRANSPORT DEVELOPMENT
A
Scattered Ports
B
Penetration Lines and Port Concentration
C
Development of Feeders
D
Beginnings of Interconnection
E
Complete Interconnection
F
Emergence of High-Priority "Main Streets"
GEOGR. REV., OCT., 1963

FIG. 1

circles P_1 and P_2. Feeder routes begin to focus on the major ports and interior centers (C). These feeder routes give rise to a sort of hinterland piracy that permits the major port to enlarge its hinterland at the expense of adjacent smaller ports. Small nodes begin to develop along the main lines of penetration, and as feeder development continues (D), certain of the nodes, exemplified by N_1 and N_2, become focal points for feeder networks of their own. Interior concentration then begins, and N_1 and N_2 pirate the hinterlands of the smaller nodes on each side. As the feeder networks continue to develop around the ports, interior centers, and main on-line nodes, certain of the larger feeders begin to link up (E). Lateral interconnection should theoretically continue until all the ports, interior centers, and main nodes are linked. It is postulated that once this level is reached, or even before, the next phase consists of the development of national trunk-line routes or "main streets" (F). In a sense, this is the process of concentration repeated, but at a higher level. Since certain centers will grow at the expense of the others, the result will be a set of high-priority linkages among the largest. For example, in the diagram the best rail schedules, the widest paved roads, and the densest air traffic would be over the P_1–I_2 and P_1–P_2 routes.

It is probably most realistic to think of the entire sequence as a process rather than as a series of discrete historical stages.[2] Thus at a given point in time a country's total transport pattern may show evidence of all the phases. Lateral interconnection may be going on in one region at the same time that new penetration lines are developing in another.

THE FIRST PHASE: SCATTERED PORTS

In both Ghana and Nigeria[3] an early period of numerous small, scattered ports and coastal settlements with trading functions may be easily identified (Figs. 3 and 4). These settlements, most of which existed or came into being

[2] It is interesting to note a few analogies to some of W. W. Rostow's stages of economic development. The scattered, weakly connected ports might be considered evidences of the isolation of Rostow's traditional society; the development of a penetration line might be viewed as a sort of spatial "takeoff"; the lateral-interconnection phase might be a spatial symptom of the internal diffusion of technology; and the impact of the auto on the latter phases of the sequence might be an expression of the emergence of certain aspects of an era of higher mass consumption in underdeveloped countries.

[3] For Ghana, the examples are based on the field data and primary statistical source material gathered by Peter R. Gould. For Nigeria, primary statistical sources are indicated where necessary. Secondary sources include Kenneth O. Dike: Trade and Politics in the Niger Delta, 1830–1885 (Oxford Studies in African Affairs; London, 1956); Gilbert Walker: Traffic and Transport in Nigeria: The Example of an Underdeveloped Tropical Territory, *Colonial Research Studies No. 27,* Colonial Office, London, 1959; "The Economic Development of Nigeria" (International Bank for Reconstruction and Development, Baltimore, 1955). The authors are also indebted to Dr. Akin Mabogunje, of University College, Ibadan, for his comments.

between the end of the fifteenth century and the end of the nineteenth, were populated by the indigenous people around a European trading station or fort. Many of the people engaged in trade with the Europeans and served as middlemen for trade with the interior, a function jealously guarded for centuries against European encroachment. Penetration lines to the interior were weakly developed, but networks of circuitous bush trails connected the small centers to their restricted hinterlands. River mouths were important, particularly in the Niger delta, but with a few exceptions during the early periods of European encroachment the rivers did not develop as the main lines of thrust when penetration began. Most of these early trading centers have long since disappeared, destroyed by the growth of the main ports, or else they linger on as relict ports, with visits of occasional tramp steamers to remind them of their former trading heyday.

THE SECOND PHASE: PENETRATION LINES AND PORT CONCENTRATION

Perhaps the most important single phase in the transportation history of an underdeveloped country is the emergence of the first major penetration line from the seacoast to the interior. Later phases typically evolve around the penetration lines, and ultimately there is a strong tendency for them to serve as the trunk-line routes for more highly developed transportation networks. Three principal motives for building lines of penetration have been active in the past: (1) the desire to connect an administrative center on the seacoast with an interior area for political and military control; (2) the desire to reach areas of mineral exploitation; (3) the desire to reach areas of potential agricultural export production. In the cases examined, the political motive has been the strongest. Political and military control dominated official thinking of the day in Africa, often as a direct result of extra-African rivalries. The second motive, mineral exploitation, is typically associated with rail penetration. It is today probably the principal motive for the building of railways in Africa, and then only after careful surveys and international agreements have virtually guaranteed the steady haul of a bulk commodity to amortize the loans required for construction.[4]

The development of a penetration line sets in motion a series of spatial processes and readjustments as the comparative locational advantages of all centers shift. Concentration of port activity is particularly important, and

[4] For example, the extension of the Uganda railway to Kasese to haul copper ore; the long northward extension of the Cameroon railway from Yaoundé to Garoua to haul manganese; and the new railway from Port-Étienne to Fort Gouraud in Mauritania to haul iron ore.

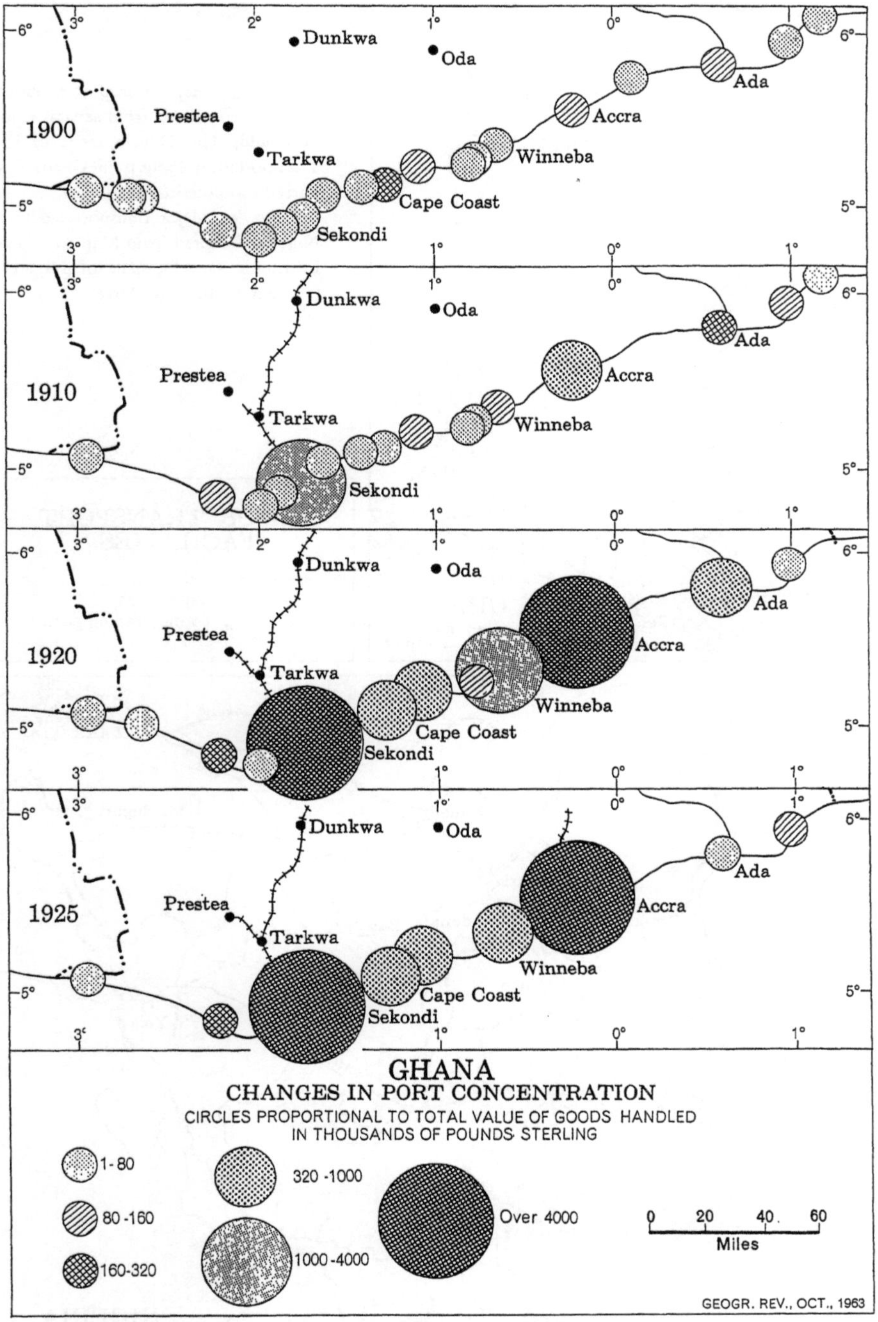

FIG. 2—Changes in port concentration, Ghana, 1900–1925. Redrawn from map in Gould, The Development of the Transportation Pattern in Ghana (see starred footnote in text), p. 45.

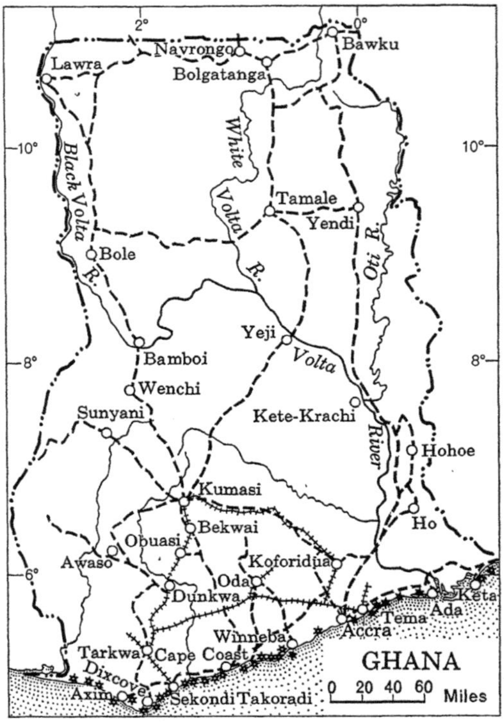

FIG. 3—Major transport facilities, Ghana. Generalized from several maps in Gould, The Development of the Transportation Pattern in Ghana (see starred footnote in text).

FIG. 4—Major transport facilities, Nigeria. Adapted from Map 1 in "The Economic Development of Nigeria" (see text footnote 3 above).

MAJOR TRANSPORT FACILITIES

- +++++++++ Railroads
- – – – – – Main roads
- * Former trading posts

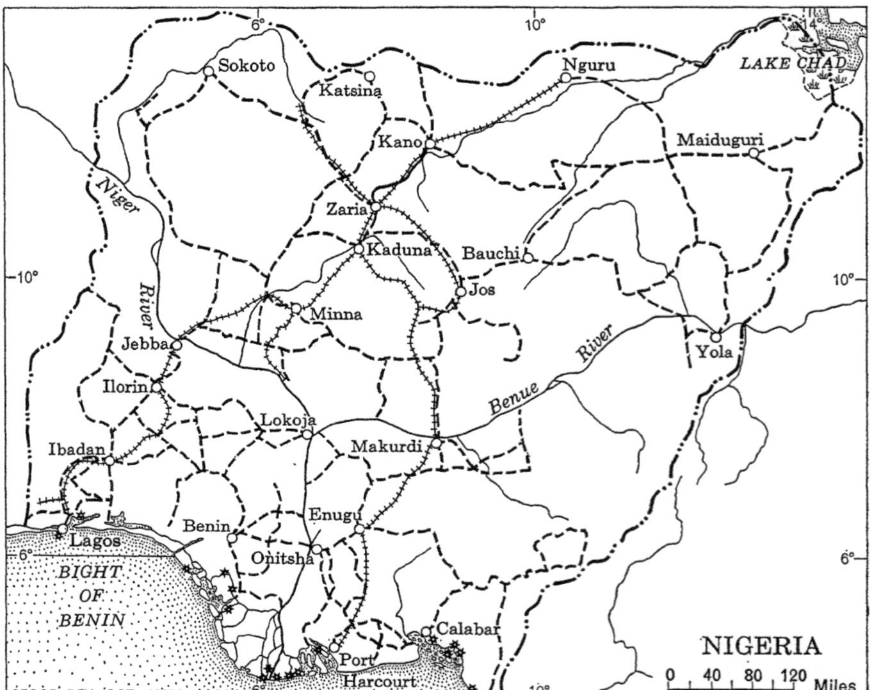

the ports at the termini of the earliest penetration lines are usually the ones that thrive at the expense of their neighbors (Fig. 2). Typically, one or two ports in a country dominate both import and export traffic, and often the smaller ports have lost their functions in external commerce.

In Ghana several interesting variations on the penetration theme appear (Fig. 3). The desire to reach Kumasi, capital of a then aggressive Ashanti, formed the essentially military-political motive for the first penetration road, which followed an old bush track, sporadically cleared whenever the local people were goaded into activity. The road was built from Cape Coast, and although it is still important as one of Ghana's main north–south links, the port function of Cape Coast declined as Sekondi increased in importance. Sekondi's great impetus came with the building of the rail penetration line to Kumasi at the turn of the century, after which adjacent ports such as Axim, Dixcove, Adjua, Shama, Komena, and Elmina suffered a rapid decline in traffic. The Pra and Ankobra Rivers, east and west of Sekondi, which were formerly of some significance as avenues of penetration, also experienced marked traffic decreases. The initial motive for the western railroad was primarily mineral production (the goldfield at Tarkwa), and secondarily provision of a rapid connection for administration between the seacoast and a troublesome internal center of population.

The eastern railroad penetration line was slower in developing, partly as a result of the interruption of the First World War and of smallpox outbreaks in the railroad camps. The link between Kumasi and Accra was not completed until 1923, twenty years after the Sekondi-Kumasi link. Connection with the rapidly expanding cocoa areas north of Accra was the immediate reason for this line, underlain by the political desire to connect the leading city, Accra, with Kumasi, the main population and distribution center in the interior. As in the case of Sekondi-Takoradi, Accra's importance increased steadily at the expense of adjacent ports as the railroad penetrated inland.

The two penetration lines forming the sides of the rail triangle were now complete, and a considerable amount of subsequent transportation development of the country was based on these two trunk lines. Penetration north of Kumasi was entirely by road, despite grand railroad plans at one time. There were no minerals to provide an economic incentive for railroads, and the barren middle zone, which separated the rail-triangle area from the densely settled north, acted as a deterrent to the continuation of the railroad in short stages.

The Great North Road is the chief line of penetration north of Kumasi, built with a strong political-administrative motive to Tamale, a town deliberately laid out as the capital of the Northern Territories in the early years of this century. Early feeder development focusing on Tamale helped to fix the position of this trunk line and its extension to the northern markets of Bolgatanga, Navrongo, and Bawku. The Western Trunk Road was built with similar motives and grew out of the extensive Kumasi feeder network to link that city with a moderately populated area north of the barren middle zone. In the building of the Eastern Trunk Road political and economic motives were mixed. This road through former British Togoland was originally extended from the Hohoe cocoa area to Yendi for transport of yams to the rapidly growing urban centers of the south, but its extension to the northern border hinged in large part on political motives that were very strong immediately before the United Nations plebiscite and Togo's resultant political affiliation with Ghana. The original road links from the rail triangle and Accra to the Hohoe cocoa area were associated with a political desire to forge a link with British Togoland, and with an economic desire to prevent the diversion of this area's cocoa traffic to the port of Lomé in then-French Togoland.

The process of penetration and port concentration in Nigeria (Fig. 4) was markedly similar to that in Ghana; the main difference lay in the greater emphasis on long-haul rail development and the subsequent higher level of economic development in the north. Again the initial motives were somewhat more political than economic; for even the early penetration to the north via the Niger River by the Royal Niger Company had imperialistic as well as economic motives. In a sense, Kano might be regarded as analogous to Kumasi. Both are important interior centers which predate European settlement and which were later connected to the main ports by rail penetration lines. The chief differences, of course, are the vastly greater distance between Kano and the coast and the greater width of Nigeria's relatively barren middle zone. Mineral exploitation was also a major motive for the building of rail penetration lines in Nigeria, particularly the eastern railroad. The line from Port Harcourt was started in 1913 and was connected with the important Enugu coalfields three years later. This port serves also as the principal outlet for the tin output of the Jos Plateau. The connecting of agricultural regions to the coast, though not a strong initial penetration motive, was apparently associated with the actual linking of the northern and southern lines.

As in Ghana, the rail penetration lines form the basis for the entire transportation network. The only area of extensive road penetration is in the northeast, from the railroad at Jos and Nguru to Maiduguri. The main motive for establishing tarred roads and large-scale trucking services was the attraction of the Lake Chad region to the northeast. However, the Maiduguri region is now being connected by rail to the main network, despite the recommendation of a mission of the International Bank for Reconstruction and Development, which felt that roads could more efficiently accommodate the expected increase in traffic. In the southeast there is no effective road or rail penetration line.

Port concentration has been marked. The decline of the delta ports began with the building of the rail penetration line from Lagos and was accentuated by the building of the eastern line and the concomitant growth of Port Harcourt. In 1958 these two major ports accounted for more than three-quarters of Nigeria's export and import trade.

THE THIRD PHASE: FEEDERS AND LATERAL INTERCONNECTIONS

Penetration is followed by lateral interconnection as feeder lines begin to move out both from the ports and from the nodes along the penetration lines. The process of concentration among the nodes is analogous to the process of port concentration; it results when the feeder networks of certain centers reach out and tap the hinterlands of their neighbors. As feeder networks become stronger at the interior centers and intermediate nodes some of them link and thereby interconnect the original penetration lines.[5]

Figures 5, 6, and 7 present a sequence of road development in Ghana from 1922 to 1958. The shading represents road-mileage density as recorded in a series of grid cells of 283 square miles superimposed on a highway map. In 1922 Ghana had just entered the phase of lateral interconnection, with east–west linkages both in the south, along the coast, and among the centers of the north, and with an extensive feeder network steadily drawing more and more of the smaller population centers into the orbit of Kumasi. Development in the southwest was weak, owing to railroad competition and to a deliberate policy of maintaining an economic road gap between Sekondi-Takoradi and Kumasi. This gap was finally filled in 1958, and only now is the southwest beginning to realize its great potential in cocoa and timber. By 1937 lateral interconnection had become more marked. The connections east and west of Tamale provide a good example of links between intermediate nodes. The 1–20-mile shading, for example, now reaches west from the

[5] The degree of interconnectedness of a transport network could be precisely evaluated by the use of such new measures as those presented in William L. Garrison: Connectivity of the Interstate Highway System, *Papers and Proc. Regional Science Assn.*, Vol. 6 (6th Annual Meeting), 1960 (Philadelphia, 1961), pp. 121–137.

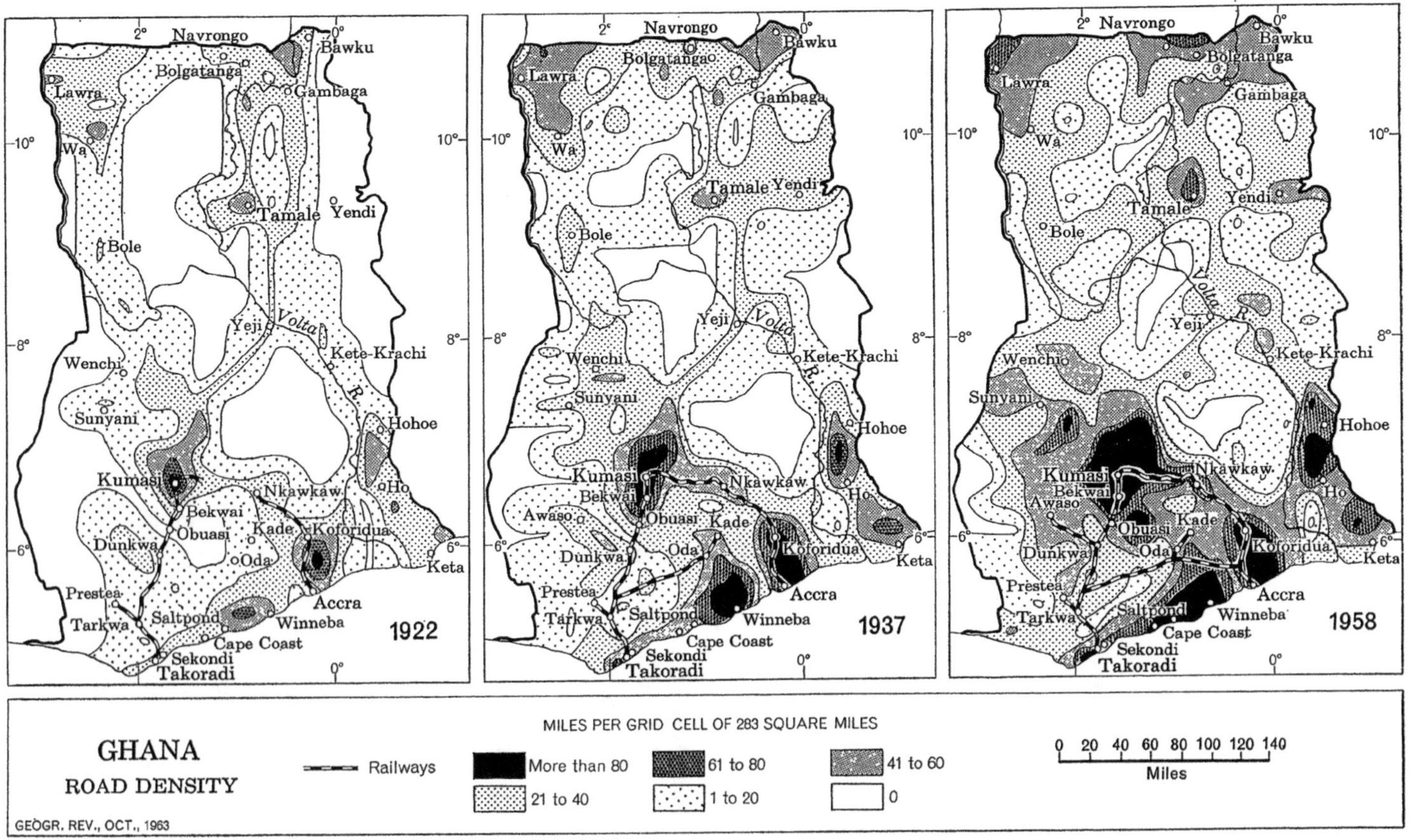

FIG. 5-7—Road density in Ghana, 1922, 1937, and 1958. Redrawn from maps in Gould, The Development of the Transportation Pattern in Ghana (see starred footnote in text), pp. 104, 107, and 109.

Tamale node on the Great North Road to the node at Bole, which was just developing on the Western Trunk Road in 1922; Yendi on the Eastern Trunk has been similarly linked. Lateral interconnection has become intensified in the north, and the 21–40-mile shading now covers the entire zone between Bawku and Lawra. In the Kumasi area feeder development has continued, and the 21–40-mile shading blankets the Wenchi–Sunyani area to provide a fairly good network of interconnections between the Western Trunk and the Great North Road in the zone where both converge on Kumasi. Urban geographers will note the strong analogy to the process of interstitial filling between major radial roads converging on a central business district.

In 1958 the lateral interconnection process is fairly well developed. Only a few areas are still without road links, formerly inaccessible areas having been tapped by the expanding road network. A new series of high-density nodes have developed since 1937 and already are reaching out toward one another. Marked examples occur in the north between Tamale, Yendi, and other northern population centers, and in the south in the developing and extending nodes around Kumasi and east of the Volta River.

It is clear from the regularity of the progression of the highway-density patterns that extrapolation of the density maps to some future date would be reasonable. In a sense the map sequence is a crude predictive device. For instance, the probability of an increase in road miles for any area between two nodes is greater than that for a comparable area elsewhere.

In Nigeria a basically similar pattern had developed by 1953 (Fig. 8), with many of the earlier nodes of high accessibility in the south linking laterally to form an almost continuous high-density strip, broken only by the Niger River near Onitsha. Lines of penetration linking the north and the south across the barren middle zone, a feature clearly brought out by the map, are relatively weakly developed, and the degree of lateral linkage is well below that of northern Ghana. Only the Kano and Zaria nodes, in areas of high agricultural production and at the center of strong administrative webs, and the Jos node, at the center of the tin and columbium mineral complex, stand out as exceptions. Heavy rail competition, which resulted in severe restraints on long-haul trucking for many years, has clearly weakened the western road penetration lines, and the similarity to southwestern Ghana is strong. Areas totally inaccessible by road are still numerous, particularly in the barren middle zone and along the periphery of the country. To the political geographer the general weakness of the linkages between the Eastern, Western, and Northern Regions will be of particular interest. There is, in fact, a clear visual impression that the general pattern of accessibility by road in Nigeria in 1953 is similar to that of Ghana in 1937—hardly surprising in view of the much larger size of Nigeria, the longer distances, and the lower per capita tax base from which the greater part of road development funds must come.

THE FOURTH PHASE: HIGH-PRIORITY LINKAGES

The phase following the development of a fairly complete and coherent network is difficult to identify, and a variety of labels might be applied to it. Certainly the most marked characteristic of the most recent phase in the cases studied is the dominance of road over railroad. A common theme throughout the evolution of the transportation system in Ghana and Nigeria, and also in the other examples studied, has been the steady rise in the importance of road traffic, which first complements the railroad, then competes with it, and finally overwhelms it. However, the evidence available seems to indicate that this occurs irrespective of the stage of transport development, and it is possible that a greater number of road penetration lines are now being built in areas which would have required rail penetration lines in the past.

The idea of a phase of high-priority linkages is based, somewhat weakly, on a logical extrapolation of the concentration processes noted in the earlier stages of transport development in Ghana and Nigeria, and is supported in part by highly generalized evidence from areas with well-developed transportation systems.

Interior centers, intermediate nodes, and ports do not develop at precisely the same rate. As some of these centers grow more rapidly than others, their feeder networks become intensified and reach into the hinterlands of nearby centers. Ultimately certain interior centers and ports assert a geographic dominance over the entire country. This creates a disproportionately large demand for transportation between them, and since some transport facilities already exist, the new demand may take such forms as the widening of roads or the introduction of jet aircraft. In general, transport innovations are first applied to these trunk routes. For example, in the United States the best passenger rates, schedules, and equipment are usually initiated over high-density routes such as New York–Chicago. In underdeveloped countries high-priority linkages would seem to be less likely to develop along an export trunk line than along a route connecting two centers concerned in internal exchange. There is some weak evidence that high-priority links may be developing in the two study countries. High-density, short-haul traffic in the vicinity of Lagos may be the forerunner of a "main street" between cities of

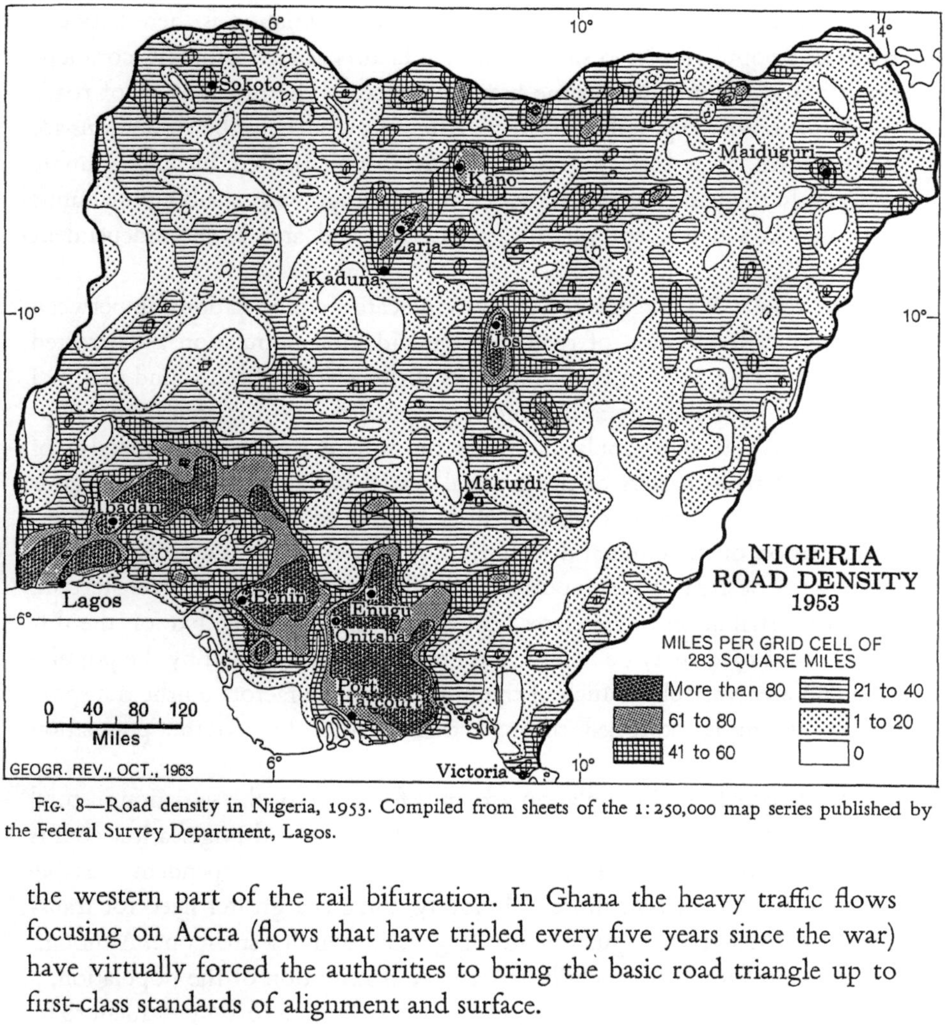

FIG. 8—Road density in Nigeria, 1953. Compiled from sheets of the 1:250,000 map series published by the Federal Survey Department, Lagos.

the western part of the rail bifurcation. In Ghana the heavy traffic flows focusing on Accra (flows that have tripled every five years since the war) have virtually forced the authorities to bring the basic road triangle up to first-class standards of alignment and surface.

RELATIONSHIPS BETWEEN ROAD MILEAGES AND SELECTED PHENOMENA

The lateral-interconnection phase of the ideal-typical sequence is the one that best depicts the current extent of transportation development in most underdeveloped countries. This phase has been accompanied by a steady increase in the importance of motor vehicles, so that at present the dominant transport characteristic of most underdeveloped countries is the expansion of the road network. Closer examination of the road networks of Ghana and Nigeria affords deeper insight into the factors that affect the spatial diffusion of roads at a time when interconnection is a prominent motive for transport development. For example, how close a relation is there between roads and population? Do such additional factors as environment, competitive transportation, and income have an effect on the distribution of roads over and above the population effect? Attempts to answer these questions are in the form of a basic regression model supplemented by cartographic analysis. In the basic regression model, road mileage within subregional units is treated as a dependent variable, population and area as the independent variables.[6]

The results of the regression analysis indicate a close relationship between the internal distribution of road mileage and total population as corrected for the differing areas of reporting units. Briefly, it has been found that in a given unit road mileage is in general proportional to the square root (approximately) of the population times the square root (approximately) of the area. Three-quarters of the internal spatial variation in road mileage is associated with these two factors alone.[7] Thus to achieve a fair first approximation of the internal distribution of road mileage at a given point in time, we look first to the population distribution. Effects of difficult terrain, unequal distribution of resources, rail competition, and the like on the distribution of roads may be regarded as being partly subsumed by the population and area variables. Much of the impact of these factors on the transportation system is expressed through their relationship to the population pattern.

As expected, total population accounts for more of the variation in total road mileage than area accounts for; in both Ghana and Nigeria it accounts for about 50 per cent. The addition of area as an independent variable accounts for 20 per cent more. Obviously, there is a greater need for transportation for a given population in a large unit than in a small one. Although the demand for roads generally reflects the distribution of the population, a large, sparsely settled unit will require a large per capita road investment to be served at all. Thus the relative weights of the two independent variables, and

[6] Sources for population figures were "Population Census of Nigeria, 1952–1953" (Lagos, 1954), and "The Gold Coast Census of Population, 1948: Report and Tables" (London and Accra, 1950); for road mileages, "Mobil Road Map of Nigeria," 1:750,000 and 1:500,000 (Federal Survey Department, Lagos, 1957), and "Road Map of the Gold Coast," 1:500,000 (Department of Surveys, Accra, 1950). Only first- and second-class roads were included. No weighting system was applied.

[7] For Ghana the regression equation was $\log Y_c = 0.1709 + 0.6285 \log X_1 + 0.4139 \log X_2$ or $Y_c = 1.482\, X_1^{0.6285} X_2^{0.4139}$, with Y_c the estimated highway mileage, X_1 the district population in thousands, and X_2 the district area. The r^2 or explained variation was 0.75 or 75 per cent. For Nigeria the regression equation was $\log Y_c = -0.44771 + 0.4458 \log X_1 + 0.4823 \log X_2$ or $Y_c = (X_1^{0.4458} X_2^{0.4823})/2.799$. The explained variation in this case was 81 per cent. In both cases the particular form of the equation was the result of normalizing the data by means of log transformations.

the closeness of the correlation, are seen to be significantly affected by variations in the size of the reporting units; hence the use of simple population densities would have been deceptive in that an understatement of road-mileage expectations for large, sparsely settled units would have resulted.

On the other hand, it is not clear that meaning may be ascribed to the area variable as a separate factor. The problem of modifiable areal reporting units is, in general, a difficult one.[8] Internal variations in size of reporting units affect the degree of apparent correlation between variables. Even if the sizes of reporting units were uniform, different correlations and different regression equations would be obtained for different levels of areal aggregation (a grid cell of 10 square miles as opposed to 100 square miles, for instance). In this case, it is best to regard the use of an area variable as a means of including the effects of any internal variations in size of reporting units on road mileage along with any effects of variations in population. Thus it can be said that three-quarters of the variation in road mileage among the areal subunits in Ghana and Nigeria is statistically associated with their combined variations in population and size. However, if the average size of the subunits were to be significantly increased or decreased, the amount of statistically "explained" variation would be affected.

This relatively simple regression analysis may now be tied directly to the maps in an attempt to uncover further possible factors that seem to be particularly relevant to the development of road transportation. In what parts of Ghana and Nigeria does there seem to be a great deal of residual variation? Where, in other words, does the population-area equation seem to give significant overestimates or underestimates of road mileage?[9] Examination of the residuals maps for Ghana and Nigeria suggests five additional factors: hostile environment; rail competition; intermediate location; income or degree of commercialization; and relationship to the ideal-typical sequence. Precise quantification of these factors did not seem to be warranted by the data. Therefore, a subjective examination was made of the relationship between each of the five factors and the distribution of regression residuals.

[8] Some aspects of the problem are discussed in Arthur H. Robinson: The Necessity of Weighting Values in Correlation Analysis of Areal Data, *Annals Assn. of Amer. Geogrs.*, Vol. 46, 1956, pp. 233–236, and Otis Dudley Duncan, Ray P. Cuzzort, and Beverly Duncan: Statistical Geography: Problems in Analyzing Areal Data (Glencoe, Ill., 1961).

[9] The use of residuals maps has been discussed by Edwin N. Thomas: Maps of Residuals from Regression: Their Characteristics and Uses in Geographic Research, *State Univ. of Iowa, Dept. of Geogr.*, [*Publ.*] *No. 2*, Iowa City [1960]. Examples of the use of residuals maps are Edward J. Taaffe: A Map Analysis of Airline Competition, Part 2, *Journ. of Air Law and Commerce,* Vol. 25, 1958, pp. 402–427, and Peter R. Gould and Robert H. T. Smith: Method in Commodity Flow Studies, *Australian Geographer,* Vol. 8, 1960–1962, pp. 73–77.

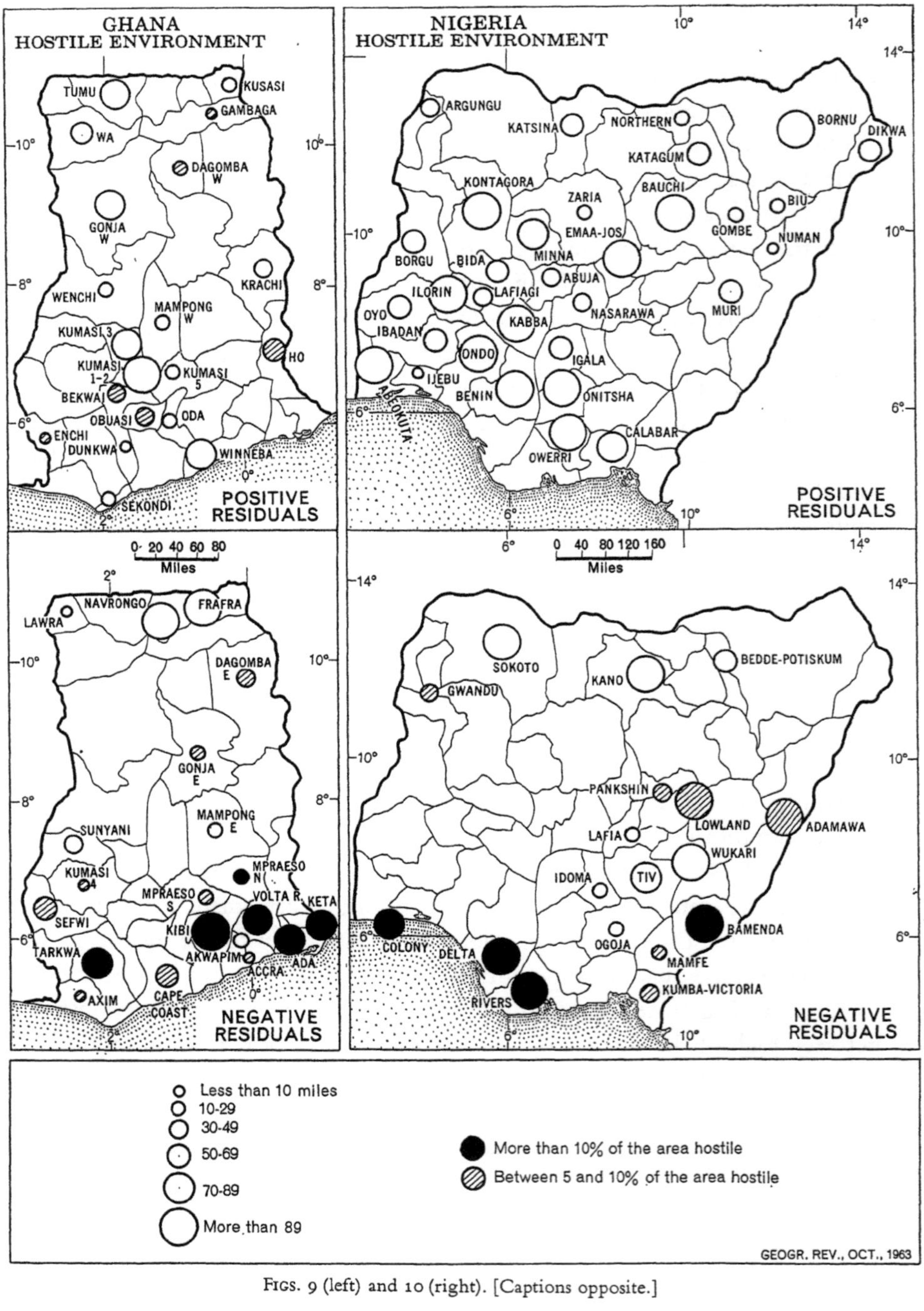

FIGS. 9 (left) and 10 (right). [Captions opposite.]

In many instances the lack of data resulted in highly generalized and arbitrary quantifications.

HOSTILE ENVIRONMENT

Consideration of the effects of the physical environment on highway mileage gives rise to some interesting speculations. The first finding is negative: namely, that the inclusion of difficult terrain in the analysis does not add greatly to one's ability to predict the road mileage in a given unit once population and area have been considered. Although visual comparison of road maps and topographic maps might suggest such a relationship, the effects of sloping land seem already to have been discounted in individual units by the fact that they are thinly populated. Comparison of slope percentages with regression residuals indicated no significant relationship. Thus there is no consistent evidence that a unit with a high percentage of sloping land has less road mileage than would be expected from its population and area as expressed in the basic equation. It should be noted, however, that this statement too is conditioned by the size of the reporting units. At the scale of observation implied by the reporting units there was little evidence of a separate effect of sloping land. But if reporting units were to become smaller, it is obvious that, at some scale, slope would have a marked effect on the selection of specific routes irrespective of population. More refined slope measures might also have produced different results at different scales.

Examination of the residuals maps (Figs. 9 and 10)[10] does indicate that,

[10] On all the residuals maps circles are proportional to the residual deviation from the regression formula. For example, on the positive residuals maps the large circles for Tumu, Ghana, and Abeokuta, Nigeria, indicate that they had considerably *more* highway mileage than would be expected from the regression formula, which takes into account population and area. Similarly, on the negative residuals maps the large circles for Kibi, Ghana, and Delta, Nigeria, indicate that they had considerably *less* highway mileage than expected. In effect, then, these units show the distribution of the variation in highway mileage attributable to factors other than population and area.

FIG. 9—Hostile environment, Ghana. Estimates were made of the area within a district which would be classified as swampy or noticeably dissected. Maps used were at the scale of 1:250,000. Circles are graded in size according to the differences between actual road mileages and road mileages estimated from the equation

$$\text{Road mileage equals } (1.482)\ (\text{Population})^{(0.6285)} \text{ times } (\text{Area})^{(0.4139)}.$$

FIG. 10—Hostile environment, Nigeria. Estimates were made of the area within a district which could be classified as swampy, noticeably dissected or containing high mountains. Circles are graded in size according to the differences between actual road mileages and road mileages estimated from the equation

$$\text{Road mileage} = \frac{(\text{Population})^{(0.4458)} \text{ times } (\text{Area})^{(0.4823)}}{(2.799)}.$$

at the scale of observation employed, two specific environmental conditions seem to be related to the tendency for road mileage to be less than would be expected from a unit's population and area.[11] These are a very steep and consistent slope, such as an escarpment, the trend of which is in general at right angles to the country's alignment of traffic, and the presence of extremely swampy land. Circles on the residuals maps have been shaded according to the amount of "hostile" land in a particular unit, and since the presence of swamps or escarpments should reduce the road mileage associated with a given population and area, we should expect to find most of the shaded circles on the negative residuals map. This proves to be the case. In Ghana fourteen of the twenty districts with more than 5 per cent of their land classified as hostile, and all six of the districts with more than 10 per cent, are on the negative residuals map. Low road-mileage figures are most closely associated with the swampy lands in the Keta, Ada, and Volta River districts. In Mpraeso North and Kibi there is strong evidence of the deterrent effect of steep slopes where the prominent Mampong escarpment has sharply curtailed the development of feeder routes.

In Nigeria low residuals are associated with swampy land in the Colony and, in the Niger Delta, Delta and Rivers districts. An important difference, however, between the swampy lands of Nigeria and Ghana is that in Nigeria the low residuals are associated not only with an environment unfavorable to road building but also with effective and long-established competitive water transportation. Waterway competition is also reflected in the low residuals in Lowland, Wukari, Lafia, Tiv, and Idoma. These districts are located along the middle and lower Benue, where waterway traffic is important in view of the absence of other effective connections to the zone of maximum economic activity. In Nigeria the only consistent relationship between slope and low road density appears where there is a steep slope over a fairly large area, such as an escarpment lying across the transport grain, or where there is a large number of steep volcanic ranges (the Cameroons districts of Kumba-Victoria, Mamfe, Bamenda, and Adamawa). In the latter example, climate may also have played an important role.

RAILROAD COMPETITION

Logically, railroad competition could affect road mileage either way.

[11] For a consideration of other specific environmental features see Benjamin E. Thomas: Transportation and Physical Geography in West Africa (Prepared for the Human Environments in Central Africa Project, National Academy of Sciences–National Research Council, Division of Anthropology and Psychology [Department of Geography, University of California, Los Angeles, 1960]). For example, Thomas documents the marked effect of seasonality on the utilization of roads in West Africa.

It could reduce the need for roads by providing an alternative form of transportation; it could increase the need for roads by promoting commercial production for interregional export. In the latter, or complementary, case one might expect a proliferation of feeder roads from nodes on rail penetration lines. Although there is some evidence of the validity of both possibilities, the stronger evidence seems to be on the side of the positive effect. It is perhaps more accurate to say that units with railroad mileage have more road mileage than would be expected from their areas and populations. Either the railroad itself promotes feeder development, or both rail and road are transport manifestations of a high level of income, urbanization, or commercialization. There are instances where low road-mileage figures are associated with the presence of railroads, but this seems to be where a deliberate government policy of protecting the railroad from competition has been in effect. It is for the last reason that the residuals maps of Ghana (Fig. 11) are difficult to interpret: railroad competition has had sharply different effects in different parts of the country. Tarkwa's large negative residual seems to be associated with the "economic gap" policy under which the building of important through roads parallel to the western line was forbidden so as to preserve the railroad from competition.[12] In Kumasi, on the other hand, and to a smaller extent along the eastern line, the building of feeder networks to the railroad appears to have had a positive effect on the residuals.

In Nigeria (Fig. 12) there is clearly no map evidence that rail competition reduces road mileage below the level expected from population and area. The concentration of high rail-mileage figures on the positive residuals map indicates that the complementary aspect of the relationship is much more evident. In Nigeria, however, it is difficult to separate the effects because of the more consistent distribution of the residuals in the zone of maximum economic activity, where such measures as railroad mileage, commercial production, income, and degree of urbanization would also be expected to be high.[13]

INTERMEDIATE LOCATION

In addition to the factors already discussed, the location of a reporting unit with respect to other distributions may have considerable bearing on its road mileage. For example, a unit with a low population density located

[12] Gould, *op. cit.* [see starred footnote above], pp. 44, 71, 102, 106, and 110.

[13] A multiple regression was also run using four independent variables (in logarithms): population, area, railway mileage, and waterway mileage. The results did not add significantly to the amount of variation in road mileage statistically "explained" by population and area. The regression coefficients for both railway and waterway mileage tested as statistically insignificant.

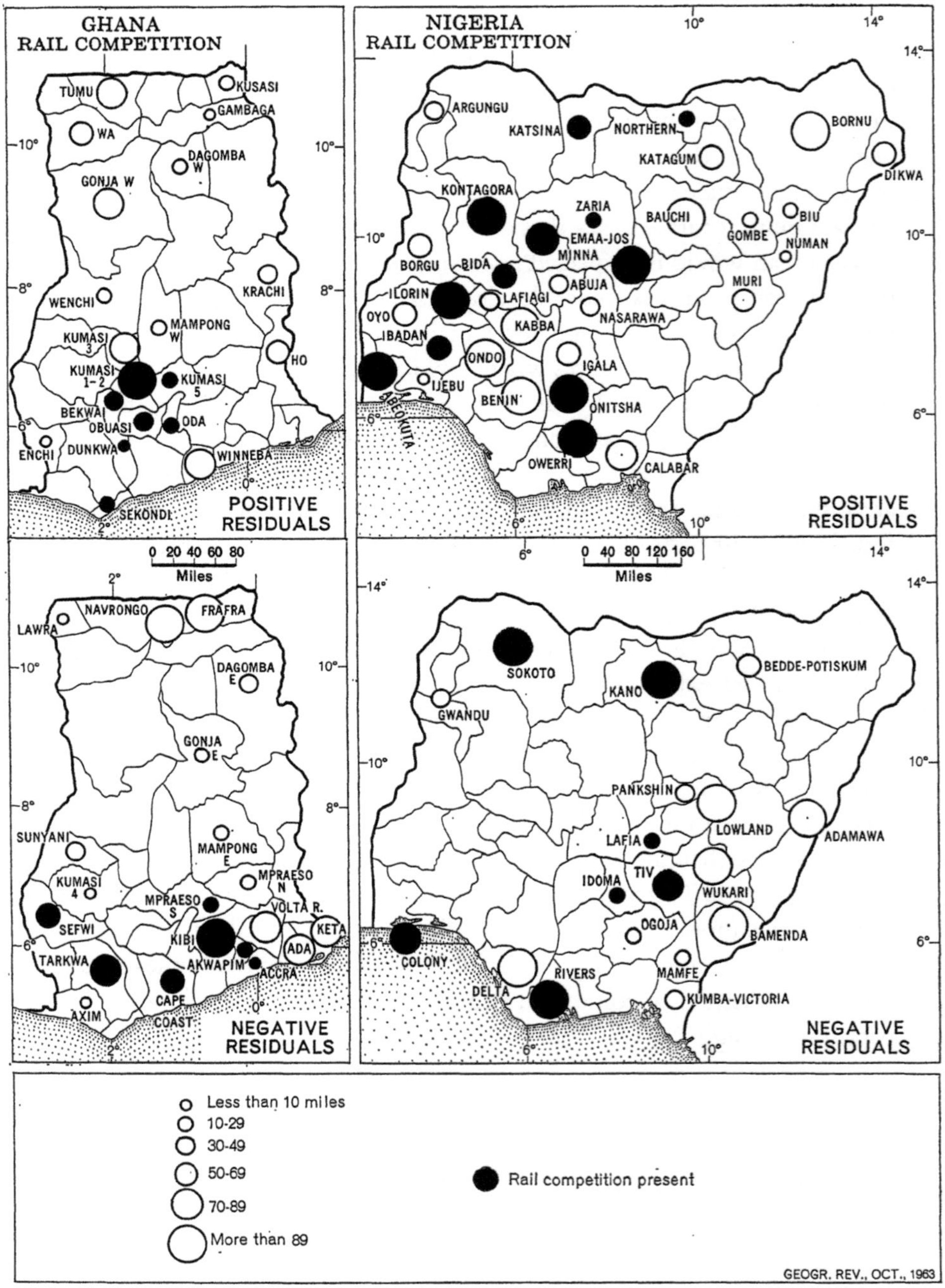

FIGS. 11 and 12—Rail competition, Ghana and Nigeria.

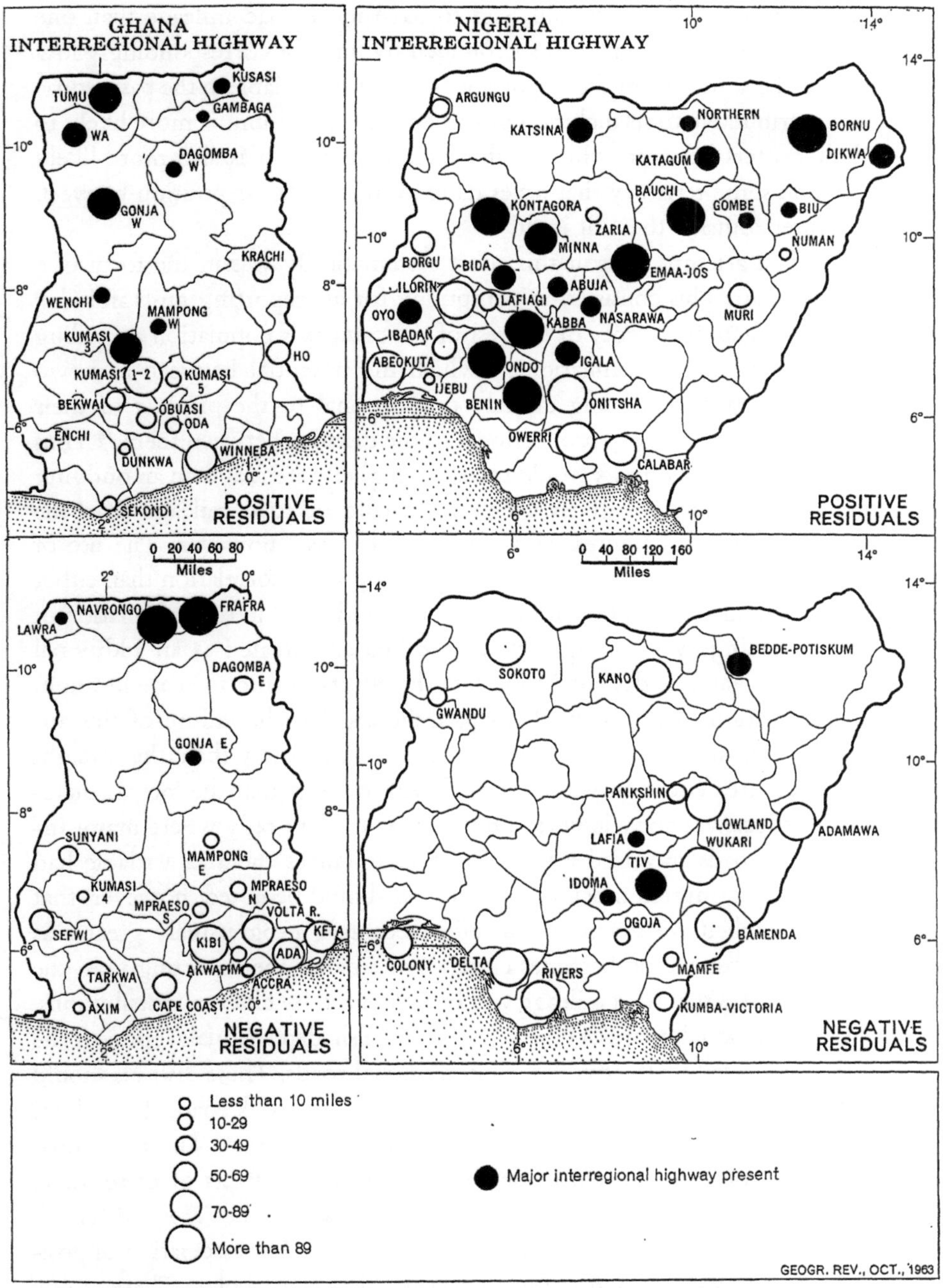

FIGS. 13 and 14—Interregional highways, Ghana and Nigeria.

between two large cities would tend to have more road mileage than one with the same density and area surrounded by units of correspondingly low densities. To some extent the importance of the area variable in the basic equation is attributable to its inclusion of this effect. A large unit is more likely to include centers between which highways are needed. It is also more likely to include some territory that owes its road mileage to its position between two centers outside the unit itself.

The first attempt to treat this effect systematically was in the form of a potential map. This consisted in computing for all reporting units an index of the aggregate proximity of the rest of the country's population according to a gravity-model formulation in which the potential between any two units was regarded as being directly proportional to the product of their populations and inversely proportional to the distance between them.[14] Thus a unit between two centers would register a higher potential than an outlying unit. This should have given a better estimate of road mileage than the population and area figures. Such was not the case, however. The use of potential figures in a regression analysis gave a poorer correlation than either population and area or population alone. Nor did it aid in the explanation of variation of highway mileage when modified and treated as an additional independent variable designed to isolate the effects of intermediate location.

Examination of the residuals maps indicates that the failure of this apparently logical index to provide a better explanation than the straight population-area equation was associated with the fact that the intermediate-location effect was not widespread but seemed evident only where major interregional roads traversed a unit. In other instances the area variable had apparently transmitted much of the effect. It should be noted, however, that need for explicit treatment of this effect becomes progressively greater as the scale of observation becomes closer. The smaller the reporting unit, the more likely it is that road mileage will be influenced by external distributions.

On the residuals maps of Ghana (Fig. 13) the shaded residuals represent the districts traversed by the two large interregional roads. These districts should have more mileage (from feeders as well as from the trunk roads themselves) than would be expected from their populations and areas. The maps provide reasonably good visual evidence of the positive effects of these interregional roads. The string of positive residuals along the Western Trunk Road from Kumasi to Tumu is particularly striking. The association of positive residuals with the Great North Road is less spectacular. There is one

[14] See John Q. Stewart and William Warntz: Macrogeography and Social Science, *Geogr. Rev.*, Vol. 48, 1958, pp. 167–184.

contradictory district (Gonja East), and none of the residuals are more than half a standard error above or below the regression estimate. The contrast among the northern districts in this respect is interesting. Tumu has a very high residual; Navrongo and Frafra have low residuals. The explanation seems to lie in the population densities rather than in the road mileages. Road mileage does not differ greatly among the three districts, but Navrongo and Frafra are much more densely populated than Tumu. Thus the negative residuals for Navrongo and Frafra may be associated with a generally lower level of feeder-road development in the north than in districts of corresponding population density in the south.

In Nigeria (Fig. 14) major interregional roads are in general parallel to the railroads. There are, however, two areas without railroad mileage where the positive residuals seem to be associated with the presence of important interregional roads. In both the use of the interregional designation may be legitimately questioned. One road, through the northeastern districts of Bauchi, Gombe, Biu, Bornu, and Dikwa, is an extension from the zone of maximum rail and economic activity to Maiduguri and the Lake Chad area outside Nigeria. The traffic density on this road is relatively high, and it represents one of the few cases where long-haul trucking is of major importance in Nigeria. In the south two roads serve as lateral interconnections between the two parts of Nigeria's bifurcated rail pattern. They are classed as interregional roads because they carry heavy long-haul traffic and serve to link the Eastern and Western Regions of a federated Nigeria.

COMMERCIALIZATION AND RELATION TO THE IDEAL-TYPICAL SEQUENCE

It can be assumed that unusually productive units have relatively high incomes and therefore have more road mileage than is called for in the population-area equation. Although the resulting relationships are somewhat ambiguous and difficult to isolate from other factors, there is a general tendency for the more productive units to have more road mileage than expected. Units focused on export agriculture appear to be more important with respect to highways than units focused on mineral production. The maps are visually inconclusive, owing chiefly to difficulties in obtaining data that realistically measured commercial production in the individual units, and in separating commercial production from population that had already been considered in the regression equation. For example, in Ghana (Fig. 15) it is probable that the highs around Kumasi and in Ho and Krachi are associated with export agricultural production. On the other hand, Tarkwa appears on the negative residuals map despite its large production for export, because much of the production consists of mineral ores (manganese and bauxite) hauled almost exclusively by rail. Thus, despite its relatively high commercialization, this district has less road mileage than would be expected from its population and area. There seems also to be an interesting time element in the relation between the cocoa districts and road transportation. Although the districts cited above show high residuals, the very new cocoa districts (Sunyani, Sefwi, and Kumasi 4) show negative residuals. This may represent a tendency for transportation development to lag behind population in rapidly growing districts.

In Nigeria (Fig. 16), as expected, the districts with a large commercial production add little to the correspondence with the positive residuals already noted on the railroad and interregional-road maps. However, the commercialization maps do point up the persistent anomaly of the two important northern districts of Kano and Sokoto. These districts register negative residuals despite a high degree of commercialization and the presence both of railroads and of interregional roads. There is an interesting analogy here to the new cocoa districts of Ghana, because Kano is also an area of rapid and recent economic expansion, and it, too, may well show a lag-and-lead pattern.

The tendency for commercialization to be intercorrelated with population and road mileage, and to occur within the general frame of the transport network, leads to a consideration of the relation between the residuals maps and the ideal-typical sequence. As a rule, high residuals, representing large road mileage, are found in units that also have large populations, a high degree of commercialization, and large railroad mileages. These units are usually in a zone of maximum activity, which includes the early penetration lines, the "main streets," and the majority of the interconnections. Outside this zone, toward the borders of the country, transportation development seems weaker, as is evidenced by negative residuals.

In Ghana the zone of maximum activity comprises the railroad triangle and some parts of the northern penetration lines. Districts peripheral to this zone that exhibit weaker road development are the new cocoa districts, the northern districts of Navrongo and Frafra, and the southeast. Ho and Krachi are conspicuous exceptions, though this may be associated with the political factors mentioned earlier.

In Nigeria there is a similarly striking concentration of positive residuals within the zone of maximum transport activity, as would be expected from the ideal-typical sequence. Most of the districts in this zone also rate high in commercialization as well as in either rail mileage or interregional roads.

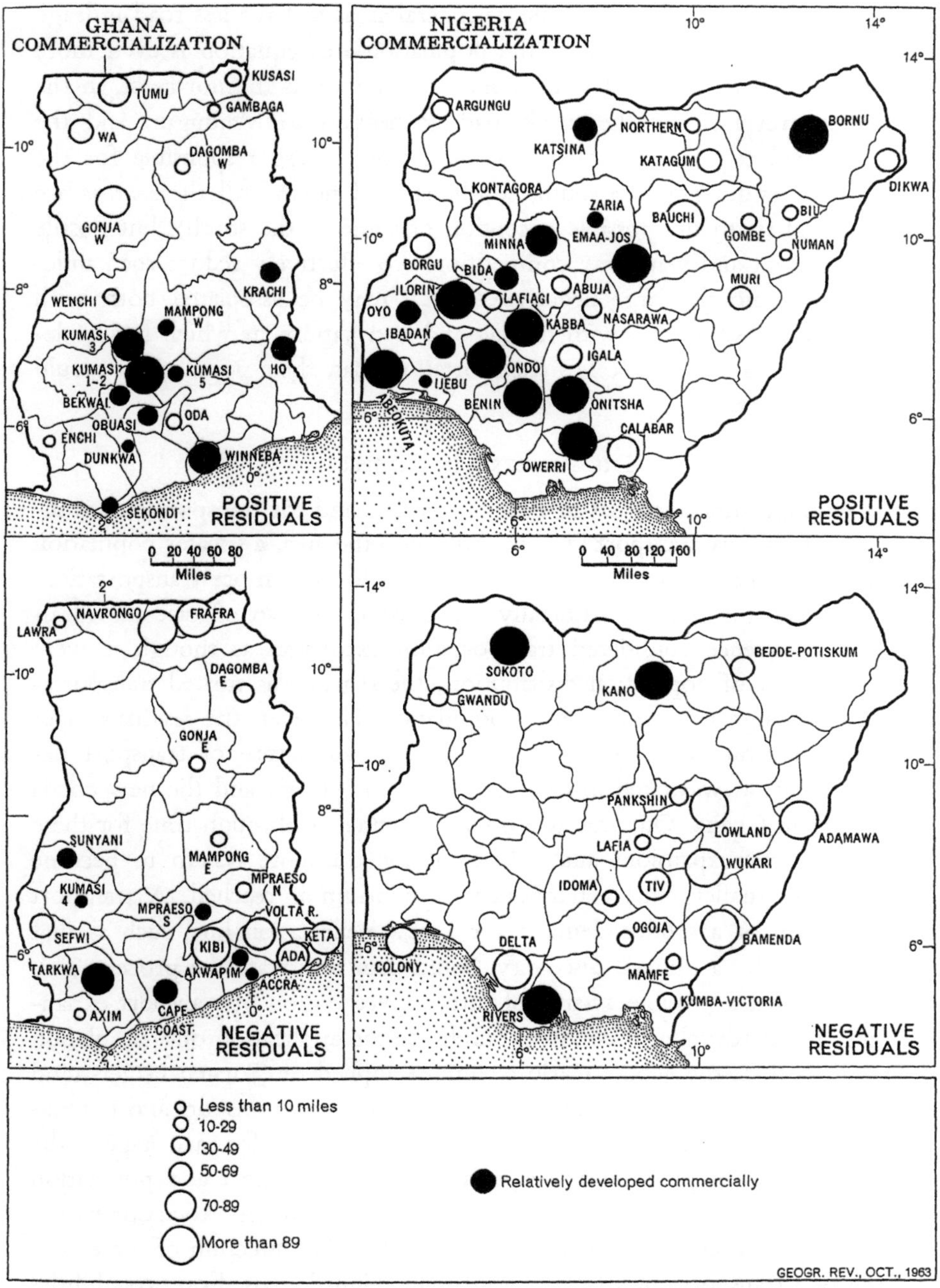

FIGS. 15 and 6—Commercialization, Ghana and Nigeria. Circles have been shaded in those districts which have been subjectively classified as being relatively urbanized or as containing a relatively large amount of commercial agriculture.

Conversely, the tendency for the peripheral areas to have less road mileage than would be expected from the population-area equation is even more marked than in Ghana. The only major exception is the northeast, in the districts traversed by the trunk-road connection to Maiduguri and the Lake Chad region. A combination of factors is also responsible for the negative residuals in the south, southeast, and north. All these areas are peripheral (with the possible exception of some in the south), but hostile environment and waterway competition also effectively reduce road mileages. In Kano the large negative residual may be associated both with peripheral position and with the lag-and-lead pattern in which newly developing areas may have less transportation than their populations would seem to warrant.

The Analysis in Perspective

As population increases in an area, the demand for transportation is intensified; as new transport lines are built into the area, a greater population increase is encouraged, which, in turn, calls for still more transportation. In a sense, the models artificially separated these two effects: the ideal-typical sequence considered transportation expansion as though it were independent of population distribution; the regressions treated transportation as though it were caused by population. However, the residuals maps provided intuitive evidence of the lag-and-lead nature of transport development, as was cited in cases of Northern Nigeria and the new cocoa districts in Ghana. One might postulate a tendency through time for these alternate overexpansions and deficits of the transport system to become gradually smaller until a temporary equilibrium is reached. A transport innovation or a sudden demand for a new penetration line, such as that occasioned by a mineral discovery, could then reactivate the process. This suggests that a possible avenue of future investigation of transport expansion in underdeveloped countries might be the application of a simulation model such as the Monte Carlo technique applied by Torsten Hägerstrand in his migration studies.[15] The spatial evolution of a transport and population pattern might be simulated through time by using for each stage in the process a set of probabilities dependent on the transport and population pattern of the preceding stage, thus bringing the essentially stochastic nature of transportation development into the model. The direction of the extension of a transport line from a given point might be based on probabilities

[15] Torsten Hägerstrand: Innovationsförloppet ur korologisk synpunkt, *Meddelanden från Lunds Univ. Geogr. Instn., Avhandl. 25,* 1953.

derived from factors similar to those noted in the discussion of penetration lines and the Ghana highway-density maps.

Finally, it should be noted that the generalizations in this study are designed to provide an initial perspective on the expansion of transportation in underdeveloped countries. At the moment, it is probable that the variations from the typical sequence and the regressions are of more interest than their explicit application.[16] It is to be hoped that future studies will bring about fundamental changes in the perspective presented here, at the same or a higher level of generalization. This may be accomplished by field investigations, by the development of more useful transportation parameters, and by the application of increasingly rigorous methods of analysis and model verification.

[16] Cursory examination of the Brazilian pattern, for instance, indicates a continued viability of some of the scattered ports in coastal commerce, due in part to weak lateral interconnection by land along the coast. Preliminary results from field investigations carried on in the state of São Paulo by Howard L. Gauthier, Jr., indicate a stronger emphasis on expansion of secondary roads toward the interior from railhead than on lateral interconnection during the period following the development of penetration lines. In former British East Africa the political boundary between Kenya and Tanganyika has apparently restricted lateral interconnection between two widely separated rail penetration lines. Population-area regressions run on road mileages for selected South American countries by Lawrence A. Brown, graduate student at Northwestern University, also resulted in explained variations of about 80 per cent. The population exponent, however, ranged from 0.41 to 0.85 and the area exponent from 0.22 to 0.37.

[8]

Mobility in development context: changing perspectives, new interpretations, and the real issues ☆

T.R. Leinbach *

Department of Geography, University of Kentucky, Lexington, NY 40506-0027, USA

Abstract

Aside from the acknowledged basic importance of transport in development, there remains the question of contemporary interpretations on its real role and the issues and concerns that must drive new research. A brief capsule of recent salient transport research is provided. The emphasis however is on the new conceptualizations of development and how mobility may be viewed and inserted in such interpretations. The paper addresses concerns such as the role of women, control of resources, institutional structures, environmental sustainability, the nature of transport enterprises and service delivery and employment and the labor process among others. Prescriptions for new research are provided. Salient development issues with transport dimensions which deserve further inquiry are de-agrarianization, the non-farm rural economy and the changing nature of the urban transition. Especially critical is our need to have a deeper understanding of personal and family mobility needs and to encourge transport policies directed to these.

1. Introduction

A critical yet neglected topic in this series of lectures and in the disciplines of geography and regional science is the theme of mobility, i.e. transport relations, in a development, but especially, third world context. My feeling is that the time is ripe to call new attention to this theme in part because development theory is in the throes of change and new interpretations abound. Indeed some would argue that development studies have entered a period of crisis. Mainstream frameworks dominant in postwar development studies seem unable to meet the developing world's most compelling challenges. Too often it was discovered that economic growth was accompanied by growing disparities and actually led to impoverishment. There was a failure of the new development to trickle down. Yet newer alternative frameworks with considerations of equitable development and human needs remain relatively undeveloped and unexplored. Whether we call these approaches popular development, sustainable development or use other labels matters not (e.g., Friedmann, 1992; Elliott, 1994; Escobar, 1995; Brohman, 1996). But it is clear in the new orientations that emphases are upon place context, power and empowerment, the role of the state, institutions, gender relations and above all people and family oriented change and strategies. Given the shortcomings of top down development efforts, community participation in decision-making has come to be recognized as critical in both alternative and mainstream development planning efforts (Brohman, 1996).

My overall goal is to situate transport more firmly in rapidly changing development analyses. More specifically I wish to first point up and clarify the prevailing wisdom of transport's role in rural development and how this has produced generally a flat and non-productive path to new understanding. Yet some intriguing recent research has been produced and these studies will be highlighted. But most important I want to look at select strands in the current thinking on development inquiry and show how the pursuit of new understanding on particular questions might benefit both the body of transport development literature and knowledge building in development studies. While transport in an urban development context is also critical, my focus as noted above is the rural Third World.

As we all know transport investment still accounts for a major share of the capital formation of less developed countries (Hirschman, 1962). In fact up to 40 percent of public expenditure is devoted to transport infrastructure

☆ Fleming lecture in Transportation Geography Association of American Geographers Annual Meeting 1999.

* Tel.: +1-606-257-1276; fax: +1-606-323-1969.

E-mail address: leinbach@pop.uky.edu (T.R. Leinbach).

investment with still additional amounts coming from the World Bank and advanced nation technical assistance programs (Leinbach and Chia, 1989; Button, 1993). These simple facts provide striking evidence once again of the prevailing recognition of the important role of transport in development. Yet the exact role of transport remains ambiguous and indeed has been subjected to periodic reappraisals. A legitimate question then in this light is: what do we really know about transport and development in the Third World? And perhaps even more important, what directions should our inquiries take? What are the questions, which should form the basis of new research?

2. What have we learned?

It is, of course, generally recognized among social scientists and development planners that transport has and will continue to play a critical role in development (e.g., Cooley, 1894; Moavenzadeh and Geltner, 1984; Owen, 1987). This relationship however is still commonly viewed largely as a permissive or supportive one; that is transport is a 'necessary but not sufficient' interactor with development. However under certain circumstances, especially where transport is a 'binding constraint' or is operating in an environment of 'prior dynamism', i.e. where factors are ripe, it may act as a catalytic agent (Wilson, 1966). But transport in the end is only one of many critical services required in order for development to move forward. Yet this simple interpretation does not yield great insight nor allow us to isolate the critical importance of transport within a variety of development contexts. Indeed in some ways the reliance on the simple supportive connection between transport and development has thwarted a deeper understanding of the richness of the connection between the two. Simple static relationships which show that wealth is highly correlated with mobility clearly do not explain the dynamic effects of providing transport in particular circumstances nor do they penetrate and illuminate the complex human fabric of development circumstances (Howe and Richards, 1984). Moreover, impact theories that emphasize separate possibilities whether positive, neutral, or negative are often based upon observations, which are too often based upon easily accessible and/or weak data. Rarely are we able to execute true ex ante and ex post analyses over significant spatial and temporal frames. What is required then is a shift from impact emphasis to an improved understanding of personal mobility. Yet there is no model which allows us to first understand the needs which drive *personal mobility*. Essentially then many time-honored approaches do not tap the rich complexity of the transport-development interface.

The fundamental weakness of these approaches is the simple and almost naive way in which they conceptualize or model human reactions and ignores elements forming the development context. The theory linking transport influences to social and economic change has not really been refined much beyond the general and aggregative levels (e.g., Wilson, 1966; Hagen, 1962; Hirschman, 1962). Few studies have addressed the matter of the distributional consequences of change nor derived comprehensive explanation to deal with this set of issues. But more critical there has been a lack of attention to transport issues which take into account family as well as individual needs. The critical importance of the family as a unit of analysis in development has become clear.

During the past decade some rather significant reassessment of the role of transport in Third World development has been taking place. This has been motivated in part by the concern over the spatial and structural maldistribution of income and inequities in delivering basic needs to dominantly rural nations. Especially now when, in many countries, austerity budgets are in effect and projects have been 'rephased' or postponed indefinitely, the exact role of transport has come under close scrutiny. Some development 'experts' indeed even maintain that overinvestment has occurred in the sector and thus future spending should be released only when a critical need can be identified. Local participation, restoration and rehabilitation are being emphasized. In addition, the involvement of the private sector (both commercial and non-commercial, i.e. NGOs) is increasingly being encouraged (e.g., Toh, 1989). Exceptions do occur especially where the target of the intended benefits are the rural and often inconspicuous and spatially, isolated poor (Chambers, 1997; Filani, 1993).

It is well documented that accessibility and inefficient transport have important impacts on the productivity and structure of agriculture as a result of, for example, the inability to obtain credit and financing as well as higher transport costs (e.g. Hine et al., 1983a,b). We suspect that transport is critically interwoven with communication and social change and moreover that access has some effect on employment searches and basic needs acquisition, including the delivery of health services. However, despite the results that document these impacts there are many situations where intended impact falls far short of our expectations. We still know all too little about the ways in which rural transport should be improved and how to deliver benefits to more needy populations. Until recently this was in part due to the consequence of equating people's needs with conventional engineered roads. Another reason for this lack of understanding of transport needs and impact is related to the biases associated with rural poverty. Essentially development needs are more easily seen near urban areas and along main roads while more remote and inaccessible areas and peoples are ignored (Chambers, 1983,1997).

In a fairly small literature several studies do stand out which provide some useful empirical evidence on the impact of rural transport. A synthesis of findings from numerous transport impact studies carried out in Africa, Asia, and Latin America through the mid-1980s (Howe and Richards, 1984) is useful even though the case studies differed widely in their characteristics. Some conclusions were drawn from ex ante predictions and others from ex post studies where data was assessed as late as a decade after construction of the facility. Nonetheless what conclusions emerge?

First, local circumstances and environment have a considerable effect on the way roads affect economic and social change. Second, the evidence cannot sustain continuing optimism over the positive impact of roads on poverty. Cited here are the rather widespread incidence of land consolidation, more landless workers, decline of local industries in the face of outside competition, and the acceleration of out-migration. Third, road improvements, in contrast to new roads, rarely lead to sharp decreases in the cost of transport, which is critical to stimulate demand. In addition, land tenancy is often a major factor in determining who benefits. If the land is unevenly distributed, the landless or land poor will receive little benefit. But if land is more evenly distributed, road projects will serve poorer households more effectively.

Some additional specific evidence comes from an insightful study by the Overseas Development Group at East Anglia which has examined the impact of several major roads in Nepal (Blaikie et al., 1979). The results show that the road development resulted in the increased penetration of local markets by Indian goods as well as some destruction of local manufacturing. In fact in these situations personal mobility expanded greatly with the new availability of bus travel. Moreover, provincial towns had acquired an administrative presence and as a result some additional income *but* the positive impacts that were expected from these new roads failed to develop. The roads served to encourage agricultural imports rather than exports. Adoption of new technologies and new crops was very limited. The move toward increasing subsistence due to population pressure outweighed any local commercial development.

Thus the roads had very little effect on the crucial prerequisites for significant development: namely increasing productivity in agriculture and industry. Why? In large part the answer is that government machinery was not organized to support peasant agriculture and poverty was so severe that farmers cannot risk innovation. The extent to which development takes place is *critically dependent upon the capacity of the local and regional economy to respond and reallocate resources.* The roads in question made little contribution in Nepal because they have not affected the major determinants of the local political economy nor do they begin to resolve the basic problems of a predominantly agrarian system. These problems are mounting population pressure, ecological decline and the meager subsistence production, which results from the peculiar circumstances of the region and its position as a dependent periphery of India.

Still other evidence supporting the differential impact of infrastructural development reports that the incorporation of rural areas into a wider transport network linking the rural economy with the urban and industrial is no guarantee of the expansion of *economic activity* in the countryside. As a positive case the illustration of Taiwan is useful. Here decentralized industrialization was powerfully influenced by the high degree of infrastructural development. But indeed the key facilitating factor in this case was the availability of a malleable, disciplined and yet cheap part-time rural labor force with a high level of education. Still another complementary factor was the specific nature of industrial activities which lent themselves to de-skilling through the vertical integration of industrial operations. Here rural women commuting daily between factory and village were a critical part of a larger set of factors which together accounted for industrial decentralization (Saith, 1992).

Rural road investments in Bangladesh show that generally the poorest groups, especially destitute women, do benefit as it is from this segment of society that workers are recruited for intensive labor construction efforts. Yet concerns for leakages in the resources made available undermine this positive benefit. From this experience it is also clear that owner operated transport and especially affordable and reliable transport services are the essential mechanism through which broader changes take place. But political influences through regulation of fares and other constraints can undermine this effect. More broadly it is clear that there exists strong biases against asset-poor villagers' realization of benefits from road investments compared with the wealthier strata of the village population (Howe, 1994).

Too often the analysis of the impact of improved accessibility has focussed only on economic criteria. Yet just as commodities flow over the linkages of a road network so too do ideas and news about innovations and new techniques. Results from Indonesia show a comparison of the impact of improved roads between urban fringe and more remote locations. The findings suggest that improved access did have an effect on the search for employment. But five times as many respondents in the urban fringe groups, as opposed to the remote rural groups, claimed that transport improvements were influential in the search for employment (Leinbach, 1983a,b).

The Taiwanese example above and other situations also show how infrastructure development may affect labor circulation. However as a result of the spatial concentration of industrial opportunities, infrastructure improvements may result in permanent rural outmigration in settings such as Malaysia and Sri Lanka which rely on export processing zones. In the electronics industry of Kerala, on the other hand, despite the existence of rural assembly units operating at considerable distances from a supplying town, prohibitively high transport costs were incurred. But this was only in part due to distance. A major factor was lack of logistical organization. As a consequence of this as well as poorer infrastructure, the result has been a decentralized approach to production in which little mobility is required of the female workforce. Such a pattern of labor circulation, of course, carries important implications for opportunities available to women. Female labor mobility is obviously related to productive and reproductive activities through the family life cycle and the gendered division of labor (Watkins et al., 1993). But how much do we really know about the way in which transport deficiencies affect the role of women in development (e.g. Barwell et al., 1987; Bryceson and Howe, 1993; Doran, 1990)?

These results raise some questions about whether indeed improved access really does enlarge the employment fields of rural villages. A major consideration here however is that still too many rural communities lie beyond the reach of an effective road network (Leinbach, 1982). Moreover, despite upgraded physical networks in some rural areas, affordable, flexible and reliable transport services are still a rarity (Replogle, 1991). The regularity, cost, and efficiency of the services is perhaps more important as an influence on employment acquisition but clearly such characteristics are critical to a wide range of needs. Given this information, what is the approach that we need to pursue to acquire further understanding about the complex interaction of transport and society?

3. New paths forward

Despite numerous useful studies detailing the role of transport in development, it is clear that traditional views of impact are not particularly useful ways of assessing development through the transport lens. We therefore almost need to reinvent transport inquiry in the context of contemporary approaches to development questions. What are required are new conceptual frameworks and indeed methodologies for analyzing the complex ways in which accessibility and mobility are embedded in development issues. The comments, which follow, are intended to provide some guidelines for the exploration of new grounds on this critical topic. What we need ultimately is a reoriented conceptual framework which effects a more meaningful approach to the analysis of transport's role especially in rural development (Jolibois, 1991). Such a framework must of course explicitly incorporate economic but equally important political and social considerations. It also must recognize that the bulk of the transport movements are related to subsistence activities (Barwell et al., 1985). It must clarify the conceptual gap between the traditional way impact, development and the use of the transport system has been viewed and the socio-economic reality of the transport systems. In this effort it is clear that micro-scale approaches are going to be much more useful than the macro-scale, holistic views of transport relationships in development. How do we begin to approach the construction of a new framework for understanding transport in development context?

As noted above, the extent to which development takes place is critically dependent upon the capacity of the local and regional economy, and especially its institutions to respond and particularly to reallocate resources. An important aspect of this is, as Donald Janelle has noted, "the social pre-conditions which are needed to facilitate positive outcomes from transport investments, including building awareness, providing education and skills and complementary resources." Individuals must be encouraged to view access means and opportunities as a way of enhancing their well being in terms of incomes, personal growth, self-reliance and other values. One way to begin to effect this is through the use of community participation in which beneficiaries influence the direction and execution of development projects and especially those which have transport elements (Edmonds, 1982; Heggie, 1989).

Transport in order to effect change must relate to the major determinants of the local political economy and be combined with efforts to resolve problems of largely agrarian economies. These include the lack of entrepreneurship, ecological decline, landlessness and mounting population pressure. A place to begin to redraw the conceptual framework for transport development knowledge building may lie in the focus on critical development issues. One especially important theme and one, which is rich in terms of potential research in Asia, Africa and Latin America, is *de-agrarianization*, its driving forces and alternatives. Non-agricultural rural employment (NARE) for households is not only desirable but also indeed a necessary alternative to farming in many communities. This is so because of constraints on land availability, contraction of urban employment as a result of economic crises (the situation in much of Asia today), and an increasingly educated population, which is much less satisfied today with the agrarian life style. Transport relations can be inserted into this theme in several ways. For example, how do various barriers (social, economic, political and physical) constrain individuals from seeking non-agricultural employment? Indeed the development of transport enterprises and employment in the transport sector are key topics worth exploring in this connection.

Another theme challenges the conventional view and wisdom of the urban transition and points to the increasingly blurred rural–urban dichotomy. The proposed hypothesis is that there has been an emergence of new zones of intense economic interaction between rural and urban activities. These zones, it is suggested, have come about as a response to the evolution of transport-communications, technological developments and economic and labor force changes among other causes (Ginsburg et al., 1991). The desakota concept (from the Indonesian terms for village and city) outlines a model of the transition of the space economy which, it is argued, applies dominantly to situations where one or more urban cores are located in densely settled rural areas but also areas of intensive agriculture. The argument focuses upon the requirements of wet rice agriculture and at the same time the urban and peri-urban employment potential for labor, adequate infrastructure to support expanded development, labor reservoirs, and the existence of 'transactive' environments. The focus on adequate transport within the changing dynamics of urban–rural relations is important as the household is restructured and becomes more diversified. Genders and generations renegotiate their respective roles (Rigg, 1997). In addition, the role of transport and supporting institutions is critical in the topic of borderless economies which clearly are taking hold across all of Asia.

One useful approach toward achieving some greater understanding of small-scale development situations and especially the role of access may be to explore the utility of variations of the political economy theme. By this I mean we should perhaps examine the interrelationship of transport with the forms of and limits to production. The objective here is to learn how individual producer incentives and accumulation could be enhanced. In this analysis the key actors and decision-making processes and institutions at the village level may be isolated as dominant elements in our new understanding. Within this broad approach many aspects deserve the attention of researchers but several topics stand out as especially worthwhile for case study analysis.

First, within this perspective it is clear that a deeper understanding must be gained of the concept of rural household travel demand and personal mobility needs. A basic deficiency of current knowledge is that reliance on market demand models has led to an undifferentiated view of the benefits and costs of work performance to household members. In this way the issue of gender bias at the intra-household level does not arise. The fallacy here is the lack of understanding of the transport needs of the subsistence-producing household, which may allocate its survival tasks on the basis of contextual factors, rather than optimizing market principles. Given a better definition of the real needs of the peasant household appropriate intervention strategies can be devised (Ellis, 1993; Bryceson and Howe, 1993).

Second, the role of transport as it affects production relations in village situations between landholders and those without land is important. What differences separate these two groups and how does transport inhibit or aid these distinctions?

We now know too little about the role of access in improved information flow. Although capital, infrastructure and education are essential, frequently it is the lack of information about a service or an opportunity which constrains change. Careful studies isolating the information impacts of accessibility would be especially useful for investment decisions and policy formulation.

A key element in both rural and urban development is employment. Yet too little information on the role of transport and accessibility in the labor process is known. Especially useful here would be studies which relate transport to labor commuting and circulation strategies as well as the search for employment. In this regard women's employment situations, given the simultaneous assumption of reproductive and productive activities, are of special interest (Ellis, 1993). But in a more basic vein, even detailed studies of women's labor time allocation have devalued the burden of inadequate transport (Bryceson and Howe, 1993). Recent findings from village surveys have revealed the dominance of human porterage and the centrality of women as load carriers in African rural transport. As household transport requirements have grown, cultural norms have remained in a traditional mould which dictates that women's responsibilities are to travel and do load carrying with regard to basic needs. But in addition they also headload commercial goods in the absence of transport assisted options. These two underlying precepts of the gendered division of labor not only stand in the way of a more rational intra-household distribution of work effort, but also thwart the equitable distribution of benefits between male and female household members in external agency transport improvement interventions aimed at replacing arduous porterage. Thus rural women's physical and economic positions could deteriorate further relative to men. Men would gain enhanced mobility with few household transport responsibilities while women would have little or no access to the innovation and remain responsible for the bulk of household transport work. International agencies have been reluctant to challenge the 'cultural preferences of communities' even though these preferences give rise to gross inequities between the sexes. In this situation and others, a shift from the almost exclusive focus on household demand to in-depth research from the perspective of women transport suppliers must take place. Here there is the need for a more comprehensive research approach which discards assumptions about the unity of household demand and welfare and considers women's decision-making and logistics as well as the multi-tasking and childcare dimensions of women's transport strategies. Moreover the matter of employment within the transport sector, especially of rural areas, deserves research attention. This general theme of course again supports the well-recognized priority in all developing areas: *employment creation.*

Finally the control of and accessibility to *transport resources* remains a largely unexplored area (McCall, 1977, 1985). Topics here range from location decisions on network placement by local level or other decision-makers to the ease of entry into transport operations. The constraints associated with the provision of transport services and infrastructure must be better understood. In this vein it is important to reiterate that many of the poor cannot afford even low cost public transportation and must walk. In rural areas, appropriate vehicles and transport services are frequently not available or affordable for the majority of people. Too frequently investments have tended to reduce the diversity of modal options forcing people and goods to conform to the few higher cost modes rather than utilize the most appropriate and affordable means. We indeed should seek a robustness in our transport systems which is produced by great diversity and differentiation. It is clear that a more diverse system will be less susceptible to inefficiency, disruption and system failure. It is obvious that transport investment policies should be reoriented to attempt to accommodate the needs of the poor and often isolated rural people. One avenue, which might be more thoroughly explored, is schemes, which provide low cost credit for affordable mobility. Moreover, and most important, researchers must begin to recognize this more informal dimension of transportation by taking on projects which probe the practical and policy implications of a broader, more diversified and equitable access provision (Replogle, 1991).

In order to improve our understanding of the dynamics of the transport relationship we must focus on the real purposes of development and orient our analyses to capture the ways in which transport interacts with and constrains the basic forces in the development change process. In these efforts economic, political and social relationships must be examined thoroughly. One new theoretical lens which may hold some promise from a policy perspective is imbedded in the relatively new area of institutional economics (Nabli and Nugent, 1989).

Most developing countries are today especially concerned with ways to produce more competitive and differentiated economies. At the same time there is a concern with cost reduction in order to effectively compete in a dynamic global economy. Physical infrastructure, and especially but, of course not exclusively transportation, looms large in these efforts. Major questions continue to focus on the forms, location and means of delivering transport capacity and services at all scales in order to eliminate constraint points and bottlenecks which result in high costs and weaken competitive position.

A major strategy by donor nations and the governments themselves has been to invest heavily in infrastructure. But time and again we see abundant illustrations and examples of situations where such investment has failed to serve as a catalyst for growth and indeed has resulted in quite unintended consequences. The matter of why these investments fail is at the heart of this paper and some insights are provided above. For example, poor locations, improper design, and competing structures are partial explanations. In addition, the ability to sustain the infrastructure once it is in place is critical for sustained development. Herein lies a problem which has been nearly intractable: how to maintain infrastructure in developing countries? Clearly a partial explanation of why infrastructure has not delivered intended benefits lies within the bounds of the lack of maintenance and the erosion of facilities. If in fact such structures can be maintained effectively a more productive transport system will result. In addition, we might ask, will a deeper understanding of the sustainability of infrastructure lead us to a more enlightened and complete view of transport's relationship to development. This is clearly plausible.

One argument which supports this view is that a major underlying cause for the failure to sustain investments in facilities lies with the *incentives* facing participants in the design, finance, construction and use of facilities. Basically when road infrastructure deteriorates after construction, it may be assumed that the actors involved in the development process were submitting to incentives that essentially rewarded them for actions that resulted in unsustainable investment (Ostrom et al., 1993). Incentives, it must be stressed, are not simply financial rewards or penalties. They basically are the positive and negative changes in outcomes that individuals perceive and which result from particular actions taken within a set of rules in certain spatial, social, political and economic contexts. This position assumes that there are reasons why individuals involved in the delivery of transport do what they do. The choices and decisions which people make regarding the use of services and facilities purposely or inadvertently benefit or harm others because of the interdependence which exists in settlement situations.

The obvious question which follows is whether we can provide and develop incentive systems which will allow us to deliver public use facilities that lead to relatively efficient and equitable outcomes? The answer is generally yes, but this is much more difficult than dealing with other goods and services. One answer, at least, lies essentially in our ability to alter individuals' decision situations by developing self governing institutions (Ostrom et al., 1993). An additional one is the way in which revenue raising devices are constructed.

The concern with economic restructuring and the simultaneous development of low cost, competitive economies has over the past decade drawn attention to the strategy of substituting public ownership with private ownership over a host of facilities and services including transport operations (Gomez-Ibanez and Meyer, 1993; Gayle and Goodrich, 1990). Under deregulation efforts this process has gained considerable momentum yet it is not without problems. The point is that transport ownership and operation must be the focus of further research as it can illuminate our understanding of the interrelationship between development, perhaps specifically in this case the changing role of the state, and transport investment.

Following this line of argument then, one possible new approach to viewing the interrelationship between transport and development lies in an enlarged understanding of the institutional bases of transport. This involves examination of the matter of provision and production of rural and regional facilities and includes questions of property rights, viability of entry and competition, regulation, rent seeking, joint use, economies of scale and the notion of excludability (Ostrom et al., 1993). But important too are deeper inquiries into the individuals influencing the facilities, the associated incentives and transaction costs. An additional obvious thrust for analysis along these lines is an inspection of institutional arrangements relevant to the structure in its spatial and socio-economic context.

4. Transport for the poor or poor transport

It would be shortsighted indeed to ignore the applied and policy aspects of the transport and development nexus. Too often development policies have tended to be based on the assumed superiority of 'motorized transport' as the sole means of meeting movement needs. There is a strong argument to be made in opposition to this position for the majority of the population in the poorest countries are in fact non-users of motorized transport and roads. Their travel needs are not addressed by such an approach. John Howe argues that prescriptively too much conventional thinking has become embedded in the *product* to the exclusion of its users and especially non-users (Howe, 1996). A number of researchers over the past decade have begun to pursue needs-based planning methods with the emphasis on understanding and meeting the travel and transport demands of rural households. This research has shown that in the poorest areas household travel is dominated by subsistence tasks which give a local orientation to most trips. The prime transport requirement is for the movement of frequent, small loads over short distances. On the other hand, social and welfare needs are the main motivations for longer distance travel for which road transport might be appropriate. Few households possess any form of vehicular transport. Instead walking, cycling and movement by animal dominate. Recognition of these transport patterns must lead to the formulation of rural transport policies and planning methodologies based upon the twin requirements of improving household access to services and the mobility of all its members, especially women who in many societies shoulder the majority of the transport tasks. Implementation of such policies implies a much wider and more heterogeneous range of investments on the part of governments and development agencies than has traditionally been the case. Policy and decision makers need to be more aware of the consequences of continued imbalanced investment in a transport system which dictates increasing social and economic differentiation because it ignores the travel and transport needs of the rural majority. Developing a more people centered transport policy climate will require fundamental changes in education in the broadest sense. This brings us back to the often mentioned and widely quoted idea of 're-inventing' the very notion of what transport is for.

5. Sustainability policies for transport reform

Finally in a policy vein, it is clear that the notion of sustainability , however we wish to define this term, has become infused in the literature on transport. The World Bank, in fact, has put sustainability at the heart of a more demanding transport policy (e.g. World Bank, 1996). The broad rationale is that rapid changes in the global economy have increased the need for flexibility and reliability in transport services; individual aspirations for more mobility have generated the need for a greater variety of transport services. Mounting social concern about the degradation of the environment has increased the need to evaluate transport strategies more carefully. The new expression of these concerns is met through (see Fig. 1):

1. Economic and financial sustainability where resources are used efficiently through competitive market structures, an enabling framework for competition and efficient use of infrastructure.
2. Environmental and ecological sustainability where external effects of transport are taken into account particularly the adverse consequences of development induced by roads and other networks on forests, wetlands and other natural habitats.
3. Social and distributional sustainability where the transport problems of the poor are targeted by

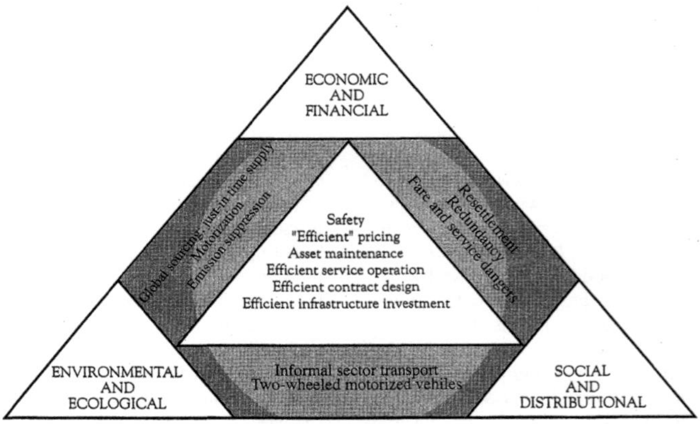

Fig. 1. Three dimensions of sustainable development: Synergies and tradeoffs.

improving access to jobs, reducing barriers to the informal supply of transport and the elimination of gender biases.

These three dimensions and specific examples are revealed in Fig. 1.

While these are magnanimous objectives there are many barriers to their implementation. Most significant, apart from the realities of the political, social and economic context of the transport situations, is the lack of real research on a variety of questions assumed in the development of these policies. It is in this regard that we as social scientists can make a real contribution (Leinbach, 1995a,b).

6. Summary

The above remarks suggest that in order to stimulate new research and understanding on transport's real role in development we need to refocus our inquiry. One approach is to explore the transport situations within development context through major problems and issues such as the process of de-agrarianization, the non-farm rural economy and the changing nature of urban transition. In addition, we need to focus on critical elements in the emerging conceptual rethinking of development. These include issues of the labor process including the mobility of workers and their search for employment, control of resources especially land, institutional structures and their efficiency, the role of women in development, the nature of transport enterprises and service delivery and a deeper understanding of personal and family mobility needs. But perhaps most critical are changes in priorities for transport policy that must meet broader human needs instead of mainly elite groups (Hine, 1982). While non-motorized and informal transport organization obviously is most pertinent to rural areas too little attention has been paid to this aspect of mobility.

References

Barwell, I., Edmonds, G.A., Howe, J.D.G.F., deVeen, J., 1985. Rural Transport in Developing Countries. Intermediate Technology Publications, London.

Barwell, I., Howe, J., Zille, P., 1987. Household time use and agricultural productivity in sub-Saharan Africa. IT Transport Ltd, Oxford.

Blaikie, P., Cameron, J., Seddon, D., 1979. The relation of transport planning to rural development. Development Studies Discussion Paper No. 50, University of East Anglia.

Brohman, J., 1996. Popular Development: Rethinking the Theory and Practice of Development. Blackwell, Oxford.

Bryceson, D.F., Howe, J., 1993. Rural household transport in Africa: reducing the burden on women. World Development 21 (11), 1715–1729.

Button, K., 1993. Transport Economics, second ed. Elgar, Aldershot.

Chambers, R., 1983. Rural Development: Putting the Last First. Longman, London.

Chambers, R., 1997. Whose Reality Counts? Putting the Last First. Intermediate Technology Publications, London.

Cooley, C., 1894. The theory of transportation. American Economics Association 9 (3), 1–148.

Doran, J., 1990. A Moving Issue for Women: Is Low Cost Transport an Appropriate Intervention to Alleviate Women's Burden in Southern Africa? School of Development Studies, University of East Anglia, Norwich.

Edmonds, G.A., 1982. Towards more rational rural transport planning. International Labour Review 121 (1), 55–65.

Elliott, J.A., 1994. An Introduction to Sustainable Development. Routledge, London.

Ellis, F., 1993. Peasant Economics, second ed. Cambridge University Press, Cambridge.

Escobar, A., 1995. Encountering Development: The Making and Unmaking of the Third World. Princeton University Press, Princeton.

Filani, M.O., 1993. Transportation and rural development in Nigeria. Journal of Transport Geography 1 (4), 248–254.

Friedmann, J., 1992. Empowerment: The Politics of Alternative Development. Blackwell, London.

Gayle, D.J., Goodrich, J.N. (Eds.), 1990. Privatization and Deregulation in Global Perspective. Pinter, London.

Ginsburg, N., Koppel, B., McGee, T.G., 1991. The Extended Metropolis: settlement transition in Asia. University of Hawaii Press, Honolulu.

Gomez-Ibanez, J.A., Meyer, J., 1993. Going Private: the International Experience with Transport Privatization. Brookings Institute, Washington, DC.

Hagen, E.E., 1962. On the Theory of Social Change. Richard Irwin, Homewood, IL.

Heggie, I., 1989. Reforming transport policy. Finance and Development 26 (1), 42–44.

Hine, J.L., 1982. Road planning for rural development in developing countries: a review of current practice. Transport and Road Research Laboratory Report 1046, Crowthorne, England.

Hine, J.L, Riverson, J.D.N., Kwakye, E.A., 1983a. Accessibility, transport costs and food marketing in the Ashanti region of Ghana. Transport and Road Research Laboratory Supplementary Report 809, Crowthorne, England.

Hine, J.L., Riverson, J.D.N., Kwakye, E.A., 1983b. Accessibility and agricultural development in the Ashanti region of Ghana. Transport and Road Research Laboratory Supplementary Report 791, Crowthorne, England.

Hirschman, A.O., 1962. The Strategy of Economic Development. Yale University Press, New Haven.

Howe, J., 1994. Infrastructure investment in Bangladesh: who really benefits and how? IHE Delft Working Paper IP-6, August.

Howe, J., 1996. Transport for the poor or poor transport? IHE Delft Working Paper IP-12, October.

Howe, J., Richards, P. (Eds.) 1984. Rural Roads and Poverty Alleviation. Intermediate Technology Publications, London.

Jolibois, S.C., 1991. Reorganizing Development: A Conceptual Framework for Efficient Rural Transport in LDCs. Institute of Transportation Studies, University of California, Berkeley.

Leinbach, T.R., 1982. Towards an improved rural transport strategy: the needs and problems of remote third world communities. Asian Profile 10 (1), 15–23.

Leinbach, T.R., 1983a. Transport evaluation in rural development: an Indonesian case study. Third World Planning Review 5 (1), 23–35.

Leinbach, T.R., 1983b. Rural transport and population mobility in Indonesia. The Journal of Developing Areas 17, 349–354.

Leinbach, T.R., Chia, L.S., 1989. Southeast Asian Transport: Issues in Development. Oxford University Press, Singapore.

Leinbach, T.R., 1995a. Transport and third world development: review, issues and prescription. Transportation Research A 29 (5), 337–344.

Leinbach, T.R., 1995b. Regional science and the third world: why should we be interested and what should we do? International Regional Science Review 18 (2), 201–209.

McCall, M.K., 1977. Political economy and rural transport: a reappraisal of transport impacts. Antipode 9 (1), 56–67.

McCall, M.K., 1985. The significance of distance constraints in peasant farming systems with special reference to sub-Saharan Africa. Applied Geography 5, 325–345.

Moavenzadeh, F., Geltner. D., 1984. Transportation, Energy and Development. Elsevier, New York.

Nabli, M., Nugent, J., 1989. The new institutional economics and its applicability to development. World Development 17 (9), 1333–1347.

Owen, W., 1987. Transportation and World Development. Johns Hopkins University Press, Baltimore.

Ostrom, E., Schroeder, L., Wynne, S., 1993. Institutional Incentives and Sustainable Development. Westview Press, Boulder.

Replogle, M.A., 1991. Sustainable Transportation Strategies for Third World Development. Transportation Research Record 1294, Transport Research Board, Washington, DC, pp. 1–8.

Rigg, J., 1997. Southeast Asia: the Human Landscape of Modernization and Development. Routledge, London.

Saith, A., 1992. The Rural Non-Farm Economy: Processes and Policies. International Labour Office, Geneva.

Toh, K.W., 1989. Privatization in Malaysia: restructuring or efficiency. ASEAN Economic Bulletin 5 (3), 242–258.

Watkins, J., Leinbach, T.R., Falconer, K., 1993. Women, family and work in Indonesian transmigration. Journal of Developing Areas. 27 (2), 377–398.

Wilson, G., 1966. Towards a theory of transport and development. In: Wilson, G.W., Bergmann, B.R., Hirsch, L.V., Kleig, M.A. (Eds.), The Impact of Highway Investment on Development. Brookings Institute, Washington, DC.

World Bank, 1996. Sustainable Transport: Priorities for Policy Reform. Washington, DC.

Equity

[9]

The BART experience—what have we learned?

MELVIN M. WEBBER

THE Bay Area Rapid Transit system (BART) has many characteristics of a huge social experiment—in vivo, as it were. Key element in a bold scheme to structure the future of the San Francisco region, BART was to stem the much-feared decline of the older metropolitan centers, while helping to give coherent order to the exploding suburbs. By offering a superior alternative to the automobile, BART was to make for congestion-free commuting. If successful, it would provide a model for rationalizing transportation and metropolitan development elsewhere.

It was experimental only in the sense that nothing quite like it had ever been tried before. Nowhere in America had a regional rail system been built in contemporary times, and nowhere in the world had such a rail system been built in an auto-based metropolitan area. Although novelty inevitably implied risk, BART was nevertheless promoted with the high confidence, if not certainty, that governmental projects seem to require. Its developers saw it as a sure bet.

They could not have chosen a better test site. The auto-ownership rate in the Bay Area is one of the world's highest, and auto-using habits are firmly entrenched, fostered by an early full-freeway system. The metropolitan region is built around a traditional European-type center with a large concentration of office employment and related service employment. Outside the small but high-density central city, which is set at the tip of the peninsula, the metropolitan settlement is organized in low-density suburban patterns, all of which were built around automobile transport. For test purposes, the Bay Area is further advantaged by having its urbanized areas topographically molded into narrow strips that parallel San Francisco Bay and follow the narrow valleys—a configuration superbly suited to the geometry of railroad lines, which also makes for long travel distances that, in turn, cry out for high-speed modes of transport. (The average Bay Area commuter lives 15.8 miles from his job; by comparison, Los Angeles commuters are only 8.9 miles away, Chicago commuters 6.6 miles, Philadelphia commuters 4.4 miles.) What better place, then, to test whether a technologically superior system could stem the decline of traditional public transit and at the same time compete with the automobile as the dominant commuter mode and shaper of metropolitan growth? If it didn't work here, the odds of its working in less suited areas would surely be low.

Following some protracted and perhaps inevitable construction and debugging delays, BART began carrying passengers in September 1972. In response, a massive effort has been launched to monitor its effects. The BART Impact Program is carefully watching a wide array of potential consequences, under the sponsorship of the Department of Transportation and the Department of Housing and Urban Development, with the local Metropolitan Transportation Commission in the key research and management roles. The aim is to learn what might be germane to other metropolitan areas and to derive some lessons from the Bay Area experience that might inform federal policy. Meanwhile, however, several other metropolitan areas have already developed transit schemes that resemble BART, without waiting for the findings of the Bay Area studies. Federal subsidies have already been allocated to Washington, Atlanta, and Miami; and several other cities are under consideration.

It will of course be some time before definitive results of the BART evaluation are in, but enough impact research has now been done to permit an early appraisal of how well BART is accomplishing the objectives that motivated its construction. In turn, enough has been done to permit some judgments concerning the wisdom of building BART-like systems elsewhere. As we shall see, BART's score is pretty low, implying that prospects for other cities are dim.

Initial objectives

The original plans for the BART system were formulated in the mid-1950's. At that time Americans perceived the major urban policy issues to be decentralization, "urban sprawl," and the decline of the central cities, particularly of their central business districts. Federal government activities were focused upon urban-renewal programs aimed at bringing middle-class suburbanites back and rejuvenating what was dolefully mourned as the "dying heart of the city." Some prophets were predicting an impending demise of metropolitan centers, as the automobile opened opportunities for free movement in all directions and combined with rising incomes to allow average families spacious suburban houses in spacious settings.

These prospects were greeted with exaggerated gloom in San Francisco. Surely, no other American city is as proud and narcissistic—no civic leaders elsewhere so obsessed by their sense of responsibility for protecting and nurturing their priceless charge. The idea that San Francisco might go the way of Newark or St. Louis was utterly abhorrent. And so it was, as the San Francisco Chamber of Commerce proudly reported in a multi-page advertisement in *Fortune*, that the civic leaders of San Francisco and their neighboring kin initiated a major effort to keep the Bay Area from going the way that cities of lesser breed were headed. The campaign was masterful in both conception and execution.

San Francisco had long been the major business and financial capital of the Western states, but it had been challenged since World War II by Los Angeles and, in the mid-1950's, was barely holding its own against explosive growth in Southern California. Probably as fundamental a motivation as any behind BART was the desire to keep "The City" as attractive to corporate headquarters, financial institutions, and upper-middle-class residents as it traditionally had been. Of course, those with personal economic interests in the central district were especially keen to promote BART. It would be simplistic, however, to ascribe their promotional fervor to their private considerations alone. Imbued with civic pride and the spirit of community enterprise, none doubted that he was involved in a happy symbiotic crusade in which everyone would be a winner.

The Golden Gateway redevelopment project was one response to the doom-sayers. Launched in 1955, it has by now successfully converted a decayed produce-market district into a spanking new cluster of high-rise office buildings, luxury residences, elaborate hotels, restaurants, and the like. Delayed for years by lawsuits, the Yerba Buena redevelopment project, the sister enterprise aimed at remodeling a 19th-century district lined with residential hotels, warehouses, and a skid row, may soon finally convert the "South of Market" street district from a low-class neighborhood to a prestige address. Between them, the two redevelopment efforts may total up to 10 million square feet of high-rise office space. More than that, they have contributed immensely to the prosperity of the San Francisco central district and to its competitive posture.

So huge an expansion in floor space and commensurate office employment obviously demanded a comparable expansion in transport capacity. BART was to provide a major increase in accessibility, supplemented by extension of the freeway system and improvements in downtown parking.

But BART was intended to do far more than bring commuters into San Francisco; it was conceived from the start as a regional system that would foster the growth of the entire Bay Area. With trains stopping on average at only 2.5-mile intervals, each station would be highly accessible, with a competitive advantage over other locations in the rivalry for apartment houses and for offices and other commercial establishments. The designers expected stations in suburban centers to attract major concentrations of offices and retail shops, while outlying stations would become surrounded by high-density housing and shopping facilities, serving commuters to the regional centers.

Simultaneously, the designers intended BART to supply the essential accessibility that would convert downtown Oakland into a major regional center. The junction of BART's four lines in the middle of Oakland's business district would indeed make that the most accessible point along BART's 71-mile route. Oakland has responded by launching its own downtown redevelopment project, which is now turning a marginal shopping area into a modern-looking business complex.

The civic leaders who promoted BART chose a rail system over additional highway improvements because they feared that the prophets of intolerable congestion might be right. The prospect that more population and more automobiles would overload the capacities of road systems seemed plausible enough to commend a system that simultaneously had a high capacity yet was conservative in its space demands. And besides, since San Francisco was a world center along with Paris, London, and New York, didn't it deserve a subway system comparable to others in its league?

Design considerations

The design criteria were thus clear: The new system had to be 1) capable of bringing increasing numbers of peak-hour commuters from near their suburban homes to within a few minutes' walk of their downtown offices, 2) attractive enough to travelers to be more than competitive with the automobile, and 3) financially viable.

The planners' response was to design a modern, electrified suburban railroad. Believing that buses could not attain the speed necessary to make them attractive to commuters, they rejected the alternative of using them as rapid-transit vehicles. Instead, they modeled their system after the New Haven and Long Island railroads, giving it much better downtown distribution with several subway stations under the main streets of San Francisco and Oakland.[1]

BART is misnamed as a "rapid transit" system. It is a hybrid among rail transportation systems, combining features of interurban electric lines, central-city subway lines, and current suburban railway lines, with the design elegance of modern aircraft and the control instrumentation of early spacecraft. As a hybrid it represents a compromise among desirable qualities of the several transport types. Unlike the interurban electric railways it cannot run in non-stop from outlying suburbs, and unlike New York subways it can offer no express service, because each station along the route is a mandatory stop. Unlike the subways of Paris, Tokyo, or London, which are interconnected networks of lines, BART offers one route in each compass direction and hence only limited distribution across the urbanized area it serves.

Because of its misnomer and because it fits none of our stereotypes, BART has been befuddling its critics, who seem unable to categorize it tidily. The important fact is that BART was designed to meet the rather specific purposes enumerated above, and those purposes happen not to be the ones by which some critics have judged it, reflecting social concerns that were not widely shared when BART was designed. For example, it was never intended to serve the kinds of short-distance trips that local buses, trolley cars, and center-city subways serve. It was not designed to carry low-income persons from central-city homes to suburban factories, even though BART officials belatedly voiced such claims.

Rightly or wrongly, BART was designed to transport peak-hour commuters from suburbs to central business districts. In turn, it was intended to generate the following effects:

to reduce peak-hour highway traffic congestion,
to reduce time expended on commuter travel,
to foster central district growth,
to generate development of subcenters throughout its region,
to raise land values,
to accommodate suburbanization of residence and centralization of employment,
to reduce land area devoted to transportation facilities.

The expectations of some enthusiasts have sometimes been rather extreme, viewing BART as a remedy for whatever snake oil fails to cure. In turn they have generated a wave of criticism in both the popular and professional media that has often been comparably extreme. Such expectations and criticism are unreasonable. BART represents a serious response to perceived problems in regional development, planned with the advice of some of the world's most accomplished engineers and designers. Because it is the first of its kind and is now being watched so closely by officials in other cities around the world, and because it is so terribly expensive, we must soberly check whether the outcomes its planners expected are being realized.

BART's patronage

Prior to the 1962 bond election that authorized BART, the key informational document was the so-called *Composite Report*, published in May 1962.[2] Among the important expectations there were hopes 1) that BART would divert 48,000 workday autos from the streets and highways by 1975 and 2) that 258,500 daily passengers would be riding BART in 1975—157,400, or 61 per cent, diverted from automobiles, 39 per cent from existing transit systems.

Although 1975 has arrived and passed, it is not yet possible to submit those predictions to the definitive test, for BART was late in getting started, and thus has not yet developed a "seasoned" patronage. Moreover, the system has been so besieged by electronic and mechanical difficulties and rising costs that it is still not operating to design standards nor offering any weekend schedule. With these qualifications in mind, we can nevertheless compare the actual volumes being experienced with the predicted patronage.

The record of BART's average daily patronage is shown in Table I, which also indicates the forecasted levels. Some of the shortfall must reflect the chronically unreliable schedules and the associated long waiting times at stations, owing to equipment

TABLE I. *Forecast as Against Actual BART Patronage, June 1976*

Route Segment	Average Weekday Trips: 1961 Forecast for 1975	1976 Actual	Actual as Per Cent of Forecast
Transbay	77,850	53,880	69
East Bay	129,493	39,725	31
West Bay	51,153	37,765	74
Total	258,496	131,370	51

failings. Some of the shortfall must also reflect sheer deficiencies in prediction; the forecasters had little prior experience to base projections upon, because there had never been a BART-like system in this kind of metropolitan area or the analytic tools for forecasting the demand for this new product.

The Concord line into San Francisco is proving highly attractive, with peak-hour standees outnumbering seated passengers. When enough cars are available to provide seats for most riders, patronage will undoubtedly exceed current levels. Other lines are doing much less well, however, and the net effect is that *total patronage is running at about half of the initial expectations.* Instead of the 258,500 weekday passengers forecasted, only 131,400 were riding in June 1976. Peak-hour riders across the bay number 8,000, a little over one fourth of the designed capacity of transbay trains—28,800 seated passengers per hour.

There are several surprises in those figures, however. Although the *Composite Report* expected that 61 per cent of riders would be diverted from private automobiles, in fact *only 35 per cent formerly made the trip by car.* In response to a BART customer survey, about a fourth of the riders said they had not made the trip at all before BART began operating. However dramatic the apparent volume of newly generated travel may be, it actually represents only a small portion of new trips BART has triggered, for it has generated additional *auto* trips as well.

BART initially reduced the number of cars on the streets and highways by 14,000 (not 48,000, as forecast). In turn, in accordance with the Law of Traffic Congestion (which holds that traffic expands to fill the available highway space until just tolerable levels of congestion are reached), other people began driving their cars on trips they would not otherwise make. Perhaps they are suburban wives who now have the family car during working hours. Perhaps people are visiting friends, now that highways are less crowded. Whoever they are, the net effect is an increase in the amount of sheer mobility within BART's region by both road and rail.

BART has brought about a rise in total transbay travel by both auto and public transit. Between 1973 and 1975, the proportion of daily transit riders on the bridge rose from 17 per cent to 22 per cent. Buses used to carry all 17 per cent; they are now down to 10 per cent, while BART is serving 12 per cent, a third of them in midday. During the peak hours, BART and buses split the transit passengers evenly, each carrying 50 per cent of them.

Half of all BART's transbay passengers formerly rode the bus. In contrast, those using *local* buses seem not to find BART as attractive, probably because of the wide station spacing. Indeed, even during the two-month transit strike, only 10 per cent of the East Bay bus riders used BART instead. Three fourths of all BART passengers travel over seven miles; half of them travel over 12 miles; a fourth over 19 miles! Clearly, BART is serving the long-distance travelers, as intended; but it is carrying only half as many travelers as intended.

The effect on highway traffic

BART's effects on auto traffic are a great disappointment. Although it has indeed attracted some 44,000 trips per day that used to be made in private cars, that is far fewer than the 157,000 forecasted. At most, the overall change in the three counties served may be a small net reduction in auto-traffic volume since BART began; but the change might also be a small net increase. The available regional data make either conclusion plausible.

There was a clear short-run reduction in auto traffic just when BART's transbay service began in September 1974, but BART may have had little to do with it. The gasoline shortage had just cut into auto use, and at the same time gas prices were rising to 60 per cent above their previous levels. Simultaneously, economic recession was increasing unemployment levels (up to 11 per cent in the Bay Area), while inflation was compelling many families to cut their spending levels and perhaps their automobile use.

Under those circumstances, we would normally expect reduced highway traffic volumes and, consequently, reduced congestion. In fact, traffic counters on the several Bay Area bridges do record a reduction in auto travel throughout the region in the period

1974-75. But the surprising fact is that the Bay Bridge, which parallels BART's transbay tube, experienced a smaller proportional reduction than any of the other bridges across San Francisco Bay. Auto travel on the bridge was sustained despite the inauguration of BART transbay service. Paradoxically, because so many commuters switched from buses to BART, the number of buses in the bridge traffic stream was reduced, thus creating more space for cars. Contrary to plan, even if only to that extent, BART has made it easier to commute by car.

It is clear that the enthusiastic expectations of dramatic reductions in auto volumes have not been realized. If BART is having any effects on overall vehicular traffic on the Bay Bridge, they are as yet so slight as to be undetectable, not exceeding two per cent or about one year's normal growth in traffic.

Pretty much the same picture emerges at the Berkeley Hills tunnels, where BART patronage has been high. The start of BART service in May 1973 was associated with a levelling of auto traffic volumes. However, a few months later traffic volumes rose again. They are now higher than ever in the auto tunnel that parallels BART's tunnel, and congestion tie-ups occur just about as frequently as they ever did.

California's Department of Transportation has made a serious attempt to measure BART's effects on highway congestion. On a few routes roughly paralleling BART's lines, peak-hour travel times were found to be reduced sharply when BART service began, demonstrating that even a few cars taken off a full freeway can convert sluggish traffic movement into a free-flowing traffic stream. It appears that where freeways ran at capacity levels, BART is making for appreciable improvement in travel times. However, the marginal autos that spell the difference between congestion and free movement are so few in number (even a two per cent reduction can be sufficient) that one cannot count on their continued diversion. On some tested routes, travel times have already crept up to pre-BART levels. On a few they were unchanged throughout the pre-BART and post-BART periods.

So the conclusions are unambiguous. 1) BART is serving large numbers of suburban commuters, as predicted. 2) People are traveling more, by both car and public transit, despite rising energy costs and undoubtedly as a direct result of the new travel capacity that BART has supplied. 3) Half of BART's transbay riders come from buses, which BART has replaced—at very high cost, as we shall see later on. 4) BART has not effected a significant change in automobile-use habits, so traffic volumes and road congestion are still at just about pre-BART levels, but will no doubt rise as auto use increases.

The effect on metropolitan development

During the 12-year period 1964-75, 35 high-rise office buildings were completed in San Francisco's central district, enclosing 18,500,000 square feet of floor space. During the next six years an additional 17 buildings are scheduled for completion, enclosing another 10,000,000 square feet. These are the structures that followed approval of BART's construction plan, each within an easy 10-minute walk of a BART subway station. Comprising 52 buildings and over 28 million square feet, they were the *cause célèbre* in the citizens' revolt against the "Manhattanization" of San Francisco, the joy of The City's regency, and the conundrum confronting transit planners elsewhere, who now wonder whether rapid-rail transit in their towns will trigger similar explosions. Whatever one's urban-design preferences—whether one loves or detests The City's new look—the question remains, what was BART's role in all that?

It is not clear to what extent BART caused the office boom, and to what extent the expected concentration of offices caused BART. Plausible explanations are obvious for either theory. BART officials like to claim credit for the spectacular change in San Francisco's skyline; they say it is a direct result of improved commuter access from the metropolitan region. However, they also argued from the outset that BART was primarily needed because forecasts of impending downtown office employment raised the specter of the ultimate traffic jam. The rub is that those forecasts were based explicitly on the assumption that BART would be built. If that reasoning were carried full circle, it should then be inferred that, without BART, there would be fewer offices concentrated downtown, less concentration of traffic downtown, more decentralized patterns of office employment, and hence no need for a BART-type system. Needless to say, that circle of reasoning is usually not closed.

Large-scale office construction in other Western and Midwestern metropolitan centers suggests that the building boom might have happened anyway, for many of them have had similar booms, although none except Chicago has had anything like BART in sight. During the 12 years following BART's successful bond election,

San Francisco's high-rise office buildings were expanded by 4,200 square feet for every 1,000 people in the metropolitan region. By contrast, Houston, the automobile city *par excellence,* added 5,500 square feet per 1,000 population, Chicago 4,500, and Dallas 3,500.

The past couple of decades have marked America's transition to the post-industrial service economy. Growing proportions of all jobs are in management and related services, new occupations that are conducted in offices rather than in factories. San Francisco has always been a center for the service occupations, and high proportions of firms coming to Northern California have traditionally sought locations in The City. Accelerating expansion of service activities during the 1950's, 1960's, and 1970's coincided nicely with the civic determination to renovate the city center. New office space was quickly filled, encouraging others to invest in rentable space; and they in turn found other waiting tenants within the growing services sector eager to be in the center of things. In effect, optimistic prophecies of blossoming central business activity became self-fulfilling. The more buildings that touched the skyline and the greater the subway-construction mess along Market Street, the more that companies wanted to be in on the action; and more engendered still more. Once the old-fashioned boom-town spirit spread, it became self-generating, building on its own image, enticing still more firms in search of the prestigious addresses, proving once again that nothing succeeds like success.

Surely BART was part of the generating force, for it was a massive piece of the big construction set. But there are no categorical answers to the question about the size of BART's role in San Francisco's reconstruction. I am inclined to think that it would have happened anyway, but that BART nonetheless made its happening bigger and quicker.

Is the story likely to be repeated elsewhere? Probably only in those cities having a comparable attraction for headquarters offices and ancillary business services. However, as the experiences of Houston, Dallas, and Denver emphatically show, a BART-type rail system is not a necessary condition for a city building boom of this sort. Unquestionably, adequate transport services are essential, but effective transport comes in many other forms as well. Those other Western cities have been undergoing explosive central office-building expansion while relying on automobiles and buses. But for the promotional attraction of BART, San Francisco might have done as well with intelligent development of its own road transport system.

Suburban sputter

The initial plan for BART was also to generate growth at selected subcenters throughout the metropolitan region. In addition to the high average speed, that was the other rationale for widely spaced stations. The planners fully expected that increased accessibility at train stations would make the surrounding areas attractive to business firms and apartment dwellers, following the model of earlier commuter railroads in the East. In turn, clusters of offices, shops, and high-density housing around these stations would visibly restructure the region, stemming the drift toward low-density dispersion and urban sprawl.

It is now 14 years since the BART project was approved, and there is, as yet, little evidence to corroborate those forecasts and hopes. Most suburban stations stand in virtual isolation from urban-development activity in their subregions, seemingly ignored by all except commuters who park their cars in BART's extensive lots.

By being heavily subsidized and charging fares well under its actual costs, BART has appreciably reduced monetary commuting expenses for outlying suburbanites who work in the central cities. *Thus, rather than deterring suburban sprawl, BART may instead be encouraging it.* The unexpected popularity of its suburban Concord line clearly signals the response of suburbanites to the bargain rates being charged. We know of no explicit empirical evidence that BART is bringing about further suburban spread, but the parallel history of freeway-induced reductions in travel costs is unambiguous. Even though BART's land-use effects may not yet have been made manifest, the longer-run expectations should be clear. If BART is to influence the future course of suburban development at all, it seems as likely that it will be an agent in spreading suburban growth as in concentrating it.

Perhaps a basic flaw in the initial planning was the failure to take into account existing high-level (and virtually homogeneous) accessibility throughout the Bay Area. The ubiquitous network of streets, highways, and freeways, combined with extremely high auto availability, made for a context that scarcely resembled the 19th-century urban settings that earlier suburban railroads were fitted into. Bay Area residents move about within the metropolitan area with great freedom; they can go from virtually anywhere to anywhere else with only occasional delays, because the road network makes all points within the region highly accessible to all others.

When BART added additional accessibility on top of existing

levels, it was proportionally only a small increment. Suburban and downtown stations are only slightly more accessible now than before BART was installed, scarcely enough to have significantly affected the location plans of many households and firms. If the rail line had been the major access route, things would have been different. But overall accessibility in this road-rich metropolitan region has not been appreciably modified, and so neither have urban development patterns.

Land use and transportation systems are highly interdependent. BART's failure to attract many riders may also be contributing to its failure to attract building investors in the areas of its stations. Moreover, seeking to reduce land-purchase costs while encouraging the park-and-ride habit, BART designers located stations at some distance from established suburban business centers and then surrounded them with parking lots. As a result, potential developers are compelled to build outside easy walking distance, and so most go to the established centers instead. It is almost as though the right-of-way agents who purchased property and the engineers who located the stations were either oblivious or opposed to the objective of fostering suburban clustering. It is also reasonable to believe that land-use effects take a long time to become manifest, especially in an auto-oriented society, and that we shall therefore have to wait another 10 years or so before we will know whether BART will affect development at all.

Who rides and who pays

Everyone expected from the outset that BART would have to be bought with solely local funds. There were no other options at the time, for BART's planning preceded by a decade federal government financial support for transit. Indeed, it may be that BART's favorable publicity may have so popularized rail transit that in some degree BART is responsible for the availability of federal subsidy funds today. But at the time it was being designed, the Bay Area was on its own.

The plan was to draw upon the traditional source for local capital improvements, general obligation bonds secured by a property tax. Those were the bonds approved in November 1962—at that time, the largest local bonding referendum ever, amounting to $792 million. The fund was to be sufficient to pay the full costs of all the capital plant except the transbay tube and the rolling stock. The tube was to be built with tolls collected from bridge users. Rolling stock and all operating expenses were to be paid from fares collected from BART users.

Subsequently, when capital costs began to exceed initial estimates, it became necessary to find other resources. Among those explored were supplements to auto-license fees, higher bridge tolls, and general highway-improvement funds. These were forcibly opposed by those who disliked a direct tax on motorists. In the end, the 1969 state legislature authorized a $150 million bond issue, financed by a one-half of one per cent addition to the general sales tax within the three-county district that was to be served by BART.

These three sources of capital funds (property taxes, sales taxes, and bridge tolls) then still turned out to be insufficient, for there was not yet enough money to buy the cars. Fortunately, the federal transit grant program was initiated in time to permit BART to purchase its rolling stock (federal grants have by now amounted to $305 million), and a parallel state-aid program was initiated that now yields an annual subsidy (about $2.6 million this year).

The 1962 *Composite Report,* circulated prior to the public vote on the bonds, had presented a favorable estimate of operating revenues and costs for 1975-76. BART's own current financial reports now present a far more gloomy picture:

TABLE II. *BART Operating Revenues and Expenses (Fiscal Year 1975-76)*

	MILLIONS OF CURRENT DOLLARS: 1962 COMPOSITE REPORT	ACTUAL (PRELIMINARY)
Gross Fare and Concession Revenue	$24.5	$23.7
Operating Expense	13.5	64.0
Net	$11.0 surplus	$40.3 deficit

When it became apparent that fares would be insufficient to cover the costs of rolling stock and operations, the 1974 state legislature authorized, as emergency aid against the impending operating deficit, a temporary extension of the one-half of one per cent sales tax to be applied to the operating deficit. This yields about $28 million per year, enabling BART to continue operating until the latter part of 1976, by which time an extension or some other operating subsidy will be needed.

As of June 1976, the overall picture of BART expenditures and sources of funds looked like this:

TABLE III. *Sources of Funds for BART*

	MILLIONS OF CURRENT DOLLARS	PER CENT OF TOTAL
Capital Funds		
General Obligation Bonds	$792	50
Sales Tax Revenue Bonds	150	10
Toll Bridge Revenues (for Transbay Tube)	180	11
Federal Grants	305	19
Miscellaneous Revenues	159	10
Total	$1,586	100
Annual Operating Funds (1975-76 Estimates)		
Fares and Concession Revenue	$23.7	37
Sales Tax Revenue (0.5 per cent)	27.7	43
Property Tax Revenue (5¢ for Operations)	5.0	8
State Aid (0.25 per cent Sales Tax)	2.6	4
Federal Aid	.5	1
Borrowings Against Capital Account	4.5	7
Total	$64.0	100

Over 60 per cent of capital costs and about 55 per cent of operating costs are being paid by all residents of the three-county district through taxes on property and retail sales, both inherently regressive modes of taxation. E. G. Hoachlander has estimated the incidences of BART's property and sales taxes and compared them with household incomes.[3] His findings indicate that both taxes fall proportionately far more heavily on households with low incomes than on wealthier ones.

In an effort to keep track of customers' characteristics and preferences, BART has been conducting a periodic survey of its riders. The last completed survey was made in May 1975, eight months after transbay service started and two-and-a-half years after East Bay trains began running. The riders' own reports on their incomes reveal that they are not a representative sample of the Bay Area population. They are drawn far more heavily from the upper sectors of the income distribution than from the lower, reflecting, no doubt, the system's attractiveness to long-distance suburban commuters. The chart below compares the income distribution with the proportions of family incomes devoted to BART taxes. The ratio of these two curves is 40 to one: The percentage of income paid to provide tax support for each ride taken is 40 times greater for an individual in the lowest income group than for one in the highest income group. Clearly, the poor are paying and the rich are riding.

Income Comparison of BART Taxpayers and BART Riders

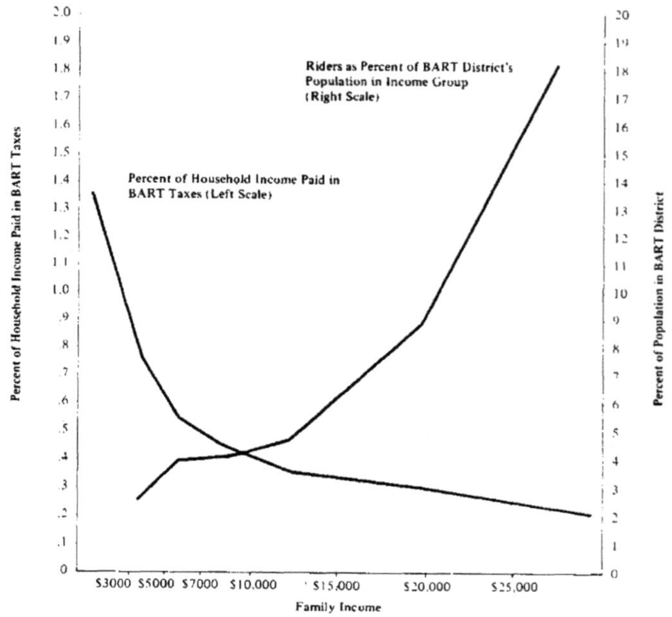

Comparative costs

Every trip calls for three kinds of expenditures. There are the dollar costs that auto owners and transit agencies pay for vehicles, fuel, roadways, drivers' salaries, interest charges, insurance, and so on. There are the time expenditures made by travelers, who seem to make sensitive assessments of the lengths of time they must spend riding vehicles, getting to and from vehicles, and especially transferring and waiting for the next vehicles to arrive. The third kinds of costs are external to the transportation systems: annoyances borne by neighbors who must suffer noise, pollution, congestion, and other nuisances that transportation systems and their patrons impose.

To compare the costs of alternative travel modes, one must tot up the monetary outlays, the time expenditures, and the social costs associated with each mode, preferably on standardized scales and in common coinage. This is obviously difficult to do, if only because each mode of travel provides somewhat different qualities of service and because some costs and some benefits are not directly traded in pecuniary terms. But it is not impossible to do.

A group of economists at Berkeley, led by Professors Theodore

Keeler and Leonard Merewitz and Dr. Peter Fisher, recently completed a careful study aimed at comparing the full costs of standardized trips on BART, local buses, and private cars.[4] Their findings indicate that *the bus is consistently more cost-efficient than even a subcompact automobile, at virtually all levels of traffic density and at current averages of 1.5 persons per car.* No surprise there. But they also find that the bus is consistently cheaper than BART for comparable qualities of service, whichever set of technical assumptions is built into their estimating formulas. *Moreover, and to everyone's surprise, their findings show that a subcompact automobile is also cheaper than BART, except when travel densities approach 20,000 passengers per hour within a single traffic corridor.*

BART's maximum design capacity on the transbay and West Bay lines is 28,800 seated passengers per hour (72 passengers per car, 10-car trains spaced at 90-second intervals). So far BART has not approached that level of patronage. In Wolfgang Homburger's October 1975 survey, BART's maximum single-hour transbay load was 8,120 eastbound passengers at the height of the evening peak period. Instead of 28,800 passengers *per hour* in each direction, BART's transbay patronage was only *28,500 per day.*

Philip Viton has computed the full costs of carrying one passenger on a representative trip by BART, bus, and auto. He concluded that *even when BART achieves its full design efficiency, it will still run on all its lines at higher costs per passenger trip than buses, and its costs may continue to be more than those of an automobile even then.*

The estimates are based on assumptions most favorable to rapid transit. They assume a hypothetically optimized system in which schedules of buses and trains minimize riders' waiting time and agencies' operating costs. Estimates for automobiles include governmental costs of building, maintaining, and policing highways, individuals' total costs of buying and operating private cars, and the external costs of pollution. For each mode, a concerted effort has been made to account for all monetary and non-monetary costs borne by governments, travelers, and neighbors.

The dollar-equivalent costs are estimated against a set of variables related to interest rates on investments, value of travelers' time while riding, value of their time while getting to a train or bus, and the average lengths of trips. BART trip-cost estimates are highly sensitive to variations in interest rates, because of the large capital investment; and they compare best against bus and car when computed for long-distance trips. As we shall see, because very few Bay Area commuters live within an easy walk of a BART station, BART trip-cost estimates are particularly sensitive to assumptions about the value of the access time.

There is some disagreement among economists on the dollar values to be assigned to passengers' time, but all agree that some dollar equivalents must be included in estimating total costs. After all, BART did go to great expense to offer high speeds, presumably because its designers believed that passengers value their time, preferring a brief trip to an extended one. Similarly bus operators work for short headways and for on-schedule performance because they believe that passengers' dollar-clocks click fastest when they are waiting for a bus to come. (If those beliefs were in error, the appropriate response would be to slow down the trains and to delay the buses. Obviously those strategies would be absurd.) The consensus among students of "modal choice" is that average passengers clock their walking-and-waiting time about three times more heavily than their riding time.

In an effort to avoid any inferences of negative bias, the estimates presented here have been deliberately biased in favor of BART. Throughout, wherever the estimating equations permit a choice of values, we have selected the combination of permissible variations to show BART trip-costs at their lowest levels when compared with the bus and the car.

The comparisons in Table IV present the full costs of carrying a hypothetical commuter 15.7 miles across the Bay from Orinda to San Francisco's Montgomery Street in the morning peak hour. We assume that he lives two miles from the BART station, which he reaches by a local feeder bus, and that he walks a quarter-mile from the Montgomery Street station to his office. Alternatively in

TABLE IV. *Full Costs of Peak-Hour Trip from Orinda to Montgomery Street*[1]

Number of Passengers in Peak Hour	Auto Standard	Auto Subcompact	Bus	BART plus Feeder Bus
1,000	$4.49	$4.05	$8.51	$34.08
5,000	4.49	4.05	3.72	9.20
8,000[2]	*4.49*	*4.05*	*3.21*	*6.77*
10,000	4.49	4.05	2.99	5.99
20,000	4.49	4.05	2.59	4.26
30,000	4.49	4.05	2.44	3.64

[1] In 1972 dollars. These are low estimates, computed to be most favorable to BART. Costs to agency plus time costs to riders plus social costs to neighbors.
[2] Current peak volume on both bus and BART.

our example, he could ride a bus from near his home to the Montgomery Street BART station, or he could drive his car and park in a public garage near his office.

Actual peak-hour volumes on the transbay routes for both bus and BART are only about 8,000 passengers per hour. At that volume, a bus ride costs $3.21 and a BART ride costs $6.77, with private cars in between. We have also computed estimates under less favorable, but nevertheless reasonable, assumptions concerning the value of commuters' time and the rate of interest on capital invested. For that same ride from Orinda to Montgomery Street at present peak-hour patronage, full costs would then be $5.83 on the bus and $11.96 on BART.

When BART has been finally debugged and operates at full schedule and full efficiency, with a fully seasoned patronage, it will surely be carrying more riders; and so both average and marginal costs will then fall. The most favorable estimate for seasoned operation on that same commuter trip from Orinda to Montgomery Street is unlikely to fall below $6.00, however. (The less favorable, but still reasonable estimate is $10.00 for that trip.) In contrast, the full cost of a nearly equivalent ride on a bus would be between $3.00 and $5.00, and the cost of driving a standard sedan would be between $4.50 and $7.50, all costs included. Under those circumstances, BART is unlikely ever to replace other modes of travel, so long as travelers are permitted free choice. Perhaps only under regulated market conditions, imposing mandatory constraints on use of buses and cars, might BART carry the volumes of passengers that were forecasted for it.

The financial picture

All estimates above are economic costs—i.e., they reflect costs to the economy, representing the opportunity costs of using resources for these purposes rather than some others.

Alternatively, we might examine accountant costs—i.e., the costs as they might be seen by BART's internal bookkeepers. That picture is somewhat different. Annualizing BART's total capital costs to find its annual "mortgage payments" at its favorable 4.14 per cent interest rate, adding in its current operating costs, then dividing by annual passengers carried, the cost per trip comes to be about $4.48. With fares averaging around 72¢, the average subsidy then comes to about $3.76 per ride.

Two major factors are of interest here to officials in other cities.

TABLE V. *BART's Annual Expense and Patronage Account (Fiscal Year 1975-76)*

ITEM	AMOUNT
Annualized Capital Costs	\$ 82,400,000
Annual Operating Cost	64,000,000
Total Cost for Year	\$146,400,000
Number of Trips during Year	32,700,000
Average Cost per Trip	\$4.48

First, direct costs to the BART district are high because the system was financed largely from local sources. With federal subsidies now available, no other metropolitan area will finance a transit expansion on its own again. Second, BART's capital-intensive plant makes it inherently expensive. Subways and elevated, exclusive, grade-separated rights-of-way cost a lot, no matter how crudely or elegantly they are constructed. Besides, BART spent a great deal of money in pioneering research and development for new systems, the benefits of which will now accrue to other cities that build subsequently. But BART had one advantage over all other descendent systems. It was built with pre-inflation dollars. However exorbitant its costs may seem, they appear cheap relative to comparable systems under construction in Washington and Atlanta.

Rail-transit systems are inherently capital-intensive. When they are also burdened with very high labor costs, as BART is, operating expenses may become excessive. BART's operating costs are running at about 15.7¢ per passenger mile, while the local bus system is costing about 13.6¢. Despite the all-out effort to reduce manpower requirements by automating train operations and assigning only one attendant per train, BART now finds itself with a wage bill only slightly less than that of the local bus system—10.5¢ per passenger mile compared to about 13.2¢ for the bus. In part because of a federal requirement that led to a costly union contract and in part because of the high-priced engineering and planning staff it maintains, BART has not yet shown how to cut the labor and operating costs that have been hurting public transit systems of late.

As a result, it has been in a virtually continuous financial crisis. Governments can continue to "throw more money at it," but the real solution is for BART to attract sufficient patronage at least to carry its operating costs, as planned. The question, then, is why haven't Bay Area travelers been flocking to this outstanding transportation system.

What went wrong?

BART was designed to be a superior way for Bay Area commuters to get to work. Its promoters believed that if only superb transit service were offered, commuters would gladly give up their private cars. Because commuters are not responding as expected, it is important that we understand why.

The clue may lie in the few, but key design decisions that were made in the early stages of BART planning and that then fixed the system's essential character. It now appears in retrospect that those design features preordained BART's failure to lure the motorist.

At the time BART's designs were being drawn, the postwar auto boom was a major factor in metropolitan growth, and freeways were accorded high priority. At the same time, public transit systems were falling into disrepair or being abandoned, and patronage was declining everywhere, projecting the prospect of a virtually all-auto transportation system. BART was seen as a lower-cost alternative to freeways and as a means for both reversing the trend and preserving the option of public transit. The design strategy was to top the automobile by producing a transit system that would incorporate and improve upon the automobile's most attractive features, thus making the transit system more than competitive.

The designers concluded that high speed, high comfort, high style, and downtown delivery were the attributes that matter most to motorists; and BART was then designed to outdo the car on those four counts. BART's management has delivered the system promised in its original specifications. Unfortunately, however, these may not be the features that will entice mass patronage.

The emphasis on high speed between stops and on comfort and overall aesthetic excellence quickly led to the decision to build a rail system, and that in turn led to a logically necessary network of decisions concerning equipment design, electronic control gear, roadbed standards, station qualities, and the like. As Randall Pozdena has insightfully noted, each of these decisions was simultaneously a decision to trade off other potentially desired qualities. Some of the more important sacrifices and compromises he identifies are in Table VI on the next page.

Of course every design for a complex system must make these sorts of trade-offs, for it is seldom possible to enjoy all advantages simultaneously. It is the tragedy of the BART story, however, that the chosen attributes compelled the sacrifice of features that were essential for attracting riders. Most important by far are the first four in the table. By choosing mainline rail, the designers created a system geometry that puts BART out of walking distance of most residents.

TABLE VI. *Design Trade-offs and Compromises*

SELECTED QUALITIES	SACRIFICED OR COMPROMISED QUALITIES
1. High average speed between stations, therefore widely spaced stations.	1. Closely spaced stations, therefore ease of access to stations.
2. Mainline system serving major traffic corridors.	2. Network of transit lines serving sub-areas of the region. Ability to complete trip in a single vehicle without having to transfer to and from feeder system.
3. Batch-type transport mode: cars in trains carrying many passengers.	3. Flow-type transport mode: smaller vehicles carrying comparable numbers of passengers at shorter headways, with branching local distribution at origin and destination.
4. Fixed rail on exclusive grade-separated right-of-way.	4. Flexible routing in response to changing travel patterns. Economy of construction. Right-of-way usable by other vehicles. Disabled vehicles do not disrupt operation of entire line.
5. Limited number of access points into system, to encourage clustered urban development.	5. Compatibility with footloose trends and low-density settlement patterns.
6. Frequent service with stops at all stations.	6. Differentiated service with both "local" and "express" operations.
7. High aesthetic and comfort standards.	7. Economy of construction.
8. Regional long-haul design.	8. Local trip-making capability.

The designers seem to have been most concerned with the attractiveness of the BART system as seen by passengers after they arrive at the station. Station decor is handsome; waiting time for trains was to have averaged a brief 45 seconds; 80-miles-per-hour speeds make for short elapsed time en route; and the station at the other end of the trip is also aesthetically pleasing. But outside BART's premises, the passengers are on their own. They must find their way to the station by bus, car, or foot, make the transfer, and then find their way at the other end after leaving the BART station. While they are BART's guests they are treated very well; outside the premises they are rather neglected.

Oddly enough, although the designers have always been explicit about the critical role of bus access to the system, the planning process was rather nonchalant about creating it. There was little effective planning for feeder-bus service until the period just preceding opening day, and it is still quite inadequate in the outlying suburbs.

Making the wrong choices

During the past 15 years, at least a dozen major studies have investigated the ways travelers assess costs when deciding how they will make intrametropolitan trips. With remarkably small variation among the cities examined, the studies all conclude that *the time spent inside vehicles is judged to be far less onerous than the time spent walking, waiting, and transferring, by a factor of up to 3 or 4 times. For commuters waiting on platforms, the factor may be as high as 10 times!*

BART designers were obviously unaware of these findings; the research was conducted after the key BART decisions were made. Indeed, their own understanding was just the opposite, as this key conclusion of the 1956 basic design report indicates:

> ... interurban rapid transit must be conceived as providing only *arterial* or *trunk-line* connections between the major urban concentrations of the region. . . . We are convinced that the interurban traveler, facing the choice between using his private automobile or using mass transportation, will be influenced in his choice more by the speed and frequency of interurban transit service than by the distance he must travel in his own car or by local transit to reach the nearest rapid transit station.[5]

Herein may lie a clue as to why their strategy erred. Their fixation on high speed meant that riders spend relatively short amounts of time in BART's vehicles, but this is the kind of time that travelers place a low cost upon. That fixation has also inevitably meant long access times, which travelers account as a high cost. The desire for high speed led to wide spacing between stations, and that, combined with the skeletal mainline-route pattern, compels most travelers to use some kind of feeder service getting from home to BART. The use of a feeder bus compounds the onus of waiting and transferring, and many potential BART patrons have therefore simply decided to ride the bus all the way through to their destinations, instead of making the transfer. Many others have simply decided to continue driving their cars. The reasons for both decisions should now be clear.

Buses and cars mixed into the traffic stream operate at slower speeds than 80-mile-per-hour trains, of course. However, trains must stop at each station every 2.5 miles or so. Meanwhile buses and cars on freeways usually move along nonstop until they reach the terminal or exit. The net effect is that *scheduled running times for some East Bay peak-hour express buses to San Francisco are just about the same as for BART trains.* Of course, whenever an accident or a breakdown clogs the freeway, traffic slows down or stops; but BART's frequent breakdowns almost even the score. Overall, where express bus service is available, the advantages of no-transfer rides and steady freeway movement make the bus competitive in route time.

The major competitive advantage of buses, however, derives from the savings they permit in access time. Buses have the capacity to thread into residential districts, collecting passengers near their homes. Automobiles do even better than that, parked in the owner's garage and available at his call. Whatever the cost of traffic congestion, access time is zero. However unpleasant the bus may be, access time is low. In these respects, the bus and the car are functional opposites of BART; they trade off high speeds en route in in favor of easy access.

It seems that BART's mistake was made at the outset, when the wrong technology was chosen. Instead of lavishing primary attention on in-vehicle travel time and physical amenities, which called for a mainline rail system on an exclusive grade-separated right-of-way, the designers would have attracted more riders by adopting more automobile-like technology. A system that could pick up passengers within a short walk of their homes and deliver them, in the same vehicle, to within a short walk of their jobs would have been far more likely to entice them out of private cars. The success of both the new Golden Gate buses from Marin County to San Francisco and the express buses from the East Bay suggest that high-quality bus service can attract significant numbers of commuters.

It is the door-to-door, no-wait, no-transfer features of the automobile that, by eliminating access time, make private cars so attractive to commuters—not its top speed. BART offers just the opposite set of features to the commuting motorist, sacrificing just those ones he values most. This was a fundamental mistake. Given commuters' propensities to weight system access and waiting time so heavily and to place much less importance on in-vehicle time, it is scarcely any wonder that BART has not lured them away.

Moreover, the error is compounded by the high construction and operating costs compelled by the insistence upon high speed. That initial standard in turn determined much of the overall and detailed designs: a new and separated guideway, automatic controls, unconventionally wide-gauge rails, a highly stable roadbed, lightweight cars of unprecedented design, a highly specialized and high-priced work force, and so on. If its stations were more closely spaced, if its routes were more extensive, if it were not so difficult to get to it, BART's patronage would certainly be far higher. The paradox is that potential passengers are not using it because it is too rapid.

Was it worth it?

Having spent $1.6 billion to avert the trend to the auto-highway system, BART is now serving a mere two per cent of all trips made within the three-county district, and about five per cent of peak-hour trips. Some 50,000 of its daily passengers have been diverted from inexpensive buses to expensive trains, 46,000 from private cars and car pools, and several thousand more from the latent pool of trips not previously made. The overall effect has been to leave highway congestion levels just about where they would have been anyway. BART may have been influential in propagating downtown building construction, but it has not yet had any visible effect on suburban development.

The most notable fact about BART is that it is extraordinarily costly. It has turned out to be far more expensive than anyone expected, and far more costly than is usually understood. High capital costs (about 150 per cent of forecast) plus high operating costs (about 475 per cent of forecast) are being compounded by low patronage (50 per cent of forecast) to make for average costs per ride that are twice as high as the bus and 50 per cent greater than a standard American car. With fares producing only about a third of the agency's out-of-pocket costs, riders are getting a greater transportation bargain than even bus and auto subsidies offer; and yet only half the expected numbers are riding. The comparative full costs of a typical transbay peak-hour commuter trip on BART are about $6.80, on a bus $3.25 and in a small car $4.00 —computing all those estimates with variables that make BART appear most competitive. The total economic costs of even a large-sized American car are still lower than BART's for a transbay commuter trip—all-day parking charges, highway construction, pollution, and all other measurable costs included.

The 50,000 passengers BART has diverted from buses could be carried in brand new luxury buses at a total capital investment of under $13 million. The BART system cost $1,600 million. The costs of buying a whole fleet of new buses sufficient to carry all BART's passengers projected to 1980 would be under $40 million, or about half of one year's worth of BART's annual mortgage payment alone. One is compelled to ask, was it worth it? Was it wise to have built so costly a system?

Rather than reverting to pre-auto technology, the designers might instead have sought a competitor to the automobile that, by incorporating similar service capabilities, might then be more likely to induce commuters to switch. Among currently available alternatives are express buses that collect passengers near their homes, or subscription buses that pick them up at their doors, then use urban freeways to speed them to job centers; jitneys and group taxis, the major transit modes in many parts of the world, which use automobiles as public transit vehicles, interlacing residential districts and delivering directly to employment places; and franchised van pools that operate as a quasi-bus and quasi-car pool, providing door-to-door service at both ends of commuters' trips, albeit with some rigidity in scheduling.

These are automobile-like modes that more nearly approximate the door-to-door, no-transfer, flexible-routing features of the private car than do 10-car trains on fixed mainline rails. Although they all surely lack the high technological glamour of BART-like systems, they can approximate BART's in-route speeds; they can greatly reduce system-access time; and furthermore, as Professor McFadden's surveys in the Bay Area indicate, where time and money costs are equal, most travelers seem not to be taken by the glamour of BART over the bus anyway. In any case, with the ratio of capital costs between BART and buses on the order of 40 to one, and with operating costs per passenger-mile essentially the same, it would seem that investment prudence is therefore on the side of the express bus.

Experience with express bus service is still rather sparse, however; and so we can have no assurance that the high levels of transit patronage that BART was aiming for are attainable with buses either. Although we do now know that express buses work, we do not know whether they work well enough. Much more experimentation with various sorts of auto-like and bus-like modes is called for, including experiments with pricing schemes that charge motorists and transit riders the full costs of travel.

There are surely many who still believe that BART will eventually make it—that only start-up problems, overly publicized equipment failures, and annoying delays dissuaded potential patrons. Given enough time, they believe, BART's trains will run on time and will then become the powerful magnets capable of attracting motorists at last. Given the uncertainties that bedevil this business, they might of course be right; only the test of time can tell. But even if BART were today carrying all 258,500 daily passengers originally projected for 1975, at present average fares it would be earning only $46,000,000 per year, or less than three-fourths of its *present* operating costs.

The power of promotion

If BART has achieved any sort of unquestionable success, it is as a public relations enterprise. BART has projected a superb image from the start: a high-speed, futuristic transport mode that would transport commuters in luxurious comfort without economic pain. It became one of the more effective signs of the Bay Area's avant-garde spirit, the very symbol of Progress.

As a result, it may be that BART's most successful effects have been felt outside the Bay Area. Urbanists and civic officials throughout the world have become intimately familiar with its promises. Many, in turn, have been encouraged to propose various sorts of modern rail systems for their own cities, some of them aided by the same consultants who mothered BART through its Bay Area gestation. Despite its problems, BART has both popularized and legitimized modern rail transit, and that much-reproduced photograph of the BART lead car has become a heraldic symbol on rail-promotion banners everywhere.

Back home in the Bay Area, the picture is rather more gloomy. Half the expected riders have chosen the convenience of their private cars or the local bus that also runs them close to their jobs. Real estate developers seem to believe that auto access is more important for their potential customers or tenants, and so they have pretty much ignored the locational opportunities BART stations opened to them. People quietly pay property and sales taxes, unable to do anything much about the BART levies, whether they like them or not, whether they benefit from them or not. Governmental officials, caught with a large investment, outstanding bonds, and rapidly rising operating costs, keep priming BART with the hope that it will eventually run on its own. According to one careful student of BART's finances, they would find it more cost-effective to abandon the train service and convert the rights-of-way into exclusive bus lanes instead. It is unlikely that any public officials will find that politically acceptable, however, especially when BART is still so new. Besides, public officials seem to be constitutionally incapable of admitting error.

The whole affair again raises some persistent questions about the analytical and ideological bases underlying local political decisions, about the locus of accountability, and about governmental capacity for learning. From the beginning, BART's planners were handicapped, because the state of transport-choice theory was so inadequate that it was impossible to simulate accurately what would happen if BART were built. They did not even have adequate descriptive data showing how people choose among travel modes, especially travelers having choices among three such first-rate systems as are now offered in the Bay Area. It was therefore virtually impossible to forecast patronage or revenues with any precision. The science of transport planning has simply been inadequate to warrant the level of confidence that has accompanied this project. However cautious the disclaimers that were attached to the forecasts, once in print the numbers somehow became reified, then accepted as facts by political leaders, voters, and bond buyers. One wonders to what degree ideological leanings affected personal beliefs when the forecasts were inherently so uncertain.

BART is the manifestation of a wide array of ideologies that must have made it attractive to a wide array of publics. It symbolizes the frontiers of science and technology and, simultaneously, the nostalgic old railways of the elder stateman's youth. It is the rationally efficient means for reducing land consumption, and it simultaneously reflects the romantic desire to capture the ethos of Parisian urban life. It is seen as sound business by both merchants and city officials, and as the instrument of sound development policy by city planners. It is the darling of the anti-auto, anti-technology ideologues and also of the engineers who admire its technological sophistication. Many people believed in BART for many reasons. The mere failure to meet its objectives is not likely to shake such faith, particularly when the appraisal is made in merely pecuniary terms.

Many are wholly pleased with the BART system. It offers perhaps the smoothest and quietest ride in the world. The cars and stations are physically handsome and pleasant. Fares are low and travel speeds are high. Save for the initial equipment failures, it is almost exactly the physical system that was promised to the 1962 voters. For those who can use it conveniently, it is superb. For others, the very fact that it was built is itself an achievement, for BART was surely one of the more spectacular civic projects, and the transbay tube one of the more daring engineering feats of our time. From those special perspectives, BART has been an unqualified success.

But the question remains whether leaders in other metropolitan areas can learn from BART's experience that these perspectives are illusory. Having been built in a metropolitan area offering what are probably the best test conditions in the country, BART is not passing the most important of those tests. It is struggling with persistent fiscal crises, with no prospect of ever becoming the self-supporting system the voters were promised. The poor continue to pay and the rich to ride, with no visible prospect that this will change. Its patronage remains low despite low fares and rising gasoline prices, and that situation seems to be stable too. Save for the possible influence on downtown San Francisco and Oakland, it has had few detectable effects on urban development patterns, and its effects on traffic congestion are similarly undetectable. Clearly, BART has not earned a passing grade on the significant tests. One wonders, then, what underlies the professional confidence of transit consultants, and what sorts of politics and ideologies in local and federal governments continue to promote BART-like systems elsewhere.

BART has been heralded as pacesetter for transit systems throughout the world. The evidence so far suggests that it may also become the first of a series of multi-billion-dollar mistakes scattered from one end of the continent to the other.

But in the long run, say in 50 years when the bonds will have been retired, when everyone will regard BART as just another built-in feature of the region, rather like Golden Gate Park, perhaps no one will question whether BART should have been either built or abandoned. It will then be regarded as a handy thing to have, a valuable facilitator of trips that would not otherwise be made by the elderly and the young, a blessing that enriches the quality of Bay Area life. And who will gainsay then the wisdom of having built a white elephant today?

FOOTNOTES

This article is the result of a collective effort by the following contributors: Seymour Adler, George Cluff, Kevin Fong, E. Gareth Hoachlander, Wolfgang S. Homburger, Leonard A. Merewitz, Mary Jo Porter, Randall J. Pozdena, David Reinke, and Philip A. Viton. The review was undertaken at the suggestion of Wilfred Owen. It was supported by the Institute of Transportation Studies at the University of California, Berkeley. Henry Bain has been our most perceptive critic.

[1] The proposal was presented in Parsons, Brinckerhoff, Hall, and Macdonald, *Regional Rapid Transit,* Report to the San Francisco Bay Area Rapid Transit Commission (January 1956).

[2] Parsons, Brinckerhoff, Tudor, Bechtel; Smith, Barney, and Company; Stone and Youngberg; and Van Beuren Stanbery; *The Composite Report, Bay Area Rapid Transit,* Reports submitted to the San Francisco Bay Area Rapid Transit District (May 1962).

[3] E. Gareth Hoachlander, *Bay Area Rapid Transit: Who Pays and Who Benefits?* (Berkeley, University of California, Institute of Urban and Regional Development, 1976).

[4] Theodore E. Keeler, Leonard A. Merewitz, and Peter M.J. Fisher, *The Full Costs of Urban Transport,* three volumes (1975); Randall J. Pozdena, *A Methodology for Selecting Urban Transportation Projects* (1975); and Philip A. Viton, *Notes on the Costs of Mass Transit in the Bay Area* (Berkeley, University of California, Institute of Urban and Regional Development, 1976).

[5] Parsons, Brinckerhoff, Hall, and Macdonald, *op. cit.*, p. 38.

[10]

GEOGRAPHY AND THE POLITICAL ECONOMY OF URBAN TRANSPORTATION

David C. Hodge
Department of Geography
University of Washington
Seattle, WA 98195

The classic view of the role of urban transportation in urban geography is based on a model of technological determinism in which changes in transportation lead inexorably to changes in urban form. This model dominates most urban geography textbooks and continues to influence contemporary accounts of the role of transportation (especially the role of the automobile) in determining urban form (Muller, 1986). While major changes in urban form and urban transportation are obviously connected to each other, there is an increasing number of scholars who argue that the significance of urban transportation runs much deeper than that of a progenitor of new urban geometries. It is the goal of this review to examine the perhaps less obvious, but more profound, sets of relations between transportation and the structure of society such as the redistribution of scarce public resources, economic development, the accumulation of capital, and the reproduction of labor.

The infusion of critical social theory into urban geography at a time of obvious structural change in urban areas has provided a rich platform from which to examine these deeper dimensions of urban transportation. It is important to note at the outset that this review derives from a view of society as a complex set of *contentious* social relations, hence the emphasis on *political economy.* To this end, I will argue that transportation is deeply embedded and implicated in the processes of production, reproduction, and legitimation in an advanced capitalist economy (Feldman, 1977; Staeheli, 1989). The paper is organized around three major themes. I begin with a consideration of urban transportation as a public service in which the welfare issues of benefits, costs, and equity dominate. The second, and major, portion of the paper is devoted to an understanding of the journey to work which involves issues related to Kain's (1968) spatial mismatch hypothesis, contemporary urban restructuring and the connections between home and work, and new studies examining gender issues. Third, I consider the role of transportation as part of urban capital accumulation processes and examine urban transportation as a private commodity, an industry subject to forces common to the production of other commodities in capitalist society.

URBAN TRANSPORTATION AS A PUBLIC SERVICE

As a public service, the provision of urban transportation is held accountable to a number of critical interests. Explicit public goals include such diverse concerns as promoting economic development, providing mobility for the disadvantaged, and reducing pollution and congestion. The rules by which this infrastructure and these services are provided are often constrained and the result of rather difficult political negotiations. While the issues of efficiency are certainly present, more overt attention is focused on issues of equity (i.e., who pays to support the service and who benefits from the service) and on indirect benefits and costs associated with the service. It is also important to distinguish between the provision of auto-related and mass transit services, since the overt political agenda differs so greatly between them.

Automobile-Related Transportation

With few exceptions urban streets and highways have long been viewed as a public responsibility. They have been considered as so fundamental to the functioning of the economic system, in fact, that since World War II at least massive state and federal support have supported their local provision. While it may be argued that most of the support has come through the dedicated use of a user fee (the gas tax), it has also been argued that the full cost of auto-oriented transportation, including all externality costs, has not been paid by those benefitting, but imposed on less powerful segments of society (Hodge, 1986; Pucher, 1988b). The increasing shortage of tax dollars in general, however, has added many new controversies and conflicts over the priorities for spending public monies as well as determining who should pay them. In many ways, the debate indirectly settles on a conflict between the central city where an aging transportation infrastructure is in critical need of support, and the outer fringes of an urban area, where the pressures of development are overtaxing the abilities of local governments to provide adequate transportation. Added to this is a reaction to increasing congestion, with the result being a growing demand for more of the cost of the transportation infrastructure to be paid by developers and, indirectly, new residents (Alterman, 1988). No-growth and controlled-growth strategies are now common responses to crowded highways and high tax bills.

Mass Transportation

Historically, mass transportation was most often a private service, provided either by a transportation company as an independent operation or in cooperation with a land development company whose land was valuable only with such transportation improvements (Jackson, 1985; Jones, 1985). Since the early 1960s, however, almost all mass transit systems have come under public ownership, with many implications for different service goals and employee relationships. Rather than generating profits or increasing the value of land, the primary goals of mass transit have become providing a means of transportation for those dependent on public transit and reducing congestion, especially in the urban cores (an additional emerging goal related to the journey to work is discussed in the next section).

Major issues and findings regarding equity and the distribution of mass transit services and expenditures are summarized in a paper by Hodge (1986). Most distributional studies have found at least a modest bias in favor of economically disadvantaged riders of the system, a finding best explained by the central location of racial minorities and low-income populations (McClafferty, 1982). More recently, Ircha and Gallagher (1985) found such a pattern for two Canadian cities, while Hodge (1988) found no systematic variation in benefits and subsidies among social groups (except for the regressive profile of nonfarebox revenues). He did, however, find substantial geographic variation, with nonriders in suburban jurisdictions generating large subsidies that were used to support central-city riders.

Although most of the funds for mass transit have come from riders and local governments, the federal government's contribution has been considerable, some 50 billion dollars since the Urban Mass Transportation Act of 1964 (Pucher, 1988a). The significance of federal subsidies, however, goes well beyond the actual dollar contributions; the money has come with "strings attached" that define critical elements of both operating conditions and preferred capital investments (Wachs, 1989). On the one hand, operating restrictions have been enormously important in requiring that resources be fairly distributed to minorities and other protected groups (Miller, 1977). On the other hand, federal involvement has defined labor conditions in a manner that has actually led to a significant increase in costs while, ironically, reducing productivity per worker (Wachs, 1989). Pickrell (1985) notes that most of the increased subsidies have gone to cover higher costs for roughly the same levels of service. In his econometric analysis, he finds that the dispersion of residences and employment into less dense suburban areas accounts for only 7.5% of increased deficits, whereas increases in labor costs accounted for about 60% of increased deficits. Such an observation leads Wachs (1989, p. 1546) to conclude, "When confronted with trends such as these, many argue that the benefits of the subsidies have accrued disproportionately to those who provide transit services rather than those who use it." Pucher (1988a) points out that subsidies are almost universally required (even in Europe), but they are highest in the United States. In general Pucher finds that productivity is lower among all countries with a significant role for central funding of transit finances.

The role of the federal government in influencing the provisions of public transit is especially pronounced in its support of capital financing, particularly for rail-based systems. The case against the rational choice of rail systems is by now overwhelming (Kain, 1988; Snow, 1986; Wachs, 1989). Yet support for rail systems persists. In part this support derives from the inherent allure of technological solutions to our problems (Eflin, 1987). Certainly, it is hard not to like the idea of it, "the rail line as art object, as status symbol, as urban artifact" (Dorschner, 1985, p. 10). Ideally, many believe, rail systems will somehow make it possible for us to shed our unfortunate dependence on the automobile and move to the future. As Snow (1986, p. 45) put it, "Smoky cars and concrete highways would give way to glass towers, monorails, and other monuments to the genius of modernity." And of course the decision was made much easier by the federal government picking up 85% of the tab. Why should not local jurisdictions pick up the rail option at 15 cents on the dollar for what, at the very least, is an effective public works project? "There is an incentive to construct new systems whose benefits fall far short of total costs yet exceed local costs" (Pucher, 1988a, p. 393). Unfortunately, downstream costs and significant local costs for rail systems (even when subsidized) usually result in the diversion of other transit funds and may have the perverse effects of reducing total transit usage (Kain, 1988) and/or making fewer resources for low-income and transit-dependent populations (Snow, 1986).

THE POLITICAL ECONOMY OF THE JOURNEY-TO-WORK

It is easy for the average citizen to view the deteriorating morning rush hour traffic as a transportation problem whose solutions lie in developing further transportation improvements. But, in reality, that congestion is merely symptomatic of a fundamental restructuring of North American cities that is not only reshaping the physical structure of cities but also many of the underlying social relations of society and thus the political economy of cities. The journey to work is singularly important in understanding these transformations for, as Feldman (1977, p. 31) perceptively notes, "Since the principal points of production and consumption are respectively the workplace and home, the transit linkage between the two provides an excellent vehicle for studying the relation between them." This theme has been voiced more recently by Hanson and Pratt (1988a, p. 299) who argue that "The link between home and work is one of the cornerstones of urban geography." Unfortunately, according to Hanson and Pratt, in the traditional study of urban geography we have tended to separate studies of economic geography (production relations) from studies of social geography (consumption relations). The realities of a restructuring metropolis as seen through the lenses of more comprehensive social theory informed by a materialist political economy are forcing a union of these interests, with the journey to work emerging as a key focal point. In this section I will first briefly review the "objective" changes occurring in the distribution of jobs and people in metropolitan America. Then I will discuss recent literature that focuses on the journey to work in general before turning to the specific issue of gender-based relations that revolve around the journey to work.

Urban Restructuring

The physical restructuring of urban areas is intimately tied to major social and economic trends that include major demographic and economic changes. The globalization of the economy, changes in family structure, and a transition to a service-oriented economy are three of the most important of these macro changes which impact local structure. At the risk of oversimplifying the impact of these changes, I will focus on one physical manifestation of these changes, viz., the changing nature of and relationships between the central city and suburbs of metropolitan areas. For a variety of reasons, academics and planners have been slow to appreciate the huge quantitative and qualitative changes that have completely outdated the formerly dominant image of a primate central city CBD surrounded by suburban bedroom communities. With the exception of some growth-related controversies, most urban transportation problems are still associated with an outdated view of the primacy of the central city and its CBD and the "uniqueness" of problems of congestion associated with the volume of commuters from suburban homes to CBD jobs. Yet no interpretation could be further from the truth.

Although most of his data are limited to changes between 1970 and 1980, Pisarski (1987) paints a particularly cogent and revealing portrait of a new suburban and commuting reality. It is clear from these data that many, if not most, suburbs are no longer bedroom communities (a point which Muller and Hartshorn have been trying to make for years), that the automobile is more dominant than ever, and that the suburb-to-central-city commute has declined dramatically in overall importance. Pickrell (1985) echoes this theme as he notes that work trips to central cities decreased by nearly 9% between 1970 and 1980 (in spite of the solid growth of office space in most CBDs), while work trips to suburbs grew by nearly 32%. Wachs (1989) even more emphatically states that the single most important change impacting urban transportation has been the suburbanization of employment, a phenomenon that has been termed America's third wave of suburbanization. In spite of increased congestion transit trips both to central cities and to suburbs declined by about one fourth (Pickrell, 1985). Increased reliance on the auto is in part a reflection of the relocation of both firms and residences to suburbs, which virtually necessitates travel by automobile.

The co-suburbanization of employment and residence at least initially permitted significant work-trip economies facilitated by shorter work trips in cross-suburb commuting (Gordon, Kumar, and Richardson, 1989). However, it would appear that these advantages may be short-lived, as Cervero (1989a), among others, notes that the average work-trip distance is steadily increasing. He argues that this is largely the result of land-use policies that separate employment from housing whose cost is consistent with wage levels offered. Also, the separation of land uses has had a major impact on the volume of nonwork travel (Cervero, 1989b). Gordon, Kumar, and Richardson (1988) find that nonwork morning peak trips in the United States increased by 42% compared to an overall growth of 16% in all urban trips. This leads them to conclude that, in spite of popular images of long lines of commuters, nonwork trips are a significant, if not the major factor in the growth of peak-hour congestion.

The Journey to Work

While the dominant public issue involving the journey to work revolves around congestion and mode choice (Pucher, 1988a, 1988b), ultimately the journey to work must be understood for its role in shaping fundamental relationships within urban areas. It is important to note here that I am not focusing on the findings that the simultaneous determination of workplace and residence in an econometric model yield better results (Simpson, 1987). Rather, I am discussing the more fundamental ways in which the home environment affects the work decision, the work environment affects home decisions (Hanson and Pratt, 1988), and urban transportation connects the two (Feldman, 1977). In one of the best papers discussing these implications, Feldman (1977, p. 30) notes the multiple and "more or less contradictory function [of urban transportation] as commodity, reproducer of labor power, social-control mechanism, and structurer of space." His basic argument is that "transit" (read urban transportation) reproduces the divisions within and between classes in society by mediating the relationship between status at the point of production (workplace) and status at the point of consumption (homeplace). Good transit can theoretically contribute to the separation of production and consumption interests (fragmenting class interests of labor) while simultaneously accommodating patterns of residential segregation that contribute to the reproduction of labor.

This home-work division provides a useful framework for further reviewing literature that addresses the impacts of urban transportation. I will look first at issues related to access by the worker from home to work and then at the issues regarding access to workers by employers.

The recognition that differential physical access to employment opportunities impacts the life chances of individuals is almost a truism in geography even though empirical research paints a more ambiguous picture. McClafferty (1982), in particular, finds that by almost all conventional accessibility measures, minorities and low-income populations are well positioned geographically because of their more central location. In his study of Chicago, Ellwood (1986) concludes that job accessibility matters only slightly for labor market outcomes for blacks. Their conclusions are challenged, however, by Aiken and Fik (1988), who find that access to an automobile is of overwhelming importance to behavioral potential; by Singell and Lillydahl (1986), who find that nonwhites have longer commutes when wage and home values are controlled for—thus suggesting that residential segregation is important in generating inequality of earnings; by Leonard (1987, p. 344), who found that "Residential segregation not only limits where blacks can live, but it also limits where they work"; and Herschberg, Burstein, Eriksen, Greenberg, and Yancey (1981), who persuasively argue that blacks in Philadelphia have been long-term historical victims of Kain's (1968) spatial mismatch hypothesis—having been in the right place at the wrong time or the wrong place at the right time.

It is especially noteworthy that the economic and spatial restructuring of metropolitan areas has made the other side of the mismatch theory more important. While most studies have focused on the commuter (worker) and his or her access to jobs, less attention has been given to the employers' access to labor. To be sure, regional variations in labor wages and qualities have long been considered important in economic studies, but intraurban variations in labor markets have received much less attention. Scott (1988), however, argues that commuting cost differentials intersect with the effects of the employment relation as a whole to produce marginal variations in wage rates which are nonetheless significant for intraurban plant location. Much of the literature pertaining to female labor markets (see below) focuses on the increasing recognition of intraurban spatial variation in female labor markets (Nelson, 1986). However, given the growing dualism of labor needs, declining numbers of young employees, less opportunity for expansion of female labor-market participation, and the increased suburbanization of employment in general, there is a growing mismatch between jobs and workers that threatens many new services in industries (Cervero, 1989b). In Seattle, for example, large hotels in the eastern suburb of Bellevue are forced to sponsor vans to South Seattle, where there are significant concentrations of new Asian immigrants, while high-tech assembly plants in other eastern suburbs go begging for labor even though they offer two dollars per hour over industry standard wages. In an even more unusual example, a national fast-food chain interviewed and hired a small group of homeless men who are bussed early each morning from their "temporary" shelter near downtown Seattle to an eastside suburb where they cook the breakfast shift. In support of Scott's argument, a recent survey in Seattle also found that employers in the Seattle CBD are forced to pay a premium to entice clerical workers to make the longer commute to downtown. In short, the issue of job access, which so long was framed from the point of view of an individual's access to opportunity, is now being recast from the point of view of employers desperate for unskilled and/or low-paid labor who are not easily accessible.

Gender and the Journey to Work

Gender issues have emerged as a critical dimension in a modern assessment of the significance of urban transportation.[1] In part the interest stems from equity issues related to fair access for female workers, in part from a business response to spatial variation in labor pools, and in part from a better understanding of the significance of macro social, economic, and demographic changes in society. It is well established that there are substantial differences between travel patterns of men and of women (Wachs, 1987). In general, most studies have found that women work closer to home, travel less time to reach their jobs, and more often use public transit (Rosenbloom, 1989). From the point of view of political economy, it is especially important to consider why those differences occur and what the implications are for women, for employers, and for urban planning and economic development.

The basic issues revolve around the extent to which women's travel behavior is structured by relations at the point of consumption and reproduction (the household), by relations at the point of production (the workplace), and/or by the geographic structure of urban places. The theoretical and empirical frameworks for these questions are rich and growing and cannot be summarized thoroughly in this review. Rather I will outline some of the key elements and point to representative work for each of these sets of relations.

There is common agreement that household responsibilities are a significant factor structuring female commuting behavior (Madden, 1981; Rosenbloom, 1989), although this conclusion has been challenged by Hanson and Johnston (1985) in their study of Baltimore. Most studies focus on the direct impact of household responsibilities, such as child care and shopping, which constrain options for a mother. Fox (1983) finds that employed married women with children have much less discretionary time than their male or unmarried female counterparts. While such explanations may account for the behavioral basis for such differences, they do not, of course, deal with the more fundamental question of how patriarchal relationships restricting women are established to begin with (Wachs, 1987). A related explanation concerns the structuring effects of residential location choices. Evidence suggests that because of the secondary status of women in the typical household, residential location decisions are made with respect to the male's work location, with the female job search then constrained by residential location (Fox, 1986; Singell and Lillydahl, 1986).

The size and persistence of income disparity for female workers has led most observers to conclude that a dual labor market exists in which females are "crowded" into occupations that are relatively low-paying, yet frequently require disproportionately high skills (Nelson, 1986). As a result, there is little incentive for longer commutes when incomes for women are relatively inelastic with respect to distance—at least as compared with men (Rutherford and Wekerle, 1988). Thus the position of women in the labor market is a major factor structuring the travel behavior of female workers.

The issue of the role of geography in all this rests on a rather complex set of relationships between the degree to which females are spatially constrained in their ability to commute, the extent to which residential patterns give rise to intraurban spatial divisions of labor, and the extent to which employers are able to respond in their location choices to exploit this labor force (Hanson and Pratt, 1988b; Nelson, 1986; Rutherford and Wekerle, 1988). The relative importance of these factors is not easy to determine since in the restructuring metropolis both jobs and population are increasing in suburbs and declining in central cities. Evidence for Scott's (1988) claim of the presence of marginal but significant intraurban variations in wage rates suggests that women working in suburbs, especially in female-dominated jobs in suburbs, have lower income or advancement potential than their central-city counterparts (Kent, 1988; Nelson, 1986; Villeneuve and Rose, 1988). On the other hand Hanson and Pratt (1988b) conclude that processes other than constraints on women's work trips could also be leading to the clustering of women's workplaces in certain parts of the city.

The most frequent response to spatial constraint is to deal with those features of women's lives which appear most directly to affect their abilities to expand commuting behavior, viz., household responsibilities and access to better transportation conditions, especially for transit-dependent female workers (Fox, 1986; Rutherford and Wekerle, 1988). While there is an obvious appeal to providing better transportation (and such features as improved daycare options) as a means to improve the lives of females, there is a real danger of such initiatives leading to perverse results. In response to Fox's (1986) list of recommendations for improving female commuting speeds, Barbara Sanford (1986, p. 18) noted that "we must ask ourselves if the quality of our working life has actually improved . . . or have we simply made our own exploitation less painful?" Similarly, Rutherford and Wekerle (1988) and Feldman (1977) both note that improving transportation options for women (which by implication increases the potential labor pool for employers) may simply lead to a lowering of wage rates. It is noteworthy that years of pushing for improvements in transportation, equal pay, and daycare options as a means to improving opportunities for women as individuals met limited success until a tightening labor market sent employers scrambling for options to enlist and retain qualified labor. As noted above, Kain's mismatch hypothesis has essentially been stood on its head and the political emphasis shifted from the needs of the employee to the needs of the employer.

URBAN TRANSPORTATION AND CAPITAL ACCUMULATION

If the *raison d'être* of cities is to provide the opportunity for interaction, few aspects of urban infrastructure match the importance of the provision of urban transportation. While the focus of transportation has been on its importance as an enabling (or determining) technology that has altered spatial form, it may prove more useful to consider urban transportation in a more contentious way as the major structural agent shaping ground rent (Feldman, 1977) and giving rise to patterns of relative (dis)advantage (Adler, 1987, 1988) for investment and capital accumulation. In this section I will briefly highlight this critical view of urban transportation as well as a second role that transportation plays in capital accumulation as an industry in its own right.

There is little disagreement about the importance of transportation to the growth and functioning of cities. There is even agreement that investments in transportation open new outlets for the investment of capital in the physical infrastructure, changing the distribution of ground rent, "creating" and coordinating space (Feldman, 1977; Gottdiener, 1985). However, there is much less agreement as to the true nature of the role played by transportation investment. Adler (1988), for example, contrasts transportation investment as either accomodationist, a somewhat benign response to other forces structuring urban areas, or developmental, leading land development (and by implication capital accumulation). Adler concludes that U.S. transportation policy has been expansionist and developmental, a position which has considerable support (Jackson, 1985; Jones, 1985). Moreover, Adler argues that most attention to urban transportation policies has stressed issues related to modal conflict and in so doing obscures "the fundamental dimension of urban transport politics, which is place conflict—*competition between places to maintain and attract capital investment*" (1988, p. 267—emphasis added). It should be added that not only is there a competition between places but also a competition between different factions of capital (Badcock, 1984; Gottdiener, 1985) and, because of social geography, different social groups.

According to Adler (1987), competition between places is a basic feature of metropolitan development processes which leads to place-based coalitions that are particularly concerned with transportation because (1) transportation facilities create locational advantage and (2) transportation facilities offer greater leverage over the location decisions of other investors. In an especially insightful comparison of San Francisco and Los Angeles, Adler (1987) argues that Los Angeles failed to get started on a rail system in the late 1940s because there were so many competing centers where the plans were viewed as an attempt to promote downtown Los Angeles. In contrast, Oakland feared competition for a more southerly crossing of the Bay and thus joined forces with San Francisco; this not only promoted the interests of San Francisco but also secured a prominent developmental position for the Oakland area. In another interesting case study Cohen (1988) investigated the deterioration of the quality of mass transit service in New York City. Contrary to popular images, Cohen argues that the deterioration did not occur because of either a preference in funding for the automobile or a lack of money because of the fiscal crisis of New York City in the mid-1970s. Rather it was the result of a deliberate policy to spend money on new construction rather than maintaining or upgrading existing capital stock. One of the key reasons for such policy was a response to the political demands for more services in developing areas. Also, such policies promoted the interests of key business and financial institutions in the region, leading Cohen to agree with Whitt and Yago's (1985, p. 59) proposition that rail construction projects are primarily "political tools for urban development."

Finally, while the emphasis above has been on urban transportation investments as a factor influencing other capital investments, it is also important to consider investment in urban transportation as a vehicle for accumulation in its own right. Certainly this was easier to do when transit companies were privately held, although the close liaisons with real estate developers, especially during the streetcar era, and the ongoing control over rates muddles the record even in that period (Jackson, 1985; Jones, 1985). The most famous chapter of the private transit industry, the alleged collusion of General Motors, Firestone, Phillips Petroleum, and Standard Oil to substitute buses for streetcars, has been carefully examined by Yago (1980) and Whitt and Yago (1985) and popularized on television's *60 Minutes* and in the movie *Who Framed Roger Rabbit?* While there seems little doubt that collusion existed and that it hastened the demise of rail transit in many cities, it is difficult to believe that the collusion amounted to much more than a relatively minor factor in the ultimate hegemony of the automobile in most cities.

CONCLUSION

In a world in which the highly visible and usually painful reality of urban congestion dominates, it is easy to lose sight of the fact that embedded in urban transportation systems are relations that are intimately and inescapably tied to the most fundamental, and often contentious, elements of cities and societies. Thus the primary focus in the political arena, in almost a sleight of hand, rests on conflicts between mode choices, carpool lanes, new highway construction, and other superficial manifestations of more basic urban processes. Yet urban transportation is important to defining which social groups, which factions of capital, which geographic areas in cities are to gain and which are to lose. It has not been the intent of this paper to answer such important questions as whether improved transportation leads to the reproduction and further exploitation of workers, or whether uneven economic development is an inevitable and appropriate process. Rather, it has been the goal of this paper to place these questions squarely in the research framework of urban transportation analysis, to ensure that the deeper relations between urban transportation and the structure of society are at least recognized, if not analyzed. As noted at the outset, the enormous spasms of restructuring impacting North American cities, combined with the infusion of critical social theory, provide a rich platform from which to understand better the political economy of urban transportation which, in many ways, is a key to unlocking the nature of even more basic urban relations.

NOTE

[1]For an excellent introduction to these issues the reader is directed to the second issue of Volume 9 (1988) of *Urban Geography*, edited by Susan Hanson, which is a special issue devoted to women and employment focusing in general on the journey to work.

LITERATURE CITED

Adler, Sy, 1987, Why BART and no LART? The political economy of rail rapid transit planning the Los Angeles and San Francisco metropolitan areas, 1945-57. *Planning Perspectives*, Vol. 2, 149–174.

_______, 1988, A comparative analysis of rail transit politics, policy and planning in Canada and the United States. *Logistics and Transportation Review*, Vol. 24, 265–279.

Aiken, Stuart, and Fik, Timothy, 1988, The daily journey to work and choice of residence. *The Social Science Journal*, Vol. 25, 463–475.

Alterman, Rachelle, editor, 1988, *Private Supply of Public Services: Evaluation of Real Estate Exactions, Linkage, and Alternative Land Policies*. New York: New York University Press.

Badcock, Blair, 1984, *Unfairly Structured Cities*. Oxford: Basil Blackwell.

Cervero, Robert, 1989a, *America's Suburban Centers: The Land Use-Transportation Link*. London: Unwin-Hyman.

_______, 1989b, Jobs-Housing balancing and regional mobility. *American Planning Association Journal*, Vol. 55, 136–150.

Cohen, James, 1988, Capital Investment and the Decline of Mass Transit in New York City, 1945-1981. *Urban Affairs Quarterly*, Vol. 23, 369–388.

Dorschner, J., 1985 (September 15), Metrofail? The Miami Herald *Tropic*, 10ff.

Eflin, James, 1987, *Technology and Social Power: Social Action, Intentional Technology, and the Social Basis of Space-Time Autonomy*. Unpublished doctoral dissertation, University of Washington, Seattle.

Ellwood, David, 1986, The spatial mismatch hypothesis: Are there teenage jobs missing in the ghetto? In R. Freeman and H. Holzer, editors, *The Black Youth Employment Crisis*. Chicago: University of Chicago Press, 149–190.

Feldman, Marshall, 1977, A contribution to the critique of urban political economy: the journey to work. *Antipode*, Vol. 9, 30–50.

Fox, Marion, 1983, Working women and urban travel. *American Planning Association Journal*, Vol. 49, 156–170.

_______, 1986, Faster journeys to suburban jobs. *Women and Environments*, Vol. 8, 15–18.

Gordon, Peter, Kumar, Ajay, and Richardson, Harry, 1988, Beyond the Journey to Work. *Transportation Research A*, Vol. 22A, 419–426.

_______, 1989, Congestion, Changing metropolitan Structure, and City Size in the United States. *International Regional Science Review*, Vol. 12, 45–56.

Gottdiener, M., 1985, *The Social Production of Urban Space*. Austin: University of Texas Press.

Hanson, S. and Johnston, I., 1985, Gender differences in work-trip length: Explanations and implications. *Urban Geography,* Vol. 6, 193–219.

_______ and Pratt, G., 1988a, Reconceptualizing the links between home and work in urban geography. *Economic Geography,* Vol. 64, 1988, 299–321.

_______, 1988b, Spatial dimensions of the gender division of labor in a local labor market. *Urban Geography,* Vol. 9, 180–202.

Herschberg, Theodore, Burstein, Alan, Eriksen, Eugene, Greenberg, Stephanie, and Yancey, William, 1981, A Tale of Three Cities: Blacks, Immigrants, and Opportunities in Philadelphia, 1850-1880, 1930, 1970. In T. Herschberg, editor, *Philadelphia: Work, Space, Family, and Group Experience in the Nineteenth Century.* New York: Oxford University Press.

Hodge, D., 1986, Social Impacts of Urban Transportation Decisions: Equity Issues. In S. Hanson, editor, *The Geography of Urban Transportation,* 301–327.

_______, 1988, Fiscal Equity in Urban Mass Transit Systems: A Geographic Analysis. *Annals of the Association of American Geographers,* Vol. 78, 288–306.

Ircha, M. and Gallagher, M., 1985, Urban Transit: Equity Aspects. *Journal of Urban Planning and Development,* Vol. 3, 1–9.

Jackson, K., 1985, *Crabgrass Frontier: The Suburbanization of the United States.* New York: Oxford University Press.

Jones, D., 1985, *Urban Transit Policy: An Economic and Political History.* Englewood Cliffs, NJ: Prentice-Hall.

Kain, John, 1968, Housing Segregation, Negro Employment, and Metropolitan Decentralization. *Quarterly Journal of Economics,* Vol. 82, 32–59.

_______, 1988, Choosing the Wrong Technology: Or How to Spend Billions and Reduce the Use of Transit. *Journal of Advanced Transportation,* Vol. 21, 197–213.

Kent, Brooke, 1988, *Central City-Suburban Variation in Female and Male Earnings in the United States.* Unpublished master's thesis. University of Washington, Seattle.

Leonard, J. S., 1987, The interaction of residential segregation and employment discrimination, *Journal of Urban Economics,* Vol. 21, 323–346.

Madden, J., 1981, Why women work closer to home. *Urban Studies,* Vol. 18, 181–194.

McClafferty, Sarah, 1982, Urban Structure and Geographical Access to Public Services, *Annals of the Association of American Geographers,* Vol. 72, 347–354.

Miller, D. R., 1977, *Equity of Transit Service,* No. DOT-UT-50029, Washington, DC: Urban Mass Transportation Administration.

Muller, P., 1986, Transportation and urban growth: the shaping of the American metropolis. *Focus,* Vol. 36, 8–17.

Nelson, Kirsten, 1986, Labor Demand, Labor Supply, and the Suburbanization of Low-Wage Office Work. In Allen Scott and Michael Storper, editors, *Production, Work, Territory: The Geographical Anatomy of Industrial Capitalism.* Boston: Allen and Unwin, 149–171.

Pickrell, D., 1985, Rising Deficits and the Uses of Transit Subsidies in the United States. *Journal of Transport Economics and Policy,* Vol. 19, 281–298.

Pisarski, A., 1987, *Commuting in America: a National Report on Commuting Patterns and Trends.* Westport, CT: Eno Foundation.

Pucher, J., 1988a, Urban Public Transport Subsidies in Western Europe and North America. *Transportation Quarterly,* Vol. 42, 377–402.

________, 1988b, Urban Travel Behavior as the Outcome of Public Policy: The Example of Modal-Split in Western Europe and North America. *American Planning Association Journal,* Vol. 54, 509–520.

Rosenbloom, Sandra, 1989, Differences by sex in the home-to-work travel patterns of married parents in two major metropolitan areas. *Espace Populations Societes,* Vol. 18, 65–75.

Rutherford, B. and Wekerle, G., 1988, Captive rider, captive labor: spatial constraints and women's employment. *Urban Geography,* Vol. 9, 116–137.

Sanford, Barbara, 1986, Reforms are Good, but . . . *Women and Environments,* Vol. 8, 18.

Scott, Allen, 1988, *Metropolis: From the Division of Labor to Urban Form.* Berkeley: University of California Press.

Simpson, Wayne, 1987, Workplace location, residential location, and urban commuting. *Urban Studies,* Vol. 24, 119–128.

Singell, L. and Lillydahl, H., 1986, An empirical analysis of the commute to work patterns of males and females in two-earner households. *Urban Studies,* Vol. 23, 119–129.

Snow, T., 1986, The Great Train Robbery. *Policy Review,* Vol. 10, 44–49.

Staeheli, Lynn, 1989, Accumulation, Legitimation, and the Provision of Public Services in the American Metropolis. *Urban Geography,* Vol. 10, 229–250.

Villeneuve, P. and Rose, D., 1988, Gender and the separation of employment from home in metropolitan Montreal, 1971–1981. *Urban Geography,* Vol. 9, 155–179.

Wachs, M., 1987, Men, women, and wheels: The historical of sex differences in travel patterns. *Transportation Research Record,* No. 1135, 10–16.

________, 1989, U.S. Transit Subsidy Policy: In Need of Reform. *Science,* Vol. 244, 1545–1549.

Whitt, J. Allen, and Yago, Glen, 1985, Corporate Strategies and the Decline of Transit in U.S. Cities. *Urban Affairs Quarterly,* Vol. 21, 37–65.

Yago, Glen, 1980, Corporate power and urban transportation: a comparison of public transit's decline in the United States and Germany. In Maurice Zeitlin, editor, *Classes, Class Conflict and the State: Empirical Studies in Class Analysis.* Cambridge: Winthrop, 296–323.

[11]

Identifying Winners and Losers in Transportation

David Levinson

The issues surrounding transportation equity, both external and internal to transportation, are explored. Several examples are provided of transportation improvements that impose transportation costs on more individuals than those who are benefited. Beyond counting the number of winners and losers, several quantitative measures of equity are suggested and applied to a test case: ramp meters in the Twin Cities, Minneapolis–St. Paul, in Minnesota. It is recommended that transportation benefit-cost analyses include an "equity impact statement," which would consider the distribution of opportunities to participate in decisions and the outcomes of those decisions (in terms of mobility, economic, environmental, and health effects) that different strata (spatial, temporal, modal, generational, gender, racial, cultural, and income) of the population receive. Policy makers would then have additional information on which to base decisions.

Social welfare includes both efficiency and equity. Transportation engineers are taught to provide for the safe and efficient movement of people and goods. They are not taught to ensure that transportation systems are equitable, in part because of the ambiguity associated with equity. Transportation textbooks seldom broach the subject, which is considered political rather than technical. [Hanson's text (*1*) is a notable exception.] In economics, the words "equity" and "fairness" do not appear in the index of Varian's standard textbook, *Microeconomic Analysis* (*2*). Still, economics does not completely ignore the topic; Baumol devoted a book to *Superfairness* (*3*).

Public-sector investment decisions are made in nonmarket forums that often suffer from a short-term viewpoint and the dominance of selected individuals (*4*). For objectivity, public-sector investments generally rely on benefit-cost (B/C) analysis to compare various proposals. The use of B/C analysis as a decision-making tool in public choices results in the separate consideration of equity and efficiency. Usually the efficiency criterion employed by decision makers for a project overrides concern for equity. A situation is considered *pareto efficient* (or *pareto optimal*) if there is no way to make all agents better off, that is, if one cannot improve person Y without worsening person Z. However, there are two problems with pareto efficiency. First, some things, such as time, are not fungible, making exchange difficult (someone cannot give you 10 minutes). Second, the exchange does not actually occur. Therefore, although the pareto criterion is important from an efficiency point of view, it is unhelpful in trying to understand equity.

Although transportation projects are formally planned on the basis of efficiency, equity criteria may affect the project. However, the concept of equity is highly subjective and changes with the individual concerned. A project that may appear equitable to the decision maker may not appear so to an individual affected by the project. This situation makes it difficult to achieve objective decisions with respect to considerations of equity, but neither does it mean that equity can be ignored.

It is therefore necessary to use an approach that considers both efficiency and equity in the evaluation of public-sector investments. The use of both quantitative and qualitative approaches to equity would help give importance to those factors that are not included in B/C analysis.

Questions of equity should be of concern to transportation analysts facing political issues whose resolution depends on perceptions of fairness. The most important problem in the analysis of equity has been the question of its definition. The term "equity" has both a descriptive (positive) and a normative use, describing the distribution of benefits and whether the distribution is for better or worse (*5*). *Horizontal equity* refers to the equivalent or impartial treatment of individuals with regard to the allocation of the benefits and costs among individuals and groups who are similar in terms of wealth and ability (*6, 7*). *Vertical equity,* however, refers to the distribution of benefits and costs among different income groups or other strata such as physical disability.

From the normative point of view, there are two additional concepts. Equality of opportunity, or *process equity,* is concerned with equal access to the planning and decision-making process. In contrast, equality of outcome, or *result equity,* examines the consequences of the product. The U.S. Constitution enshrines the first, whereas the Declaration of Independence only posits the right to pursue happiness, not happiness itself. In contrast to the utilitarian aim to maximize total welfare, the egalitarian view would maximize the welfare (or opportunities) of the least advantaged member of society, and thus move society toward greater equity, as championed by the environmental justice movement (*8, 9*). Compared with the wealthy, the poor spend a larger portion of their income on transportation (as well as a variety of other goods). Furthermore, the poor and disadvantaged have historically borne the burden of transportation investments and improvements, which are often sited in their neighborhoods (*10*, pp. 190–234).

The recent slow pace of additions to the transportation network, as illustrated in Figure 1, has in part been blamed on increasing opposition to new construction. The neighboring opponents, often termed NIMBYs, for Not In My Back Yard, are accused of selfishness, people who would place their own needs above those of the rest of the community. An addition to the transportation network that moves people from A to B often must cross C to get there. If the project harms or disadvantages the people living near C, and if they are uncompensated, it is perfectly reasonable for them to oppose expansion. Their opposition is sometimes claimed to be because of the negative externalities of transportation (noise, air pollution, etc.), which are increased in the locality of the project. However, the opposition may also reflect a sense that the project will make them worse off in mobility terms. It is seldom acknowledged (though

Department of Civil Engineering, University of Minnesota, 500 Pillsbury Drive SE, Minneapolis, MN 55455.

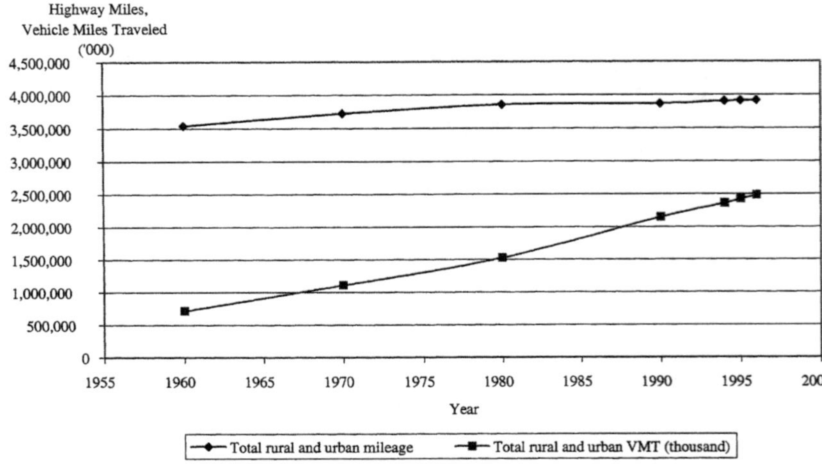

FIGURE 1 Growth in network and traffic, 1955–1997.

occasionally claimed to be obvious) that many transportation projects and policies increase the travel times of some travelers in order to decrease the times of others. Those travelers who lose from the improvement may also be automobile commuters making local trips, or they may be pedestrians now forced to use an overpass or tunnel, or transit users placed on a more circuitous route.

Any new transportation project or policy creates both winners and losers from the standpoints of mobility, accessibility, and environmental and economic concerns. In some cases, an improvement does not even make society better off as a whole; the gains of the winners do not exceed the losses of the losers. Whether society gains overall depends on both the shape of the network and the elasticity of demand. Thus there may be a great deal of conflict around new construction or policy changes, in which the losers attempt to use the political process to stop projects that may have an overall net benefit to society. In this paper, methods for measuring gains and losses are developed. It is hoped that this information can be used to create situations in which the losers are compensated rather than excluded from the process.

The issues surrounding transportation equity are explored. The external effects are investigated first. Next some examples are presented to illustrate how transportation improvements may create winners and losers. These results are due to the coupling inherent in transportation networks. Changes to capacity on one link change the patterns of demand (and the implicit supply) on others. Quantitative measures of equity are suggested and applied to a case study of ramp meters in the Twin Cities of Minneapolis–St. Paul, in Minnesota. These measures can be used in an equity impact analysis to determine explicitly the distribution of gains and losses across subpopulations.

EXTERNAL INEQUITIES

Costs borne directly by consumers are considered internal costs. The costs of a transaction that are borne by third parties are termed external costs. The definitions distinguish harm committed between strangers, which is an external cost, and harm committed between parties to an economic transaction, which is an internal cost. When external costs are estimated, one can use either the estimated amount of economic damage (including both market and nonmarket costs) produced by the externality or the cost of preventing that damage in the first place. Rational economic actors would choose the lower of prevention costs or damage costs when costs are internalized. If there were a cheaper prevention measure, it could be used, but if prevention were more expensive, the actors would accept damages.

Levinson and Gillen (*11*) proposed a full-cost model including user costs, infrastructure costs, and external costs, without double-counting transfers (e.g., user taxes for infrastructure or automobile insurance). Estimates of that model for a specific set of assumptions for intercity highway travel in California are given in Table 1. (The costs, particularly for travel time, shown in Table 1 can be expected to be lower than those found in urban regions.) Although internal costs are much larger than external costs, the external costs are significant and measurable. These costs are only a portion of total external costs, since there are some costs that cannot be easily monetized. *Social severance* is the cost of dividing communities with infrastructure. It is a real externality, but its valuation is difficult.

TABLE 1 Long-Run Average Costs by Category (*11*)

Cost Category	Long Run Average Cost ($/vkt)
User	0.13
Infrastructure	0.0174
Free Flow Time	0.15
Congestion	0.0045
Accidents	0.031
Noise	0.006
Air Pollution	0.0056
Total	0.34

Another important cost is *ecosystem severance,* the environmental cost from placing a highway amidst native ecologies. Costs such as the defense of the Persian Gulf, parking, or urban sprawl have been suggested by other researchers, but whether these costs should be counted is subject to significant disagreement.

From the perspective of investigating winners and losers, clearly the recipients of those external costs are losers; they need to be either persuaded (or required) to take the loss for the good of society or compensated through some type of side payment or bargain in order to obtain their acquiescence to new infrastructure that would impose such heavy costs. In principle those external costs could be offset by external benefits from transportation, but no good quantification of those external benefits has yet been made (*12*).

MOBILITY INEQUITIES

Generally, in the case of fixed (inelastic) demand, the addition or expansion of a link will make the travelers on that link better off and nobody else worse off. However, even for that situation there is a famous counterexample. Braess's paradox (*13*) shows that in a simple network serving one origin and one destination, an additional link can worsen overall welfare if travelers behave according to the Wardrop user equilibrium principle (i.e., they travel on the shortest-time path, given that everyone else is doing likewise). Since research has shown that demand is induced by capacity expansions (*14, 15*), the assumption of inelastic demand is clearly unreasonable.

Here several additional examples are developed of how transportation improvements create both winners and losers. These cases are not a complete set but simply are intended to illustrate some of the issues involved and to suggest where the resentment toward new transportation projects may come from.

In dealing with multiple markets, the issue of coupled supply and demand arises. In transportation, markets can generally be thought of as origin-destination pairs. A *coupled market* is one in which a change in the supply or demand in one market changes the supply or demand curve for another market. In coupled markets, a careful examination is required before conclusions about net and specific benefits can be deduced. Coupling implies a nonzero cross-elasticity of demand (or price) with respect to a change in price (or demand) in another market.

Example 1: Y-Network

To begin, a very simple network, shown in Figure 2*a*, is assumed. An improvement to link J–D_1 may have several effects, depending on the assumptions that are made about demand elasticity. In all cases it is assumed that the travel time functions on links J–D_1 and J–D_2 are an increasing function of demand.

1. Inelastic trip production and attractions: If the number of trips produced at O is fixed and the number of trips attracted to D_1 and D_2 is also fixed, nothing happens except that travelers from O to D_1 see an improvement in travel time. An example of this effect in the short term is when O represents work trips originating at home and D_1 and D_2 represent job sites.
2. Inelastic trip production and elastic trip attractions: If trip origins remain fixed but there is no constraint on destinations, some or all travelers from O to D_2 will switch to O to D_1. An example of this effect is when O represents shopping trips originating at home and

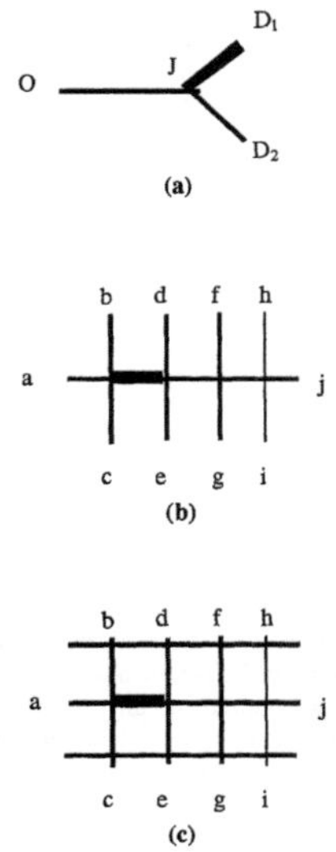

FIGURE 2 Example networks.

D_1 and D_2 represent otherwise identical supermarkets, and there is no capacity constraint at the supermarkets. O–D_1 has a relative accessibility improvement over O–D_2, though in fact both improve in absolute terms compared with before the improvement.

3. Elastic trip productions and attractions: If it is assumed that demand is sensitive to supply for both origins and destinations, the improvement will increase demand from O to D_1 as the price for those trips declines. This effect will inevitably increase flows on the unimproved link O–J. This additional flow will worsen travel times (and reduce demand) from O to D_2.

Example 2: Network Bridge

The degree of substitution and complementarity will determine to what extent losers are created by a change relative to the number of winners. In the Example 2 network (Figure 2*b*), with $n = 10$ origins and destinations, there are $n(n - 1) = 90$ markets (ignoring internal trips). Assuming fixed demand, improving the bridge link (heavy dark line) will improve all of the trips using the bridge link (42) and leave the other 48 markets unchanged. Assuming elastic demand, the 42 improved markets will see growth in traffic, which will worsen travel times in the other 48 markets (as seen in Table 2) because of downstream congestion. Although there may very well be a net benefit, the number of beneficiaries is potentially less than the number of losers.

Example 3: Grid Network

In contrast to the bottleneck bridge, expanding a link on a grid will have a larger share of winners, as illustrated in Figure 2*c*. Revising Example 2 and assuming elastic demand, each of the forty-two

TABLE 2 O-D Pairs for Example 2

	a	b	c	d	e	f	g	h	i	j
a	x	0	0	1	1	1	1	1	1	1
b	0	x	0	1	1	1	1	1	1	1
c	0	0	x	1	1	1	1	1	1	1
d	1	1	1	x	0	0	0	0	0	0
e	1	1	1	0	x	0	0	0	0	0
f	1	1	1	0	0	x	0	0	0	0
g	1	1	1	0	0	0	x	0	0	0
h	1	1	1	0	0	0	0	x	0	0
i	1	1	1	0	0	0	0	0	x	0
j	1	1	1	0	0	0	0	0	0	x

1 indicates improved, 0 otherwise, x unaffected, 42 improved OD pairs

directly improved markets will still see growth in traffic, though others may as well. In this example, traffic is now able to reroute, so other markets, which do not use the improved link, may benefit as well. For instance, trips from *d* to *f* and from *d* to *h*, which in Example 2 mixed with traffic from *b* to *j* and *b* to *g* and *b* to *i*, may benefit because some fraction of the *b*-originating traffic shifts southward to take advantage of the improvement. In this particular example, the number of beneficiaries now exceeds the number of losers.

MEASURING EQUITY

Beyond a simple count of the number of winners and losers, richer, more systematic measures of equity can be developed. The Gini coefficient of concentration and the Lorenz curve have historically been used to measure the equality in the distribution of a good. These tools have been used in economics to analyze income inequality. The distribution of the total income is represented by the Lorenz curve, and the statistical analysis is done using the Gini coefficient, which is hence a measure of the degree of inequality.

The Lorenz curve is the line separating the A_1 and A_2 areas in Figure 3. The Lorenz curve relates the proportion of the population receiving a given proportion of income. Whereas the bottom 100% of the population gets 100% of the income by definition, the bottom 50% may only get 30% of the income. The Gini coefficient is the ratio $A_1/(A_1 + A_2)$. A Gini coefficient of 0.0 indicates perfect equality, and 1.0 indicates perfect inequality (the richest person gets all). For the income in the United States, according to 1994 census results, the Gini coefficient was 0.456.

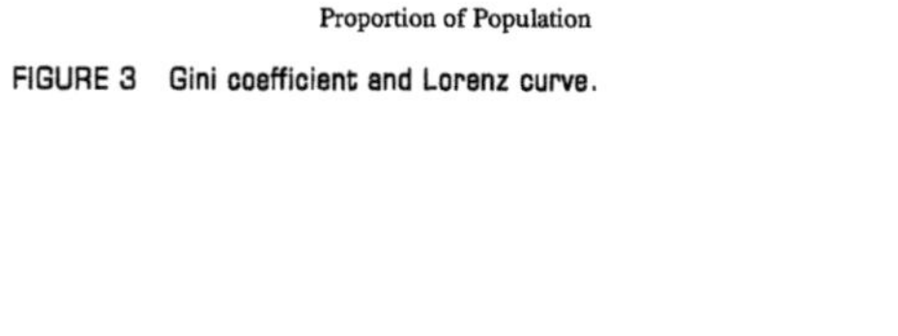

FIGURE 3 Gini coefficient and Lorenz curve.

The Lorenz curve concept can be used to measure other kinds of equity, not just income, by changing what is shown on the *Y*-axis. For instance, rather than income, it could be frequency of bus service on the nearest route or delay at ramp meters (though here the best outcomes are at the left of the *X*-axis rather than to the right). In general, any type of quantifiable impact, market or otherwise, can be assessed with this type of measure. Although it may not be efficient to aim for a Gini coefficient of zero, the change in the coefficient before and after a project gives some sense of whether the project is equity enhancing or equity subtracting.

The *entropy statistic* (*H*) focuses on the evenness of the distribution. An alternative is said to be equitable the closer it is to an equal distribution of net gains:

$$H = -\sum_j (y_j \cdot \log_k y_j)$$

where

H = entropy statistic,
y_j = proportion of average net gains to *j*th class, and
k = log base.

To analyze traffic data, y_i equals the proportion of total delay accrued by each individual [or on each link or for each origin-destination (OD) pair, etc.].

The computed *H*-statistic approaches zero as the distribution approaches complete inequality. The computed *H*-statistic gets larger as the distribution moves toward equality. For historical reasons arising from its emergence in information theory, log base 2 has been the most widely used in the computation of the entropy statistic. Another measure that has been used is the redundancy statistic, *R*, which calculates the bias in the distribution as opposed to the index of equality measured by the entropy statistic, *H*.

An *R*-value of 0% represents complete equality in distribution, or lack of bias. An *R*-value of 100% represents complete inequality, or bias in favor of one class. The equality in the average marginal utility of net change in real income is the third measure of equity:

$$R = 1 - \frac{H}{H_{max}}$$

where

R = measure of redundancy,
H = calculated entropy, and
H_{max} = maximum possible entropy.

RAMP METERS

In fall 2000, the ramp meters in the Twin Cities were turned off for 8 weeks so that an assessment of their effectiveness could be made. Although this assessment focused on the efficiency of the system, considering mobility and safety particularly, a transportation equity analysis of the delay distribution across space could also be made. In this section the relationship is estimated between mobility and equity for O-D pairs on Route 169, a suburb-to-suburb limited-access highway connecting the north and south legs of the region's beltway, with and without ramp meters. In order to make the results comparable, the data used for the analyses (ramp metering on and off) were collected on Tuesday, March 21 and November 7. November 7 is the third Tuesday after the ramp meters were shut down. The calculation methodology is described more fully elsewhere (*16*). It is assumed that an equilibrium was established and that traffic patterns were stable after 3 weeks.

In this discussion the equity and mobility for specific trips are considered. The mobility measures include total travel time, speed, and delay. All the measures are computed for each 5-min interval for each OD pair. Since the number of trips on each OD pair is not available, all the averaging is done without weighting by number of trips between origin and destination. The analysis is performed for 105 OD pairs.

What ramp meters bring in terms of mobility and equity can be shown by the comparison of the two cases. Previous research indicates that ramp meters can increase the mobility of freeway networks, which is confirmed by the findings here. With ramp metering, the average travel speed (taking ramp delay into account) of the highway increases from 60 mph to 100 mph (37 km/h to 62 km/h); travel delay per mile decreases from 136 s to 112.5 s and the average travel time for one trip decreases from 610 s to 330 s.

No previous results can be relied on to guide this analysis of equity. In contrast with the consideration of freeway segments in the previous section, when one looks at trips, one finds a drop in the Gini coefficient in the absence of metering. This result suggests that the system becomes more fair when meters are removed. This drop is observed for three primary measures: travel time per mile, travel speed, and travel delay per mile.

Figure 4 illustrates the trends in the change of mobility and equity with and without metering for trips. It should be noted that in Figure 4, the shortest trips (those on the right side of the graph) actually are hurt in mobility terms by ramp metering, whereas the longest trips (those on the left side) benefit the most.

EQUITY IMPACT STATEMENT

Environmental justice is a good beginning, but it only considers "fair treatment for people of all races, cultures, and incomes" regarding the development of environmental laws and policies (*17*). It thus only examines environmental outcomes and only addresses a few strata.

There are ways of grouping the population to determine the fairness of the distribution of gains and losses to specific subpopulations. Different groupings of the population will result in different assessments of a project's fairness. Because there is no right way of grouping, multiple groupings should be considered. To that end, transportation B/C analyses should include an "equity impact statement." [The Applied Research Center (*18*) has also developed what they call an equity impact statement; however, that document is a qualitative approach to assessing equity and can be seen as complementary to what is being suggested here. The city of Toronto, Canada (*19*), has issued an equity impact statement but again of a more qualitative and less systematic type than that suggested here.] This document would specifically consider the winners and losers for a project. In particular, a set of specified subgroups would be identified. Then the outcomes of the project (e.g., travel time and delay, accessibility, consumer's surplus, air pollution, noise pollution, accidents) would be assessed for each of the population groups. Although inequity across some dimensions is almost inevitable, it is crucial both for fairness and for political expediency, given the grow-

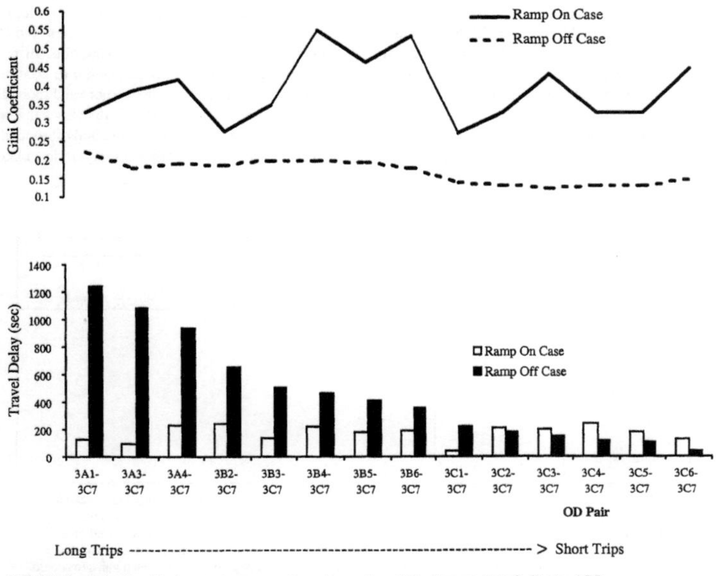

FIGURE 4 Relationship between temporal equity and mobility (travel delay), Route 169.

ing environmental justice movement, to acknowledge the inequity and its relative magnitude before a project is implemented.

Chen (*20*) argues that the principles of social equity and environmental justice can be realized only when the conventional top-down approach to decision making ends. The only way that this can be done is by including all the groups of the community in the decision-making process. Social equity can be realized only when the needs of all groups are adequately represented. This argument calls for an inclusion of *opportunity to participate* as a key criterion in an equity impact statement. For each group, identification of whether that group had equal opportunity to affect the project would be made. Questions would be raised such as "Was the group included among the analysts and decision makers in proportion to its share of the affected population?" Although state departments of transportation and metropolitan planning organizations are attempting to involve minority and low-income populations to a larger degree, historical biases remain.

The equity impact statement, a checklist for which is given in Table 3, would thus consider the inputs (the opportunity to participate in decision making) as well as the outcomes (mobility, economic, environmental, health, and other) for transportation projects. The strata are worth discussing in some detail:

- The population stratification just looks at the population as a whole and investigates how equally distributed are both the opportunities to participate and the outcomes.
- The spatial (or jurisdictional) stratification would examine how different areas (from small areas like census blocks or traffic zones to larger areas like census tracts, jurisdictions, or metropolitan areas) are affected by the project. For example, the U.S. Congress has a House of Representatives, whose seats are allocated in proportion to population, and a Senate, which has two seats for every state. One ensures population equity, the other a type of spatial equity.
- The temporal stratification would consider the benefits and losses to current residents in comparison with those of (potential) future residents. Many transportation and land use policies, such as impact fees, have significant temporal effects (*21*).
- Modal equity considers whether users of different modes (e.g., drivers, pedestrians, transit riders) receive different gains or losses from a project and had equal input into the decision.
- Generational equity differentiates individuals by age: do the elderly or middle-aged benefit at the expense of the young?
- Gender equity contrasts men and women. Because there are known differences in the transportation use patterns by sex, distinguishing the effects on the two groups is important.
- Ability compares the fairness accorded to those without any physical or mental disability with the fairness to those facing such challenges.
- Racial and cultural equity consider the effects on different races, ethnic groups, religions, and cultures. Insufficient research has to date examined the transportation uses by these groups, but if only because of historic spatial segregation, transportation investments will have differential impacts.
- Similarly, some investments that serve certain vehicle types and certain areas will inevitably favor the rich over the poor, an issue addressed by examining income equity.

It must be recognized that collecting such data may be difficult or costly. Some data, such as income, race, or sex, may be known geographically (from census data) but not according to network use (which can only be estimated with models). Further, there will inevitably be the need to forecast when land use changes are anticipated. There may also be privacy concerns if some of these data are collected on facilities. Nevertheless, reasonable attempts can be made to estimate this information in a consistent way across alternatives so that general trends can be assessed.

CONCLUSIONS

A healthy skepticism by concerned citizens toward transportation projects is warranted on the basis of both the transportation and the external effects such projects have. It is no longer enough to apply the pareto maxim that so long as the losers could be compensated by the winners, the project is worthwhile. In the absence of such compensation, political opposition will continue to rise and new construction will continue to be more and more difficult.

Philosopher John Rawls (*22*) discusses the conditions for a fair outcome. He imagines two individuals shrouded in a "veil of ignorance"; they know what they prefer but don't know things like their

TABLE 3 Equity Impact Statement Checklist

	Process	Outcomes				
Stratification	Opportunity to Engage in Decision-Making	Mobility	Economic	Environmental	Health	Other
Population						
Spatial						
Temporal						
Modal						
Generational						
Gender						
Racial						
Ability						
Cultural						
Income						

NOTE: This checklist is suggestive of the considerations that should be explicitly taken. To apply this checklist, each cell should be considered. E.g. taking spatial/opportunity cell, we can ask a set of questions: Was the opportunity to engage in decision making fair across all jurisdictions (locales within a jurisdiction)? Did each place have a say in the planning, engineering, public meetings, financing, and final decision process? To what extent were small places given a voice equal to large? To what extent were populous areas given a voice in accord with their population?

social class. They must agree to divide some spoils (political rights, money, etc.) but don't know which side of the spoils they will get. Rawls argues that they will come to a fair agreement because each has an equal possibility of receiving either side of the division. Rawls's approach is just a sophisticated version of the pie-cutter problem. Imagine that there is a pie and several (N) people: how do you ensure that each gets an equal share? The solution is to let one person cut the pie into N pieces, but the person who cuts the pie gets the Nth piece. He will ensure that the pieces are as equal as possible in order to get that last piece. However, the pie-cutter problem assumes a zero-sum world, whereas often there are gains from trade. Solutions to equity problems include ideas such as bundling improvements, so that not only is there a net benefit (when all projects are considered together), but the number of winners exceeds the number of losers by a significant amount.

Because of the sensitivity of equity analysis to the units of measurement and the definition of the groups, it becomes difficult to select a single, right method of evaluation. The method of equity analysis must be based on the community aspirations and needs. Leaving aside what is the "right" thing to do, consideration of equity is the efficient thing to do in a political environment that empowers many disparate groups. An equity impact statement or its equivalent could help to clarify the impacts of a policy or infrastructure proposal and to test alternative strategies. Equity considerations should be given consideration just as efficiency has traditionally been considered in transportation. Decisions are not made by society based on equity alone or efficiency alone, but rather some mix. Improving the measurement of both equity and efficiency can only lead to better decisions.

ACKNOWLEDGMENT

The author would like to thank Pavithra Parthasarathi and Lei Zhang for their research assistance on this paper. This research was conducted as part of the project Measuring the Equity and Efficiency of Ramp Meters, funded by the Minnesota Department of Transportation.

REFERENCES

1. Hanson, S. *Geography of Urban Transportation.* Guilford Press, New York, 1995.
2. Varian, H. *Microeconomic Analysis,* 3rd ed. W.W. Norton & Company, New York, 1992.
3. Baumol, W. *Superfairness.* MIT Press, Cambridge, Mass., 1986.
4. Sagner, J. Benefit/Cost Analysis Efficiency-Equity Issues in Transportation. *Logistics and Transportation Review,* Vol. 16, No. 4, 1980, pp. 339–388.
5. Lee, D. B., Jr. Making the Concept of Equity Operational. In *Transportation Research Record 677,* TRB, National Research Council, Washington, D.C., 1978, pp. 48–53.
6. Khisty, C. J. Operationalizing Concepts of Equity for Public Project Investments. In *Transportation Research Record 1559,* TRB, National Research Council, Washington, D.C., 1996, pp. 94–99.
7. Litman, T. What is Transportation Equity? *Community Transportation Reporter,* Vol. 14, No. 7, 1996, pp. 22–23.
8. Stolz, R. Race, Poverty and Transportation. *Poverty and Race,* March/April, 2000. Reprinted at www.communitychange.org/transportation/raceandtransportation.asp. Accessed Oct. 18, 2001.
9. Bullard, R., and G. Johnson, eds. *Just Transportation: Dismantling Race and Class Barriers to Mobility.* New Society Publishers, Gabriola Island, B.C., Canada, 1997.
10. Altshuler, A. The Intercity Freeway. In *Introduction to Planning History in the United States* (D. A. Krueckeberg, ed.), Center for Urban Policy Research, Rutgers University, New Brunswick, N.J., 1983.
11. Levinson, D., and D. Gillen. The Full Cost of Intercity Highway Transportation. *Transportation Research,* Vol. 3D, No. 4, 1998, pp. 207–223.
12. Rothengatter, W. Do External Benefits Compensate for External Costs of Transport? *Transportation Research,* Vol. 28A, No. 4, 1994, pp. 321–328.
13. Murchland, J. D. Braess's Paradox of Traffic Flow. *Transportation Research,* Vol. 4, 1970, pp. 391–394.
14. Hansen, M., and Y. Huang. Road Supply and Traffic in California Urban Areas. *Transportation Research,* Vol. 31A, No. 3, 1997, pp. 205–218.
15. Noland, R. Relationships Between Highway Capacity and Induced Vehicle Travel. *Transportation Research,* Vol. 35A, No. 1, 2001, pp. 47–72.
16. Levinson, D., L. Zhang, S. Das, and A. Sheikh. Measuring the Equity and Efficiency of Ramp Meters. Presented at the 81st Annual Meeting of the Transportation Research Board, Washington, D.C., Jan. 2002.
17. *Executive Order 12898 About Environmental Justice.* Office of Solid Waste and Emergency Response, Environmental Protection Agency, 1994. www.epa.gov/swerosps/ej/aboutej.htm. Accessed Oct. 18, 2001.
18. *Developing an Equity Impact Statement: A Tool for Local Policy Making.* Applied Research Center, Grass Roots Innovative Policy Program, 1999. www.arc.org.
19. *Diversity Our Strength: Access and Equity Our Goal,* Appendix G: Access and Equity Impact Statements. Reports to Council Committees. Toronto City Council, 2001. www.city.toronto.on.ca/getting_around/index.htm.
20. Chen, D. *Social Equity, Transportation, Environment, Land Use, and Economic Development: The Livable Community.* www.fta.gov/library/policy/envir-just/backcf.htm. Accessed Oct. 18, 2001.
21. Levinson, D. Financing Infrastructure over Time. *Journal of Urban Planning and Development,* ASCE, Vol. 127, No. 4, 2001, pp. 146–157.
22. Rawls, J. *A Theory of Justice,* rev. ed. Belknap Press, Cambridge, Mass., 1999.

The opinions and errors in this paper remain those of the author.

Publication of this paper sponsored by Committee on Transportation Economics.

Costs Associated with Transport

[12]

Time Pollution

by

John Whitelegg

Although time-savings provide the principal economic justification for new road schemes, the expansion of the road network and the increase in traffic does not seem to have given people more free time. This is because pedestrian time is not evaluated, because cars are deceptively time-consuming, and because people tend to use what time savings they do gain to travel further.

Time is money, we are told; and increasing mobility is a way of saving time . But how successful are modern transport systems at saving time?

Michael Ende's novel *Momo*[1] describes the changes which took place in the daily lives of a small community when "time thieves" persuaded the residents to save time rather than "waste" it on idle conversation, caring for the elderly and similar social activities. The effects were dramatic: as the traditional café was converted into a fast-food outlet and other changes took place, people were too busy saving time to find any time for each other. The village barber found that:

> "he was becoming increasingly restless and irritable. The odd thing was that, no matter how much time he saved, he never had any to spare; in some mysterious way, it simply vanished. Imperceptibly at first, but then quite unmistakably, his days grew shorter and shorter. Almost before he knew it, another week had gone by, another month, and another year, and another and another."

Ende's novel compresses into a few months the process of community disintegration that has been taking place over the last few decades in Europe. The observation that "no one has any time for each other any more" is a commonplace, particularly among older people; yet there are few attempts to examine why this should be so. How can we explain the Momo effect, the paradox that the more people try to save time, the less they seem to have? In other words, what do people do with the time they save?

More Speed, Less Access

The work of Torsten Hagerstrand over the last thirty years is an important but neglected contribution to the understanding of people's use of space and time.[2] He suggests that the ability to make contact with places and other people is the central organizing feature of human activity and that it is ease of access to other people and facilities that determines the success of a transportation system, rather than the means or the speed of transport.

> *"The odd thing was that, no matter how much time he saved, he never had any to spare."*

It is relatively easy to increase the speed at which people move around, much harder to introduce changes that enable us to spend less time gaining access to the facilities that we need.

On this important matter there are very few indicators which can reveal how well our transportation systems are performing in the 1990s, by comparison (for example) with the 1920s. What is without doubt is that facilities are sited further apart and that people have to travel further than they did 70 years ago to reach them. In their home territories, they must travel further to supermarkets or leisure facilities and often must cover some distance while looking for somewhere to park. In their work, they must be prepared to commute further afield to find jobs. In their leisure time people contemplate day trips to Brussels, Paris or Stockholm when previously they would have thought the idea ridiculous.

C. Marchetti has shown that the amount of time each person devotes to travel is roughly the same regardless of how fast or how far they travel. "When people gain speed they use it to travel further and not to make more trips. In other words most individuals treat their territory the same way whatever size it is."[3] Those who use technology to travel at greater speeds still have to make the same amount of contacts — still work, eat, sleep and play in the same proportions as always. They simply do these things further apart from each other.

Do they do so by choice or through obligation? A circular logic operates here. While the distances between hospitals, schools, shopping centres and the like have risen, nothing can be done to increase the number of hours in the day. Speed must therefore be increased, and investments are made in quicker forms of transport — families buy faster cars, governments build faster roads and railways. But the time savings promised by new motorways and high speed trains appear to release time for more travel and thus spur the consumption of distance to ever higher levels of achievement. When people save time, they use it to buy more distance.

Social Speeds

The suggestion that people spend about the same amount of time travelling, whatever their mode of transport, does not, however, explain the Momo effect: many people feel they have *less* time than they had before, despite faster means of transport.

There is another hidden time factor in the equation. Motor cars and other high speed vehicles do not save as much time as they appear to, as Ivan Illich pointed out in 1974:

> "The typical American male devotes more than 1,600 hours a year to his car. He sits in it while it goes and while it stands idling. He parks and searches for it. He earns the money to put down on it and to meet the monthly instalments. He works to pay for petrol, tolls, insurance taxes and tickets."[4]

Elaborating on Illich's observations D. Seifried[5] has coined the term "social speed" to signify the average speed of a vehicle, once a number of these hidden factors have been taken into account (*see*

Dr. John Whitelegg is Head of the Geography Department at Lancaster University and an international transport and environment consultant.

Table 1). According to Seifried, the social speed of a typical bicycle is 14 kilometres per hour (kph), only three kph slower than that of a small car. If other external costs (air and noise pollution, accident costs, road construction costs and so on) are taken into account as well, then the small car is one kph slower than the bicycle.

Thus the owner of a small car who spends 30 minutes per day driving 20 kilometres may feel that she is travelling faster than a bicyclist who spends the same time covering seven-and a half kilometres. But when the social speed is taken into account, it emerges that the car owner is likely to be spending 70 minutes per day while the bicyclist is spending only 32. *Ecce* Momo!

Space Pollution

Whereas speed consumes distance, a mode of transport occupies space — and the faster the mode of transport the more space it requires. According to a 1985 Swiss study,[6] a car travelling at 40 kph requires over three times as much as space as a car travelling at 10 kph (see Table 2, p.134). Furthermore the "bodywork" often associated with high speed vehicles demands space even when the vehicle is travelling slowly: a single person in a car travelling at 10 kph requires six times as much space as a person riding a bicycle at the same speed.

Space therefore has to be consumed in large quantities to provide the infrastructure for high speed travel, as can be witnessed in the land requirements for new motorways, high speed rail routes and airports. Roads designed to carry traffic at speeds over 120 kph take up more land than roads designed for lower speeds, and the same is true for high speed rail — fast cars and trains cannot take tight bends. Urban motorway and "relief" road construction is the ultimate expression of space sacrificed for speed.

When the demand for space is not met at certain points in the network, the result is congestion — the familiar situation where cars costing up to £20,000 and designed to travel at 175 kph cannot average speeds much above 20 kph. The current enthusiasm for charging motorists for their use of road space through toll roads and electronic road pricing arises out of a hope that it will ease congestion. Traffic flow on these roads can be regulated by adjusting the level of the toll. This will save time for one group (wealthy motorists) at the expense of other groups (such as poor car-owners or pedestrians) and at the expense of greater levels of space inefficiency. Table 2 shows that in terms of space efficiency, the car is extremely wasteful. Paying for that space does not alter this equation.

Calculated speeds taking into account the total amount of time spent in transport

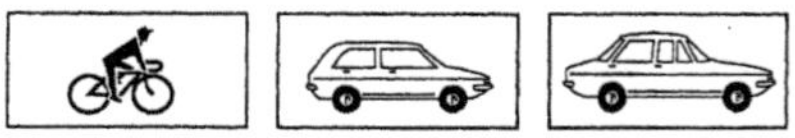

Average speed over all journeys over a year	15 Kph	40 Kph	60 Kph
Annual Cost	120 DM	4700 DM	16000 DM
Monthly net income of vehicle owner	1600 DM	1600 DM	6400 DM
Amount of time which must be spent at work to earn the yearly cost of that mode	15 h	470 h	400 h
Average 'social' speed	14 Kph	17 Kph	21 Kph
External cost per Km		15 Pf/Km	30 Pf/Km
Average social speed when taking into account external costs	14 Kph	13 Kph	18 Kph

Table 1: *Average social speeds of the bicycle and the car*

Time Thieves

As higher speeds lead to greater distances between facilities, people overcome this distance either by allocating more time to travel or by gaining access to modes of transport with higher speeds. The result of both has been an accentuation of social differences. While those with access to high-performance cars and regular transcontinental air flights have seen their radius of activity expand immeasurably over the last few decades, that of an unemployed black resident of London or an elderly person in Montgomery, Alabama, for instance, may be no greater than that of an urban resident 100 years ago. The poor and unemployed, whose time is valued very low, are expected to find the time to devote to travel; the rich have the money to buy travel and are more likely to do so because their time is considered more valuable. The more emphasis put on time savings, the more the whole transport system becomes skewed to serve a wealthy élite.

Transport policies and policies which influence location and accessibility of basic facilities steal time from different groups in society and reallocate it to (usually) richer groups. The relocation of shops, hospitals and schools at a greater distance from the community that needs them imposes serious time penalties on other users. Those without cars (still about 35 per cent of the UK population) and those without access to them during the day must spend more time searching for other facilities, waiting for buses, waiting for friends to give them lifts, or walking.

Among the groups particularly affected in a male- and car- dominated planning system are women, children, the elderly and the infirm. For women travelling alone after dark, there are potentially serious consequences arising from waiting at bus stops or for late trains or for using another device designed to maximize vehicle

Left: *A Wiltshire village in the early 20th century. The sign on the far left says "Danger Motorists Caution". The veranda, indicates time to spare, and a congenial street life.*

Below: *The same view in 1988. The motor car has usurped both space and time: the veranda has been replaced by a parking place.*

Nikki Houghton

convenience at the expense of pedestrians: the underpass. Women are more likely to be bus users than men, more likely to be in charge of young children in dangerous pedestrian environments and more likely to be involved with escorting duties arising from the unacceptability of letting children walk unsupervised in environments rendered lethal by traffic. In Britain, women spend many thousands of hours escorting children in an environment rendered unsafe for children, mainly by men. Using Department of Transport (DOT) methods of valuation, the cost of this escorting has been estimated at over £10 billion.[7] If this cost had been taken into account the planning process would have produced a different outcome.

The Price of Time

The provision of high quality urban roads, large car parks and (soon) in-car navigation is dependent upon a high valuation of the time of the car occupant. Road schemes in Britain are justified by assigning a monetary value to the time they will save for motorists. The author of one study[8] describes an urban road construction and improvement scheme in Leicester where time savings made up 96.4 per cent of the gross benefits in the DOT's cost-benefit evaluation (COBA). The average time savings over several projects was 90 per cent of the value of the benefits. Where the proposed road might block pedestrian movements or require an increase in the time devoted to escorting children, this was not offset against the time gained. Nor was attention given to the question of how this newly-won time might be reallocated in an economically productive way to justify the assignment of monetary values.

The Leicester study also revealed that most of these predicted time savings for motorists were very small, in the order of five minutes or less. It calculated that when the value given by COBA to each time-saving of less than three minutes was reduced by 75 per cent, the estimated first year rate of return of the scheme fell from 20 per cent to five per cent — a rate of return that would cast severe doubts upon the financial viability of the project. Time savings of three minutes are likely to fall within the routine variability of any journey and cannot be easily be reallocated to "useful" time. Furthermore, in any road scheme there will be innumerable other repercussions which take up three minutes — the time taken by pedestrians to make a detour through an underpass, for example. The monetarization of motorists' time savings is a convenient fiction that enables the evaluation process to come up with the desired answer — build the road.

If putting high values on the time of drivers, even down to very short periods, leads to more road building, putting a high value on the time of cyclists and pedestrians would restructure present transport systems. Traffic would have to give way to pedestrians so as not to delay them, purpose-built pedestrian and cycle facilities would win new investment, and proposals that encouraged pedestrians to linger and make use of space whilst slowing down traffic would gain precedence. This is encouraging a "waste of time" and might be seen to imply that motorists' time should have a negative value. But allocating a negative value to motorists' time-savings is no less ridiculous than current practices and would encourage cities and villages to develop as social, productive, enjoyable and secure places.

Maintaining Community

Jane Jacobs account of city life in the US some thirty years ago[9] shows how important ordinary but diverse contact is to people's well-being. Maintaining a sense of community needs an investment of time and energy in contact with neighbours and local groups. The opportunities for such contact depend on time available and thus on priorities. The decision to travel longer distances (and save time at higher speeds) means that little time is available for interaction with neighbours and so there is less chance of a genuine community developing or maintaining itself.

Motorists not only restrict their own lives in this respect, but also those of other people. Detailed studies on the effect of traffic volumes upon different street communities in San Francisco[10] showed, unsurprisingly, that streets with heavy traffic have relatively little social interaction; residents of streets with light traffic had three times as many local friends and twice as many acquaintances as did residents of busy streets

Transport Mode	Speed	Space required per person
Pedestrian		0.8 M² per Person
Cyclist		3 M² per Person
Fully Occupied Car		6.2 M² per Person
Fully Occupied Car		20 M² per Person
Car with 1 Person		18.7 M² per Person
Car with 1 Person		60 M² per Person
Bus - Full and 1/3 Full		3.1 M² per Person (full) 9.4 M² per Person (1/3full)
Bus - Full and 1/3 Full		9.4 M² per Person (full) 28.1 M² per Person (1/3full)
Light Rail/Metro - Full and 1/3 Full		1.5 M² per Person (full) 4.6 M² per Person (1/3full)
Light Rail/Metro - Full and 1/3 Full		2.2 M² per Person (full) 6.9 M² per Person (1/3full)

Table 2: *Consumption of space by different modes of transport.*

Time is central to notions of sustainability. A sustainable city or a sustainable transport policy or a sustainable economy cannot be founded on economic principles which, through their monetarization of time, orientate society towards higher levels of motorization, faster speeds and greater consumption of space. The fact that these characteristics produce energy intensive societies and pollution is only part of the problem. They also distort value systems, elevate mobility above accessibility, associate higher speeds and greater distances with progress and dislocate communities and social life.

Sustainability involves significant changes in the way markets operate and the ways individuals behave. Time valuation is one area ripe for change. Current methods of valuation provide an economic rationale for more travel and more pollution and justify the poor conditions for cyclists and pedestrians. They also explain why solutions such as catalytic converters and road-pricing and even improved public transport are irrelevant. None of these agents in themselves will alter significantly the economic trajectory that is now in place.

This article is a shortened and adapted version of Chapter V of *Transport for a Sustainable Future: The Case for Europe*, by John Whitelegg, published by Belhaven Press, London, 1993, price £12.99.

References

1. Ende, M., *Momo*, Penguin, London, 1984.
2. Hagerstrand, T., "Space, Time and the Human Condition", in Karlquist, A., Lundquist, L. and Snickars, F., (eds.) *Dynamic Allocation of Urban Space*, Saxon House, Lexington, MA, 1975.
3. Marchetti, C., *Building Bridges and Tunnels: the Effect on the Evolution of Traffic*, Document SR-88-01, International Institute for Applied Systems Analysis, Vienna, 1981.
4. Illich, I., *Energy and Equity*, Marion Boyars, London, 1974.
5. Seifried, D., *Gute Argumente: Verkehr, Beck'sche Reihe*, Beck, Munich, 1990.
6. Navarro, R.A., Heierli, U. and Beck, V., *Alternativas de Transporte en America Latina: la Bicicleta y los Triciclos*, SKAT, Centro Suizo de Technologia Apropiada, St Gallen, Switzerland, 1985.
7. Hillman, H., Adams, J. and Whitelegg, J., *One False Move: a Study of Children's Independent Mobility*, Policy Study Institute, London, 1990.
8. Sharp, C.H., *Transport Economics*, Macmillan, London, 1973.
9. Jacobs, J., *The Life and Death of American Cities*, Pelican, London, 1961.
10 Appleyard, D., Livable Streets, University of California, Berkley, 1981.

[13]

A Review of the Literature on the Social Cost of Motor Vehicle Use in the United States

JAMES J. MURPHY
Department of Agricultural and Resource Economics
University of California, Davis

MARK A. DELUCCHI
Institute of Transportation Studies
University of California, Davis

ABSTRACT

Over the past five years, analysts and policymakers have become increasingly interested in the "full social cost" of motor vehicle use. Not surprisingly, there is little agreement about how to estimate the social cost or why, with the result that estimates and interpretations can diverge tremendously. In this situation, policymakers and others who wish to apply estimates of the social cost of motor vehicle use might find it useful to have most of the major estimates summarized and evaluated in one place. Toward this end, we review the purpose, scope, and conclusions of most of the recent major U.S. studies, and summarize the cost estimates by individual category. We also assess the level of detail of each major cost estimate in the studies.

INTRODUCTION

Over the past five years, analysts and policymakers have become increasingly interested in the full social cost of motor vehicle use. Researchers have performed social cost analyses for a variety of

James J. Murphy, Graduate Student, Department of Agricultural and Resource Economics, University of California, Davis, CA 95616. Email: jjmurphy @ ucdavis.edu.

reasons, and have used them in a variety of ways, to support a wide range of policy positions. Some researchers have used a social cost analysis to argue that motor vehicles and gasoline are terrifically underpriced, while others have used them to downplay the need for drastic policy intervention in the transportation sector. In any case, social cost analyses excite considerable interest, if only because nearly all of us use motor vehicles.

Interest in full social cost accounting and socially efficient pricing has developed relatively recently. From the 1920s to the 1960s, major decisions about building and financing highways were left to "technical experts," chiefly engineers, who rarely if ever performed social cost-benefit analyses. Starting in the late 1960s, however, "a growing awareness of the human and environmental costs of roads, dams, and other infrastructure projects brought the public's faith in experts to an end" (Gifford 1993, 41). It was a short step from awareness to quantification of the costs not normally included in the narrow financial calculations of the technical experts of the past.

Today, discussions of the social costs of transportation are routine. In most accounts, the social costs of transportation include external, nonmarket, or unpriced costs, such as air pollution costs, as well as private or market costs, such as the cost of vehicles themselves. Government expenditures on motor vehicle infrastructure and services usually are included as well.

Purposes and Uses of Social Cost Analyses

By itself, a social cost analysis does not determine whether motor vehicle use on balance is good or bad, or better or worse than some alternative, or whether it is wise to tax gasoline or encourage alternative modes of travel. A social cost analysis can provide cost data, cost functions, and cost estimates, which can help analysts and policymakers evaluate the costs of transportation policies, establish efficient prices for transportation services and commodities, and prioritize research and funding. Let us examine these uses more closely[1]:

Use #1: Evaluate the costs of transportation projects, policies, and long-range scenarios. In cost-benefit analyses, policy evaluations, and scenario analyses, analysts must quantify changes to, and impacts of, transportation systems. The extent to which a generic national social cost analysis can be of use in the evaluation of a specific transportation policy or system depends, of course, on its detail and quality. At a minimum, a detailed, original social cost analysis can be mined as a source of data and methods for cost evaluations of specific projects. Beyond this, if costs are a linear function of quantity, and invariant with respect to location, then estimates of national total or average cost, which any social cost analysis will produce, may be used to estimate the incremental costs for specific projects, policies, or scenarios. Otherwise, analysts must estimate the actual nonlinear cost functions for the project, policy, or scenario at hand.

Use #2: Establish efficient prices for, and ensure efficient use of, those transportation resources or impacts that at present either are not priced but in principle should be (e.g., emissions from motor vehicles) or else are priced but not efficiently (e.g., roads). Again, at a minimum, the data and methods of a detailed social cost analysis might be useful in analyses of marginal cost prices. Beyond this, the average cost results of a social cost analysis might give analysts some idea of the magnitude of the gap between current prices (which might be zero, as in the case of pollution) and theoretically optimal prices, and inform discussions of the types of policies that might narrow the gap and induce people to use transportation resources more efficiently. And to the extent that total cost functions for the pricing problem at hand are thought to be similar to the assumed linear national cost functions of a social cost analysis, the average cost results of the national social cost analysis may be used to approximate prices for the problem at hand.

Use #3: Prioritize efforts to reduce the costs or increase the benefits of transportation. The total cost or average cost results of a social cost analysis can help analysts and policymakers rank costs (e.g., whether road dust is more damaging than ozone), track costs over time (e.g., whether the cost of air pollution is changing), and compare the costs and benefits of pollution control (e.g., whether expenditures on motor vehicle pollution control devices are more or less than the value of the pol-

[1] See also Lee (1997).

lution eliminated). This information can help people decide how to fund research and development to improve the performance and reduce the costs of transportation.

Overview of the Debate in the Literature

Not surprisingly, there is little agreement about precisely which costs should be counted in a social cost analysis, which costs are the largest, how much the social cost exceeds the market or private cost, or to what extent, if any, motor vehicle use is "underpriced." On the one hand, many recent analyses argue that the "unpaid" costs of motor vehicle use are quite large—perhaps hundreds of billions of dollars per year—and hence that automobile use is heavily "subsidized" and underpriced (e.g., MacKenzie et al 1992; Miller and Moffet 1993; Behrens et al 1992; California Energy Commission 1994; Apogee Research 1994; COWIconsult 1991; KPMG 1993; Ketcham and Komanoff 1992; Litman 1996). Others have argued that this is not true. For example, the National Research Council (NRC), in its review and analysis of automotive fuel economy, claims that "some economists argue that the societal costs of the 'externalities' associated with the use of gasoline (e.g., national security and environmental impacts) are reflected in the price and that no additional efforts to reduce automotive fuel consumption are warranted" (NRC 1992, 25). Green (1995) makes essentially the same argument. Beshers (1994) and Lockyer and Hill (1992) make the narrower claim that road-user tax and fee payments at least equal government expenditures related to motor vehicle use, and Dougher (1995) actually argues that road-user payments exceed related government outlays by a comfortable margin.

We could cite other examples. This extraordinary disagreement exists because of differing accounting systems, analytical methods, assumptions, definitions, and data sources. The root of the problem is that there are few detailed, up-to-date, conceptually sound analyses. With few exceptions, the recent estimates in the literature are based on reviews of old and often superficial cost studies. Moreover, some of the current work confuses the meaning of *externality, opportunity cost*, and other economic concepts. And, because there is no single, universally accepted framework for conducting a social cost analysis of motor vehicle use, it is often difficult, if not impossible, to make meaningful comparisons of the results from different studies.

In this situation, policymakers and others who wish to apply estimates of the social cost of motor vehicle use might find it useful to have most of the major estimates summarized in one place. This is the purpose of our paper: to review much of the present literature on the social cost of motor vehicle use in the United States as an aid to those who wish to use the estimates. Although we are not able to provide a simple evaluation of the overall quality of the studies, we do offer, as a partial indicator of quality, an evaluation of the degree of detail of the cost estimates in each study.

Our Review

The studies reviewed are presented in chronological order. Generally, we review the purpose, scope, and conclusions, and summarize the cost estimates by individual category. We also assess the level of detail for each major cost estimate in the studies.

In each review, the definitions and terms are those of the original study. For example, we report as an "external cost" what each study calls an external cost; we do not define external cost ourselves and then categorize estimates of each study with respect to this definition. This of course means that what may appear in different studies to be estimates of the same cost—the external cost of accidents, for example—might actually be estimates of different costs. Because of this, and because of differences in scope, timeframe, and so on, one must be careful when comparing estimates.

The bulk of the paper consists of a set of relatively detailed reviews, with tabulations of the estimates of cost and level of detail of some of the more frequently cited studies. In the main set of detailed reviews, we include only U.S. studies whose primary purpose is to estimate some significant part of the social cost of motor vehicle use. We do not include studies where the use, review, or development of estimates is secondary to application or theoretical discussion. Also, we do not review here studies of a single cost category, such as air pollution or noise (these studies are reviewed in the appropriate report of the social cost series of

Delucchi et al 1996). Although this literature review focuses specifically on U.S. studies, European studies are summarized at the end of the paper.

KEELER AND SMALL (1975)

Keeler and Small is one of the most influential and widely cited studies of the costs associated with automobile use. It was one of the first attempts to quantify the nonmarket costs of automobile use, such as time and pollution, as well as the direct costs, such as operation and maintenance. Although most of the costs in this report are now outdated, and many of the methods have been improved, we summarize Keeler and Small because of its influence on subsequent research.

Goals and Methodology

This report develops estimates of the costs of peak-hour automobile transportation in the San Francisco Bay Area. To facilitate intermodal comparisons, the authors also develop similar cost estimates for bus and rail work trips. They divide automobile trips into three main components, and estimate costs associated with each: 1) residential collection (i.e., going from a residence to the freeway interchange), 2) line-haul trip (i.e., travel by freeway to the edge of the central business district), and 3) downtown distribution. They evaluate two alternative trip lengths: 1) a 6-mile line-haul trip with an average feeder distance of 1 mile, and 2) a 12-mile trip with an average feeder distance of 2 miles. For both trips, the downtown distribution is assumed to be about 0.75 miles in length.

Capital and Maintenance Costs

To estimate highway capacity costs, Keeler and Small develop statistical cost models for construction, land acquisition, and maintenance for 1972. The data used in the three models covers all state-maintained roads in the Bay Area, including expressways, arterials, and rural roads. The construction cost model, which accounts statistically for the effects of urbanization and economies of scale on expressway construction costs, allows them to estimate the cost of a lane-mile of freeway under different degrees of urbanization and road widths. Land acquisition costs are modeled in a similar manner. Finally, maintenance costs per lane-mile are expressed as a function of the average annual vehicles per lane on the relevant stretch of road.

User Benefits and Costs of Speed

Keeler and Small recognize that there is a tradeoff between highway traffic speed and capacity utilization: faster speeds save travel time, but result in lower capacity utilization and increased fuel consumption.[2] This tradeoff is represented by speed-flow curves. They develop a model that calculates optimal tolls and volume-capacity ratios for each period as a function of time values and lane capacity costs. To develop the model, the authors adjusted the results of a study by the Institute of Transportation and Traffic Engineering that estimated speed-flow curves for the Bay Area. On the basis of a literature review, they assume the value of time in the vehicle is $3 per hour per person. Finally, they use data on hourly vehicle flows to determine the peaking characteristics of traffic.

Public Costs

For Keeler and Small, public costs include environmental costs, the costs of police and supporting social services (e.g., city planning, fire department, courts), and any maintenance costs related to the number of vehicles that use the road (as opposed to the capacity of the road). To estimate police and social service costs, the authors cite an earlier, unpublished paper (Keeler et al 1974), in which they estimated the average costs of police and supporting social services was about 4.5 mills[3] per vehicle-mile in the Bay Area. They assume that the marginal and average costs are about the same.

Their estimate of the environmental costs (i.e., noise and pollution) are also drawn from a previous paper (Keeler and Small 1974). They argue that marginal noise costs are likely to be low, no more than one or two mills per vehicle-mile, because costs are high only on quiet residential streets where an extra vehicle is likely to be

[2] However, fuel consumption is not by any means a simple linear function of speed, and in some cases a increase in the overall average speed reduces fuel consumption.

[3] One mill equals one-tenth of a cent.

noticed. They estimate that composite pollution (the average from all vehicle types) in 1973 cost about 0.92¢ per vehicle-mile. They note that this is a conservative figure, because it assumes that the cost of human illness and death is only equal to hospital bills and foregone wages. On the other hand, they expect that this cost will decline as more rigorous standards come into effect.

Accidents and Parking Costs

To estimate accident costs, Keeler and Small first compute a national average accident cost figure, and then use the results of two earlier studies (May 1955; Kihlberg and Tharp 1968) to allocate costs among the different highway types and locations. Parking costs are derived by combining the results of two engineering cost studies (Meyer et al 1965; Wilbur Smith and Associates 1965). From this, they derive estimates of the annual cost per parking space for five types of facilities (lot on central business district—CBD—fringe, lot in low land value CBD, and garage in low-, medium-, and high-value CBD). They compare the results of these studies with actual rates at privately owned parking facilities in San Francisco and find that they are consistent.

Related Work

The work of Keeler and Small spawned additional work by Small (1977) on air pollution costs. The objective of Small (1977) is "to provide some rough and aggregate measures of the economic costs imposed on society by air pollution from various transport modes in urban areas." Small uses the work of Rice (1966), Lave and Seskin (1970), and the Midwest Research Institute (1970) to estimate the total health and materials costs of air pollution. He then disaggregates the total pollution cost by specific pollutant and geography. Finally, he estimates the motor vehicle contribution to each pollutant and hence to air pollution damages. The result is an estimate of $1.64 billion in air pollution damages by automobiles, and $0.55 billion by trucks, in 1974.

FEDERAL HIGHWAY ADMINISTRATION (1982)

Goals and Methodology

In the introduction, the authors state:

> This report . . . responds to the [congressional] request for: (1) an allocation of Federal highway program costs among the various classes of highway vehicles occasioning such costs; (2) an assessment of the current Federal user charges and recommendations on any more equitable alternatives; and (3) an evaluation of the need for long-term monitoring of roadway deterioration due to traffic and other factors (p. I-1).

Although the primary focus of the report is the allocation of federal highway expenditures, appendix E of the report contains a discussion of some of the social costs and provides estimates of efficient highway user charges for some of these costs in 1981.[4] The authors focus solely on costs that vary with vehicle-miles of travel (VMT). Of 11 cost items mentioned in the report, the authors attempt to estimate costs on a VMT basis for 6: pavement repairs, vibration damages to vehicles, administration, congestion, air pollution, and noise. Costs associated with the first two items are significant for trucks, but negligible for automobiles on a VMT basis. The authors note that of the five costs not estimated in cents per VMT, "accidents looks to be the only category that might lead to a substantial increase in user charges if more were known about causal relationships. Other marginal costs may be large in the aggregate but small in relation to VMT" (p. E-52). In their conclusion, they estimate that "efficient user charges could raise almost $80 billion annually (ignoring collection costs and assuming revenues from different types of charges are additive), in contrast to the $40 billion currently spent on highways by all levels of government or the $22 billion now raised by user fees" (p. E-7). In addition, appendix E also contains a fairly detailed discussion of the standard economic theory on which their analysis is based.

[4]"Efficient highway user charges are those which will lead to the greatest surplus of benefits over costs, for a given stock of capital facilities" (p. E-17).

KANAFANI (1983)

Kanafani is a review of published estimates of the social costs of motor vehicle noise, air pollution, and accidents. As he puts it:

> the purpose of this report is to review and assess recent attempts at the evaluation of the social costs of road transport. It is intended to provide a comparative evaluation of the economic magnitude of the social costs of road transport in selected countries, particularly as occasioned by the environmental and safety impacts of motor transport" (p. 3).

He defines social costs as "those costs that are incurred by society as a whole, not solely by the users as direct costs, nor those that are incurred solely by the nonusers (pp. 2–3). He discusses the key cost components for each of these categories, and summarizes the results from other studies.

Kanafani reviews studies from several different countries, including the United States, France, and West Germany. Based on a literature review, he estimates that the social cost of noise in the United States is between \$1.3 billion and \$2.6 billion (0.06% to 0.1% of GDP), the social cost of air pollution ranges between \$3.2 billion and \$9.7 billion (0.14% to 0.36% of GDP), and accidents cost between \$33.0 billion and \$37.0 billion (about 2% of GDP). (The year of these estimates varies, because Kanafani reports estimates from the literature without converting or updating the dollars to a base year.)

FULLER ET AL (1983)

Background and Scope

Fuller et al was prepared in conjunction with the Federal Highway Administration (FHWA) Cost Allocation Study (1982). Although the FHWA report discusses external costs, its primary focus is on allocating government outlays. Fuller et al, on the other hand, focuses exclusively on external costs. The costs identified in this report are: congestion or interference (including accidents), air pollution, and noise damages. The analysis was performed using data for 1976 to 1979, with forecasts for 1985.

Although the report "does not undertake to develop new techniques for the measurement of damages," and instead performs "a comprehensive review of the literature and data available for each type of damage" (p. 4), it does in fact use detailed models to estimate marginal and total costs, particularly for noise.

The work of Fuller et al was incorporated into the FHWA study, and has been cited in a number of others.

Congestion and Accident Costs

Fuller et al model traffic interference and marginal accident rates as a function of the volume-to-capacity ratio on several different functional classes of roads (interstates, arterials, collectors, and local roads in rural and urban areas). They combine these functions with estimates of the value of time by functional road class, and the injury, fatality, and property damage costs of accidents, to produce marginal cost curves for the different functional classes of roads.

Air Pollution Costs

Fuller et al estimate air pollution costs in three steps. First, they review and analyze the literature on the health, vegetation, and materials damages of air pollution (e.g., Small 1977; Lave and Seskin 1970) in order to estimate dollar damages per ton of each pollutant. Second, they multiply the dollars per ton estimates by the U.S. Environmental Protection Agency estimates of grams per mile of emissions, for each pollutant, and sum across all of the pollutants, to obtain dollars per VMT. Finally, they "correct" the dollars per VMT estimates for "microscale" differences in exposure, meteorology, and other factors.

Noise Costs

Fuller et al calculate the dollar cost of motor vehicle noise in residential areas as the product of three factors:

1. the number of housing units in each of up to three distance/noise bands along roads: moderate exposure (55 to 65 dBA), significant exposure (65 to 75 dBA), and severe exposure (more than 75 dBA);
2. excess dBA of noise, equal to the noise level at the midpoint of each distance/noise band minus the threshold noise level (assumed to be 55 dBA);

3. the dollar reduction in property value per excess dBA (estimated to be $152 per excess dBA in 1977 dollars).

They use a 1970s-vintage noise generation equation to delineate the distance/noise bands, and national average data on housing density, housing value, and traffic volume. They do not consider noise costs outside of the home.

MACKENZIE ET AL (1992)

Goals and Methodology

The goal of this report is to quantify the costs of motor vehicle use in the United States that are not borne by drivers. Because it is one of the more widely cited studies on the social cost of motor vehicle use in the United States, we provide some additional comments on the derivation of their cost estimates.

Two types of costs are identified in this study: market and external. "Market costs are those that are actually reflected in economic transactions . . . (They) represent the direct, ordinary, expected costs of owning and operating a motor vehicle" (p. 7). Examples of this include vehicle purchase, fuel and maintenance costs, and road construction and repair. External costs, or externalities, are those costs, such as global warming and illnesses resulting from pollution, that are not incorporated into market transactions. Social costs are the sum of market and external costs.

The results of this study are summarized in table 1. MacKenzie et al estimate that the annual market costs not borne by drivers in 1989 was about $174.2 billion, and that the annual external costs not borne by drivers totaled $126.3 billion, for a total of approximately $300 billion.

TABLE 1 Annual Social Costs of Vehicle Use not Borne by Drivers: 1989
MacKenzie et al 1992

Market costs	$ billion
Highway construction and repair	13.3
Highway maintenance	7.9
Highway services (police, fire, etc.)	68.0
Value of free parking	85.0
Total market costs	*174.2*
External costs	
Air pollution	10.0
Greenhouse gases	27.0
Strategic petroleum reserve	0.3
Military expenditures	25.0
Accidents	55.0
Noise	9.0
Total external costs	*126.3*
Total social costs	*300.5*

Most of the cost estimates provided by MacKenzie et al are direct citations from another work or simple extrapolations from someone else's analysis. In the following sections, we discuss some of the estimates derived by MacKenzie et al. The costs in table 1 that we do not discuss below are essentially direct citations from other studies.

Highway Services

In this category, MacKenzie et al mean to include police motorcycle patrols and details for auto theft, parking enforcement, accident aid, fighting garage fires, and various public works expenses, such as traffic and road engineering. Their estimate of the cost of these services is from Hart (1986), which in turn is based on Hart's earlier, more detailed analysis (Hart 1985).

Hart's (1986) estimates of the national cost of highway services is an extrapolation of his detailed estimate for the city of Pasadena. This extrapolation is questionable. Moreover, it appears that some of the costs that Hart (1985; 1986), and hence MacKenzie et al, count as highway service costs are actually highway capital and operating costs in FHWA (1990), and hence are double counted in MacKenzie et al.

Employer-Paid Parking

MacKenzie et al assume that 86% of the workforce commute by car, and that 90% receive free parking, and use this to calculate that 85 million Americans receive free parking at work. Assuming that the average national value of a parking space is $1,000 (Association for Commuter Transportation 1990),[5] MacKenzie et al estimate that the annual parking subsidy for workers is about $85 billion.

[5] This estimate probably is too high (Delucchi et al 1996).

MacKenzie et al note that their estimate is for the cost of free parking at work, and therefore it does not include the cost of free parking for other kinds of trips. Because commuting to work constitutes only 26% of all vehicle trips, the cost of free parking for nonwork trips is probably not trivial.

Climate Change

Because there is so much uncertainty about the magnitude, effects, and costs of climate change, MacKenzie et al assume that "it is not possible to accurately estimate the actual costs of the current buildup of greenhouse gases" (p. 14). Instead, they develop an "imperfect" estimate, based on Jorgenson and Wilcoxen (1991), that a phased-in carbon tax that reached $60 per ton of carbon emissions (about 20¢ per gallon of gasoline) in the year 2020 would reduce emissions to 80% of the 1990 level by 2005. By assuming that motor vehicle fuel consumption would continue at roughly 1990 levels, MacKenzie et al estimate that a phased-in tax of 20¢ per gallon would eventually cost motorists about $27 billion per year, which they use as an estimate of the cost of climate change. We emphasize that this is not an estimate of the damage cost of global warming at all, but rather an estimate of the aggregate revenue from a somewhat arbitrarily assumed carbon tax on gasoline.

KETCHAM AND KOMANOFF (1992)

Goals and Methodology

Ketcham and Komanoff are concerned about the inefficient use of New York City's transportation infrastructure. They believe that the compactness of New York City creates an opportunity to provide people with a greater variety of transportation alternatives, but that public policies are skewed toward motor vehicle use and prevent these opportunities from materializing. They argue that New York City's "transportation and air pollution problems are solvable, through an approach that systematically charges motorists for a fair share of the fiscal and social costs of driving and invests much of the revenues in transit and other non-motorized modes" (p. 3). Their paper explains this approach, and how it can "benefit the vast majority of residents in the region" (p. 3).

In their report, costs are divided into four categories. 1) The costs that motorists pay when they drive are called "the direct costs of roadway transportation borne by users." Examples of these direct costs include vehicle purchase, fuel, insurance, and maintenance and repair. 2) The costs of building and maintaining roads, net of user fees such as tolls and taxes, are called "the direct costs of roadway transportation borne by non-users." 3) The portion of motor vehicle externalities, such as congestion, noise, and accidents, that is borne by motorists in the act of driving is called "the externality costs borne by users." 4) Finally, environmental damages and other external costs that are borne by society as a whole are called "externalities borne by non-users."

Much of the paper is devoted to public policy issues that focus primarily on New York City. However, a portion of the paper provides an analysis of the social costs of motor vehicle use for the whole United States. Our review focuses on Ketcham and Komanoff's national estimates, most of which they derived from their review of other published studies, particularly FHWA (1982), Eno Foundation (1991), and MacKenzie et al (1992). The results of their study are shown in table 2, and discussed in more detail below.

Direct Costs of Roadway Transportation

Ketcham and Komanoff's estimates of the direct costs borne by drivers—vehicle ownership, taxi services, school bus transport, and freight movement by truck—are from the Eno Foundation (1991). They do not estimate the national costs associated with off-street parking. Their estimates of the direct costs not borne by drivers—costs associated with roadway construction, maintenance, administration and services—are calculated from FHWA data on highway finances (FHWA 1990).

Externalities of Roadway Transportation

Accidents and congestion. In Ketcham and Komanoff, the two largest external costs are congestion ($168 billion) and accidents ($363 billion), which combined represent almost 75% of their total estimated external costs of roadway transport. To estimate congestion costs they use the cost factors in the FHWA Cost Allocation Study (1982), adjusted to

TABLE 2 Costs of Roadway Transportation in the United States: 1990
Ketcham and Komanoff 1992

Direct costs of roadway transportation borne by users	$ billion
Personal transportation (auto)	510.8
Taxi/limousine services	7.5
School bus transport	7.5
Freight movement by truck	272.6
Roadway construction and maintenance	48.1
Off-street parking	n.e.
Total direct costs of roadway modes (A)[1]	*798.4*
Direct costs of roadway transport borne by non-users	
Roadway construction, maintenance, admin. services	16.0
Parking	n.e.
Total direct costs not borne by users (B)	*16.0*
Externality costs borne by users	
Congestion costs	142.8
Air pollution: health and property costs	1.5
Accident costs	290.4
Noise costs	1.1
Pavement damage to vehicles	15.0
Total externality costs borne by motorists (C)	*450.8*
Externality costs borne by non-users	
Congestion costs	25.2
Air pollution: health and property costs	28.5
Accident costs	72.6
Noise costs	21.1
Vibration damage to buildings and infrastructure	6.6
Land costs	66.1
Security costs	33.4
Climate change	25.0
Total externality costs borne by non-users (D)	*278.5*
Total cost of roadway transport (A+B+C+D)	*1,544*
Direct cost of roadway transport (A+B)	*814*
External cost of roadway transport (C+D)	*729*
Roadway costs borne by everyone (B+D)	*295*

n.e. = not estimated.

[1] It is unclear why Ketcham and Komanoff did not include the cost of "Roadway construction and maintenance" in this total. It probably was an oversight. In any case, we report the totals as they are shown in the original source.

1990 dollars, but not to 1990 congestion levels. Their estimate of the national cost of motor vehicle accidents is from the Urban Institute (1991). The bulk of these two external costs is borne by users.

Land costs. According to Ketcham and Komanoff, the land cost of motor vehicle use is one of the largest external costs borne by nondrivers. They estimate the land cost nationally by scaling the estimated cost in New York City. They estimate the cost in New York City on the basis of three assumptions: that street space is one-third of the city's land area; that half of the street space is needed for movement of public vehicles, bicycles, and pedestrians (and therefore is not to be assigned to motor vehicle use); and that the value of the land in New York City is 45% of the city's $26 billion budget derived from property taxes. They then estimate the national land cost by scaling up the cost in New York City on the basis of population and labor force.

One can question all three of the assumptions that Ketcham and Komanoff use to estimate the value of land devoted to motor vehicle use in New York City. Certainly, one can question the basis for scaling the result from New York City to the entire country. Beyond that, however, it is not clear to us why they consider all of the estimated land value to be an external cost: FHWA's estimates of the cost of road construction (FHWA 1990), which Ketcham and Komanoff use in their national analysis, include the cost of acquiring rights-of-way for roads. Hence, at least some of the cost of the land is counted as an infrastructure cost, and is partially recovered from users through user fees.

Air pollution and noise. Ketcham and Komanoff derive their estimate of the cost of air pollution ($30 billion) from the estimates in the FHWA Cost Allocation study (1982), which the authors say are consistent with the ranges published in other studies. (Actually, on basis of these other studies, the authors feel that their estimate of $30 billion is conservative.) Noise cost estimates are derived from a 1981 study for FHWA by the Institute of Urban and Regional Research at the University of Iowa (Hokanson et al 1981), which estimates the nationwide costs of noise in 1977. Ketcham and Komanoff make some adjustments to this figure to account for differences between 1977 and 1990.

HANSON (1992)

Goals and Methodology

Hanson's article "delineates the nature and magnitude of automobile subsidies in the United States and considers their significance for transportation and land use policy. The central argument . . . is that the U.S. transportation system, based on and designed largely for the automobile, has been systematically subsidized in a way that produces a more dispersed settlement pattern than would have otherwise evolved" (p. 60).

In Hanson, an automobile subsidy is any direct cost of providing for and using the automobile system that is not paid for privately or through a transportation fee. Hanson uses data provided by the state of Wisconsin, supplemented with a review of existing studies, to estimate these subsidies. Wisconsin is used because it is near the national average for the percentage of state highway user revenues shared with local governments, and because Wisconsin is unique in its extensive reporting requirements.

Direct Costs

Hanson divides direct costs into three major categories. Highway construction includes right-of-way acquisition, engineering, signage, and construction costs for pavement, bridges, culverts, and storm sewers. Highway maintenance includes maintenance of pavements, bridges, culverts, storm sewers, and traffic control devices, and snow plowing. Other highway infrastructure includes machinery and vehicles, buildings, debt service payments, and street lighting. Hanson analyzes government data to make these estimates. After estimating the gross direct costs, Hanson nets out offsetting user revenues to calculate the subsidy to motor vehicle use.

Externalities and Other Indirect Subsidies

Hanson estimates the external costs of air pollution, water pollution resulting from road salt use, personal injury and lost earnings associated with accidents, land-use opportunity costs for land removed from other sources, and petroleum subsidies. Hanson points out that there are a number of other external costs, such as noise and community disruption, that he has not attempted to quantify.

In order to estimate air pollution costs for Madison, Wisconsin, he notes that the midpoint estimate in the studies of national costs that he reviewed was $7 billion. To allocate a share of this to Madison, he multiplied this midpoint figure by the ratio of the population of Madison to the population of the United States.

To estimate the personal injury costs associated with accidents, Hanson multiplies the number of accidents in 1982 (1,628 according to the Wisconsin Department of Transportation, WDOT) by the personal injury cost per accident ($7,700). He also uses a WDOT estimate of the cost of lost earnings, $1.6 million. These estimates do not include a value for fatalities.

Hanson also uses WDOT data to generate an estimate of the value of property damages resulting from accidents. However, in quantifying the amount that should be considered a subsidy, he assumes that "because a substantial portion of property damage is insured by automobile users via separate insurance coverage, and to a lesser degree by direct payments, those costs are mostly internalized and, therefore, not included."

Hanson assumes that "a land opportunity cost occurs when land, used for roads, could have been used for some other purpose." A subsidy will result if more than the "optimal" amount of land is used for highways. To provide a rough estimate of this subsidy, Hanson assumes that one-third of the surface area of highways in Madison is unnecessary. This is based on two assumptions. First, according to Cervero (1989), local roads provide 80% of the lane-miles, but only 15% of the vehicle-miles. Second, he assumes that higher travel costs would reduce travel demand and alter land use in the long run. He uses foregone property tax revenues to estimate the cost of land, and calculates that, with the existing property tax rates, Madison would gain $1 million in revenues if the area of roadways was reduced by one-third.

Hanson notes that air emissions from motor vehicles contribute to water pollution and acid rain, but believes there are few reliable published estimates of the damages. As a result, he focuses only on damages from road salt. He begins with the estimates provided by Murray and Ernst (1976), adjusts their figures to avoid double counting, converts their estimate to 1983 dollars, and finally allocates a portion of the cost to Madison on the basis of the population in the snowbelt "salt zone."

To estimate petroleum subsidies, Hanson uses Hines' (1988) estimates of the depletion allowances and other tax breaks received by the petroleum industry in 1984. This is allocated to Madison by combining gasoline consumption for personal travel in Madison with the subsidy level per British thermal unit (Btu).

CONGRESSIONAL RESEARCH SERVICE (BEHRENS ET AL 1992)

Goals and Methodology

Congress asked the Congressional Research Service (CRS) to summarize for the U.S. Alternative Fuels Council what is known about monetary estimates of the side effects (external costs) of oil used in highway transportation. In its analysis, CRS considers three kinds of costs: economic costs stemming from the dependence on world oil markets, national defense costs, and health and environmental impacts. They review previously published studies, and develop what they believe are reasonable low- to mid-range estimates of the monetary value of these external costs.

The results of this study are summarized in table 3. Note that CRS, like Kanafani (1983), reports estimates directly from the literature without converting or updating the dollars to a base year.

TABLE 3 Estimated External Costs of Oil Used in Transport
Congressional Research Service
(Behrens et al 1992)
(Billions of dollars)

Cost category	Low	High
Risk of supply disruption	3.2	4.9
Monopsony effects	11.3	13.0
Military expenditures	0.3	5.0
Air pollution—human health	3.6	3.6
Air pollution—crop damages	1.1	1.1
Air pollution—material damages	0.3	0.3
Air pollution—visibility	0.8	0.8
Oil spills	n. e.	n. e.
Total with monopsony effects[1]	*10.5*	*17.0*
Total without monopsony effects[1]	*21.8*	*30.0*

n.e. = not estimated.
[1] The estimates in each category and the totals shown here are those reported in Behrens et al and are based on a review of the literature. The authors did not convert the dollar estimates in the literature to a single dollar base year. The totals are the overall estimates, not the sum of the individual estimates.

Economic Costs of Oil Dependence

CRS considers two effects on the economy due to oil dependency: the risk of a supply disruption, and the market power or monopsony effect. The former is the result of exposure to "possible market manipulation or disruption by exporting nations" (p. 7). Some of the potential adverse impacts include higher inflation and unemployment, as well as possible balance of payments and exchange rate effects. The range of estimates of the costs associated with this are from zero to $10 per barrel. Multiplying the results of a mid-range estimate by U.S. oil imports for 1990, the authors estimate a $6 billion to $9 billion cost to the economy due to the risk of disruption.

According to CRS, "the market power or monopsony component reflects the influence on the world price that a large importer such as the United States causes" (p. 7). The economic cost of this is the failure to transfer wealth to U.S. citizens by reducing U.S. oil imports (which would result in lower world oil prices). Based on a literature review, the authors use a mid-range estimate between $21 billion and $24 billion for not exploiting this power.

National Defense Costs

CRS considers military expenditures that could be avoided if the United States and other industrialized countries did not need to import oil from the Persian Gulf (p. 23). In developing its estimate, CRS reviews the estimates provided by the U.S. General Accounting Office (1991), Ravenal (1991), and Kaufmann and Steinbruner (1991).

CRS discusses at length some of the difficulties of attributing military expenditures to the defense of Persian Gulf oil interests. As a result of these difficulties, cost estimates can range from a few cents to a few hundred dollars per barrel. The authors conclude that attempts to reduce U.S. dependence on imported oil will probably have little effect on the amount spent in the Persian Gulf. They also note that attempts to internalize these costs may not have a significant impact on reducing the costs.

Health and Environmental Impacts

CRS estimates the impacts of motor vehicle pollution on human health, crop yields, certain species in forests, materials, and visibility, but not climate change. The authors acknowledge that there will be damage to ecosystems resulting from oil spills, but believe that there are no "defendable estimates of the monetary value of the external costs associated with oil spills" (p. 55).

CRS emphasizes that "the effects on the environment and health . . . are imperfectly understood. And how these environmental and health damages can be approximated in monetary terms is controversial" (p. 10).

On the basis of a literature review, the authors conclude that a "reasonable estimate of the lower range of health and welfare damages resulting from transportation-related pollution is between $5 and $6 billion per year" (p. 52). We note that this is one of the lower estimates in the literature.

MILLER AND MOFFET (1993)

Through a survey of existing literature, Miller and Moffet attempt to develop estimates of the full cost of transportation in the United States in 1990. In addition to estimating the costs associated with automobile transportation, they also estimate these costs for bus and rail transportation. Table 4 summarizes their estimates of the costs of automobile use.

They consider three categories of costs. "Personal costs," which include the costs to purchase, register, maintain, and operate a car, are borne solely by the vehicle owner. "Government subsidies" include direct construction and maintenance expenditures plus other government expenses directly associated with providing transportation services. Miller and Moffet's estimate of these costs are net of user fees. "Societal costs" include all other indirect costs, or what is often referred to as externalities. Examples of this include energy dependence, pollution, and congestion.

Miller and Moffet estimate that the full annual costs of automobile transportation were between $1.1 trillion and $1.6 trillion in 1990. They estimate that $72 billion of this was government subsidies, between $310 billion and $592 billion were societal costs, and the remaining $775 billion to $930 billion were personal costs incurred by the vehicle owners. However, one must be cautious in interpreting their estimate of the full annual costs of automobile transportation. The bulk of this estimate is comprised of personal costs entirely borne by vehicle users, and it can be somewhat confusing when this figure is added to net government expenditures, rather then gross government expenditures. Total unpaid social costs, that is, net government subsidies plus societal costs, totaled between $378 billion and $660 billion in 1990.

TABLE 4 The Full Cost of Transportation in the United States: 1990
Miller and Moffet 1993

Category	$ billion
Personal costs	
Ownership and maintenance	775–930
Government subsidies	
Capital and operating expenses	64.0
Local government expenses	8.0
Total government subsidies	*72.0*
Societal costs	
Energy dependence	45–150
Congestion	11.0
Parking	25–100
Accidents	98.0
Noise	2.7–4.4
Building damage	0.3
Air pollution	120–220
Water pollution	3.8
Total societal costs	*310–592*
Unquantified costs	
Wetlands lost	n.e.
Agricultural land lost	n.e.
Damage to historic property	n.e.
Changes in property value	n.e.
Equity effects	n.e.
Urban sprawl	n.e.
Total government and societal costs	*378–660*
Total costs	1,153–1,590

n.e. = not estimated.

Note: This table is reproduced directly from Miller and Moffet. Note that both the total government and societal costs and the total costs do not add up, presumably due to rounding.

KPMG PEAT MARWICK, STEVENSON, AND KELLOGG (1993)

As part of a long-range transport planning initiative, the Greater Vancouver Regional District and the Province of British Columbia hired KPMG Peat Marwick, Stevenson, and Kellogg to analyze the full costs of various modes of passenger transportation in British Columbia. The resulting study estimates the total cost of transporting people in the Lower Mainland in 1991, and calculates the average cost "per unit" of travel, by different modes, for urban peak, urban off-peak and suburban travel (p. iii). There are three specific goals of the analysis: 1) estimate the total economic costs of different modes of passenger transport in the region; 2) determine how much is paid by users of different transport modes and how much is paid by non-users; and 3) provide a broad basis for assumptions and recommendations regarding the future levels and methods of pricing the movement of people in the region.

The authors utilize a computer model to estimate these costs for five different modes of private transport (average car, fuel-efficient car, car pool, van pool, and motorcycle), four modes of public transport (diesel bus, trolley bus, SkyTrain, and SeaBus), and three modes of nonmotorized transport (bicycle, walking, and telecommuting). The costs are evaluated for travel in urban areas during peak and off-peak hours, as well as for suburban travel. They find that the total subsidy for automobile transport in 1991 was \$2.7 billion Canadian dollars (C\$2.7 billion).

The authors estimate that the total cost associated with the transportation in the Lower Mainland of British Columbia in 1991 was approximately C\$13.6 billion. The five modes of transport via private motor vehicles accounted for C\$11.7 billion (86%) of the total cost, and were subsidized at approximately C\$2.7 billion, or 23% of the total cost of private transport.

CALIFORNIA ENERGY COMMISSION (1994)

Purpose

In the aftermath of the 1991 Persian Gulf War, the California legislature passed, and Governor Pete Wilson signed into law, Senate Bill 1214, which provides, in part, that:

> it is the policy of this state to fully evaluate the economic and environmental costs of petroleum use . . . including the costs and value of environmental externalities, and to establish a state transportation energy policy that results in the least environmental and economic cost to the state (CEC 1994, 1).

The task of developing a "least environmental and economic cost scenario," including the costs and values of environmental externalities and energy security, was assigned to the California Energy Commission (CEC), as part of its biennial report. To fulfill this charge, CEC analyzed the social costs and benefits of several state and national energy policies, relative to a base case. The policy measures included increasing fuel taxes, increasing fuel economy standards, and subsidizing the price of alternative fuels and vehicles. For each policy, CEC estimated the differences in travel, emissions, fuel use, and so forth, relative to the base case. The value of the differences was the net social cost or benefit of the policy.

Estimates of Avoidable Costs

CEC quantified several kinds of social costs: travel time, accidents, infrastructure maintenance and repair, governmental services, air pollution, carbon dioxide, petroleum spills, and energy security.

Travel time. CEC used the "Personal Vehicle Model," a demand forecasting model that projects vehicle stock, VMT, and fuel consumption for personal cars and trucks, to estimate that congestion costs, including the disutility of aggravation, are \$10.60 per hour (1992\$). CEC also estimated the actual net change in travel time in Los Angeles under the various policy scenarios.

Accidents. The cost of accidents is estimated by multiplying the cost per injury or death by the number of injuries or deaths, for several kinds of injuries. Ted Miller, lead author of the much-cited Urban Institute (1991) study of the cost of highway crashes, developed California-specific unit costs for the Commission. CEC uses the Urban Institute study to allocate costs to different vehicle classes.

Air pollution. To calculate the cost per mile of air pollution, CEC multiplied the change in total emissions (estimated using California's mobile source emissions inventory models, EMFAC and

BURDEN), by the dollar-per-ton value of emissions, and then divided by the change in travel. The dollar-per-ton values, estimated for nitrogen oxides, sulfur oxides, reactive organic gases, particulate matter, and carbon monoxide, are from the Air Quality Valuation Model, a damage function model that estimates the cost of air pollution from powerplants in California air basins. CEC acknowledges that damage values for powerplants might not apply to motor vehicles.

Carbon dioxide. Because, according to CEC, "reliable data on damage functions are not available . . . the Energy Commission uses carbon emission control costs alone to represent carbon values" (p. 3G-1). CEC adopted its own control cost estimate of $28 per ton-carbon from its 1990 electricity report. To estimate the total carbon dioxide cost of different policies, CEC multiplied the cost per ton by the change in carbon emissions under the different scenarios. Carbon emissions rates for different fuel cycles were taken from reports by CEC, the U.S. Environmental Protection Agency, and DeLuchi et al (1987).

APOGEE RESEARCH (1994)

The report by Apogee Research, prepared for the Conservation Law Foundation, presents the results of case studies of intraurban passenger transportation in Boston, Massachusetts, and Portland, Maine. The report "attempts to develop a framework for comparing transportation costs and to provide specific quantification of the costs of passenger transportation" in the two regions analyzed. The methodology developed by the authors was constructed such that it could be adapted for other case studies.

The study evaluates nine "sub-modes" of transportation: single-occupancy vehicles (SOVs) on expressways, SOVs off expressways, high-occupancy vehicles (HOVs) on expressways, HOVs off expressways, commuter rail, rail transit, bus, bicycle, and walking. It also distinguishes between high, medium, and low population densities, and between on-peak and off-peak travel. Table 5 summarizes the cost categories used in their report.

Their report is divided into four main sections. The first is a comprehensive literature review that provides background information for the analytic framework. The next section describes the methodology used in the case studies, and defines the costs and travel parameters studied. The analytic framework is then applied to estimate the costs in Portland and Boston. Finally, the report presents the results of the case studies and suggests some policy responses.

Apogee Research focuses primarily on developing original estimates for user and governmental costs, and relies on existing estimates for the societal costs. Wherever possible, they try to use data from the relevant agencies to develop their cost estimates. This is supplemented by literature reviews when data were unavailable. The cost estimates derived from these data are primarily the result of relatively simple, yet intuitively reasonable, analysis, rather than the product of more complex and rigorous statistical models. The authors acknowledge this, stating that "while additional research and analysis on particular costs would undoubtedly lead to more refined results, we believe that these case studies provide a good sense of the magnitude of the various costs of transportation" (p. 59).

The policy recommendations provided in the report are common to most analyses: reduce trip length, favor lower cost modes, increase vehicle occupancy, explore single occupancy vehicle pricing, and educate the public on transportation costs.

LEE (1994)

This draft paper examines the debate about the extent to which drivers pay the costs associated with motor vehicle use. Lee uses a "full cost pricing" approach to analyze this issue. "Full cost pricing is a policy strategy based on the idea that the economy would benefit from imposing the discipline on each enterprise that all its costs should be recovered from consumers, i.e., total user revenues should equal total cost for each activity" (p. 1).

Lee is concerned more with theoretical issues than with estimates of costs. After discussing the fundamental economic issues pertaining to full and marginal cost pricing, Lee outlines a strategy for estimating these costs. His focus is not so much on estimating costs as developing an appropriate analytical structure. He discusses which costs should be included in a social costs analysis, why they

TABLE 5 Transportation Cost Categories
Apogee Research 1994

User costs[1]	Governmental costs[2]	Societal costs[3]
Vehicle purchase and debt	Capital investment—land, structures, vehicles	Parking—free private
Gas, oil, tires[4]	Operations and maintenance	Pollution—health care, cost of control, productivity loss, environmental harm
Repairs, parts	Driver education and DMV	Private infrastructure repair—vibration damage, etc.
Auto rentals	Police, judicial system, and fire	Accidents—health insurance, productivity loss, pain and suffering
Auto insurance	Parking—public, tax breaks	Energy—trade effects
Tolls[4]	Energy—security	Noise
Transit fares[4]	Accidents—public assistance	Land loss—urban, crop value, wetlands
Registration, licensing and annual taxes[4]	Pollution—public assistance	Property values and aesthetics
Parking—paid		Induced land-use patterns
Parking—housing cost		
Accidents—private expense		
Travel time		

DMV = Department of Motor Vehicles.
[1] User costs are the costs borne by vehicle owners: the direct ownership and operating costs, such as gas, oil and parts; the indirect costs, such as garage parking and accident risks.
[2] Governmental costs include expenditures that are not explicitly for the purpose of transportation, but which nevertheless are necessitated by vehicle travel.
[3] Societal costs of transportation are those paid by neither the traveler nor the government, but rather are spread across the economy.
[4] These items are, or include, dedicated taxes that fund governmental transportation expenditures and must be deducted from costs in

should be estimated, and important theoretical issues on how they should be calculated. However, Lee does make some estimates of unpaid costs, primarily on the basis of a literature review. Table 6 summarizes his estimates.

COHEN (1994)

The goal of this study is to "update and extend the analysis of the external costs of highway operations that was reported in appendix E of the final report on the 1982 Federal Highway Cost Allocation Study" (p. 1). The present report actually is an interim report. It summarizes the literature on estimating external costs, assesses recent efforts to develop national estimates of these costs, and recommends procedures that should be used to develop cost models and estimate the monetary value of external costs.

When the final report is completed, it will contain three primary elements. First, it will provide estimates of the external costs due to congestion delay, highway crashes, noise, and air pollution. Second, the report will include a simple computer model to reproduce these results in future analyses. Third, it will include a detailed discussion of institutional barriers, equity implications, and political consideration that affect marginal cost pricing and other methods to charge highway users for external costs.

For the most part, the literature review in the interim report refers to studies that we have reviewed here. And, because this is an interim report, there are no actual cost estimates for us to report. However, it appears that the authors are in the process of developing a useful framework for making original estimates of these costs. Recent unpublished manuscripts from this project indicate that they are using external cost estimation methods similar to those summarized in Delucchi et al (1996).

TABLE 6 Estimates of Highway Costs not Recovered from Users: 1991
Lee 1994

Cost group	Cost item	$ billion
Highway capital	Land (interest)	74.7
	Construction, capital expenditures	42.5
	Construction, interest	26.3
	Land acquisition and clearance	n.e.
	Relocation of prior uses and residents	n.e.
	Neighborhood disruption	n.e.
	Removal of wetlands, aquifer recharge	n.e.
	Uncontrolled construction noise, dust, runoff	n.e.
	Heat island effect	n.e.
Highway maintenance	Pavement, right-of-way, and structures	20.4
Administration	Administration and research	6.9
	Traffic police	7.8
Parking	Commuting	52.9
	Shopping, recreation, services	14.9
	Environmental degradation	n.e.
Vehicle ownership	Disposal of scrapped or abandoned vehicles	0.7
Vehicle operation	Pollution from tires	3.0
	Pollution from used oil and lubricants	0.5
	Pollution from toxic materials	0.0
Fuel and oil	Strategic petroleum reserve	4.4
	Tax subsidies to production	9.0
Accidental loss	Government compensation for natural disaster	n.e.
	Public medical costs	8.5
	Uncompensated losses	5.9
Pollution	Air	43.4
	Water	10.9
	Noise and vibration	6.4
	Noise barriers	5.1
Social overhead	Local fuel tax exemptions	4.3
	Federal gasohol exemption	1.2
	Federal corporate income tax	3.4
	State government sales taxes	13.2
	Local government property taxes	16.0
	Total cost	*382.1*
	Current user revenues	*52.1*
	Profit (loss)	*(330.0)*

n.e. = not estimated.

LITMAN (1996)

The purpose of Litman's analysis is to establish a foundation for analyzing transportation costs. After estimating the costs for the United States in 1994, primarily through an extensive literature review, he discusses the implications of these costs with respect to efficiency, equity, land use, stakeholder perspectives, and future policy options.

Litman classifies transportation costs into three dichotomies, as shown in table 7: internal (users) or external (social) costs, market or nonmarket costs, and fixed or variable costs. He estimates these costs for 11 different modes of transportation. In order to estimate the costs, Litman conducted a literature review, and from this information, generates his "best guess" at the true cost.

TABLE 7 Motor Vehicle Transportation Costs
Litman 1996

		Variable	Fixed
Internal	*Market*	Fuel Short-term parking Vehicle maintenance (part)	Vehicle purchase Vehicle registration Insurance payments Long-term parking facilities Vehicle maintenance (part)
	Nonmarket	User time and stress User accident risk	
External	*Market*	Road maintenance Traffic law enforcement Insurance disbursements	Road construction "Free" or subsidized parking Traffic planning Street lighting
	Nonmarket	Congestion delays Environmental impacts Uncompensated accident risk	Land-use impacts Social inequity

His estimates of costs for motor vehicles are summarized in table 8. In 1994, internal costs were about $1.6 trillion, and accounted for about two-thirds of the total costs. External costs amounted to about $0.8 trillion.

In Litman's analysis, the value of user time alone accounts for over 20% of the total cost of the average automobile used during peak times in urban areas. As a basis for deriving the costs, Litman uses a 1992 value of time schedule for British Columbia because it is "current and comprehensive." That study assumes that the value of the personal vehicle driver's time is 50% of the current average wage, which Litman assumes to be $12 per hour. He calculates total costs assuming average speeds of 30 mph (urban peak), 35 mph (urban off-peak), and 40 mph (rural), and an hourly cost premium of 16.5% in congestion.

In Litman's analysis, land-use impacts and parking costs are the largest external costs associated with an average car. On the basis of a review of the literature, Litman assumes that the average automobile off-street parking cost is around $3 per day.

TABLE 8 Motor Vehicle Costs in the United States: 1994
Litman 1996
(Billions of 1994 dollars)

	Internal costs	External costs	Total costs
Urban peak	327	281	607
Urban off-peak	653	313	966
Rural	589	184	773
Total	*1,569*	*778*	*2,347*

According to Litman, "a primary conclusion of this research is that a major portion of transportation costs are external, fixed, or non-market . . . This underpricing leads to transportation patterns that are economically inefficient and inequitable . . ." (p. vi).

LEVINSON ET AL (1996)[6]

Goals and Methodology

The goal of this report is to compare the costs of intercity passenger travel by air, automobile, and high-speed rail in the California Corridor (i.e., between San Francisco and Los Angeles). The policy question they address is whether the full costs of developing a high-speed rail line are comparable to the costs of expanding the air or highway transportation systems. To accomplish this, they develop long- and short-run average and marginal cost functions for each of the three modes of travel. Our discussion of this report will be limited to their analysis of the highway costs.

[6]In this review, we refer to the pair of 1996 papers by Levinson et al (1996a and 1996b) as Levinson et al (1996). The later paper, 1996b, is a condensed journal article that summarizes the more detailed research report, 1996a.

TABLE 9 Long-Run Full Costs of the Highway System
Levinson et al 1996
(Dollars per vehicle-kilometer)

Cost category	Short-run costs		Long-run costs	
	Marginal	Average	Marginal	Average
Infrastructure costs				
Construction and maintenance	0.0055	0.0008	0.0180	0.0174
External costs				
Accidents	0.0350	0.0310	0.0350	0.0310
Congestion	0.0330	0.0680	0.0330	0.0068
Noise	0.0090	0.0060	0.0090	0.0060
Pollution	0.0046	0.0046	0.0046	0.0046
Total external costs	*0.0816*	*0.1096*	*0.0816*	*0.0484*
User costs				
Fixed + variable	0.0490	0.1300	0.0490	0.1300
Time	0.5000	0.5000	0.1500	0.1500
Total user costs	*0.5490*	*0.6300*	*0.1990*	*0.2800*
Total costs[1]	*0.2861*	*0.3292*	*0.2986*	*0.3458*

[1] This table is reproduced directly from Levinson et al without changes. Note that the total for the short-run costs do not add up properly.

They identify three types of costs associated with automobile use: infrastructure costs, user costs, and social (or external) costs.[7] For the most part, Levinson et al develop their own econometric models to estimate these costs. Each of these is discussed in more detail below. A summary of their estimates of the long-run full costs of the highway system is provided in table 9.

Infrastructure Costs

Infrastructure costs include the capital costs of infrastructure construction and debt servicing, and operations and maintenance costs. Levinson et al develop an econometric model that predicts total expenditures as a function of the price of inputs (interest rates, wage rates, and material costs), outputs (miles traveled per passenger vehicle, single unit truck, and combination truck), and network variables (the length of the network and the average width of the links). The data used for the model come from a variety of sources, such as FHWA data on maintenance and operating costs, and Gillen et al (1994) data on capital stock, among others. Costs are allocated among the different vehicle classes on the basis of an engineering analysis of the amount of damage caused by each vehicle type.

[7] Note that Levinson et al (1996) use a different definition of social costs than we do in our own analyses (Delucchi et al 1996). In their report, they limit the definition of social costs to negative externalities, or external costs.

User Costs

Levinson et al estimate the cost of gas, oil, maintenance, tires, and depreciation for an intermediate-size automobile, the most popular vehicle type in 1995. (They omit insurance costs, license and registration fees, and taxes on the grounds that they are transfers.) For most of their estimates, they use data from the American Automobile Association (AAA). However, to estimate depreciation, they regress the posted price (not the actual transaction price) in an Internet classified ad for Ford Taurus and Honda Accord against the age of the vehicle and the distance traveled multiplied by the vehicle age. From this, they estimate depreciation costs of $1,351 per year and 2.3¢ per vehicle-mile of travel, which, assuming 10,000 miles per year, translates to an annual depreciation of about $1,581, as compared with the AAA estimate of $2,883 in 1993. To estimate the cost of user time, the authors assume that travel time costs $10 per hour and vehicles travel at 100 km per hour.

External Costs

Levinson et al identify four external costs, which they also refer to as social costs: accidents, congestion, noise, and air pollution. Their estimates for each of these costs are based on simple models and an analysis of existing work.

Accidents. Their estimate of accident costs is developed by combining an accident rate model by Sullivan and Hsu (1988) with the work of the Urban Institute (1991). The accident cost is obtained by determining the value of life, property, and injury per accident, and multiplying this by an equation that represents accident rates. They estimate that a crash on a rural interstate costs about $120,000 (in 1995 dollars), and a crash on an urban interstate costs about $70,000. The disparity is largely attributable to the higher death rate associated with accidents on rural highways due to the higher speed of travel.

Congestion. Assuming a modest average traffic flow of 1,500 vehicles per hour per lane, a $10 per hour value of time, and 1.5 passengers per vehicle, the authors estimate that the average congestion costs are $0.005 per passenger-kilometer of travel. This is based on a simple analysis of the relationship between traffic volumes and time delay.

Noise. For noise costs, they develop a simple analytical framework and use the results of previous research to derive their estimates. Essentially, this involved translating noise production rates into economic damages using total residential property damage costs per linear-kilometer of roadway.

Air pollution. The authors identify four types of air pollution (photochemical smog, acid deposition, ozone depletion, and global warming), which generate three types of damages (health effects, material and vegetation effects, and global effects). Their estimate of the total cost of air pollution is derived by combining the results of a number of other studies.

Costs Excluded from the Analysis

Levinson et al (1996) do not include U.S. defense expenditures in the Middle East or the costs of parking in their analysis. They dispute the notion that a significant share of U.S. defense expenditures are directly related to the transportation sector. They exclude parking on the grounds that it is a local cost that is unlikely to be avoided by switching intercity travel modes.

DELUCCHI ET AL (1996)

In a series of 20 reports, Delucchi et al (1996) estimate the annualized social cost of motor vehicle use, as:

- 1990 to 1991 periodic or "operating" costs, such as fuel, vehicle maintenance, highway maintenance, salaries of police officers, travel time, noise, injuries from accidents, and disease from air pollution;

  plus
- the 1990 to 1991 value of all capital, such as highways, parking lots, and residential garages (items that provide a stream of services), converted (annualized) into an equivalent stream of annual costs over the life of the capital.

This annualization approach essentially is an investment analysis, or project evaluation.

They classify and estimate costs in six general categories: personal nonmonetary costs, motor vehicle goods and services priced in the private sector, motor vehicle goods and services bundled in the private sector, motor vehicle goods and service provided by government, monetary externalities, and nonmonetary externalities.

Personal Nonmonetary Costs

In Delucchi et al, personal nonmonetary costs are those unpriced costs of motor vehicle use that a person imposes on him or herself as a result of the decision to travel. The largest personal costs of motor vehicle use are personal travel time in uncongested conditions and the risk of getting into an accident that involves nobody else. Delucchi et al perform detailed analyses of travel time costs in this category.

Motor Vehicle Goods and Services Priced in the Private Sector

The economic cost of motor vehicle goods and services supplied in private markets is the area under the private supply curve: the dollar value of the resources that a private market allocates to supplying vehicles, fuel, parts, insurance, and so on. To estimate this area, Delucchi et al subtract producer surplus (revenue in excess of economic cost) and taxes and fees (mainly noncost transfers) from total price-times-quantity revenues. The cost items in this category include those in the "transportation" accounts of the Gross National Product (GNP), and several others. For several of these costs, Delucchi et al use the same primary data and methods used in GNP accounting.

Motor Vehicle Goods and Services Bundled in the Private Sector

Some very large costs of motor vehicle use are not explicitly priced separately. Foremost among these are the cost of free nonresidential parking, the cost of home garages, and the cost of local roads provided by private developers. However, all of these costs are included in the price of "packages," such as homes and goods, that are explicitly priced.[8] Delucchi et al use a variety of primary data sources to estimate national parking and garage costs in detail.

Motor Vehicle Goods and Services Provided by the Public Sector

Government provides a wide range of infrastructure and services in support of motor vehicle use. The most costly item is the highway infrastructure. Delucchi et al analyze survey data from FHWA, the Bureau of the Census, the Department of Energy, the Department of Justice, and other government departments to estimate these infrastructure and service costs. They note that, whereas all government expenditures on highways and the highway patrol are a cost of motor vehicle use, only a portion of total government expenditures on local police, fire, corrections, jails, and so on is a cost of motor vehicle use.

Monetary Externalities

Some costs of motor vehicle use are valued monetarily yet are unpriced from the perspective of the responsible motor vehicle user, and hence are external costs. Examples of these are accident costs that are paid for by those *not* responsible for the accident, and congestion that displaces monetarily compensated work. Delucchi et al estimate that the largest monetary externalities are those resulting from travel delay.

Nonmonetary Externalities

Delucchi et al follow Baumol and Oates (1988) and define a nonmonetary externality as a cost or benefit imposed on person A by person B but not accounted for by person B. Environmental pollution, traffic delay, and uncompensated pain and suffering due to accidents are common examples of externalities.

Environmental costs include those related to air pollution, global warming, water pollution, and noise due to motor vehicles. Delucchi et al use damage functions to estimate air pollution and noise costs. They find that by far the largest environmental externality is the cost of particulate air pollution.

The authors' estimates of the total social costs in each of the six cost categories are summarized in table 10.

STUDIES OF THE SOCIAL COSTS OF MOTOR VEHICLE USE IN EUROPE

Although this paper focuses on U.S. studies, there are a number of good studies of the social costs of motor vehicle use in Europe. Quinet (1997) provides the most comprehensive and up-to-date summary of European studies of the external cost of traffic noise. In Quinet, the range of noise cost estimates is between 0.02% and 2.0% of Gross Domestic Product (GDP); the range of local pollution costs, between 0.03% and 1.0% of GDP; and the range of accident costs, between 1.1% and 2.6% of GDP.

Verhoef (1994) also summarizes many estimates of the external cost of noise (0.02% to 0.2% of GDP), air pollution (0.1% to 1.0% of GDP), and accidents (0.5% to 2.5% of GDP) attributable to road traffic, and Kageson (1992) and Ecoplan (1992) summarize estimates of the damage cost of air pollution caused by the transport sector (0.01% to 1.0% of GDP). These ranges indicate that

[8] Delucchi et al note that this bundling is not necessarily inefficient: in principle, a producer will bundle a cost, and not price it separately, if the administrative, operational, and customer (or employee) cost of collecting a separate price exceed the benefits.

TABLE 10 **Summary of the Costs of Motor Vehicle Use: 1990–91**
Delucchi et al 1996

Category	Low	High	Low	High
	(billion 1991$)		(percent)	
1. Personal nonmonetary costs of motor vehicle use	$584	$861	30	26
2. Motor vehicle goods and services priced in the private sector (estimated net of producer surplus, taxes, fees)	$761	$918	40	28
3. Motor vehicle goods and services bundled in the private sector	$131	$279	7	8
4. Motor vehicle infrastructure and services provided by the public sector	$122	$201	6	6
5. Monetary externalities of motor vehicle use	$55	$144	3	4
6. Non-monetary externalities of motor vehicle use	$267	$885	14	27
Grand total social costs of highway transportation	*$1,920*	*$3,289*	*100*	*100*
Subtotal: monetary cost only (2+3+4+5)	*$1,069*	*$1,543*		

European estimates of air pollution and accident costs are somewhat lower than recent detailed U.S. estimates (e.g., Delucchi et al 1996).

Several recent, detailed studies are not included in the reviews by Quinet (1997), Verhoef (1994), or Kageson (1992). Eyre et al (1997) estimate the effects of fuel and location on the damage cost of transport emissions. Bickel and Friedrich (1995; 1996) use a damage function approach to estimate the external costs of accidents, air pollution, noise, land use, and "dissociation effects" (e.g., roads as barriers or dividers in communities) of passenger vehicles, freight trucks, passenger rail, and freight rail in Germany in 1990. Otterström (1995) uses a detailed damage function approach, similar to the method of Delucchi et al (1996, Report #9), to estimate the external cost of the effect of traffic emissions on health, crops, materials, forests, and global warming in Finland in 1990. Maddison et al (1996; summarized in Maddison 1996) use a variety of methods to estimate the marginal external costs of global warming, air pollution, noise, congestion, road damage, and accidents attributable to road transport in the United Kingdom in 1993. Mayeres et al (1996) develop marginal cost functions, again similar to those of Delucchi et al (1996, Report #9), to estimate the marginal external cost of congestion, accidents, air pollution, and noise attributable to cars, buses, trams, metro rail, and trucks in the urban area of Brussels in the year 2005.

SUMMARY AND CONCLUSION

Our Rating of the Level of Detail

A review of the study summaries, in tables 1 to 10, indicates that in most cost categories, there is a very wide range of estimates. These ranges result from differences in every conceivable facet of the analysis: scope, accounting system, analytical methods, assumptions, and data sources. Because of this, it is not possible to give a simple summary of the *overall* quality of each analysis, or of the sources of discrepancies between analyses. However, it is possible and we hope useful to evaluate the studies according to one partial indicator of quality: the level of analytical detail.

Tables 11a to 11d identify some of the major cost categories included in these studies. For each cost category, we give a rating of A through F, which is our assessment of the level of analytical detail underlying each estimate in the studies reviewed. These ratings are explained in more detail in table 12. We emphasize that they are not necessarily assessments of the *overall* quality, because there is more to quality than detail, and a review and analysis of sound and pertinent literature is preferable to a poorly done detailed, original analysis. Nevertheless, it is useful for policymakers to know who has done a detailed original analysis, and who has done a combination of literature review and detailed analysis, and who has simply cited the work of others.

TABLE 11a Summary of Social Cost Items and Level of Detail in the Studies Reviewed[1]

Author	Keeler and Small (1975)	FHWA (1982)	Kanafani (1983)	Fuller et al (1983)
Geographic region	San Francisco	USA	USA	USA
Year(s) of estimates	1972–73	1981	Varies	1976–79
Primary purpose or objective	Efficient resource use; compare travel modes	Cost allocation	Compare estimates for different countries	Cost allocation
Cost categories[2]				
Accidents	B	F	C	A1/B
Air pollution	A1/B	B	C	A1/B
Congestion/time	A1	B		A1/B
Energy dependence[3]				
Equity				
Global warming/climate change				
Military expenditures				
Noise pollution	A1/B	B	C	A1
Parking	C			
Pavement damage to vehicles		E		
Roadway construction	A1/A2			
Roadway maintenance	A1/A2	A2		
Highway services[4]	A1	C		
Strategic petroleum reserve				
Urban sprawl/land use				
Vehicle ownership and operation		F		
Vibration damage to buildings				
Water pollution		F		

FHWA = Federal Highway Administration.
[1] The ratings A through F are defined in table 12.
[2] This list of cost categories is not meant to be all-inclusive. Instead, it represents some of the costs that are commonly estimated in these studies. The category definitions in this table necessarily are generic, because each study uses its own specific definitions. It is possible that some of the studies include other costs that are not identified in this table.
[3] Energy dependence may include such costs as macroeconomic effects of monopsony power, threats of supply disruption, trade effects, and petroleum subsidies.
[4] Highway services include such costs as police services, fire protection services, the judicial system, and paramedics.

Of course, there is a fair bit of judgment in our assessment here. What one person might consider a combination of literature review and detailed analysis of primary data (our "B" rating), another might consider a detailed analysis of the literature (our "C" rating). Although we tried to assess the studies consistently and evenhandedly, we recommend that readers consult the original studies to fully understand their level of detail as well as their overall quality.

Table 11 shows that the range in the level of detail is quite broad. For example, most of the estimates of MacKenzie et al (1992)—one of the most widely cited analyses—are based on a straightforward literature review. Miller and Moffet (1993) provide a significantly more detailed discussion of the issues, but still derive most of their estimates from the literature. Litman (1994) conducts a rather extensive literature review, and uses this as a basis for generating his "best guess" of the costs. By contrast, Levinson et al (1996) derive their estimates of the marginal and average costs from econometric models, and Delucchi et al (1996) primarily use original data analysis for their figures.

Conclusion

This review, and the ratings in tables 11a to 11d, indicate that many of the current estimates are based on literature reviews rather than detailed analysis. Of course, this in itself is not *necessarily* bad. The real problems are: 1) many of the reviews rely on outdated, superficial, nongeneralizable, or otherwise inappropriate studies; and 2) many of the cost-accounting systems are not fully articulated, or else are a mix of economic and equity crite-

TABLE 11b Summary of Social Cost Items and Level of Detail in the Studies Reviewed[1]

Author	MacKenzie et al (1992)	Ketcham and Komanoff (1992)	Hanson (1992)	Behrens et al (1992)
Geographic region	USA	USA	Madison, WI	USA
Year(s) of estimates	1989	1990	1983	Varies
Primary purpose or objective	Equity; efficient resource use	Efficient resource use	Equity; efficient resource use	Estimate external costs; compare alternative fuels
Cost categories[2]				
Accidents	D	D	D	
Air pollution	C	D, C	C	C
Congestion/time	C	D		
Energy dependence[3]			D	C
Equity				
Global warming/climate change	C	D		F
Military expenditures	D	D		C
Noise pollution	D	D		
Parking	D			
Pavement damage to vehicles		D		
Roadway construction	A2	D	A2	
Roadway maintenance	A2	D	A2	
Highway services[4]	D/E	D	A2	
Strategic petroleum reserve	D	D		C
Urban sprawl/land use			B	
Vehicle ownership and operation	D	D		
Vibration damage to buildings	E	D		
Water pollution			D	

TABLE 11c Summary of Social Cost Items and Level of Detail in the Studies Reviewed[1]

Author	Miller and Moffett (1993)	KPMG (1993)	CEC (1994)	Apogee (1994)	Lee (1994)
Geographic region	USA	British Columbia	California	Boston; Maine	USA
Year of estimates	1990	1990	Varies	1993	1991
Primary purpose or objective	Efficient resource use; compare travel modes	Efficient resource use; compare travel modes	Efficient resource use	Efficient resource use; compare travel modes	Efficient pricing and resource use
Cost categories[2]					
Accidents	B/C	A1/B	B	B	C
Air Pollution	B	B	B	B	C
Congestion/time	C	A1/B	A1/B	B/D	F
Energy dependence[3]	C		C	D	
Equity	F				
Global warming/climate change	C	B	D		
Military expenditures	C				
Noise pollution	C	A1/A2		D	C
Parking	C	A1/A2		A1	B
Pavement damage to vehicles					
Roadway construction	A2	A2		A2	A2
Roadway maintenance	A2	A2	A2	A2	A2
Highway services[4]	D	A2/E	D	A2	C
Strategic petroleum reserve	C				B/C
Urban sprawl/land use	F	E			F
Vehicle ownership and operation	D	B		A1/B	C
Vibration damage to buildings	D				
Water pollution	B	D	B/C		C

See the notes in table 11a.

TABLE 11d Summary of Social Cost Items and Level of Detail in the Studies Reviewed[1]

Author	Cohen (1994)[5]	Litman (1996)	Levinson et al (1996)
Geographic region	USA	USA	California
Year(s) of estimates	1990	1990	1995–96
Primary purpose or objective	Cost allocation	Equity; efficient resource use and pricing; compare travel modes	Compare travel modes
Cost categories[2]			
Accidents	F (A1/B)	B/C	A1/B
Air pollution	F (A1/B)	C	B
Congestion/time	F (A1)	B	B
Energy dependence[3]		C	
Equity		E	
Global warming/climate change			
Military expenditures			F
Noise pollution	F (A1)	C	B
Parking		B/C	F
Pavement damage to vehicles			
Roadway construction		C	A1/A2
Roadway maintenance		C	A1/A2
Highway services[4]		C	
Strategic petroleum reserve			
Urban sprawl/land use		E	
Vehicle ownership and operation		C	B
Vibration damage to buildings			
Water pollution		C	

See the notes in table 11a.
[5] Cohen (1994) is an interim report; the ratings in parentheses refer to expected level of detail of the final estimates when the research is completed.

ria. Thus, with a few exceptions, the recent literature on national social costs in the United States, taken at face value, is of limited use.

There is, however, a good deal of excellent work focusing on particular costs or localities, and it is to these, rather than generic summaries, that analysts and policymakers should turn. For example, there now are at least three detailed, original, and conceptually sound analyses of air pollution costs in the United States (Delucchi et al 1996, Report #9; Krupnick et al 1997; Small and Kazimi 1995, for Los Angeles), and several good European analyses (see discussion above). These analyses supersede previous work. Similarly, the noise cost estimates of Delucchi et al (1996, Report #14) supersede the older and heretofore widely cited estimates of Fuller et al (1983). The recent volume edited by Greene et al (1997) summarizes state-of-the-art estimates of accident costs, congestion costs, travel time costs, air pollution costs, and parking costs. As analysts continue to develop detailed marginal social cost models and sound cost-benefit evaluation tools, policymakers will begin to have more reliable cost information to consider in the complex task of making transportation policy.

TABLE 12 The Level of Detail Rating System

A1: Estimate based on detailed analysis of primary data

This designation was used if the author performed a detailed, original analysis based mainly on primary data, or developed detailed cost models, such as damage-function models of the cost of air pollution. Primary data include, but are not limited to: original censuses and surveys of population, employment and wages, government expenditures, manufacturing, production and consumption of goods and services, travel, energy use, and crime; financial statistics collected by government agencies, such as the Internal Revenue Service and state motor vehicle departments; measured environmental data, such as of ambient air quality and visibility; surveys and inventories of physical infrastructure, such as housing stock and roads; and the results of empirical statistical analyses, such as epidemiological analyses of air pollution and health.

A2: Estimate based on straightforward analysis of primary data

This designation was used if the author made relatively straightforward use of primary (or "raw") data published (typically) by a government agency. An example of this that appears in many studies is the use of Federal Highway Administration data (e.g., FHWA 1990) to estimate highway construction and maintenance costs. (See above for other examples of primary data).

Difference between A1 and A2 ratings: A1 work is more detailed and extensive than A2 work.

B: Estimate based on a combination of original data analysis and literature review

This designation was used if the author took published estimates and then adjusted them by changing some of the variables used to derive the estimates, or if the author combined published results from various sources to develop his own estimate. For example, in the FHWA Cost Allocation Study (FHWA 1982), the authors estimate the costs of air pollution by combining vehicle pollutant emissions rates published by the U.S. Environmental Protection Agency with an estimate of air pollution damage cost rates for each pollutant.

Difference between A2 and B ratings: A2 work is based mainly on primary data, such as from government surveys or data series or physical measurements; whereas B work is more dependent on the secondary literature. However, the calculations in B work can be more extensive than those in A2 work, which can involve direct use of relevant primary data.

C: Estimate based on a review and analysis of the literature

This designation was used if estimates were based on a review and analysis of literature, with perhaps some simple calculations. Some studies, such as Kanafani (1983), simply provide tables listing the results of other studies. Other studies, such as Behrens et al (1992) and Litman (1996), conduct a literature review and then make their own estimate on the basis of the review.

Difference between B and C ratings: B work involves some primary data (e.g., data from government surveys, from physical measurements, or primary economic analyses), whereas C work by and large does not; correspondingly, B work requires more calculation than C work.

D: Estimate is a simple extrapolation, adjustment, or citation from another study

This designation was used if the author did some simple manipulation or update of a previously published result. For example, in estimating congestion costs, Ketcham and Komanoff (1992) adjusted FHWA's (1982) congestion factors to reflect 1990 data. Similarly, MacKenzie et al (1992) cite the results of a study by the Urban Institute (1991). They adjust the constant dollar year to 1989, but make no significant adjustment to the published estimate.

Difference between C and D ratings: C work involves more sources and analysis than D work.

E: Estimate is based mainly on supposition or judgment

This designation was used for estimates or simple, illustrative calculations based ultimately on supposition or judgment. For example, Ketcham and Komanoff's (1992) found no reliable estimates of vibration damage to buildings, and so used their judgment to develop their own.

Difference between D and E ratings: D work cites a substantive analysis or estimate of the cost under consideration; E work is based on judgment without reference to any direct estimate of the cost or its major components.

F: Cost item is discussed, but not estimated

This designation was used for those costs that the authors acknowledge as important, but do not attempt to quantify. For example, Lee (1994) discusses, but does not estimate, the costs of vehicle use. Miller and Moffet (1993) provide estimates for most costs, but do not estimate others due to insufficient data.

REFERENCES

Studies Reviewed or Referenced in this Paper

Apogee Research, Inc. 1994. *The Costs of Transportation: Final Report.* Boston, MA: Conservation Law Foundation.

Behrens, C.E., J.E. Blodgett, M.R. Lee, J.L. Moore, and L. Parker. 1992. *External Costs of Oil Used in Transportation,* 92-574 ENR. Washington, DC: Congressional Research Service, Environment and Natural Resources Policy Division.

Beshers, E.W. 1994. *External Costs of Automobile Travel and Appropriate Policy Responses.* Washington, DC: Highway Users Federation.

Bickel, P. and R. Friedrich. 1995. *External Costs of Transport in Germany.* Stuttgart, Germany: Institut fur Energiewirtshcaft und Rationelle Energieanwendung.

____. 1996. External Costs of Transport in Germany. *Social Costs and Sustainability: Valuation and Implementation in the Energy and Transport Sector.* Edited by O. Hohmeyer, R.L. Ottinger, and K. Rennings. Berlin, Germany: Springer-Verlag.

California Energy Commission (CEC). 1994. 1993–1994 California Transportation Energy Analysis Report, draft staff report and technical appendices, P300-94-002.

Cohen, H. 1994. *Incorporation of External Cost Considerations in Highway Cost Allocation,* Interim Report, BAT-93-009. Washington, DC: U.S. Department of Transportation, Federal Highway Administration.

COWIconsult. 1991. *Monetary Valuation of Transport Environmental Impact: The Case of Air Pollution.* Lyngby, Denmark: Danish Ministry of Energy, The Energy Research Programme for Transport.

DeCicco, J. and H. Morris. 1996. The Costs of Transportation in Southeastern Wisconsin, draft report. American Council for an Energy Efficient Economy.

Delucchi, M.A. et al. 1996. *The Annualized Social Cost of Motor Vehicle Use in the United States, Based on 1990–1991 Data,* UCD-ITS-RR-96-3, in 20 reports. Davis, CA: University of California, Institute of Transportation Studies.

Dougher, R.S. 1995. *Estimates of Annual Road User Payments Versus Annual Road Expenditures,* Research Study #078. Washington, DC: American Petroleum Institute.

DRI/McGraw-Hill. 1994. *Transportation Sector Subsidies: U.S. Case Studies.* Washington, DC: U.S. Environmental Protection Agency, Energy Policy Branch.

Ecoplan. 1992. *Damage Costs of Air Pollution, A Survey of Existing Estimates,* T&E 93/1. Brussels, Belgium: European Federation for Transport and the Environment.

Eyre, N.J., E. Ozdemiroglu, D.W. Pearce, and P. Steele. 1997. Fuel and Location Effects on the Damage Costs of Transport Emissions. *Journal of Transport Economics and Policy* 31:5–24.

Federal Highway Administration (FHWA). 1982. *Final Report on the Federal Highway Cost Allocation Study.* Washington, DC: U.S. Department of Transportation.

Fuller, J.W., J.B. Hokanson, J. Haugaard, and J. Stoner, 1983. *Measurements of Highway Interference Costs and Air Pollution and Noise Damage Costs,* Final Report 34. Washington, DC: U.S. Department of Transportation, Federal Highway Administration.

Gifford, J. 1993. Toward the 21st Century. *The Wilson Quarterly* 17:40–47.

Green, K. 1995. *Defending Automobility: A Critical Examination of the Environmental and Social Costs of Auto Use,* Policy Study No. 198. Los Angeles, CA: Reason Foundation.

Greene, D.L., D.W. Jones, and M.A. Delucchi, eds. 1997. *Measuring the Full Social Costs and Benefits of Transportation.* Heidelberg, Germany: Springer-Verlag.

Hanson, M. 1992. Automobile Subsidies and Land Use: Estimates and Policy Responses. *Journal of the American Planning Association* 58:60–71.

Kageson, P. 1992. *External Costs of Air Pollution: The Case of European Transport,* T&E 92/7. Brussels, Belgium: European Federation for Transport and the Environment.

Kanafani, A. 1983. *The Social Costs of Road Transport: A Review of Road Traffic Noise, Air Pollution and Accidents.* Berkeley, CA: University of California, Institute of Transportation Studies.

Keeler, T. and K.A. Small. 1975. *The Full Costs of Urban Transport, Part III: Automobile Costs and Final Intermodal Comparisons,* Monograph 21. Berkeley, CA: University of California at Berkeley, Institute of Urban and Regional Development.

Ketcham, B. and C. Komanoff. 1992. Win-Win Transportation: A No-Losers Approach to Financing Transport in New York City and the Region, draft. Transportation Alternatives.

KPMG Peat Marwick, Stevenson, and Kellogg. 1993. *The Cost of Transporting People in the Lower British Columbia Mainland,* prepared for Transport 2021. Burnaby, British Columbia, Canada.

Krupnick, A.J., R.D. Rowe, and C.M. Lang. 1997. Transportation and Air Pollution: The Environmental Damages. *Measuring the Full Social Costs and Benefits of Transportation.* Edited by D.L. Greene, D.W. Jones, and M.A. Delucchi. Heidelberg, Germany: Springer-Verlag.

Lee, D.B. 1994. Full Cost Pricing of Highways, draft. U.S. Department of Transportation, Research and Special Programs Administration.

____. 1997. Uses and Meanings of Full Social Cost Estimates. *Measuring the Full Social Costs and Benefits of Transportation.* Edited by D.L. Greene, D.W. Jones, and M.A. Delucchi. Heidelberg, Germany: Springer-Verlag.

Levinson, D., D. Gillen, A. Kanafani, and J.M. Mathieu.

1996a. *The Full Cost of Intercity Transportation—A Comparison of High Speed Rail, Air and Highway Transportation in California*, UCB-ITS-RR-96-3. Berkeley, CA: University of California at Berkeley, Institute of Transportation Studies.

Levinson, D., D. Gillen, and A. Kanafani. 1996b. *The Social Costs of Intercity Transportation: A Review and Comparison of Air and Highway*. Berkeley, CA: University of California at Berkeley, Institute of Transportation Studies.

Litman, T. 1996. *Transportation Cost Analysis: Techniques, Estimates and Implications*. British Columbia, Canada: Victoria Transport Policy Institute.

Lockyer, B. and B. Hill. 1992. Rebuttal to Argument Against Proposition 157 (Toll Roads and Highways, Legislative Constitutional Amendment). *California November 1992 Election Ballot Pamphlet*. Sacramento, CA: California Legislative Analysts Office.

MacKenzie, J.J., R.C. Dower, and D.D.T. Chen. 1992. *The Going Rate: What It Really Costs to Drive*. Washington, DC: World Resources Institute.

Maddison, D. 1996. The True Cost of Road Transport in the United Kingdom. *Social Costs and Sustainability: Valuation and Implementation in the Energy and Transport Sector*. Edited by O. Hohmeyer, R.L. Ottinger, and K. Rennings. Berlin, Germany: Springer-Verlag.

Maddison, D., O. Johansson, E. Calthrop, D.W. Pearce, T. Litman, and E. Verhoef. 1996. *The True Cost of Road Transport*, Blueprint #5. London, England: Earthscan.

Mayeres, I., S. Ochelen, and S. Proost. 1996. The Marginal External Costs of Urban Transport. *Transportation Research D* 1D:111–130.

Miller, P. and J. Moffet. 1993. *The Price of Mobility: Uncovering the Hidden Costs of Transportation*. New York, NY: Natural Resources Defense Council.

Morris, H. and J. DeCicco. 1996. *A Critical Review of API's Estimates of Road User Payments and Expenditures*. Washington, DC: American Council for an Energy Efficient Economy.

National Research Council. 1992. *Automotive Fuel Economy: How Far Should We Go?* Washington, DC: National Academy Press.

Otterström, T. 1995. *Valuation of the Impacts from Road Traffic Fuel Emissions: Summary of Results and Conclusions*, MOBILE207Y. Helsinki, Finland: Ekono Energy, Ltd.

Qin, J., J. Weissman, M.A. Euritt, and M. Martello. 1996. Evaluating the Full Costs of Urban Passenger Transportation, Paper 961131, presented at Transportation Research Board 75th Annual Meeting, Washington, DC.

Quinet, E. 1997. Full Social Costs of Transportation in Europe. *Measuring the Full Social Costs and Benefits of Transportation*. Edited by D.L. Greene, D. Jones, and M.A. Delucchi. Heidelberg, Germany: Springer-Verlag.

Royal Commission on National Passenger Transportation. 1992. *Directions: The Final Report of the Royal Commission on National Passenger Transportation, Summary*. Ottawa, Canada: Minister of Supply and Services.

Small, K.A. 1977. Estimating the Air Pollution Costs of Transport Modes. *Journal of Transport Economics and Policy* 11:109–133.

Small, K.A. and C. Kazimi. On the Costs of Air Pollution from Motor Vehicles. *Journal of Transport Economics and Policy* 29:7–32.

Verhoef, E. 1994. External Effects and Social Costs of Road Transport. *Transportation Research A* 28A:273–287.

Studies Referenced in the Papers Reviewed and Mentioned Here

Association for Commuter Transportation and the Municipality of Metropolitan Seattle. 1990. *Proceedings of the Commuter Parking Symposium*. Seattle, WA: Association for Commuter Transportation.

Baumol, W.J. and W.E. Oates. 1988. *The Theory of Environmental Policy*, 2nd edition. New York, NY: Cambridge University Press.

Cervero, R. 1989. Jobs-Housing Balance and Regional Mobility. *Journal of the American Planning Association* 55: 136–150.

DeLuchi, M.A., R.A. Johnston, and D. Sperling. 1987. *Transportation Fuels and the Greenhouse Effect*, UERG-180. Berkeley, CA: University of California at Berkeley, University Energy Research Group.

Eno Transportation Foundation, Inc. 1991. *Transportation in America*, 9th edition. Washington, DC.

Federal Highway Administration (FHWA). 1990. *Highway Statistics 1989*, FHWA-PL-90-003. Washington, DC: U.S. Government Printing Office.

Hart, S. 1985. An Assessment of the Municipal Costs of Automobile Use, unpublished manuscript. 14 December.

____. 1986. Huge City Subsidies for Autos, Trucks. *California Transit*: 4, July-September. California Transit League.

Hines, L.G. 1988. *The Market, Energy, and the Environment*. Boston, MA: Allyn and Bacon.

Hokanson, B., M. Minkoff, and S. Cowart. 1981. *Measures of Noise Damage Costs Attributable to Motor Vehicle Travel*, Technical Report 135. Iowa City, IA: University of Iowa, Institute of Urban and Regional Research.

Jorgenson, D.W. and P.J. Wilcoxen. 1991. Reducing U.S. Carbon Dioxide Emissions: The Cost of Different Goals, CSIA Discussion Paper 91-9. Kennedy School of Government, Harvard University.

Kaufmann, W.W. and J.D. Steinbruner. 1991. *Decisions for Defense Prospects for a New Order*. Washington, DC: The Brookings Institution.

Keeler, T.E., G.S. Cluff, and K.A. Small. 1974. *On the Average Costs of Automobile Transportation in the San Francisco Bay Area*. Berkeley, CA: University of California at Berkeley, Institute of Urban and Regional Development.

Keeler, T.E. and K.A. Small. 1974. *On the Environmental Costs of the Various Transportation Modes*. Berkeley, CA: University of California at Berkeley, Institute of Urban and Regional Development.

Kihlberg and Tharp. 1968. *Accident Rates as Related to Design Elements of Rural Highways*, National Cooperative Highway Research Program Report #47. Washington, DC: Highway Research Board.

Lave, L.B. and E.P. Seskin. 1970. Air Pollution and Human Health. *Science* 169:723–733.

May, A.D., Jr. 1955. Economics of Operation on Limited-Access Highways. *Highway Research Board Bulletin* 107:49–62.

Meyer, J., J. Kain, M. Wohl. 1965. *The Urban Transportation Problem*. Cambridge, MA: Harvard University Press.

Midwest Research Institute. 1970. Systems Analysis of the Effects of Air Pollution on Materials, revised edition, prepared for the U.S. National Air Pollution Control Administration.

Murray, D.M. and U.F.W. Ernst. 1976. *An Economic Assessment of the Environmental Impact of Highway Deicing*, EPA-600/2-76-105. Washington, DC: U.S. Environmental Protection Agency.

Pisarski, A.E. 1987. *Commuting in America*. Washington, DC: Eno Foundation for Transportation.

Ravenal, E.C. 1991. *Designing Defense for a New World Order*. Washington, DC: The Cato Institute.

Rice, D. 1966. *Estimating the Cost of Illness*, P.H.S Publication No. 947-6. Washington, DC: U.S. Department of Health, Education, and Welfare, U.S. Public Health Service.

Sullivan, E.C. and C-I. Hsu. 1988. *Accident Rates Along Congested Freeways*, UCB-ITS-RR-88-6. Berkeley, CA: University of California at Berkeley, Institute of Transportation Studies.

Urban Institute. 1991. *The Costs of Highway Crashes*, FHWA-RD-91-055. Washington, DC: U.S. Department of Transportation, Federal Highway Administration.

U.S. General Accounting Office. 1991. *Southwest Asia: Cost of Protecting U.S. Interests*, GAO/NSIAD-91-250. Washington, DC.

Wilbur Smith and Associates. 1965. *Parking in the City Center*. New Haven, CT.

Part II
Individual Behaviour in Urban Spatial Context

[14]

THE DETERMINANTS OF DAILY TRAVEL-ACTIVITY PATTERNS: RELATIVE LOCATION AND SOCIODEMOGRAPHIC FACTORS[1]

Susan Hanson
Clark University

In an effort to explore the determinants of urban travel behavior, this paper examines the effects of two sets of variables on different aspects of individuals' complex travel-activity patterns: (1) sociodemographic variables (attributes of the individual and household) and (2) spatial variables (the location of the individual relative to potential destinations). The paper reviews earlier work that has investigated the relationships between these factors and travel and finds that (1) studies investigating the impact of sociodemographics on travel have generally not controlled for variation in spatial constraints and (2) few empirical studies have examined the relationship between travel and spatial constraints. Using the Uppsala travel-diary data, this study assesses the role of spatial and sociodemographic factors in explaining each of a number of different aspects of travel (e.g., frequency of travel, dispersion of destinations visited). The significance of the regression coefficients in stepwise regression establishes which independent variables are important in explaining each travel measure. The results indicate that the operative variables differ as a function of the particular aspect of travel in question. Spatial constraints are important to several travel measures, and sociodemographics remain significantly related to most aspects of travel even after the effects of spatial constraints have been controlled.

Hardly anyone who has contemplated spatial behavior has failed to advance the idea that observed behavior patterns are somehow related to the personal characteristics of the individual actors and to the nature of the environment within which the behavior takes place (Anderson, 1971; Chapin, 1974; Cullen and Godson, 1975; Davies, 1969; Hägerstrand, 1974; Horton and Hultquist, 1971; Horton and Reynolds, 1971; Huff, 1960; Isard, 1956; Kutter, 1973; Lloyd and Jennings, 1978; Marble, 1959; Marble and Bowlby, 1968; Nystuen, 1967; Pahl, 1970; Rushton, 1971; Westelius, 1973). The operative word here, of course, is "somehow" because this assertion, though often made, remains relatively unexamined. The contention that spatial behavior patterns might be explained by sociodemographic and environmental variables has been clothed in an assorted wardrobe of terminology. For example, sociodemographic characteristics have been referred to as representing the "demand" side of the equation, while the spatial environment has been viewed as incorporating the "supply" side (Chapin, 1974). Current lexicons favor use of the term, constraints, for any nonchoice variable used to explain travel.[2] Regardless of the terminology used, most researchers, after acknowledging that behavior is likely to be a function of these two sets of variables, have set for themselves some research task other than that of assessing the impact of either one or both of these sets of variables on travel (e.g., Nystuen, 1967; Rushton, 1971).

The motivation for this paper is, then, neither novel nor startling; yet it is fundamental. If we are to advance our understanding of the determinants of urban travel-activity patterns, we need to begin examining systematically the spatial behavior of people with different options and different constraints. For their potential impact on travel, the two factors that merit the most careful scrutiny are the set of sociodemographic descriptors that reflect the personal circumstances of the individual and the set of spatial variables that describe the location of the individual relative to urban activity sites. If a comprehensive theory of travel is to emerge, or if, in fact, urban transportation policy is ever to be effective, we need to know more about the impact of these two sets of variables on different aspects of travel—for example, travel for different purposes. Some would no doubt argue that there is already in the geography and transportation literature a large body of work that deals squarely with this important issue. One purpose of this study is to review this earlier work and to point out what we have and have not been able to learn from it about the determinants of travel behavior.

Examining the impact of social and spatial factors on travel means grappling with a number of difficult problems in data collection, measurement, and analysis. Another goal of this present paper is to sketch out these problems, to suggest some appropriate measurement procedures, and to illustrate their use in an empirical analysis.

Empirical work that has explored the relationship between travel and sociodemographics has done so, by and large, without regard to the spatial configuration of opportunities; yet urban social geography suggests that location relative to activity sites is likely to vary for different social groups. The third purpose of this paper is, therefore, to examine empirically the following questions: (1) Do the sociodemographic characteristics of individuals help to explain travel even when the individual's location relative to opportunities is held constant? and (2) What aspects of the individual's travel-activity pattern are most influenced by spatial, rather than sociodemographic, variables?

The potential effect of spatial constraints on daily travel-activity patterns has received a great deal of attention recently (Burnett, 1979; Burns, 1979; Ellegård, Hägerstrand, and Lenntorp, 1977; Heggie and Jones, 1978; Lenntorp, 1976). Burnett (1979), for example, has criticized spatial choice models for not adequately or explicitly handling spatial constraints insofar as such models assume that all individuals face the same set of alternatives and that all decision makers actually do have a choice between at least two options. But any model incorporating detailed spatial constraints (describing the location of the individual vis-a-vis opportunities) will be exceedingly data hungry; before spatial constraints are routinely incorporated into spatial choice models, we need evidence that spatial variables do improve our ability to explain travel, and we need to know which aspects of travel patterns are most dependent upon the spatial configuration of the environment relative to the location of the individual. This need is particularly acute at the intraurban scale: Is there enough interpersonal variation in spatial constraints within cities to account for a significant amount of variation in travel? If the goal is to understand observed behavior (rather than to devise some models of normative behavior), then we must focus on both sets of constraints believed to affect behavior. Spelling out the constraints and learning about their respective roles is especially imperative in light of the fact that, as Pirie (1976), Gray (1975), and Olsson (1975) have observed, overt behavior patterns are not necessarily globally optimal nor do they necessarily reflect individuals' preferences.

PREVIOUS WORK

Certainly aggregate spatial interaction models (where the units of analysis are zones, not individual actors) of trip generation and trip distribution have incorporated sociodemographic and environmental variables in predictive models of aggregate flows. Sociodemographic characteristics of zones have appeared primarily as predictors of trip generation rates while environmental measures have been used in trip distribution models to specify the attractiveness of a destination zone in terms of the density of opportunities occurring there (e.g., Wilson, 1975). Similar aggregate measures of attractiveness have been adopted by disaggregate modellers attempting to explain the spatial distribution of nonwork trips (e.g., Kern and Lerman, 1978). Measures of a zone's attractiveness, therefore, have been used, mainly to explain the choice of a particular destination zone (e.g., the proportion of a sample choosing zone j) rather than to explain the characteristics of an individual's complex travel-activity pattern or some selected aspect of that pattern. One study that does focus upon travel pattern, rather than simply choice of destination zone, is Wiseman's (1975) aggregate-level analysis in which he examined the importance of a zone's intraurban location vs. its socioeconomic characteristics in explaining travel frequency and travel distances. Using standard transportation planning data gathered in the mid-1960s (which presumably include vehicular travel only), he concluded that "... location within the city, here measured in reference to the CBD, is generally more important than the socioeconomic characteristics of zones in predicting travel patterns. To a lesser degree, variations in travel patterns are also associated with certain socioeconomic characteristics of zones" (p. 176). Travel patterns were measured by (1) zonal trip frequencies and (2) distances between origin and destination zones.

Those who have investigated the determinants of travel behavior at the individual level have focused primarily on the role of sociodemographics and have conducted empirical analyses that either ignore the environmental variables altogether (e.g., Chapin, 1974; Hanson and Hanson, 1981b; Tardiff, 1975) or examine the impact of one or two demographic variables while attempting to hold constant the effects of the spatial environment by using spatially clustered samples (e.g., Davies, 1969; Horton and Reynolds, 1971; Lloyd and Jennings, 1978). As an example of the former approach, Chapin (1974) developed a rich conceptual framework that emphasized the importance of both social and spatial factors only to execute an empirical analysis that completely disregarded all constraints imposed by the configuration of the spatial environment. Another problem is that most previous studies have considered only a handful of sociodemographic characteristics simultaneously. For example, Davies (1969) focused on the role of income; Potter (1977) considered social class, age, and family size; Marble, Hanson, and Hanson (1973) examined the impact of age; Hanson and Hanson (1980, 1981a) studied the impact of traditional gender roles; and Lloyd and Jennings (1978) investigated the impact of income and race. Finally, previous studies assessing the role of sociodemographics in travel have not attempted to capture the impact of these constraints on many different aspects of the individual's complex travel-activity pattern. With a few exceptions (Hanson and Hanson, 1981b; Horton and Hultquist, 1971; Kansky, 1967), studies have used isolated travel variables such as trip frequency (Tardiff, 1975) or distances traveled (Hathaway, 1975) as the dependent variable. There is some evidence that the set of sociodemographic variables that affect travel depends upon the particular aspect of travel in question; that is, the influence of the sociodemographic descriptors varies, for example, according to trip purpose (Hanson and Hanson, 1981b). The same principle should hold when spatial measures are included as independent variables.

Although the importance of the spatial environment has been widely acknowledged in conceptual discussions of individual behavior, relatively little empirical work has investigated at the disaggregate level the relationship of environmental variables to travel. Simple geometric facts indicate that, in a monocentric city at least, people living at different distances from the city's center face different spatial situations (see, for example, Moore (1971)); Sheppard (1978) and Curry (1978) have suggested that the nature of the spatial environment itself is likely to shape the individual's utility function. Hägerstrand and his colleagues (Ellegård et al., 1977; Lenntorp, 1976) and researchers at the Transportation Studies Unit in Oxford, England (Heggie and Jones, 1978; Jones, 1977) have discussed at considerable length the nature of different constraints and the ways in which constraints might be expected to affect the individual's travel-activity pattern. Neither Hägerstrand's group nor the Oxford group, however, has looked systematically at the role of spatial variables relative to sociodemographic ones in explaining travel. They have contributed a rich conceptualization and have directed thinking towards the role of space-time constraints, but they have not yet operationalized this concept for the analysis of large samples. Consequently, it is not yet altogether clear whether the interpersonal variation in spatial constraints is sufficient to produce an impact on travel, particularly at the intraurban scale.

The few empirical studies that have examined the role of spatial constraints, either alone or in conjunction with sociodemographic constraints, report mixed results. Some provide support for the notion that the spatial circumstances of the individual do affect travel; others disclose no relationship between travel characteristics and variables describing the location of the individual vis-a-vis opportunities. In a 1965 study of out-of-home activity patterns in Osnabrück, West Germany, Von Rosenbladt (1972) examined the frequency of travel and distances traveled by people living in four different distance rings from the center of the city (0–1 km, 1–1.5 km, 1.5–2.7 km, 2.7–5 km). He found that travel frequency was lower among people living farther from the city center. Similarly, Westelius (1973) demonstrated with data from Stockholm, Sweden that as an individual's distance to the nearest shopping center increased, the number of trips made to the shopping center decreased, but the number of errands conducted per trip increased. In a study of urban grocery shopping in Buffalo, N.Y., Recker and Kostyniuk (1978) included one spatial variable in their model of destination choice; opportunity, specified as the number of food stores of a given type within a prescribed travel-time interval of home divided by the total number of food stores of all types within that travel-time interval, was found to be important to the individual's choice of food store.

Studies in which the results call into question the importance of spatial variables include those by Marble (1959) and Fawcett and Downes (1977). Marble's 1959 study was an early inquiry into the roles of both sociodemographic and spatial variables. Using the 1949 Cedar Rapids travel-diary data (see, for example, Horton and Hultquist, 1971; Hanson and Marble, 1971), Marble examined the relative importance of socioeconomic as opposed to locational variables in explaining household travel frequency and travel distances. The location variables consisted of four distance measures, each measuring the distance from home to the nearest shopping center of a given size in the retail hierarchy. He found only socioeconomic variables to be significantly related to travel frequency, indicating that "... trip frequency is not significantly affected by the location of the residence relative to the retail structure of the city" (p. 263). However, while none of the socioeconomic variables were significantly related to the total distance traveled, three of the locational ones were, suggesting that the total miles traveled is sensitive to variations in residential location. Fawcett and Downes also set out to ascertain the effect of a household's location with respect to opportunities on household trip generation. Households were cross classified according to size, car ownership, and location in one of three concentric rings of different travel time to the town center of Reading, England (less than 6 minutes, 6 to 10 minutes, and over 10 minutes travel time). Variations in trip rates were tested by mode and purpose, and the authors concluded that, with the exception of walking trips to work in the central area, household location had little effect on trip rates: "The main conclusion is that there was sufficient spatial uniformity in household trip rates in Reading in 1971 to make household location generally unimportant in the modelling of trip generation by household classification techniques, particularly in the case of vehicular trips" (p. 6). In sum, the few studies that have examined the role of spatial constraints on travel have yielded inconsistent results.

The prevailing preoccupation with the sociodemographic correlates of travel, to the neglect of spatial variables, no doubt reflects in part the data-collection and measurement problems that empirical work on this topic poses. The following section considers some problems in and approaches to measuring both sets of constraints as well as complex travel-activity patterns.

MEASUREMENT ISSUES

Measuring Sociodemographics

Measuring the sociodemographic characteristics of a person, while not problem-free, is considerably more straightforward than measuring the spatial constraints facing an individual. The list of demographic descriptors is comfortably familiar and enjoys being part of a long tradition of attention to careful measurement procedures in sociology. Relevant variables include measures of both socioeconomic status (e.g., occupation, education, income) and the individual's role within the household and the larger society (e.g., stage in the life cycle, marital status, sex, employment status). In addition, auto ownership or auto availability, involving both status and role variables, is invariably included in transportation studies.

Collecting data even on these factors is of course not without difficulties. For example, the number of autos owned by the household is often not in itself an accurate index of the mobility potential for members of that household. First, all adults might not have the driver's license; second, one adult might take the car to work all day, leaving at home alicensed but car-less driver. Third, the household might have regular access to a firm or company car. One crude but expedient way of handling these problems is to collect data on possession of the driver's license and on the number of autos regularly available to members of the household, and to create for each household member an auto availability variable, which is the number of cars in the household divided by the number of driver's licenses in the household. The index is automatically zero if the person is not licensed to drive. This simple example indicates how complex measuring even apparently basic sociodemographic variables can be.

Measuring the Relative Location of the Individual

Measuring relative location is akin to measuring accessibility; such measurement is patently complex and demands simplifications and generalizations that risk rendering the resultant measures impotent. (For a discussion of some of the trade-offs involved in measuring accessibility, see Weibull (1980).) Past efforts to ensure that a group of sample individuals was located similarly with respect to the opportunity set have employed spatially clustered samples (Bowlby, 1972; Davies, 1969; Hanson, 1976; Horton and Reynolds, 1971; Lloyd and Jennings, 1978). The drawback to this approach in the present context is of course the difficulty in observing sufficient variation in socioeconomic status when a spatial cluster is used. Given the social geography of most urban areas, the members of a spatially clustered sample are not likely to display a broad spectrum of sociodemographic characteristics, but they are not likely to be completely homogeneous either. Furthermore, it is difficult to isolate two or more cluster samples with similar spatial constraints yet different social characteristics. The use of zonal or clustered aggregates as a way of handling spatial constraints is, in short, clumsy. Moreover, even if an areally-defined population group were homogeneous with respect to every important sociodemographic descriptor, everyone in the same zone would not necessarily face the same set of spatial alternatives (McCarthy, 1969; Pirie, 1979). For example, not everyone can live next door to a corner grocery or be two blocks from the nearest park; moreover, employed people are likely to work in different parts of the city, each of which has a unique relationship to the spatial distribution of activity sites. Given these obstacles, it is not surprising that the measurement difficulties involved in separating locational from sociodemographic influences on travel have been widely recognized (Anderson, 1971; Horton and Hultquist, 1971; Rushton, 1971).

For a sample of individuals randomly located throughout an urban area, disaggregate measures of individuals' locations vis-a-vis potential activity sites must employ techniques that permit a certain level of abstraction or generalization. Such measures of spatial constraints can be either objective (i.e., taken from a map showing the distribution of opportunities vis-a-vis the individual) or subjective (i.e., generated by interviewing individuals to learn about the options perceived as being available) (Hanson and Burnett, 1979). Objective measures include distances that separate the home and/or workplace location from different activity sites and transportation routes (e.g., bus lines) in the city. These distances can be measured either in absolute distance (miles or kilometers) or in travel times, but in opting for the latter, one must choose between on-peak and off-peak travel times. The activity sites encompassed in such measures are conceptualized as points in urban space, and therefore an extremely fine scale of geocoding is required in which coordinates are assigned to street addresses. Generating these measures requires information on the location of the bus stops, the establishments (e.g., shops, doctors' offices, banks), and other activity sites (e.g., parks, meeting halls) within the urban area. Taking such an establishment inventory can be extremely tedious and time consuming, but a few such data files are becoming available in this country (e.g., for Baltimore). The advantage of measures such as these is that they can be used to describe the relative location of a random (not spatially clustered) sample of individuals.

Measuring Travel

An observed travel pattern is the outcome of an individual's participation in different activities at certain places and at certain times. Travel is thus conceptualized as the movement required in order to participate in daily activities (Jones, 1977). This means that, by definition, a travel pattern is complex; it is something a good deal more complicated than a simple link from one origin zone to one destination zone. Measuring travel means collecting spatial and temporal data on the entire out-of-home activity pattern of the individual over some time period (a day, a week, or a month).

A fundamental question asks what is the appropriate time period over which to observe travel? There is something to be said for the argument that because constraints do not change much from day to day, a one-day (weekday) record of the person's travel is sufficient. However, the one-day record does not permit insight into a number of questions, such as the level of variety in the individual's activity pattern or the overall frequency of participation in different types of activities. Nor is there currently any basis for judging how typical or atypical a particular one-day segment is (Hanson and Huff, 1982).

The focus on observed (as opposed to normative or prototypical) travel-activity patterns[3] means grappling with the data collection and measurement problems inherent in trying to delineate travel in all its complexity. Travel consists of a series of stops made over the observation period and each of these stops has a number of dimensions—for example, purpose, timing, mode of travel, location of destination (Burnett and Hanson, 1979). The complexity of observed travel has proven elusive when attempts have been made to capture it in suitable measures. In devising operational measures of complex travel, two approaches are possible (see Hanson and Burnett, 1979, for a more detailed discussion of this point). One is to treat the complex pattern *as a whole* and devise methods to classify these patterns such that an individual's travel pattern can be assigned to one distinct class. An alternative approach is to generate a *suite of measures*, which, taken together, catch all the important aspects of the complex travel-activity pattern. At this time, given our lack of knowledge about the determinants of travel, pursuing both strategies appears prudent.

Measuring travel patterns wholistically precludes a detailed inquiry into a number of issues. Heggie and Jones (1978) have suggested that there are distinct classes (or domains) of travel, each having a different level of complexity and each involving a different process in its formation. It is altogether likely that for different aspects, purposes, or domains of travel, different sociodemographic and/or spatial factors will be the key explanatory variables. In other words, different independent variables are likely to prove salient depending upon the aspect of travel in question. Only a design that (1) identifies distinct aspects of travel and (2) examines the impact of independent variables on each aspect will permit us to assess the validity of this proposition. The approach taken here is to generate a suite of measures or an individual's travel-activity pattern over an extended period of time and to relate each of these measures in turn via stepwise regression to the independent variables comprising the two sets of independent variables outlined above.

TABLE 1.—MEASURES OF TRAVEL-ACTIVITY PATTERN USED IN THE REGRESSION ANALYSIS

Percent of variance explained* by factor represented by variable	Variable name	Transformation
23.5	Number of trips	square root
17.3	Average distance between destinations and home	natural logarithm
11.0	Number of shopping stops	natural logarithm
8.3	Number of work stops	none
6.4	Number of social stops	square root
5.2	Time spent in recreation	square root
3.8	Kilometers traveled on weekdays	$\log_{10}$
75.5		

*In matrix of travel measures.
Source: Hanson and Hanson (1981b).

AN EMPIRICAL EXAMPLE

In an earlier paper (Hanson and Hanson, 1981b), the Uppsala travel-diary data were used to generate a comprehensive set of indices measuring individuals' travel-activity patterns observed over a 35-day period.[4] A matrix of 51 travel measures was reduced via Principal Components Analysis to seven dimensions that explained 75.5% of the variation in the matrix. The present study uses one variable for each of these seven dimensions as the measures of travel-activity pattern and relates each one, in turn, to the variables comprising sociodemographics and relative location.

Method

Table 1 shows the measures of travel that are used subsequently as dependent variables in the stepwise regressions. Each of these variables is a measure of all out-of-home travel by all modes over a 35-day period, and each represents one dimension of complex travel-activity patterns in that it was the variable most closely correlated with a particular component in the Principal Components Analysis mentioned above. The variables are listed according to the proportion of variance explained by the component associated with that variable. The first variable, number of trips (where a trip is a home-to-home circuit involving one or more stops), is the highest-loading variable on a component identifying overall travel frequency and frequency of simpler (one- and two-stop) trips. This variable (and its associated dimension) is akin to standard measures of trip generation. The second measure, the average distance[5] between destinations visited and home, loads highest on a dimension describing the spatial dispersion of the places the individual visited during the survey period. This is a weighted average because each destination point is weighted by the frequency of contact. This variable is essentially a trip distribution measure at the disaggregate

TABLE 2.—LOCATIONAL VARIABLES

1.	Distance from home to work.
2.	Distance from home to CBD.
3.	Distance from home to nearest bus stop.
4.	Number of establishments within 0.5 km of home.
5.	Number of establishments within 1.0 km of home.
6.	Number of establishments within 2.0 km of home.
7.	Number of establishments within 3.0 km of home.
8.	Number of establishments within 4.0 km of home.
9.	Number of establishments within 5.0 km of home.
10.	Number of different land use types within 0.5 km of home.
11.	Number of different land use types within 1.0 km of home.
12.	Number of different land use types within 2.0 km of home.
13.	Number of different land use types within 3.0 km of home.
14.	Number of different land use types within 4.0 km of home.
15.	Number of different land use types within 5.0 km of home.
16.	Percent of all food stores within 0.5 km of home.
17.	Percent of all food stores within 1.0 km of home.
18.	Percent of all food stores within 2.0 km of home.
19.	Percent of all food stores within 3.0 km of home.
20.	Percent of all food stores within 4.0 km of home.
21.	Percent of all food stores within 5.0 km of home.

Note: Variables 2 through 21 are repeated for the workplace if individual is employed.

level. The third dimension of travel is described by the number-of-shopping-stops variable, and is a combination of more complex, multiple-stop (3-, 4-, and more than 4-stop) trips, shopping-related travel, and the level of variety present in the person's travel-activity pattern. Each of the next three variables in Table 1 loads highest on a dimension describing travel for a particular purpose (work, social, recreation), and the final travel measure is an index of overall distances traveled. This set of measures of complex travel-activity patterns is considerably more comprehensive than the single measure, trip frequency, used by Von Rosenbladt (1972), Westelius (1973), and Fawcett and Downes (1977) or than the trip-frequency and distance-traveled measures used by Marble (1959). Where necessary, the variables used here have been transformed to normalize their distributions prior to being entered in the regression analysis. (See Table 1 for the functions used in the transformations; note that all are monotonic.)

The variables used to measure the individual's relative location or access to opportunities appear in Table 2. These variables were computed from data collected in a complete inventory of all establishments and activity sites in Uppsala, undertaken as part of the Uppsala Household Travel Survey. Each establishment or activity site was assigned a land use code and a geocode (x–y coordinates) corresponding to a street address. Perhaps the surrogate most commonly used in the literature to measure access to opportunities is distance from home to CBD (see, for example, Fawcett and Downes, 1977; Hanson and Hanson, 1980, 1981a; Von Rosenbladt, 1972). The rationale for this easily obtained proxy variable is that, in monocentric cities at least, the density of opportunities generally declines with increased distance from the center. In addition to home-to-CBD distance, the variables used here include

TABLE 3.—AGGREGATED LAND-USE CODES/
UPPSALA HOUSEHOLD TRAVEL SURVEY

01	Food store
02	Bakery-cafe
03	Kiosk
04	Liquor store
05	Flower shop
06	Sports shop and toys
07	Clothing and jewelry
08	Drug and illness supplies
09	Book store
10	Photography shop
11	Department store
12	Restaurant, hotel, motel
13	Local meeting place
14	Indoor recreation
15	Outdoor recreation
16	Library
17	Barber shop
18	Dry cleaner, shoe repair
19	Movie theater
20	Bank
21	Insurance, architect, real estate, law offices, etc. (professionals' office)
22	Art, antiques, home furnishings, appliances
23	Hardware, building materials, paints, etc.
24	Train, bus station
25	Parking place
26	Post office, telegraph office
27	Car repairs, accessories
28	Gas station
29	Destination outside Uppsala, summer home
30	Hospital
31	Doctor, dentist office
32	School
33	Public office
34	Church, cemetery
35	Own home
36	Other's home
37	Workplace

one home-based measure of access to the public transportation system (distance from home to nearest bus stop) and several other measures of the home location vis-à-vis opportunities (potential destinations): measures of the density of all establishments within the different distances of home, measures of the variety of land use types within differing distances of home, and measures of the proportion of one type of establishment (food stores) within different distances of home. The codes used to classify land use types in deriving the set of measures, "number of different land use types within x km of home," are shown in Table 3. Because the workplace is an important node in the travel-activity patterns of persons employed outside the home (Hanson, 1980),

TABLE 4.—RESULTS OF PRINCIPAL COMPONENTS ANALYSIS OF LOCATIONAL MEASURES RELATIVE TO HOME VARIMAX ROTATED FACTOR MATRIX

Loadings	Variables	% of variance explained
	Factor 1: Far from home	
.964	Number of establishments within 4.0 km of home	
.943	Number of land use types within 3.0 km of home	
.943	Number of establishments within 5.0 km of home	
.924	Number of land use types within 2.0 km of home	
.921	Percent of all food stores within 4.0 km of home	
.812	Number of establishments within 3.0 km of home	
-.801	Distance from home to CBD.	
.774	Percent of all food stores within 3.0 km of home	73.0
	Factor 2: Near to home	
.948	Number of establishments within 1.0 km of home	
.917	Number of establishments within 0.5 km of home	
.915	Percent of all food stores within 1.0 km of home	
.904	Percent of all food stores within 0.5 km of home	
.814	Number of land use types within 0.5 km radius of home	
.783	Number of establishments within 2.0 km of home	
.764	Percent of all food stores within 2.0 km of home	
.724	Number of land use types within 1.0 km radius of home	14.0

the entire set of location variables is computed for both the home and the workplace for employed people. Doubtless there is considerable redundancy among these variables measuring relative location; Principal Components Analysis was used to reduce each complete set of measures (one for workers, one for nonworkers) to a smaller number of factors for use in further analysis.

The results confirm that the original measures are highly intercorrelated. For nonworkers, 87% of the variance in locational variables is accounted for by two factors (Table 4). The first identifies the number and variety of opportunities within distances that are relatively distant from home, that is, within mostly 3.0 to 5.0 km of home. It is interesting to note that a large number of establishments and land use types available within 5.0 km of home is positively related to one's proximity to the CBD. This makes sense in that in Uppsala in 1971, the farther one's residence was from the CBD, the less likely the CBD (with its density and variety of establishments) was to fall within a 3.0, 4.0, or 5.0 km annulus of home. In the Uppsala case, at least, home-to-CBD distance does appear to be a reasonable surrogate for the number and variety of opportunities within a considerable distance of home; the home-to-CBD distance variable is not, however, the leading variable on this factor, nor is home-to-CBD-distance related to the second factor, which describes the number and variety of opportunities closer to home (within 0.5 km to 2 km of home).

Four factors account for 86.4% of the variance in the matrix of variables describing the relative location of employed people (Table 5). The second and third factors—summarizing the number and variety of opportunities relative to the residence—are essentially the same as the two factors described above for nonworkers. The first and fourth factors describe the opportunity surface within near (0.5 to 2.0 km) and far

TABLE 5.—RESULTS OF PRINCIPAL COMPONENTS ANALYSIS OF LOCATIONAL MEASURES RELATIVE TO HOME AND WORKPLACE VARIMAX ROTATED FACTOR MATRIX

Loadings	Variables	% of variance explained
	Factor 1: Near to work	
.972	Number of establishments within 1.0 km of work	
.966	Percent of all food stores within 1.0 km of work	
.944	Number of land use types within 0.5 km of work	
.928	Percent of all food stores within 0.5 km of work	
.919	Percent of all food stores within 2.0 km of work	
.901	Number of establishments within 0.5 km of work	
.886	Number of establishments within 2.0 km of work	
.841	Number of land use types within 1.0 km of work	
-.819	Distance from workplace to CBD	
.778	Percent of all food stores within 3.0 km of work	
.699	Number of establishments within 3.0 km radius of work	40.6
	Factor 2: Far from home	
.965	Number of establishments within 4.0 km of home	
.952	Number of land use types within 3.0 km of home	
.946	Number of establishments within 5.0 km of home	
.931	Percent of all food stores within 4.0 km of home	
.930	Number of land use types within 2.0 km of home	
.833	Number of establishments within 3.0 km of home	
-.807	Distance from home to CBD	
.791	Percent of all food stores within 3.0 km of home	
-.600	Distance from home to workplace	29.8
	Factor 3: Near to home	
.940	Number of establishments within 1.0 km of home	
.907	Percent of all food stores within 1.0 km of home	
.901	Number of establishments within 0.5 km of home	
.893	Percent of all food stores within 0.5 km of home	
.806	Number of land use types within 0.5 km of home	
.768	Number of establishments within 2.0 km of home	
.747	Percent of all food stores within 2.0 km of home	
.707	Number of land use types within 1.0 km of home	8.9
	Factor 4: Far from work	
.941	Number of establishments within 4.0 km of work	
.927	Number of establishments within 5.0 km of work	
.918	Number of land use types within 3.0 km of work	
.895	Percent of all food stores within 4.0 km of work	
.823	Number of land use types within 2.0 km of work	7.1

(3.0 to 5.0 km) distances of the workplace, respectively. Thus, while two factors are sufficient to describe the spatial distribution of opportunities available to nonworkers, four factors are required to summarize the spatial situation of the worker. The similarity among the components for the two samples is, however, striking. For neither sample were the distance-to-bus-stop variables significantly related to any factor with an eigenvalue greater than one.

TABLE 6.—SOCIODEMOGRAPHIC VARIABLES USED IN THE REGRESSION ANALYSIS

Variable name	Variable	Transformation
Occupation	Treiman's International Occupational Prestige Score (see Lin, 1976) for household head	none
Education	Years of formal education	none
Income	Household income before taxes for 1971 in Swedish crowns	square root
Auto availability	Auto availability: number of autos owned by household ÷ number of persons in household with valid driver's license; index is 0 if person does not hold driver's license or if household owns no cars.	none
Hours worked per week	Number of hours worked per week in the paid labor force	none
Age	Age in years	none
Sex	0 = Female; 1 = Male	none
Household size	Number of people in the household	none
Marital status	Marital status 1 = single, divorced, widowed 2 = married	none

In the following regression analysis, the location variable with the highest loading on each factor is used as an independent variable and is a surrogate for that dimension of the individual's location relative to opportunities. "Number of establishments within 4.0 km of home" and "number of establishments within 1.0 km of home" are the two spatial variables entered in regressions for nonworking people, and in addition to these two, "number of establishments within 1.0 km of work" and "number of establishments within 4.0 km of work" are entered in regressions for employed people. The highest correlation among these variables is r = .53 between the two variables describing opportunities relative to home for the sample of nonworkers.

The variables used to measure the individual's sociodemographic characteristics include measures of socioeconomic status (occupation, education, income), auto availability, and role-related variables (employment status, age, sex, household size, marital status) (Table 6). Among nonworkers, three pairs of sociodemographic variables have correlation coefficients exceeding |.50|: household size and age (r = -.74); marital status and age (= -.58); and marital status and household size (r = .60). For the sample of employed people, no pair of sociodemographic variables has a correlation coefficient over |.50|. For neither sample are the sociodemographic and spaital variables strongly related; the highest correlation between any sociodemographic variable and spatial variable is r = .25. In general, relationships among the independent variables are weak enough that multicollinearity should not be a serious problem.

An earlier investigation (Hanson and Hanson, 1981b) found several sociodemographic variables to be significantly related to each dimension of travel; spatial variables, however, were not considered in that earlier analysis. The questions addressed here are (1) do the spatial variables contribute significantly to an explanation of any of the dimensions of travel given in Table 1, and (2) are sociodemographic variables significant in explaining travel after the individual's relative location is controlled. The model used to investigate these questions is

$$Y_r = a + B_l X_l + \ldots + B_j X_j + B_k X_k + \ldots + B_m X_m + e \qquad (1)$$

where Y_r is one of the travel measures given in Table 1; X_i (i = 1, 2 for nonworkers and i = 1, . . . , 4 for workers) are the variables measuring relative location; X_i (i = k, . . . , m) are the sociodemographic variables that are significantly related to Y_r when relative location is held constant; and B_i (i = 1, . . . , m) are the standardized regression coefficients. Standardized regression coefficients are reported here because the interest is in identifying the relative importance of the different variables rather than in using the equations to predict values of the dependent variables. In order to control for relative location, the spatial variables were forced into each equation before any sociodemographic variable was permitted to enter. Then only those sociodemographic variables with regression coefficients significant at the $p < .05$ level were included in the final equation. This strategy allows one to examine the effect of sociodemographics on travel after the effects of individuals' differing spatial constraints have been removed. Because the spatial constraints of nonworkers are different from those of workers and because the independent variables affecting travel are likely to differ for the two groups, the analysis examines workers and nonworkers separately.

Expectations

There are a number of ways in which a person's travel might be affected by the level of access to facilities, defined in terms of having many and varied opportunities available close as opposed to far from home (see, for example, Von Rosenbladt, 1972). As distances to facilities increase, the individual might reduce the frequency of engaging in out-of-home activities so that the total distance traveled over an extended time period would not increase. Alternatively, overall travel distances would increase with decreasing accessibility if the frequency of participation in out-of-home activities remained unchanged. A third possible response is for the individual to reduce travel frequency but increase the duration of stay at destinations such that the amount of time spent in an activity is not affected by reduced access to facilities.

The ways in which sociodemographic factors can be expected to affect travel have been reviewed in detail elsewhere (Hanson and Hanson, 1981b). Briefly, those with higher social status and those with the fewest constraints imposed by household and societal roles are expected to engage more frequently in and to spend more time in activities outside the home, to visit destinations farther from home, and, therefore, to travel greater distances. Those with more expansive travel-activity patterns are, therefore, expected to have higher income, education, occupation, and auto availability levels; to be male, young, and unmarried; and, if employed, to work fewer hours per week. Because employment imposes a major time constraint on travel-activity patterns, the significant determinants of travel are expected to differ for employed and unemployed people.

Results

The results of the stepwise regressions are presented in Table 7. The independent variables are listed in the order in which they entered each equation, and the adjusted coefficient of determination (multiple R^2) at each step is also given. The standardized regression coefficients describe the final equation and, with the F-ratio from the final equation, give an indication of the relative importance of each independent variable in explaining each dimension of travel. Because stepwise regression searches at each step for the variable with the largest F (rather than inserting a preselected variable), the tabled F-value does not provide the basis for a standard significance test (Dixon and Brown, 1979, p. 404, p. 855). Nevertheless, in the absence of other guidelines, the tabled F-value at $p < .05$ for 1, ∞ *d.f.* (3.84) was used here to determine which variables would be included in the final equation; only variables with an F exceeding this level are included in the equation and shown in Table 7.

In general, the regression results show that for most dimensions of travel-activity pattern, sociodemographic descriptors explain more variance than do the spatial descriptors, which, in most of the equations, account for a very small proportion of the total variance. Furthermore, with the exception of one travel variable (average distance between destinations and home for nonworkers), there is at least one sociodemographic variable in the equation after the effects of relative location have been removed. In addition, it is worth noting that, for workers and nonworkers, different although overlapping sets of explanatory variables are important in explaining travel. Finally, it is clear that the relative importance of spatial variables to travel is different for different aspects of travel and that the degree to which locational and sociodemographic conditions can together explain travel varies considerably among the different aspects of travel-activity pattern. Many of the multiple coefficients of determination are disappointingly small, which indicates that perhaps variables other than those included here need to be considered.

The spatial variables, although forced to enter the equation first, are of minor importance in explaining the two trip-generation aspects of travel-activity patterns: (1) number of trips, representing travel frequency and especially the frequency of one- or two-stop trips, and (2) number of shopping stops, representing the frequency of longer multistop journeys. As expected, especially because walk trips are included here, travel frequency is positively related to a high density of opportunities close to home, corroborating the findings of Westelius (1973) and Von Rosenbladt (1972). Among workers, sex, income, and age are all important in explaining number of trips, with trip frequency higher among males, higher-income people, and younger people. Among nonworkers, three-quarters of whom are women, trip-making increases with household size (indicating that household needs, not just personal needs, generate travel), with auto availability, and with being married rather than single. For nonworkers, then, role variables rather than social status variables are important to an explanation of number of trips. Spatial variables add even less to an explanation of number of shopping stops than they do to one of number of trips. For both workers and nonworkers, sex and education appear in the final shopping equation; women and those with more education shop with greater frequency. Perhaps the education variable indicates that highly educated people are more likely to engage in comparison shopping, a practice that would increase the number of shopping

TABLE 7.—RESULTS OF STEPWISE REGRESSION ANALYSIS

	Workers, $N = 240$				Non-workers, $N = 110$			
Dependent variables	Independent variables	Standardized regression coefficient	F-value	Multiple R^2	Independent variables	Standardized regression coefficient	F-value	Multiple R^2
Number of trips	Establishments within 1.0 km of home	.15	9.2	.04	Establishments within 1.0 km of home	.29	7.0	.06
	Sex	.22	11.9	.10	Household size	.23	23.2	.24
	Income	.17	5.2	.12	Auto availability	.25	6.2	.28
	Age	-.14	5.0	.14	Marital status	.22	4.4	.31
Average distance between destinations and home	Establishments within 1.0 km of home	-.37	80.8	.25	Establishments within 1.0 km of home	-.42	53.3	.33
	Establishments within 4.0 km of home	-.18	7.6	.28	Establishments within 4.0 km of home	-.28	9.6	.39
	Establishments within 1.0 km of work	-.10	6.9	.30				
	Sex	.22	29.7	.38				
	Auto availability	.16	10.3	.41				
	Education	-.14	5.4	.42				
	Income	.11	4.0	.43				

Number of shopping stops	Establishments within 1.0 km of work	.06	4.7	.02				
	Sex	-.27	31.0	.13	Sex	-.28	9.8	.24
	Education	.23	13.7	.18	Education	.21	5.6	.28
	Hours worked per week	-.16	5.5	.20	Auto availability	.36	15.1	.17
Number of work stops	Establishments within 1.0 km of work	.31	20.2	.08				
	Hours worked per week	.28	42.6	.23				
	Sex	.20	9.6	.26				
	Income	.15	6.2	.28				
Number of social stops	Age	-.43	48.5	.19	Age	-.30	10.3	.10
	Household size	-.21	9.5	.23				
	Hours worked per week	-.27	7.3	.25				
	Sex	.21	11.5	.29				
	Education	.12	4.2	.30				

stops made. For workers, regardless of sex, spending many hours at work means making fewer shopping stops.

There are two travel measures that describe the spatial aspect of travel-activity pattern: (1) average distance between destinations and home and (2) kilometers traveled on weekdays. The individual's relative location vis-a-vis activity sites is extremely important in explaining the former of these two. In fact, for nonworkers, the two spatial variables are the only variables in the final equation. A high density of establishments within 1.0 km of home is strongly and positively related to the selected destination set being close to home. This is not only intuitively plausible but also indicative of some rationality (in terms of distance-minimizing behavior) on the part of the individual traveler. For workers, the spatial distribution of activity sites vis-a-vis home is more important than the distribution vis-a-vis the workplace, but variables encompassing both spatial distributions are included in the equation explaining the location of the destination choice set relative to home. Sociodemographic variables also play a role in determining the average distance from destinations to home for workers. Sex again explains a significant amount of variation with men's destination sets being more far-flung than women's. As expected, auto availability and income are positively related to this aspect of travel, but curiously, with sex and auto availability already in the equation, education is inversely related to average distance between destinations and home. Perhaps education yields distance-minimizing behavior via higher levels of information about the nature of the spatial environment.

For workers, three spatial variables are significant after being forced to enter the equation explaining kilometers traveled on weekdays; sensibly enough, the greater the density of establishments close to home and work, the fewer kilometers traveled. In light of the fact that living in high-density surroundings is associated with a higher level of trip making, this result is not as trivial as it might appear. People living and/or working in high-density environments make more trips and still manage to travel fewer total kilometers. Given that many of these trips were made on foot,[6] the implied fuel savings are considerable. Sex again emerges as the most important sociodemographic variable among workers, with males traveling farther than females, a finding that corresponds to results of other studies (e.g., Hathaway, 1975). For workers, greater travel distances are also associated with auto availability, youth, and higher income. Auto availability is clearly the most important determinant of this aspect of travel for nonworkers, with marital status also playing a role.

The remaining travel variables all have to do with travel for a particular purpose: work, socializing, and recreation. The density of establishments in the vicinity of the workplace contributes significantly to an explanation of the number of stops at work. This probably is because workers are more likely to visit other establishments (e.g., shops, banks, restaurants) on the lunch hour if such places are within walking distance of the workplace. Because part-time as well as full-time and workaholic workers are represented in the group of workers, the number of hours worked per week also is important in explaining the number of work stops. Even with this employment-status variable held constant, sex and income enter the equation. True to their breadwinner role, men make more work stops than women, and, as proof that the work ethic does pay off, making more work stops is associated with higher income.

Neither social nor recreation travel is apparently affected by the overall density of establishments within the city. For both groups, age is the most important determinant of social travel, with the number of social stops declining with age, and it is the only variable included in the nonworkers' equation. Among workers, the number of social stops is also related inversely to household size and to the amount of time spent in the labor force, while it is positively related to being male and well educated. Recreation travel (time spent in recreation) by workers is influenced by gender and time spent in the labor force, and for nonworkers by age and gender. The importance of gender confirms the results of time-budget studies that have found men spending more time than women in recreation (Chapin, 1974; Michelson, 1975).

Taken as a whole, these results from the regression analyses demonstrate that sociodemographics and particularly the role-related factors (sex, household size, employment status, marital status, and age) remain important in explaining most of the aspects of travel even when the effects of relative location have been controlled. It should be stressed again that because one purpose of the regression analysis was to assess the importance of sociodemographics to travel-activity patterns when relative location is held constant, the spatial variables were forced into the regression equations before the sociodemographics were allowed to enter. It is possible that multiple regression equations that maximize the explained variance would include fewer spatial and more sociodemographic variables.

Of the sociodemographics included in these equations, role-related factors, which have until recently been relatively unexamined in the context of travel behavior, stand out as being particularly important. Among the role-related factors, sex stands out as the single most important determinant of travel-activity patterns, especially among employed persons. Interestingly, marital status, while unimportant to the travel-activity patterns of workers, enters two of the equations for nonworkers. The findings by and large confirm Kutter's (1973) hypothesis that sociodemographic factors are more important than locational factors in explaining travel behavior.

Nevertheless, spatial variables were found to be significant in explaining a number of travel dimensions, and the initial notion that spatial constraints would be of varying importance to different dimensions of travel was also confirmed. The results here tend to support the notion advanced by Marble (1959) and by Fawcett and Downes (1977) that the individual's location within the city plays a relatively minor role in explaining travel frequency but plays a more important role in explaining travel distances. The implication is that while it is probably unnecessary to include detailed measures of relative location in all models of travel behavior, explanatory models of trip distribution would certainly benefit from the inclusion of spatial variables.

CONCLUSION

A number of people have suggested the need to examine the role of supply constraints (i.e., the location of the individual relative to potential destinations) in spatial choice modeling (e.g., Sheppard, 1980; Burnett, 1980). Early in this paper, I reviewed the empirical work that has addressed the role of spatial variables in travel behavior and found no coherent pattern of findings reported in the literature. The inconsistency in findings quite likely reflects the variety of approaches and particularly of measurement techniques that have been employed in the few studies that have examined the impact of relative location on travel behavior. In particular, the level of disaggregation at which spatial constraints are measured and the amount of interpersonal variability present in the measures of relative location are likely to affect results, as is the use of objective versus subjective spatial-constraints measures.

Using highly disaggregated measures of the individual's location vis-a-vis the objectively defined opportunity set, this study has shown that relative location does affect certain aspects of the individual's travel-activity pattern. That the relationships uncovered were not, however, extremely strong raises several possibilities. Additional or different disaggregate measures of relative location might be more closely related to travel than were the measures of relative location might be more relative location might be more successful at explaining travel; or perhaps subjective, rather than objective, spatial constraints are the more important spatial variables to include. One fascinating area for future research concerns the relationship between objectively defined and subjectively defined spatial constraints or choice sets; that is, what is the nature of the relationship between the spatial distribution of opportunities in the objectively defined choice set and the set of opportunities that the individual perceives as viable options? The results of the analysis reported here also suggest that at the intraurban scale at least, the interpersonal variation in distance to opportunities may be insufficient to affect certain aspects of travel, such as travel frequency. That is, spatial choice models or models of trip generation at the intraurban scale will not be seriously under-specified if they omit spatial variables. Future research should investigate this question at larger scales, for example at the metropolitan scale or at the county scale in rural areas.

At the intraurban scale investigated here, sociodemographic factors outweigh spatial ones in explaining overall trip frequency and travel for particular purposes, but spatial factors cannot be overlooked in explaining patterns of trip distribution. It is also interesting to note that at this scale there is only a weak relationship between the two sets of independent variables—sociodemographic factors and the objective measures of relative location. This finding contradicts the notion that certain social groups are located differentially with respect to opportunities in urban areas.

A great deal of work remains to be done before we can be assured of having a firm grasp on the role of spatial factors in travel or on the relative importance of spatial versus sociodemographic variables in explaining travel-activity patterns. Given the importance of these issues to theory building, the need for additional research is clear.

NOTES

[1]The support of NSF grant #SOC-7820530 is gratefully acknowledged. The author also appreciates the comments of the anonymous reviewers.

[2]Inasmuch as both (1) the spatial configuration of the opportunity set relative to the location of the individual and (2) the individual's sociodemographic characteristics impinge on travel behavior, these two sets of variables may be considered constraints. The particular ways in which social or spatial variables can act as constraints on travel-activity patterns are discussed briefly later in this paper and have been considered in detail elsewhere (e.g., Burnett and Hanson, 1979; Chapin, 1974; Heggie and Jones, 1978; Jones, 1979; Lenntorp, 1976; Pirie, 1976). Here I shall use the term "constraints" to refer to both sets of variables.

[3] At some future point, we may wish to investigate (propose) normative models of individuals' travel-activity patterns. However, even these will rely on insights acquired from descriptive studies of observed activity patterns within different types of constraints. For example, one might establish normative travel-activity patterns for different socioeconomic groups with a certain level of spatial constraints by working out the requirements of certain social (household) roles. Alternatively, one could establish normative patterns by looking at the activity chosen when certain opportunities are available and use the group with the fewest constraints as a benchmark against which to establish disadvantaged groups. At this point, we set aside the value question of which groups evidence an "adequate" mobility level or which groups display perhaps an overzealous level of mobility.

[4] Briefly, the Uppsala Household Travel Survey collected detailed travel data on about 300 individuals resident in Uppsala, Sweden in 1971. For a 5-week period, each panel member kept a travel diary in which he/she recorded information on all out-of-home activities. A home-to-home circuit was defined as a *trip*, which could consist of one or more *stops*. For each stop on each trip, individuals recorded the time of arrival and departure, the mode of travel, the location (street address) of the destination, and the purpose of the stop. All origins and destinations were subsequently coded to X-Y coordinates corresponding to street addresses. The survey also collected information on the sociodemographic characteristics of the individual and the household. The Uppsala Household Travel Survey is described in greater detail elsewhere (e.g., Hanson and Hanson, 1980, 1981a).

[5] All distances in the Uppsala data are airline distances in kilometers, measured between origins and destinations.

[6] On average, slightly over half of the individual's movements were made on foot.

LITERATURE CITED

Anderson, J., 1971, Space-time budgets and activity studies in urban geography and planning. *Environment and Planning*, Vol. 3, 353-368.

Bowlby, S., 1972, Spatial variation in consumers' information levels. Household Travel Behavior Study, Report No. 4, The Transportation Center, Northwestern University, Evanston, Illinois.

Burnett, P., 1979, Choice and constraints-oriented modeling: Alternative approaches to travel behavior. White Paper, Federal Highway Administration, Washington, D.C.

_______, 1980, Spatial constraints-oriented approaches to movement, micro-economic theory and urban policy. *Urban Geography*, Vol. 1, 53--67.

_______ and Hanson, S., 1979, A rationale for an alternative mathematical paradigm for movement as complex human behavior. *Transportation Research Record*, 723, 11-24.

Burns, L. D., 1979, *Transportation, Temporal, and Spatial Components of Accessibility*. Lexington, Mass.: Lexington Books.

Chapin, F. S., 1974, *Human Activity Patterns in the City*. New York: John Wiley and Sons.

Cullen, I. and Godson, V., 1975, Urban networks: the structure of activity patterns. *Progress in Planning*, Vol. 4, 1-96.

Curry, L., 1978, Demand in the spatial economy: II Homo-Stochasticus. *Geographical Analysis*, Vol. 10, 309-344.

Davies, R. L., 1969, Effects of consumer income differences on shopping movement behavior. *Tijdschrift Voor Economische en Sociale Geographie*, Vol. 60, 111-121.

Dixon, W. T. and Brown, M. B., 1979, *BMDP-79*. Los Angeles: University of California Press.

Ellegård, K., Hägerstrand, T., and Lenntorp, B., 1977, Activity organization and the generation of daily travel: two future alternatives. *Economic Geography,* Vol. 53, 126-152.

Fawcett, F. and Downes, J. D., 1977, The spatial uniformity of trip rates in the Reading area, 1971. *Transport and Road Research Laboratory Report 797,* Crowthorne, England.

Gray, F., 1975, Non-explanation in urban geography. *Area,* Vol. 5, 228-235.

Hägerstrand, T., 1974, *The impact of transport on the quality of life.* Paper presented at the European Conference of Ministers of Transport, Athens, Greece.

Hanson, S., 1976, Spatial variation in the cognitive levels of urban residents. In R. G. Golledge and G. Rushton, editors, *Spatial Choice and Spatial Behavior.* Columbus, Ohio: Ohio State University Press.

_______, 1980, The importance of the multi-purpose journey to work in urban travel. *Transportation,* Vol. 9, 225-248.

_______ and Burnett, P., 1979, *Understanding complex travel behavior: measurement issues.* Paper presented at the Fourth International Conference on Behavioral Travel Modeling, Munich, Germany.

_______ and Hanson, P., 1980, Gender and urban activity patterns in Uppsala, Sweden. *Geographical Review,* Vol. 70, 291-299.

_______ and _______, 1981a, The impact of married women's employment on household travel patterns. *Transportation,* Vol. 10, 165-183.

_______ and _______, 1981b, The travel-activity patterns of urban residents: dimensions and relationships to sociodemographic characteristics. *Economic Geography,* Vol. 57, in press.

_______ and Huff, J. O., 1982, *Assessing day-to-day variability in complex travel patterns.* Paper presented at the annual meetings of the Transportation Research Board, Washington, D.C.

_______ and Marble, D. F., 1971, A preliminary typology of urban travel linkages. *East Lakes Geographer,* Vol. 7, 49-59.

Hathaway, P. J., 1975, Trip distribution and disaggregation. *Environment and Planning,* Vol. 7, 71-97.

Heggie, I. and Jones, P. M., 1978, Defining domains for models of travel demand. *Transportation,* Vol. 7, 119-125.

Horton, F. and Hultquist, J., 1971, Urban household travel patterns: definition and relationship to household characteristics. *East Lakes Geographer,* Vol. 7, 37-48.

_______ and Reynolds, D., 1971, Effects of urban spatial structure on individual behavior. *Economic Geography,* Vol. 47, 36-48.

Huff, D., 1960, A topological model of consumer space preferences. *Papers and Proceedings of the Regional Science Association,* Vol. 6, 159-173.

Isard, W., 1956, *Location and Space-Economy.* Cambridge, Mass.: MIT Press.

Jones, P. M., 1977, New approaches to understanding travel behavior: the human activity approach. In D. A. Hensher and P. R. Stopher, editors, *Behavioral Travel Modelling.* London: Croom Helm.

_______, 1979, HATS: A technique for investigating household decisions. *Environment and Planning A,* Vol. 11, 59-70.

Kansky, K. J., 1967, Travel patterns of urban residents. *Transportation Science,* Vol. 1, 261-285.

Kern, C. and Lerman, S., 1978, *Models for predicting the impact of transportation policies on retail activity.* Paper presented at the annual meetings of the Transportation Research Board, Washington, D.C.

Kutter, E., 1973, A model for individual travel behavior. *Urban Studies*, Vol. 10, 235-258.

Lenntorp, B., 1976, *Paths in Time-Space Environments: A Time Geographic Study of Movement Possibilities of Individuals.* Lund: CWK Gleerup.

Lin, N., 1976, *Foundations of Social Research.* New York: McGraw-Hill.

Lloyd, R. and Jennings, D., 1978, Shopping behavior and income: comparisons in an urban environment. *Economic Geography*, Vol. 54, 157-167.

Marble, F. F., 1959, Transport inputs at urban residential sites. *Papers of the Regional Science Association*, Vol. 5, 253-266.

Marble, D. F. and Bowlby, S., 1968, Shopping alternatives and recurrent travel patterns. In F. Horton, editor, *Geographic Studies of Urban Transportation and Network Analysis.* Northwestern University Studies in Geography Number 16. Evanston, Ill.: Northwestern University Press.

_______, Hanson, P., and Hanson, S., 1973, Intra-urban mobility patterns of elderly households: a Swedish example. *Proceedings, International Conference on Research*, Bruges, Belgium, 655-664.

McCarthy, G. M., 1969, Multiple regression analysis of household trip generation—a critique. *Highway Research Record* 241, 31-43.

Michelson, W., editor, 1975. Time Budgets and Social Activity. Toronto: Centre for urban and community studies, University of Toronto.

Moore, E. G., 1971, Some spatial properties of urban contact fields. *Geographical Analysis*, Vol. 3, 376-386.

Nystuen, J. D., 1967, A theory and simulation of intraurban travel. In W. Garrison and D. Marble, editors, *Quantitative Geography Part 1: Economic and Cultural Topics.* Evanston, Ill.: Northwestern University Press.

Olsson, G., 1975, *Birds in Egg.* Ann Arbor, Mich.: Department of Geography, University of Michigan.

Pahl., R. E., 1970, *Whose City?* London: Longmans.

Pirie, G., 1976, Thoughts on revealed preference and spatial behavior. *Environment and Planning A*, Vol. 8, 947-955.

_______, 1979, Measuring accessibility: a review and proposal. *Environment and Planning A*, Vol. 11, 299-312.

Potter, R. B., 1977, The nature of consumer usage fields in an urban environment: theoretical and empirical perspectives. *Tijdschrift Voor Economische en Sociale Geographie*, Vol. 68, 168-176.

Recker, W. and Kostyniuk, L., 1978, Factors influencing destination choice for the urban grocery shopping trip. *Transportation*, Vol. 7, 19-33.

Rushton, G., 1971, Behavioral correlates of urban spatial structure. *Economic Geography*, Vol. 47, 49-58.

Sheppard, E. S., 1978, Theoretical underpinnings of the gravity hypothesis. *Geographical Analysis*, Vol. 10, 386-402.

_______, 1980, The ideology of spatial choice. *Papers of the Regional Science Association*, Vol. 45, 197-213.

Tardiff, T., 1975, Comparison of effectiveness of various measures of socio-economic status in models of transportation behavior. *Transportation Research Record*, 534, 1-9.

Von Rosenbladt, B., 1972, The outdoor activity system in an urban environment. In A. Szalai, editor, *The Use of Time.* The Hague: Mouton Press.

Weibull, J. W., 1980, On the numerical measurement of accessibility. *Environment and Planning A*, Vol. 12, 53-67.

Westelius, O., 1973, *The Individual's Way of Choosing Between Alternative Outlets.* Stockholm: National Swedish Building Research.

Wilson, A. G., 1975, Some new forms of spatial interaction model: a review. *Transportation Research*, Vol. 9, 167-179.

Wiseman, R. F., 1975, Location in the city as a factor in trip making patterns. *Tijdschrift Voor Economische en Sociale Geographie*, Vol. 66, 167-177.

[15]

Space–time budgets, public transport, and spatial choice

P C Forer, Helen Kivell
Department of Geography, University of Canterbury, Christchurch 1, New Zealand
Received 11 October 1979, in revised form 16 July 1980

Abstract. This paper addresses the problem of access to urban facilities for housewives without cars, and the methodology of the Lund School is used to investigate the spatial constraints affecting access to and choice between a selected group of urban facilities in the city of Christchurch, New Zealand. To do this, the characteristics of the public transport system are investigated, and time-budget data used to specify typical windows of free time during a housespouse's day. From there the potential action and activity spaces of individuals in four suburbs are delimited, and these are used in assessing the variations in access to and choice between facilities in these suburbs. Finally, the social impact of the current bus provision in the context of the social structure of the city is raised as a policy issue.

Space–time budgets and activity patterns

The importance of the interaction of time and space in determining the activity patterns of individuals at the microlevel has gained increasing recognition through the work of two main schools. The earlier and less quantitative work of Professor Hägerstrand and other geographers at Lund (compare Carlstein, 1977; Pred, 1977; Thrift, 1977) has been matched with later work by planners such as Cullen and Godson (1975), and Tomlinson et al (1973). This latter school has tended to place a stronger emphasis on aggregate behaviour and mathematical modelling, but both schools have founded their work on two basic tenets: (1) that the individual in society has a budget of necessary activities to undertake in a period of time, and (2) that his scheduling of these activities is constrained by locational requirements of activities both in time and in space.

Hägerstrand (1966; 1975) has most clearly illustrated this with his concept of the space–time 'aquarium' in which the individual's life can be traced as a line through two spatial dimensions (as on a map grid) and one temporal one (chronological time). As he points out, workers such as Szalai (1972) and Chapin (1970; 1974) have tended to confirm that there are large-scale regularities across major groups of the population in the apportionment of time between activities. All activities, however, have certain requirements for their being undertaken, and these are often of the nature of socially-defined restrictions either in time or in space. Television watching, for instance, requires a location near a receiver (at home or in a social institution) and by social custom is available only in certain time slots of the day. A more formal example might be the location and opening hours of a day-care centre. In this light, activity patterns can be seen as an attempt to schedule an individual's movements in space and time to achieve optimally or suboptimally a desired balance of activities.

Palm and Pred (1974), in a largely theoretical paper, discuss this concept with respect to the American housewife. They draw heavily on the notion of a generally applicable set of constraints and opportunities to which suburban housewives are subject. Constraints are seen largely as those of access in time and space and include the need to be at home at meal times for primary school children or to attend regular child-health clinics. Working from this observation they highlight the degree to which the suburban housewife, with apparently considerable leisure time during the day, is in fact unable to utilise the time to any effect because of its partitioning into small interludes within the day. These interludes are so short that any housewife seeking to utilise them for any activity outside the home inevitably finds herself unable to do so. A major constraint is the time spent travelling to facilities she wishes to use. Typically she may have a two-and-a-half-hour free-time period bounded by highly structured activity blocks or 'markers', that is, activities which are highly restricted both in time and in space. The need to be with school children at lunch and after school would be a case in point where time and location (home) are strictly governed. Trying to leave the house for activities elsewhere and be back by the end of this 'window' of leisure time frequently means that the time available at the new location after travelling is not sufficient to be of any practical significance. A simple time budget which may appear well-provided with leisure can thus, when set in a spatial context, stand revealed as deficient in real leisure opportunities. Access to transport is a key factor in this phenomenon.

Palm and Pred's paper concentrates on *presence* or *absence* of leisure opportunity. The current paper seeks to explore the opportunities available to housewives in terms of *choice* within a set of routine activities outside of the home, and seeks to provide empiric evidence of variations in opportunity through space. It thus parallels the recent work of Hägerstrand (1974) and Tivers (1977), but with less emphasis on authority constraints.

The space-time budget of the housewife

The housewife portrayed here is typical of New Zealand housewives in the early stages of child-rearing. We have adopted a housewife in a household with one car, that car being used by the breadwinner during the day. As only 24% of *all* New Zealand households are households with two or more cars, there are potentially upwards of 80% of families with young children in such a situation. Even with some housewives having access to the car during the day, a reliance on public transport for many housewives is a normal state. Our hypothetical housewife is further restricted here by assuming both a preschool and school-age child. Having responsibilities for school children actually affects 51% of housewives. For no cars and a preschooler, we are now talking about roughly 9% of New Zealand households, possibly 8500 in Christchurch alone.

Although such a housewife must suffer extreme constraints to her activities, the preschool only mother (18% of households) is in fact often no less hindered because of the everyday physical needs of her children. Similarly, the mother of primary school children may well feel strongly constrained by the major markers in the day.

As an example, the typical activity pattern of two housewives is shown in figure 1. Here space is stylised into specific locations on the horizontal axis while time is shown on the vertical. The five locations represent typical needs in a day. The markers within this day are a need to be at home until 8.30 am, between noon and 1 pm, and after 3.00 pm. This leaves two periods of 2-3 hours for 'flexible' activities. In the case in point, housewife 1 visits shops and playcentre in the morning and the park and clinic in the afternoon. She is at any one of these locations while the line of her day is vertical. The nonvertical portions represent travel times. It is clear that a second housewife located at a different point and needing to travel longer to get to some facilities would need to modify her activities (dotted line). Hägerstrand has labelled these travelling-activity-travelling sequences which are bounded by definite 'marker' episodes as *prisms*. The effective activity choice open to any individual operating from a given home will depend on these prisms. How many facilities for any activity exist near to her home and the speed at which the individual can travel are clearly the major determinants of access in space.

In order to allow some quantitative analysis of access to facilities one needs to specify a 'typical' New Zealand housewife. In the discussion above we have translated what in reality may be a variety of hazy and individualised activity patterns into a practicable stereotype. We have assumed, amongst other things, particular patterns of response by carless housewives (short walking or bus usage) and particular periods of free time bounded by strong markers. How realistic are these assumptions?

The main avenue for testing them stems from a 1975 recreational survey conducted by the New Zealand National Council for Recreation and Sport. This survey included the collection of 3000 twenty-four-hour time budgets with related spatial information on activity location. The data yielded that some 200 women who lived within the main cities, were nonworking mothers of young children, and completed the diary on a weekday.

Analysis of the survey data has only just begun, but results to date indicate that the stereotype used here to investigate opportunities has grounding in actual behaviour. Although the car was used by those who had access to it, walking and public transport were equally popular as second alternatives. (In spite of Christchurch's ideally flat terrain the number of persons in this group using bicycles for mobility was predictably small.) The use of public transport to this degree contrasts somewhat with the findings of Hillman et al (1973), perhaps as a result of the more extensive nature of New Zealand urban centres. The tendency to eschew public

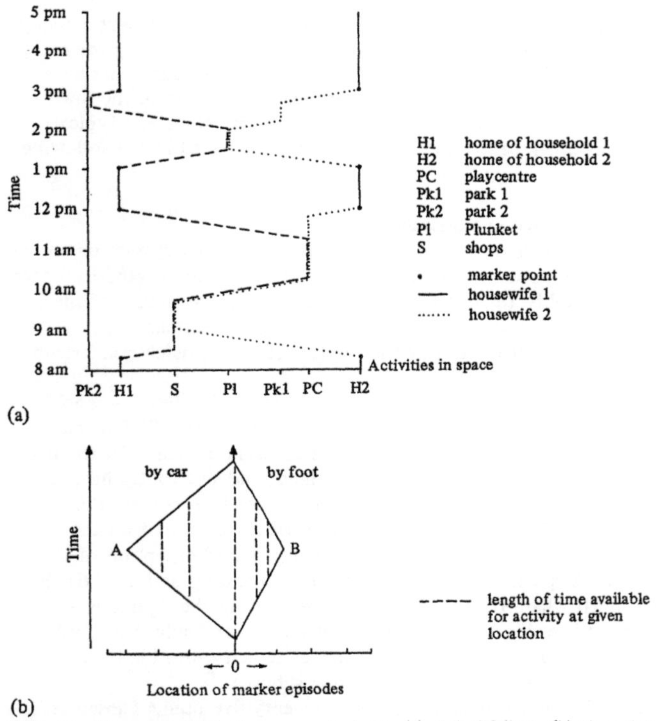

Figure 1. The housewife's time-space budget: (a) daily lifeline; (b) the prism and access to transport.

transport when deprived of one's car is typical of the population *at large*, however, as one report on transport choices reveals (Elliott et al, 1980).

In terms of the time periods available for travel to facilities the two-and-a-half-hour maximum also appears tenable. For instance the pattern of leisure activity through the day has three distinct peaks. Two of these fall in the morning (9.30–11.30 am) and afternoon (1.30–4.30 pm) when, respectively, 30% and 50% of the women are engaged in leisure at any given time. These periods also include some nonleisure components such as shopping and housework which may be reasonably flexible in timing. These periods are bounded by three short but distinct periods when leisure is minimal, and activities based on the home assume a high significance (8%, 9·5%, and 15% of the women engage in leisure in the three periods 8.30–9.30 am, 11.30–12.30 pm, and 4.30–5.30 pm). Travel and fetching children from school can also be seen to characterise the margins of the morning and afternoon periods. It seems that a shorter and less-used time window for extradomiciliary activities is typical of the morning period, whereas a longer, more popular one typifies the afternoon period. The general results suggest that our assumptions of a two-and-a-half-hour window may be an acceptable basis for investigating general levels of access to facilities.

The following sections pose four questions. First, in a society where many activities run on the assumption of car ownership what general level of choice of certain target facilities is available to a typical housewife. Second, to what degree does this choice vary over space. Third, does any variation in choice relate to socioeconomic status. In particular is access to facilities by public transport actually better in higher-status areas where greater car ownership enhances mobility for many households and depresses need. Last, what is the effect of the current structure of public transport in ameliorating any isolation or transport deprivation in the lower-status or more peripheral suburbs. By using the city of Christchurch as a typical medium-size Australasian urban centre these questions are addressed in the following sections.

Public transport and accessibility in Christchurch

The housewife in our example is dependent for mobility on the bus system or on her feet. Compared to a car-user she is at a strong disadvantage, as is indicated by Forer (1978). Not only is her ability for fast travel restricted, but the structure of the bus system considerably influences her ability to travel in certain directions. It is necessary, first, to look at the properties of the services of the Christchurch Transport Board (CTB).

Bus services in Christchurch are not poorly provided for, in that few households would be more than 0·5 km from a bus route. In 1977–1978 some 253·6 route kilometres were in existence and some 13·7 million passengers carried. On the main routes, frequencies during the day may average 6 an hour, and few routes have less than half-hourly buses. However, a combination of historical factors and falling levels of patronage has seen the persistence of a route structure centred on the central business area around the Square. Apart from the limited exception of the bus to the International Airport, all routes originate in one suburb and pass via the Square to another suburb where they terminate. Consequently the Square is a dominant interchange and readily reached from most locales, and suburban dwellers find themselves able easily to reach diametrically *across* the city, but hindered in attempts to move circumferentially into neighbouring suburbs.

Figure 2 illustrates the major routes and presents seventy-five points chosen as sample locations within that system. These points cover all timetabled stages on the CTB system, along with certain junction points. The time to travel between any pair of the seventy-five points was derived in the following manner. Initially all timetabled linkages were extracted from the 1977 winter timetables. To these times were added 'waiting time' based on the frequency of service and, hence, on the likely average delay to the user. To arrive at the waiting time the average time between buses was calculated and divided by three. [Dividing by two as for MacGregor (1953) would, of course, give the average waiting time if attempts to use the bus were randomly generated through time. The value three is used to allow for adaptation to the timetable.]

Once these total times between pairs of points were calculated they were fed into a standard minimum-path programme to produce a matrix of travel times between all points (compare Kissling, 1969). Advantage was taken here to allow for the interchange times in the Square by obtaining the average changeover time between all services from the timetables. This figure was then added to any linkages involving interchange in the Square. Finally it was realised that for some journeys the dog-leg imposed by the bus network might mean that walking was faster than going by bus. (This is most noticeable for two points close together but joined by bus only indirectly through the Square, for example, the Museum and Christchurch Public Hospital.) Therefore street mileage was used to calculate walking times, based on a walking speed of 6 km/h. For a mother with a preschool child an upper limit of 1 km walking was imposed. The walking times were compared with bus times and the lower time taken to give our final matrix of travel times between points. When one sums along the rows of this matrix in the manner of Shimbel (1953) one gets the total time taken to travel from any single point (the row) to all other points, a standard measure of access to a transport system.

The accessibility pattern that emerges is shown in figure 3 where the 100 isoline represents average accessibility in the system. Given the distribution of points the pattern is not exceptional, although one can notice certain areas where access is better or worse than one might expect. The general pattern is naturally of roughly concentric rings around the Square, but a tongue of good accessibility stands out

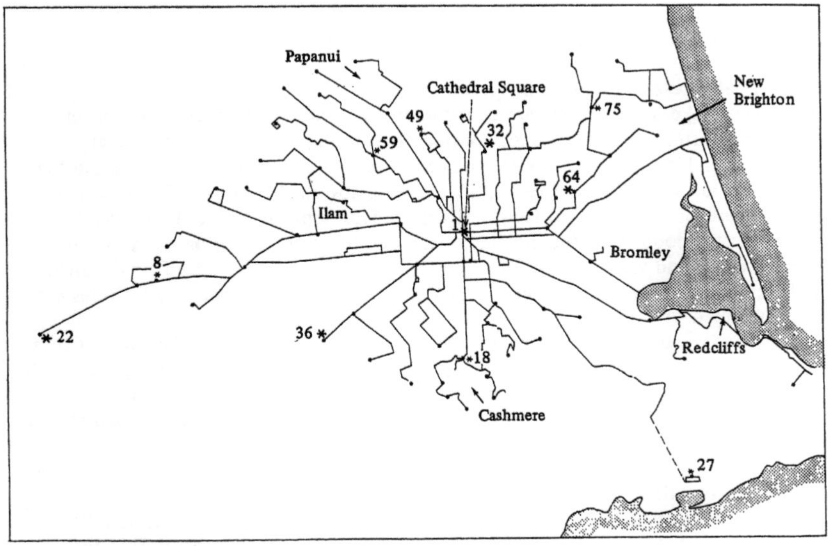

Figure 2. The structure of public transport in Christchurch.

towards the northeast. It is clear that the Square (1) is well served, needing only 70% of the average time to contact all points, whereas peripheral points, such as Lyttelton (27) or Hornby (8), pay for their peripherality. In general, the bus seems largely to mirror the natural accessibility of the various locations, even if the mirror is slightly a distorting glass.

We can now return to the housewife's time budget in order to see the effective activity range of individuals in certain locations and choice of facilities open to them.

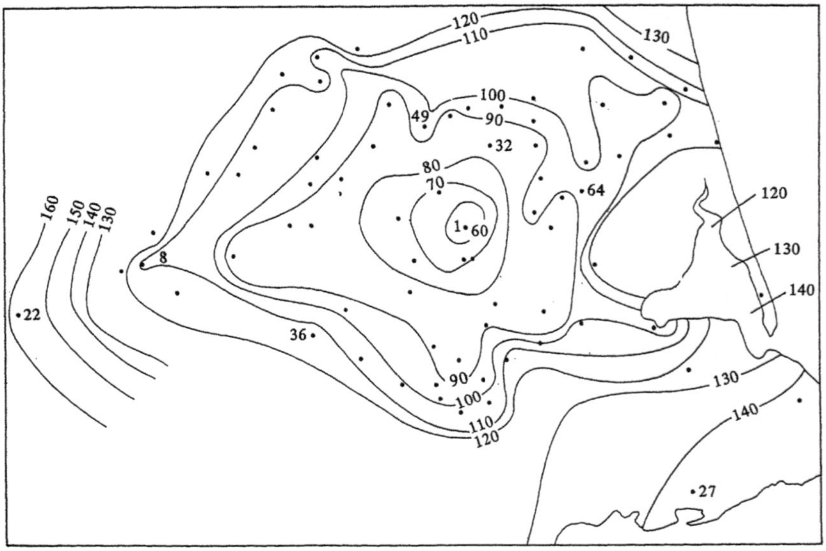

Figure 3. Public transport accessibility in Christchurch (surface in percent).

Differential access to facilities in space

Housewives, like most people, are constrained by a variety of gender and authority constraints, often, in their case, of considerable consequence. Accepting that spatiotemporal constraints are thus just one class amongst many, one can argue that the ability to use facilities at certain locations is governed by the ability to get to those places for a meaningful length of time. This depends on the time window available, the mobility of the individual, and the location of one or many facilities relative to the individual's own location. We have theoretically specified the first, and empirically derived the second. Finally, the location of various key facilities has been obtained.

The facilities selected were those likely to be of everyday or regular interest to a young housewife: middle-range shopping (post office present), central shops, doctor's surgery, Plunket (infant welfare) rooms, and playcentres. The next section examines the degree to which mobility and relative location interact to influence the real choices available to the housewife in a standard day.

Mobility: variations in activity space

The four maps of figure 4 provide an insight into the great variation in mobility between different suburban areas. We employ in each case the notion that the housewife has a two-and-a-half-hour window of time bounded by two inflexible markers.

By means of the derived travel times between nodes, the trade-off between time and distance can be seen to be quite different in the four cases covered. This will be seen to combine with the better facility provision of some areas to provide marked contrasts in access to facilities and, particularly, in the degree of choice.

The first map [figure 4(a)] deals with node 32, St Albans. This middle-status node is readily accessible to much of the system, having an 81% accessibility figure. The isolines around the node indicate the time available from the two-and-a-half-hour window after travelling to any point and back again. By staying at home there is a full two-and-a-half hours available, whereas at the margin of the isolines one is arriving at a point merely to return home immediately. Thus, although the zero-time line represents the extreme for the search space of the housewife, the area within the one-hour isoline is probably the maximum practical area for many activities. It represents the effective *activity* rather than *action* space of Brown et al (1970). Here the ratio of time wasted in transit to time spent usefully in the activity is 3 : 2. Even those activities which *could* be undertaken in less time are likely therefore to be increasingly irksome beyond this point because of the redundant time involved. Roughly 22% of the area of the city falls within an area in which one hour's free activity time would be available and a potential two hours can be spent at the central shops.

Figure 4(b) shows the extreme contrast in a satellite village, Templeton (22), where the only available direction for travel is in towards the city centre by a limited bus service. Many basic facilities are available only in Hornby which means only one to one-and-a-half hours of activity time being available. The total action space is a mere 17% of the city, the activity space only 4%. Templeton can conceivably be dismissed as an extreme, semi-rural example. However, the maps of Hendersons Road (36) and Bromley (64) [figures 4(c) and 4(d)] are clearly not. Again a picture of depressed provision of service emerges compared to St Albans. Given that St Albans has not been especially chosen as the best-provided area for transport facilities there

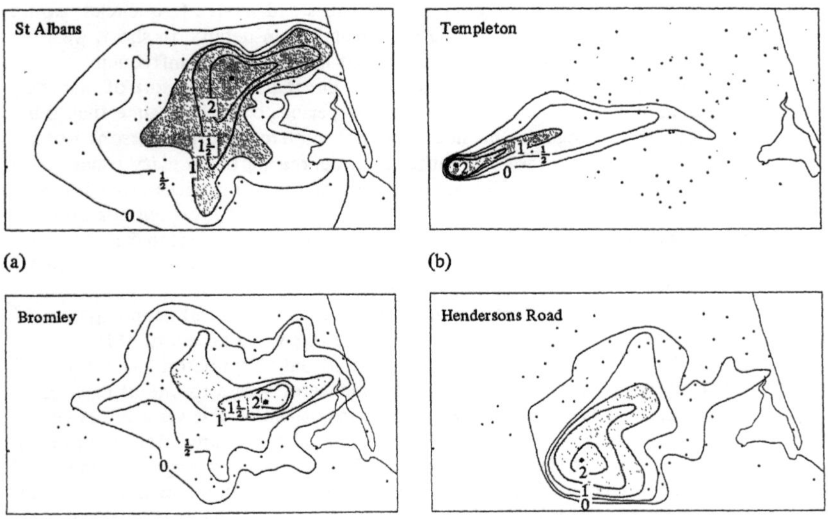

Figure 4. Activity-time zones from four sample nodes (isolines by half-hour intervals).

is clearly considerable variation in the measure of spatial mobility afforded housewives at different locations. This is underlined by the graph of figure 5.

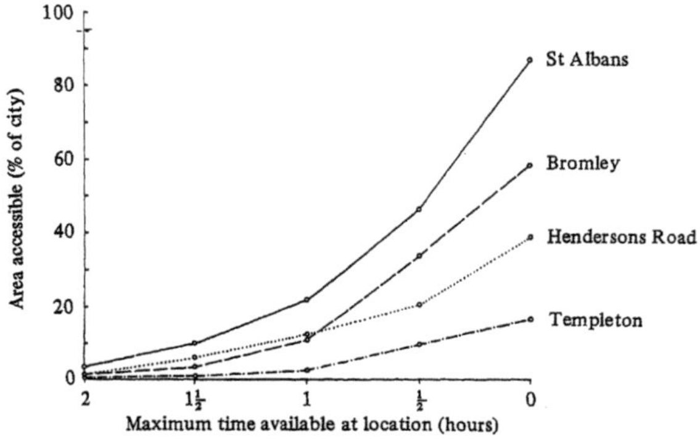

Figure 5. Extent of areas with selected minimum activity times from four nodes.

Facilities: location in action space

Figure 6 links the differences of access in space into the more pertinent one of access to facilities and choice within facilities. The differences in action and activity spaces crystallised in figure 5 translate into far wider variations in facility access. In figure 6 access to a range of facilities is shown. Time available at the specialist services of the central business district is shown by the bar across the top of each graph. The other lines depict the *log* of the number of certain facilities that can be reached for a given activity period within a two-and-a-half-hour window. (It could at this point be argued that the application of an aggregated measure of accessibility throughout the day, based essentially on journey times and service frequencies, would present a different picture than examining the timetables in detail. In short, we could suggest that the bus scheduling was such that in the cases of infrequent services the departure times were so arranged as to dovetail with the sort of markers we have proposed and so would in fact give longer periods of activity time than our maps suggested. We investigated this notion with Templeton and Hendersons Road and found that our aggregate measure, rather than depressing the activity times available in fact extended them. The actual timing of the bus services made it even harder for our housewife to achieve a worthwhile visit to other locations than our aggregate measure suggested. Upon examining other routes this appeared not to be an isolated occurrence and if anything the aggregate measure may overstate the real mobility of the housewife under current timetabling.)

The four services used in the analysis are Plunket (infant welfare) rooms, playcentres, post offices, and doctors' surgeries. As far as getting access to at least one site of each facility for a workable time it seems that only Templeton has any real problems, namely with a playcentre. The convenience of access to one of each of these four facilities clearly varies however. In terms of *choice* this variation is accentuated greatly. It is arguable that only a jaded palate requires more than ten doctors to choose from, or that choice between Plunket rooms is a questionable need. However, *some* choice between playcentres may be desirable or options in the doctor one attends may be valued. In the case of post offices we may consider these as surrogates for minor shopping centres, some of which may have unique retail outlets associated with them. A wider choice of small shopping areas may well reflect a wider range of shops as well as wider choice within a few basic goods.

Thus the three post offices yielding a one-and-a-half-hour visit from Hendersons Road may indicate a poorer living environment than the five from Bromley or the twelve for St Albans.

Equally, one may view these four facilities as indicators of choice and access to a wider range of many urban facilities. In either case it seems that although little absolute deprivation of access to facilities exists the costs of access in time and opportunity vary greatly. Choice of facilities in some areas is muted by the dictates of time and space whereas elsewhere there is considerable freedom of choice. These variations can also be linked to work on the prosperity of housewives engaged in the workforce, a prosperity clearly related to access to employment (Andrews, 1978).

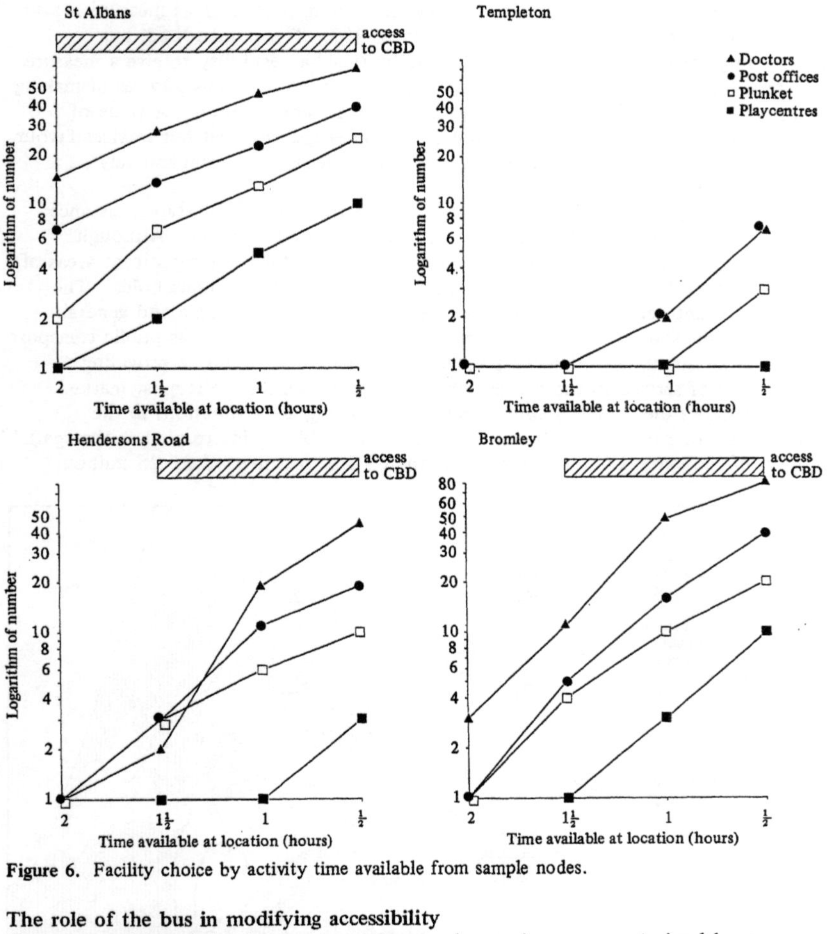

Figure 6. Facility choice by activity time available from sample nodes.

The role of the bus in modifying accessibility

We have established the degree to which some housewives are constrained by a combination of facility location and day-time mobility. In some instances this constraint is most marked. It is apparent that facility location is far from evenly spaced. What role does the bus play in ameliorating or aggravating this imbalance?

The picture painted by figure 3 suggested that the 'natural' centrality of the Square was somewhat exaggerated by the bus, and that some peripheral areas seemed better served than others. It is clear though that the pattern of accessibility on any map of that kind reflects in part the sample points chosen in compiling the map. Even with a generally even selection of points such as is used here the pattern of isolines will be to some degree a product of selected data points. Figure 7 is an attempt to show more clearly the relative impact of the variation in bus-service levels.

The map is derived from the accessibility measure described in figure 3. The standardised pattern here is taken and compared to the accessibility surface associated with simple mileage through the road system. This latter can be seen as the accessibility possessed by the point by virtue of its location in physical space and clearly can be used as a control for the pattern of data points. This measure is also standardised to an average nodal accessibility of 100. The standardised 'bus' accessibility for each point is divided by the 'mileage' accessibility to give a measure of 'stress' in access. This index indicates the relative effect of bus services in making areas of the city more or less accessible than their natural location. A value of 1·0 indicates no comparative change, whereas low values indicate that bus services favour a given point. Values in excess of 1·0 mean that the point is comparatively disadvantaged.

Private transport provision varies markedly as a function of the status, distance from the city centre, and household income of the various suburbs. Although distance from the centre encourages two-car ownership, there are significant areas of new housing in the north and east with high levels of single-car households. The more important determinant of car ownership appears to be income and general socioeconomic status. Given this, it might be argued that, insofar as public transport is seen as an instrument of social policy, the bus network should be providing a higher level of service to those areas where transport deprivation is most marked. Figure 7 shows the degree to which this is so. Failing great variations in the profitability of particular CTB routes, it can conceivably be interpreted as the spatial impact of the current annual support for the CTB of roughly NZ $1·25 million.

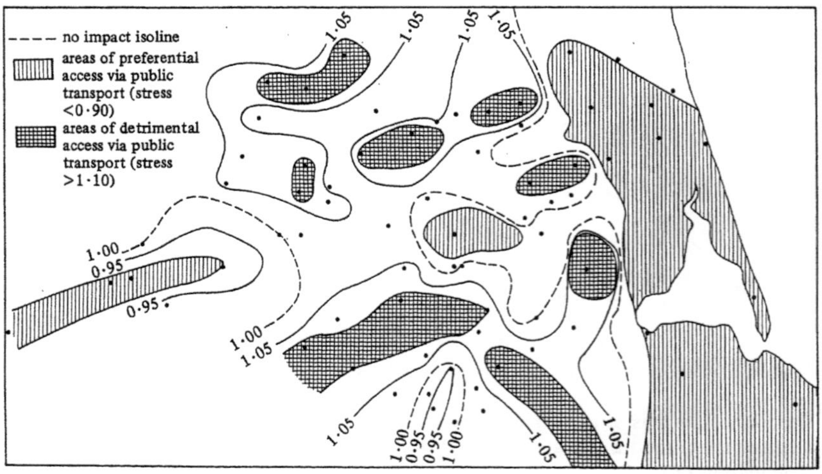

Figure 7. The spatial pattern of the stress index.

The actual range of values obtained is not as great as for some other transport media examined (Forer, 1975). The Square has a minimal stress value of 0·69, indicating the effect of the interchange concentrated there. The highest value is 1·39 recorded by Huntsbury, at the end of the 26 service. This maximum effect of 30% either side of zero change reveals certain regularities. In comparison to any idea of access one might arrive at by contemplating the normal map, it is apparent that three areas are in fact assisted by the bus. Apart from the Square and areas immediately surrounding it, the whole of the east of the city and the extreme western suburbs of Templeton and Islington are favoured. In contrast, the areas of disadvantage are in a mid-suburban belt especially to the north and west, with a few outliers in the south. The main reason for this is level of service provision. However, this is embroidered on by the effects of network structure. As the main effect of the bus system is to discourage circumferential links, the points away from the immediate centre, which are too far apart for going by foot, are involved with trips into the centre and back out again. Not only that, but the waiting time component, which is usually constant along one route, has to be carried on a short journey time. By contrast the suburbs on the outskirts lose out to a lesser degree. Templeton's loss of cross-linking clearly affects it far less than say Brett's Road (49), as the latter has so many more potentially beneficial links denied to it. Templeton's trips being lengthy are also comparatively less affected by the waiting time component.

Investigating further the relationship between stress and status proves a difficult task. Using socioeconomic status measures obtained from a factorial ecology of Christchurch (Newton, 1976) and plotting these against stress yields a weak but perceptible relationship between lower status and better bus service provision. However, the relationship is marked by many, significant deviations from the central trend. A fundamental problem which in part suggests caution with any detailed comparative exercise lies in the differing spatial characteristics of the transport and household data. The former relates to a set of nodes which frequently lie on major roads which themselves are the borders of the census divisions used in the measurement of status. As some of these roads divide areas of quite considerable

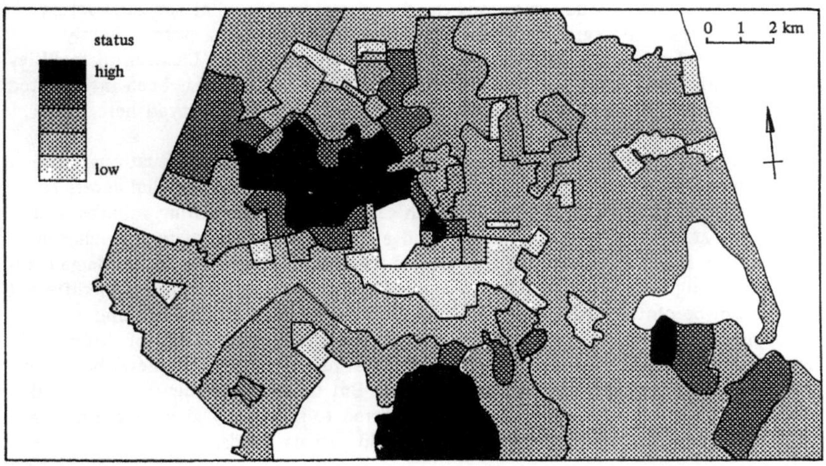

Figure 8. Social status in Christchurch.

difference in status (both of whom use the same bus service and experience the same relative level of provision) there is room for some distortion. In addition, service provision on a given route generally extends the full length of that route. A bus route predominantly serving a peripheral low-status suburb may traverse a high-status suburb en route. Both suburbs benefit equally, but without detailed traffic data the finer analysis of reasons for and impact of a given level of service must remain undone.

In the light of this deficiency only gross relationships may be drawn but a visual comparison of figures 7 and 8 suggests that, to a small degree, the effect of the current bus system is socially beneficial. It does tend, weakly, to be of benefit to those areas most needing transport subsidy. It is also plain, however, that such influence is negligible in terms of real access to facilities. The absolute accessibility of Templeton is still restricted, even if the bus-oriented housewife is relatively better served there than geography suggests.

Conclusions

We have shown that, within the constraints of a time–space budget, differential access to facilities exists for the bus-dependent housewife with young children. In some cases choice between alternative facilities is nonexistent, and fulfilling the needs of a young family without access to a car becomes problematical. The role of public transport in differentiating access to facilities was investigated in some detail. It was plain that the residents of certain suburbs enjoyed far larger expanses of the city in which to seek out and visit facilities, and that in some of these cases abundant choice of facilities was available because of an above-average provision of facilities in the surrounding neighbourhood. In other cases both action space and facility provision were restricted with consequent circumscribing of opportunities. The interaction of transport and location of facilities thus provides considerable variation in the quality of life at various locations. It seems that public transport has a socially beneficial effect, but one that is marginal in the final determination of the housewife's available options.

In this paper we have considered one aspect of access to facilities, namely the spatiotemporal one. Policy questions clearly arise regarding what is a satisfactory degree of access and degree of intraurban departure from such a norm. Purely spatiotemporal solutions to any disparity include revision of the location of facilities and provision of alternative transport arrangements. The latter has been investigated for Christchurch based on a similar methodology to the one employed here (Forer, 1980).

The same methodology also permits analysis of a wide range of situations. The impact of a proposed urban expressway which improves suburban radial access to CBD facilities, but amputates a facility-rich sector from neighbouring suburbs, is a case under study at present. The time budgets of other groups of the population may also be taken to investigate their access to relevant facilities. As the demands for services through the life cycle change so variations in access to facilities for different groups can be plotted, based on their different time-budget possibilities and a 'shopping basket' of required facilities. The impact on quality-of-life of different mobility scenarios is a further area that can be studied. The results described above focus on one particular, disadvantaged group. But the methodology has a far wider applicability in understanding the options offered to society by given combinations of transport technology, built environment, and activity needs.

References

Andrews H F, 1978 "Journey to work considerations in the labour force participation of married women" *Regional Studies* **12** 11-21

Brown L A, Horton F E, Wittich R I, 1970 "On place utility and the normative allocation of intraurban migrants" *Demography* 7 175-183

Carlstein T, 1977 "An annotated bibliography of time geographic studies" Rapporter Och Notiser 30, Institutionen fur Kultivgeografi och Ekonomisk Geografi Vid Lunds Universiteit, Lund, Sweden

Chapin F Jr, 1970 "Activity analysis, or the human use of urban space" *Town and Country Planning* **38** 28-30

Chapin F Jr, 1974 *Human Activity Patterns in the City: What People Do in Time and Space* (John Wiley, New York)

Cullen I G, Godson V, 1975 *Urban Networks and Activity Patterns* (Pergamon Press, Oxford)

Elliott J, Fletcher I, Lauder G, 1980 "The impact of the carless day scheme on the behavioural and attitudinal aspects of car-usage in Christchurch" report to the Ministry of Energy, Christchurch, by Department of Geography, University of Canterbury, Christchurch, New Zealand

Forer P C, 1975 "Relative space and regional imbalance: domestic airlines in New Zealand's geometrodynamics" *Proceedings of the IGU Regional Conference and Eighth NZ Geography Conference Palmerston North*, New Zealand Geographical Society, Christchurch, NZ, pp 53-62

Forer P C, 1978 "Time-space and area in the city of the plains" in *Timing Space and Spacing Time, Volume 1: Making Sense of Time* Eds T Carlstein, D Parkes, N Thrift (Edward Arnold, London) pp 99-118

Forer P C, 1980 "Public transport and access to urban facilities for the car-less: the case of Christchurch" in *Proceedings of Section 21 (Geographical Sciences) Australia and New Zealand Association for the Advancement of Science* Ed. W Moran (New Zealand Geographical Society, Christchurch) pp 31-36

Hägerstrand T, 1966 "The spatial structure of social communication" *Papers, Regional Science Association* **16** 27-42

Hägerstrand T, 1974 "The impact of transport on the quality of life" Rapporter och Notiser 13, Department of Geography, Lund University, Lund, Sweden

Hägerstrand T, 1975 "Space, time and human conditions" in *Dynamic Allocation of Urban Space* Eds A Karlqvist, L Lundqvist, F Snickars (Farnborough, Saxon House) pp 3-12

Hillman M, Henderson I, Whalley A, 1973 "Personal mobility and transport policy" Political and Economic Planning Broadsheet 542 (PEP, London)

Kissling C, 1969 "Linkage importance in a regional highway network" *Canadian Geographer* **13**(2) 113-127

MacGregor R D, 1953 "Daily travel" *Scottish Geographical Magazine* **68** 117-124

Newton P, 1976 *Residential Structure and Intra-urban Migration* PhD thesis, Department of Geography, University of Canterbury, Christchurch, New Zealand

Palm R, Pred A, 1974 "A time-geographic perspective on problems of inequality for women" WP-26, Institute of Urban and Regional Development, University of California-Berkeley, Berkeley, Calif.

Pred A, 1977 "The choreography of existence: comments on Hägerstrand's time-geography and its usefulness" *Economic Geography* **53** 207-221

Shimbel A, 1953 "Structural parameters of communication networks" *Bulletin of Mathematical Biophysics* **15** 501-507

Szalai A (Ed.), 1972 *The Use of Time* (Mouton, The Hague)

Thrift N, 1977 *Catmog 13: An Introduction to Time Geography* (Geo Abstracts, Norwich)

Tivers J, 1977 "Constraints on spatial activity patterns: women with young children" OP-6, Department of Geography, Kings College, London

Tomlinson J, Bullock N, Dickens P, Steadman P, Taylor E, 1973 "A model of students' daily activity patterns" *Environment and Planning* **5** 231-266

[16]

Spatial Knowledge Acquisition by Children: Route Learning and Relational Distances

Reginald G. Golledge,* Nathan Gale,** James W. Pellegrino,*** and Sally Doherty****

*Department of Geography, University of California at Santa Barbara, Santa Barbara, CA 93106, FAX 805/893-3146

**Robert D. Niehaus, Inc., Santa Barbara, CA 93117, FAX 805/681-7307

***Learning Technology Center, Peabody College, Vanderbilt University, Nashville, TN 37703

****Salk Institute, San Diego, CA 92138

Abstract. Our purpose is to examine spatial knowledge acquisition via route learning processes. To gain as much control as possible, yet study learning in actual environments, we conducted field experiments using an unfamiliar suburban neighborhood and a subject population of 9- to 12-year-old children whose activity spaces are most highly oriented to the neighborhood level. In addition to multiple-trial route learning, we gave subjects a battery of tasks including cue recognition, cue sequencing, and interpoint distance estimation, using a variety of route segment scenarios. The children generally achieved route learning after three trials and successfully completed on and off-route cue location and sequencing tasks. Subjects had difficulty with the distance estimation tasks, particularly when conditions of segment inclusion were violated. Sketch mapping and interpoint judgment tasks clearly indicate that this subject group found it difficult to integrate knowledge acquired from the two separate but partially overlapping routes. Results suggest that the term "route knowledge" is ambiguous, for subjects can learn to navigate routes without having definable procedures for unpacking the basic spatial knowledge contained in routes. They can acquire the ability to learn and follow a route between an origin and destination while procedures for integrating that knowledge into an understanding of spatial layout may be absent. This suggests that the development of configurational knowledge structures is not a simple consequence of learning declarative and procedural knowledge systems. We emphasize the implications of these results for the cognitive mapping process and the need to understand spatial reasoning and its developmental phases.

Key Words: spatial knowledge, landmarks, route knowledge, configurational knowledge, wayfinding behavior, integrating knowledge, route segmentation, route integration, cognitive map, cognitive distance, field experiments, learning tasks.

OUR purpose is to examine the process of acquiring spatial knowledge by learning routes. We proceeded by giving subjects repeated trials of well-defined route learning experiences in conjunction with a battery of tasks designed to test various components of the spatial knowledge thus acquired. To focus the subjects' attentions on the route rather than on obvious landmarks or major environmental changes, we defined the routes to be learned in a relatively homogeneous suburban residential neighborhood. Our interest was in examining how children recognize, recall, and use the spatial properties of routes in route-related spatial tasks. Because the neighborhood is the prime focus for most of the activities of fifth and sixth grade elementary school children, and since the neighborhood knowledge structure of many adults is governed by the need to move through rather than within neighborhoods, we selected a sample of nine- to twelve-year-old children, equally counterbalanced by gender.

To accomplish our purpose, we first recorded the subjects' actual navigational behavior over prescribed routes as descriptive measures of task-oriented environmental learning. Next, we gave subjects scene recognition tests to evaluate sensitivity for different types of scenes and locations. In a third task, subjects were asked to draw sketch maps, as a less structured means of testing a variety of knowledge components. We assumed the features indicated on these maps reflected subjects' recall for important choice points and environmental features that could, at this scale, be interpreted as "landmarks." The representation of the route itself was presumed to contain segments and subdivisions. We also designed sequencing tasks to test for knowledge of ordinal relations among cues along the routes. Finally, we employed a distance judgment experiment in an attempt to examine the effects of route segmentation on spatial distance estimates. This final task was based on Allen's (1981) hypothesis that sections between choice points are chunked (organized cognitively as subdivisions of a route) and this affects judgment of distance. We specifically designed this task to examine relations between chunking, distance judgments, and the integration of knowledge gained from learning separate routes.

This study falls within the area of spatial cognition, an interdisciplinary field in which geographers, environmental and cognitive psychologists, architects, environmental designers, and cognitive scientists are active participants. Using the eminently spatial task of learning about and navigating a route as its focus, the paper draws on a variety of multidisciplinary concepts to examine the problems of how wayfinding takes place in unfamiliar environments. Not only do we examine the elemental spatial properties of location, spatial sequencing, connectivity, pattern and interpoint distance estimation, but we also speculate on how piecemeal information might be integrated into a complete spatial knowledge structure.

Since much of the relevant theoretical and empirical literature on this spatial task is found in cognate disciplines rather than in geography, we begin with an overview of related research. For more comprehensive reviews of spatial knowledge acquisition or cognitive mapping generally, see Golledge (1987), Golledge and Timmermans (1990), Gärling and Golledge (1989), Heft and Wohwill (1987), and Aiello (1987).

Background

The literature on both cognitive mapping and spatial knowledge acquisition has strong ties to psychological developmental theory (Piaget and Inhelder 1967). As interpreted by Hart and Moore (1973) and Siegel and White (1975), individuals pass through stages of egocentric to allocentric knowledge, from topological to fully metric comprehension of space, and from knowledge structures dominated by *landmarks* to those where landmarks and connections between them are proceduralized into *routes*, to full *configurational* (*survey*) understanding of specific environments. Over the last decade in particular, traditional developmental theory has been questioned by life-span theorists and others interested in spatial knowledge acquisition (Liben et al. 1981; Evans et al. 1981), who suggest that progression through developmental stages may only partially account for progressive changes in spatial abilities. Increasing quantities of evidence indicate that the spatial abilities formerly hypothesized to exist only at particular ages have been found at earlier ages (Acredolo 1976, 1978, 1981, 1987; Liben 1982). Research also suggests that when one learns about an unfamiliar environment, one does not automatically obtain understanding at a level equivalent to one's developmental stage. Rather, one goes through all stages from egocentric to allocentric and topological to metric as part of the local learning process (Golledge 1977a, b; Evans et al. 1981). Kuipers (1977) provides a reason for this phenomenon. He suggests that a person acquires knowledge by a piecemeal "bottom-up" process dependent on experiencing an environment. Bits and pieces of declarative knowledge are absorbed experientially. This information is collected from a human-level perspective rather than a birds-eye perspective and must be integrated by a set of procedural rules. Such rules develop a "common sense" understanding of an environment, allowing one to use the sensed bits of information in an integrated way to solve problems of spatial movement.

Anderson (1976, 1982) conceptualized the importance of proceduralizing piecemeal knowledge to facilitate spatial problem solving and provided the general framework for knowledge acquisition processes used for investigating how different types of learning occur. Since this model was not specifically designed for spatial learning, there is question as to whether or not it can account for all stages or types of spatial understanding. One problem we researched here is whether higher-order concepts involving associations and relations of phenomena can be achieved via proceduralizing bits of information learned in a wayfinding context. Others have often raised this question (Carey and Diamond 1977; Golledge et al. 1988; Freundschuh 1989; Gale et al. 1990), and we investigate it further as part of this research.

Related Literature on Route Knowledge

In earlier studies, we examined spatial knowledge acquisition over time via the mechanism of repeated learning trials on selected routes within familiar and unfamiliar environments (Doherty 1984; Doherty and Pellegrino 1986; Doherty et al. 1989; Gale et al. 1990; Golledge et al. 1990; Pellegrino et al. 1990). Our research confirmed findings of Allen (1981, 1985, 1987) and Allen and Kirasic (1985) that, when learning routes, individuals tended to subdivide or chunk them to help identify segment sequence and to help locate specific places along the route. This chunking procedure also was important in route composition to simplify the process of sequentially ordering the total sum of information collected during navigation. Given a more differentiated inner-city environment, Allen found that subjects, when learning a route, tended to segment the route and to use segmentation as a means of organizing "along-the-route" information. In relatively undifferentiated environments, however, such as the uniform tract residential neighborhood that we chose for our experiment, natural segmentation of the route in terms of land-use types or dominant features is impossible, for there are none to use. We have designed various tasks in our study to examine whether or not chunking takes place under these conditions and how it facilitates the development and use of intra- and inter-segment spatial information.

In other research that can be related to route learning, Kuipers (1978) and Anderson (1982) advanced hypotheses suggesting that route learning is a "piecemeal process" and its success depends on how experiential data are stored and proceduralized. For example, Anderson suggests that different forms of "if-then" production systems can be applied to any declarative knowledge base (or internally stored list of environmental cues and features) in such a way that correct route following behavior takes place. This theory has been most useful for the building of computational process models (CPMs) of environmental knowledge acquisition (e.g., CRITTER, Kuipers 1985; TURTLE, Zimring and Gross forthcoming; TOUR, Kuipers 1977; TRAVELLER, Leiser and Zilberschatz 1989; and NAVIGATOR, Gopal et al. 1989).

An integral part of the route learning literature is the concept of cognitive distance. Geographers such as Golledge et al. (1969), Briggs (1973a, b, 1976), Cadwallader (1979), and Lowry (1973) have made major contributions as have other researchers such as Lee (1964), Canter and Tagg (1975), and Gärling and Gärling (1987). In addition, an important sequence of papers has been produced by Burroughs and Sadalla (1979), Sadalla and Staplin (1980a, b), and Sadalla et al. (1980) who suggested that (a) cognitive distances are essentially asymmetric, (b) perceived effort is a critical component in the estimation of cognitive distance, (c) information integration theory provides a sound basis for examining recall and retrieval of cognitive distance information, and (d) segmentation of routes using features such as intersections is a critical part of the route learning process. Smith et al. (1982) and Golledge et al. (1985), who undertook detailed studies of route learning in familiar environments, achieved similar results. Their results reinforce the importance of choice points in the route segmentation process. Evans and Pezdek (1980), in discussing the concept of cognitive mapping, also have investigated the relationship between objective and subjective distances and the impact it has on locational specificity and the linking of locations in a route context. In one paper (Evans et al., 1984) a research team specifically examined the effect of pathway configuration on landmark selection and route segmentation as well as off-route and on-route cue identification using simulated environments constructed on the Berkeley Environmental Simulator.

Another group of researchers at the University of Umeå also had a long-term project examining different components of distance in the context of navigation and wayfinding behavior. For example, Gärling and Böök (1981) examined the spatiotemporal sequencing of activities to determine if people were conscientious shortest-path travellers in urban environments. The results indicated that people were somewhat less than optimizers in terms of activity sequencing and path selection. Gärling and Gärling (1987) continued examining distance minimizing procedures in downtown pedestrian traffic while coworkers (Böök and Gärling 1980, 1981; Lindberg and Gärling 1981, 1982) have respectively tried to model wayfinding and to define the essential structure of route knowledge by comparing blindfolded and sighted locomotion activities. More recently, Montello (1991) has reviewed the literature on cognitive distance and has pointed to the comparative effectiveness of reproduction tasks as compared to the traditional distance estimation tasks used in most previous studies. He also summarizes the importance of distance compression in subjective distance estimation, emphasizing the importance of Stevens's power law for comparing subjective and objective distances. Psychologists such as Maki (1981) and Potts (1974) have combined the ideas of chunking and sequencing in examination of processes of storing and retrieving spatial linear orders and ordinal relationships among components of sequenced data. Reitman and Reuter (1980) have discussed how organization of information, spatial and otherwise, is revealed by recall order and by pauses in response latencies between chunked segments of information. Herman (1980) examined the importance of exploration direction and repeated experience in route learning, while recently Hirtle and Hudson (1991) have examined the question of precisely *what* environmental information is obtained during the process of route learning. It is at once obvious that the mental integration of bits and pieces of environmental knowledge into cognitive maps incorporates distance distortions, asymmetries, and simplifications, probably accounting for much of the error in cognitive maps and the decisions and behaviors that are based on them.

Thus attention has been focused on *what* is learned during wayfinding and *how* it is learned. The research tasks discussed later in this paper similarly concentrate on both the *content* of learning and the *process* of learning information via repeated experiences with different routes. While much of the route learning literature is self-contained, focusing largely on learning itself and how successful behavior might consequently develop, our research goes beyond this by asking questions about how information learned via route experience can be integrated into a larger, more comprehensive understanding of the environment in which the routes are embedded. We now turn to a discussion of the problems and procedures involved in our specific experiments.

Research Procedures

According to both life-span and developmental theories, the ability to understand routes and to learn them should be well developed as early as preteen years. With this in mind we chose subjects of age 9–12 years. Our previous experiments (Doherty et al. 1989; Gale et al. 1990) showed that preteenage girls and boys acquired more information about their local neighborhood when traversing through it than did adults. Possible reasons for this included the neighborhood as the focal point of daily activity patterns for the children whereas the city-at-large was more the domain of adult daily activities. The attentional processes of preteenage child subjects was thus more focused on the route learning task and less on relating the route being learned to the domain of a larger adult cognitive map. Since we were interested in discovering details of what was learned along the route and whether fundamental spatial concepts such as location, spatial order or sequencing, direction, orientation, and inter-point distances were embedded in the information learned, we deliberately chose a subject population that was most appropriate for our test situations.

Since we had chosen to have our subjects learn routes through an environment with which they were unfamiliar, it also seemed more likely that taking children outside their home neighborhood to a more distant one would reduce the chances of a subject having prior knowledge of the task environment. Although there were no dominant landmarks nearby that could possibly have provided the spatially aware adult with contextual or frame-of-reference information that might aid their recall and recognition processes, we again felt that the choice of preteenage children as subjects would minimize the possibility that this might occur.

Methods

Since previous research (Gale 1985; Lloyd 1989) had indicated that active learning in the field produced a quicker understanding of both routes and the general environment in which they are embedded, we designed this study as a field experiment. Subjects actively participated in the learning process and had the opportunity to observe phenomena not only along the routes learned but in the general vicinity of each route. We assumed that access to a field site would help subjects develop a general sense of direction and orientation in terms of a "neighborhood" frame of reference.

The specific task environment for the field study was a residential neighborhood (Fig. 1) selected for route learning experiments in part because it provided a quite regular yet sufficiently complex configuration of streets and intersections to add difficulty to the learning task. The streets included longer thoroughfares, as well as much shorter connected streets and cul-de-sacs. Two partially overlapping routes, both between 0.7 and 0.8 miles in length, were selected. Both routes incorporated seven changes of direction occurring at intersections. Distinguishing the second route from the first were two street crossings that do not occur at natural intersections; these were included to examine the effect of arbitrary decision points.[1]

Subjects

Subjects were sixteen children—eight females and eight males, aged 9–12 years—all of whom were unfamiliar with the study area. All the children lived in another residential neighborhood approximately four miles from the study area. While their home area had a few differentiating landscape features (e.g., elementary schools, a 7-11 convenience store, a gas station, and a high school) scattered throughout, housing styles were not remarkably different, and street arrangements included the same type of patterns including thoroughfares, dead-end streets, cul-de-sacs, and curvilinear sections. The study area was, therefore, not so different in nature from their home area as to produce concern, but was sufficiently different in terms of structure and organization to warrant a need to learn the neighborhood in order to navigate within it successfully. Equal numbers of male and female subjects were chosen as a counterbalancing condition of the experimental design and to allow us to check a hypothesis of the lack of difference between genders at this age as far as spatial task performance was concerned.[2] Since the route learning experience was carried out strictly in the field, there were two experimental conditions. Subjects in Condition I learned Route 1 followed by Route 2, whereas subjects in Condition II learned the routes in reverse order.

Materials

To facilitate testing in a mobile field laboratory, 3 by 5-inch photographic prints of locations and scenes in the neighborhood were developed (Fig. 2). Using these prints, we constructed a scene recognition task for each route similar in nature to that used by Golledge et al. (1980) and Allen (1981). Distance judgment tasks were developed using photographs of on- and off-route neighborhood scenes. The structuring of the tasks was based on the assumption that choice-points chunk a route into segments and that such chunking affects judgments of distance (e.g., Allen 1981). Thus, given a referent location, R, a near comparison scene, C_n and a far comparison scene approximately three times more distant, C_f, four classes of estimation were defined as follows (Fig. 3):

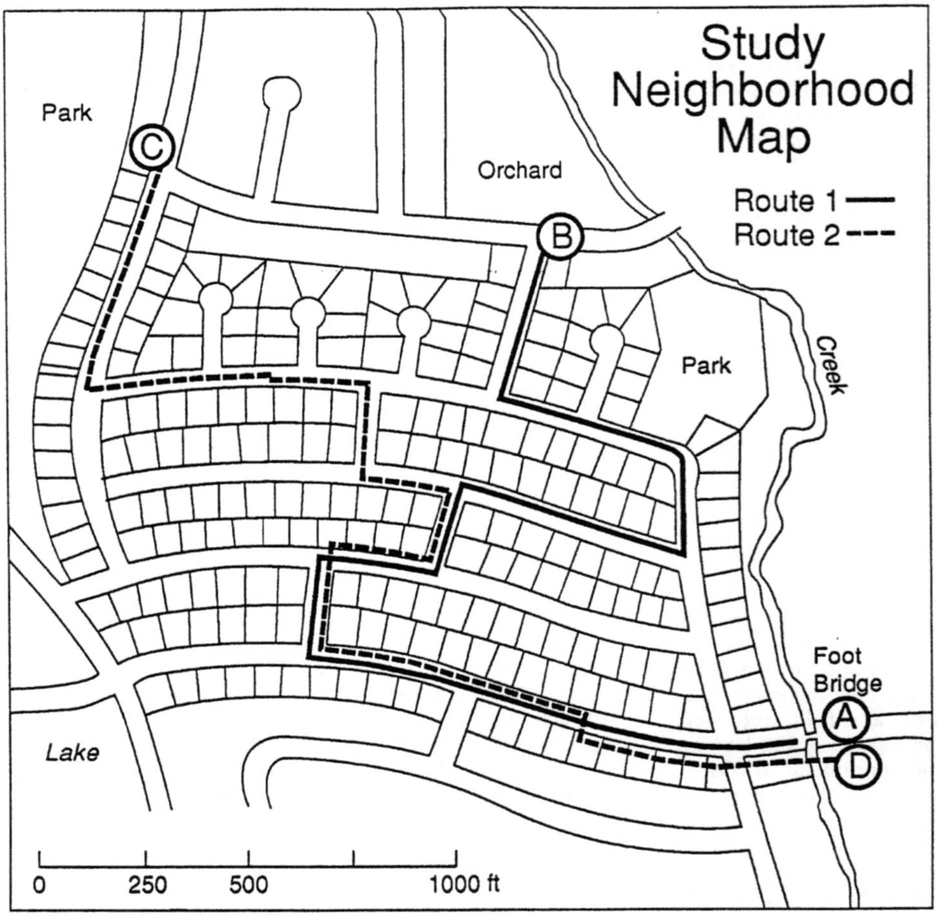

Figure 1. Study neighborhood. Source: adapted from Golledge, et al. (1991, 3) and Gale, et al. (1990, 7).

1. Segment Assignment: R and C_n in the same segment, C_f in the following or consequent segment (a correct response depends on segment assignment only);
2. Segment Assignment and Order: R, C_n and C_f each in sequential segments (a correct response depends on segment assignment and segment order);
3. Segment Assignment and Distance Estimation: R in one segment, C_n and C_f both in the same different segment (a correct response depends on segment assignment and actual distance);
4. Segment Order/Inclusion Violations: R and C_f in the same segment, C_n, in another segment (a correct response depends on actual distance overriding segment commonality).

For each route, we selected six referent locations and, for each referent, four pairs of comparison scenes. The ratio of distances along the route between the two comparison scenes and the referent was held constant at 3:1, with the type of comparison varying across the categories outlined above. In addition to the set of four (target) comparisons, one off-route dis-

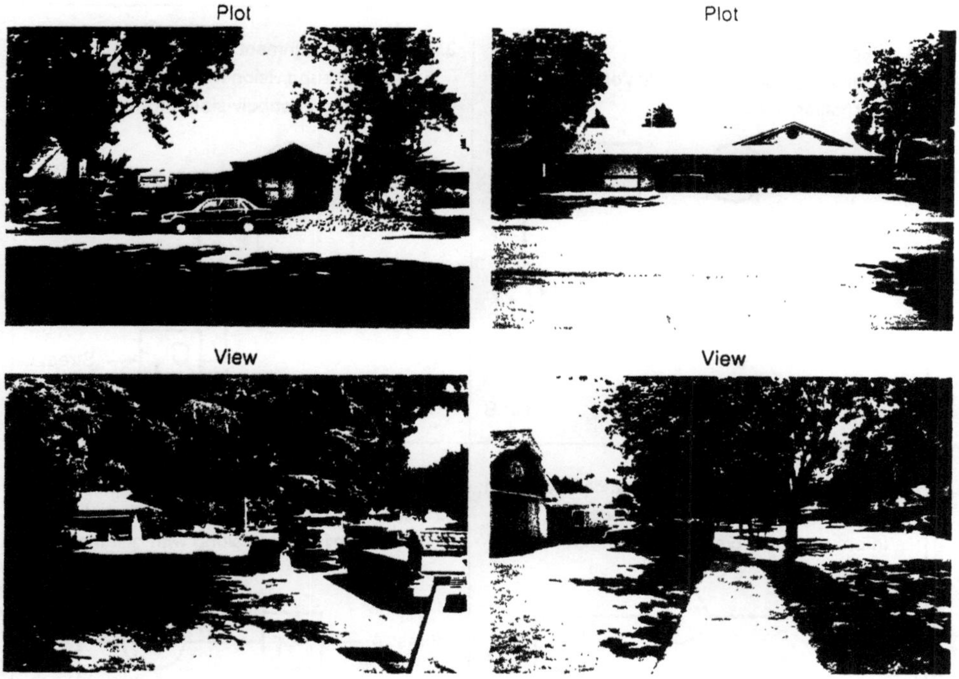

Figure 2. Photos of sample "plots" and "views" used for scene recognition and distance judgment tasks.

tractor comparison also was included. The task was then duplicated using scenes of both "views" (i.e., along streets), and "plots" (i.e., frontal shots of individual houses) at each designated location; thus, in the complete task there were 12 referent scenes (a view and a plot for six locations), each with a corresponding set of five pairs of comparison scenes (four on-route, one off-route).

To test whether or not subjects could integrate information over the two partially overlapping routes, we created a combined segment chunking and distance judgment task wherein one comparison was located on Route 1 and the other on Route 2. Four referent locations were used in this task. Again, distance ratios were held constant at 3:1, but types of comparisons were split between those that were route-compatible (i.e., closer comparison scene on the same route as referent), and route-incompatible (i.e., closer comparison scene not on the same route as referent).

Finally, a set of 2 by 3-inch photos showing views of the salient choice points along each route were produced for use in a scene sequencing task. Ten scenes were used for the Route 1 task, while twelve scenes were used for Route 2—the two arbitrary crossings on this route account for the additional scenes.

Experimental Procedures

Each subject participated in ten sessions over a two-week period, with daily sessions typically lasting from 60 to 90 minutes, and was paid an hourly rate as compensation. In all phases of the experiment, we tested subjects individually. Each trial carried out a different combination of navigation, sketch mapping, scene recognition, and distance-judgment testing

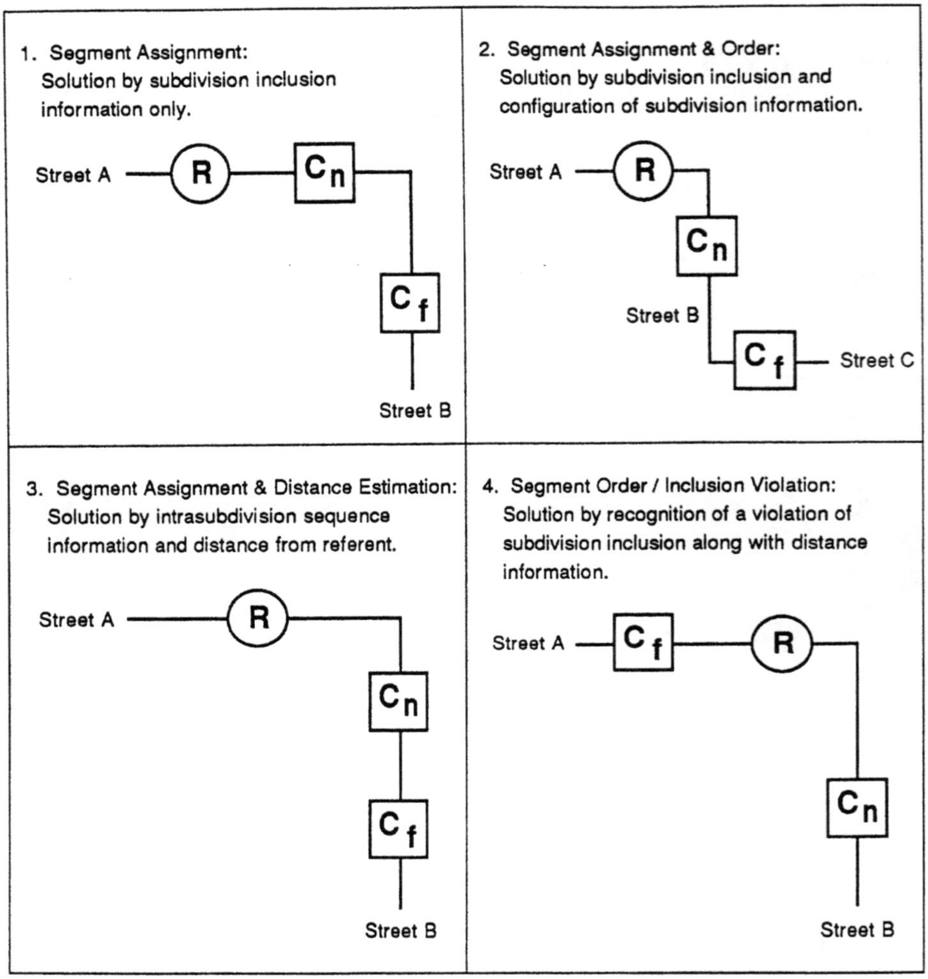

Figure 3. Schematic of reference/comparison configurations used for problem construction of the distance judgment task.

procedures. Each route was traversed in both directions and the scene recognition testing was done following Trials 1, 2, and 3. Other tasks in this study examined route integration and included a combined sketch map of both routes, inter-route distance judgments, and route sequencing, which were completed on the final day of testing. The full sequence of tasks for the ten-day course of the experiment is outlined in Table 1.

To carry out the field testing, each subject was first driven to the start of one of the routes via streets that were outside the study area. Their instructions before beginning the walk were to follow the experimenter and attend to the route.[3] We made clear that, after the first day, the roles would be reversed and the experimenter would follow the subject. Furthermore, we told subjects about the scene recognition and map drawing tasks to be completed

Table 1. Experimental Tasks in Field Study

Week #1	Day #1	Day #2	Day #3	Day #4	Day #5
Cond. I Rt. 1 Cond. II Rt. 2	Show route Sketch map Scene recognition	Navigation Scene recognition	Navigation Sketch map Scene recognition Distance judgments	Navigation Distance judgments	Navigation Sketch map Distance judgments Sequencing
Week #2	**Day # 1**	**Day #2**	**Day #3**	**Day #4**	**Day #5**
Cond. I Rt. 2 Cond. II Rt. 1	Show route Scene recognition	Navigation Sketch map Scene recognition	Navigation Scene recognition Distance judgments	Navigation Sketch map Distance judgments	Navigation Sketch map Combined sketch map Distance judgments Combined distance judgments Sequencing

after traversing the route. During the walk, the experimenter neither initiated nor encouraged conversation, so as not to aid or distract the subject in any way. A single trial consisted of one forward and one reverse traversal of the route.

After the first experimenter-guided trial, the subjects undertook successive trials without experimenter help except to correct an unrealized navigation error. Where errors of different types occurred, the experimenter recorded them on a map of the study area. For example, if the subject made a wrong turn, he or she was allowed some time to realize that a mistake had been made—a distance of one or two house plots from the point of error. If the mistake was recognized and corrected before this end point, a "realized mistake" was recorded on the map; if the mistake was not realized, the subject was informed of her/his error and returned to the choice point to make a different decision. The latter produced an "unrealized mistake" recorded on the map at the location where the wrong decision was made. Given this limited form of experimenter interference and opportunity to correct realized mistakes, it was inevitable that subjects would eventually make correct navigational choices.

Following each trial we administered various tests to each subject. These included a map drawing task in which the subject was given a piece of paper with the start and finish of the route marked with Xs. Subjects were instructed to work as accurately as possible to reproduce a map of the route they had just followed. They were urged to put on the sketch everything they could remember but no clues were given as to starting orientation or hints as to what type of information should be placed on the sketch. The sketch map task was carried out on the first, third, and fifth days of each week of testing.

On the fifth day of the second week—the final day of testing—each subject was given a piece of paper with the start and finish of *both* routes marked. In an effort to test how well the spatial information acquired over the two-week period could be integrated and represented, each subject was asked to draw a map that combined both routes.

On the first three days of each week, the scene recognition test was given. Each subject was shown 80 photographs and asked to judge whether the scenes were on or off the route.

On the third, fourth, and fifth days, we gave the distance judgment task, consisting of presenting each subject with a referent and a pair of comparison scenes. As each scene was presented, the subject was asked the scene recognition question: i.e., if the scene was on or off the route. With three photos on the table, subjects were then instructed to indicate which of two comparison scenes was judged to be closer to the referent. Then the next pair of comparison scenes was presented. The combined route-distance judgment task was also administered on the final day of testing.

After the fifth and final trial of each route, we asked subjects to complete a sequencing task. First the subject laid out photos on the table in the order in which the scenes were thought to appear on the route. Then he or she was given a piece of paper on which a line segment had been drawn to represent the route. On this line segment, the subject was asked to place a hatch mark where each of the scenes occurred.

Results

Navigation

We tabulated behaviors according to three hierarchically defined categories: (1) Recorded Behaviors, including all mistakes, stops, hesitations, tentativeness, and verbal expressions of disorientation; (2) Total Mistakes, including turns in the wrong direction, crossings to the wrong side of the street, and failures to turn, whether realized by the subject without the aid of the experimenter or not; and (3) Unrealized mistakes, including only those mistakes not realized and corrected by the subject, thus necessitating experimenter intervention. As a final measure of navigation behavior, the lapsed time from start to finish of the routes also was recorded. The mean number of recorded behaviors by trial is presented in Table 2.

The majority of route learning was completed by the end of the third trial, with little apparent change between the fourth and fifth trials. Likewise, the lapsed time showed a similar leveling off after the third trial. Thus, with respect to the variables collected from the field navigation experiment, there appears strong evidence for relatively rapid and systematic learning across all measures. Before the end of the week of testing, learning of the navigation task was virtually complete.

The data on each of the three behavioral measures and lapsed time were analyzed in turn, using a 2 × 2 × 4 × 2 (*condition* × *route number* × *trial* × *direction*) ANOVA design. No main effect was found for the grouping factor (*condition*), which separated the subjects according to the order of route presentation; there were, however, significant effects found for the within-subject factors. In all cases there was a significant main effect of *trial*, reflecting progressive learning over the course of the experiment (Recorded Behaviors, $F(3, 24) = 55.25$, $p < .001$; Time, $F(3, 42) = 6.65$, $p < .001$). In addition, for all the observed navigation behavior measures except for Time, there was a significant main effect of *direction* (Recorded Behaviors, $F(1, 14) = 33.69$, $p < .001$). Since the forward direction always preceded the reverse, this effect is clearly due to within-trial learning.

In all measures, Route 2—the route with the arbitrary crossings—appeared to be the more difficult. A main effect for this factor was found to be statistically significant for the sum of all recorded navigational behaviors ($F(1, 14) = 9.54$, $p < .01$). Subjects who learned the easier route first were able to use this experience in learning the more difficult route, and the performance on both routes was essentially equivalent. On the other hand, subjects who learned the more difficult route first did exceptionally

Table 2. Mean Number of Recorded Behaviors from Field Navigation Trials

Recorded behaviors	Route 1			Route 2			Combined		
	Forward	Reverse	Total	Forward	Reverse	Total	Forward	Reverse	Total
Trial 2	3.38	1.00	4.38	5.06	3.06	8.13	6.22	2.03	6.25
Trial 3	0.69	0.56	1.25	1.88	0.75	2.63	1.28	0.66	1.94
Trial 4	0.13	0.13	0.25	0.50	0.31	0.81	0.31	0.22	0.53
Trial 5	0.14	0.29	0.43	0.27	0.07	0.34	0.21	0.17	0.38

Units are *mean total counts* of all realized mistakes, stops, hesitations, and verbal expressions of disorientation recorded by the supervisor during each trial.

well later with the easier route; their performance on the two routes differed considerably until the fourth trial. This differential learning pattern was evidenced in two significant interactions—condition × route, and condition × route × trial. In sum, the route navigation results emphasize the orderly manner in which the skills necessary for route traversal are acquired. The implication is that, even in a completely unfamiliar environment, children can quickly adopt successful strategies for the achievement of a specific wayfinding goal.

Scene Recognition

Nonparametric signal detection measures for sensitivity and bias, calculated using the formulas given in Grier (1971), were used to analyze the scene recognition data collected on the first three trials. The sensitivity measure (ranging from .5 to 1.0) is derived from the probability of a correct recognition (or "hit") corrected by the probability of an incorrect positive response (or "false alarm"). Bias reflects the propensity to give either a negative or a positive response regardless of the stimulus. Both measures were analyzed using a 2 × 2 × 3 × 2 × 2 (*condition* × *route number* × *trial* × *location* × *scene*) ANOVA design. With respect to scene recognition sensitivity, significant main effects were found for *trial*, ($F(2, 28) = 7.15$, $p < .01$), and for *scene*, ($F(1, 14) = 26.16$, $p < .001$). There was also a significant two-way interaction between *location* and *scene*, ($F(1, 14) = 5.95$, $p < .03$), as well as a three-way interaction between *route number*, *trial*, and *location*, ($F(2, 28) = 4.39$, $p < .03$).

The main effect of *trial* reflects an increase in sensitivity with experience from an initial value of .71 on the first trial, to .74 on the second trial, to .76 on the third trial. The pattern of increase in sensitivity over time was quite constant across all other factors, yielding no significant two-factor interactions involving *trial*. The stronger main effect of *scene* was due to much higher sensitivity for views (.77) than for plots (.71). This result likely stems from the fact that the stimulus pictures of views along a street more closely represented the task environment as perceived by the subjects during navigation than did the pictures of house plots.

The lack of a main effect of *location* appears, at first, to be problematic since it seems to

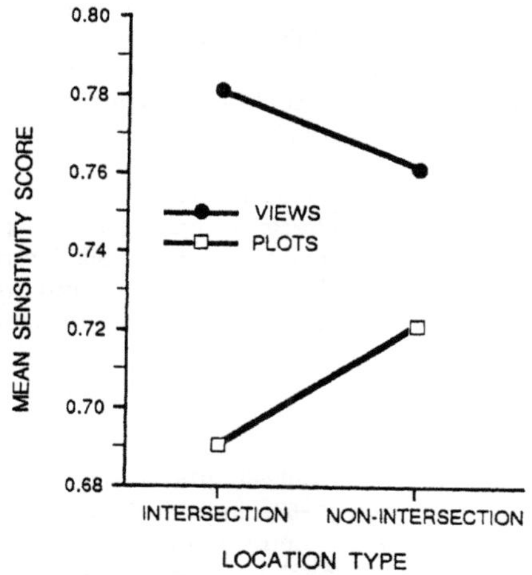

Figure 4. Scene recognition interaction—mean sensitivity scores as a function of scene type and location type.

contradict both theoretical arguments and empirical findings suggesting that choice points become important loci for the organization of spatial knowledge (Smith et al. 1982; Golledge et al. 1985). Nevertheless, an explanation for this result can be found in light of the significant interaction between *location* and *scene*. As depicted in Figure 4, the highest sensitivity was found for views at intersections, but the reason for no main effect of *location* is that plots at intersections had the lowest overall sensitivity. It appears then that, similar to the results of a previous study reported in Gale et al. (1990), there may have been a trade-off at intersections between views and plots. At intersections where navigational decisions are made, subjects remembered wide-angle views, rather than more narrowly focused images of particular houses.

The significant interaction between *route number*, *trial*, and *location* was due primarily to the differences between the two routes, assumed to be caused in large part by the inclusion of arbitrary choice points on Route 2. This more difficult route, as evidence from the navigation results suggests, presented more of a challenge to the subjects and, at the same time, led to higher sensitivity scores. Presumably, the higher sensitivity can be attributed to greater levels of attention associated with the more difficult task. In addition, the reason for relatively higher sensitivity for nonintersection locations on Route 2 could be explained by the fact that two such locations were actual decision points.

Analysis of the bias measures yielded no significant main effects or interactions. It appears, therefore, that there was no significant propensity for either a negative or a positive response associated with the scene recognition task.

Sketch Mapping

As expected (Wohlwill 1976; Tversky 1981; Moar and Bower 1983), individual differences in graphic abilities and styles led to considerable variation in the sketch maps drawn following route navigation. Even with such variations, it was possible to define appropriate quantitative measures describing key elements of the sketches in terms of orientation, segmentation, and features. Orientation measures were based on an evaluation of whether or not the beginning and end points of the route were oriented correctly both with respect to each other and with respect to a surrounding frame of reference (a standard north line). Segmentation consisted of a count of the number of segments recorded on the maps compared to the actual number of segments experienced on the route. Feature recognition consisted of a count of the number of features identified and correctly located along each route. Figure 5 shows examples of sketches of individual routes after the fifth trial by a 12-year-old girl and a 12-year-old boy respectively. It also shows the difficulty experienced when attempting to integrate both routes in a single map. The data for each of these map elements were analyzed, in turn, using a 2 × 2 × 3 (*condition* × *route number* × *trial*) ANOVA design.

Orientation

A simple binary classification (correct/incorrect) was used to code the orientation of the beginning and ending route segments on each of the sketch maps. Combining the results of both segments together, a single map could thus be scored 0.0, 0.5, or 1.0 (i.e., having both beginning and end route segment incorrectly oriented, one correct, and one incorrect, or both correct). Analysis of these data yielded significant main effects for *trial*, ($F(1, 28) = 5.48$, $p < .01$), and for *condition*, ($F(1, 14) = 5.39$, $p < .05$). No interaction effects were significant at the .05 (or even the .10) level. The *trial* effect indicates significant improvement over the course of the experiment, with the largest increase coming between the first and third trials. Overall percent correct representations for the three trials were .50, .67, and .69 for Trials 1, 3, and 5, respectively.

If the four cardinal directions are considered equally likely outcomes, the probability of a correct representation for the beginning and ending route segments if chosen at random would be .25. Since some information was provided to the subject by indicating the start and finish of the route, a more conservative assumption is that the probability of being correct is .50. Assuming independence among events, the results of the first trial are no different than chance, but those on the third and fifth trials would be significant at the .05 level.

The main effect found for *condition* was due to much higher performance levels with respect to orientation for subjects in Condition II (.78) than for those in Condition I (.46). A possible explanation for this finding is that Route 2 has only short sections that double back; the direction of travel is generally toward the destination for all but a short segment of about four house plots. On the other hand, navigation of Route 1 requires reversing direction for much longer route segments. Such significant reversals, faced during the first week of field experience, may have tended to make the task of orientation to a larger framework more difficult.

Segmentation

Results of the analysis on the number of segments represented on the sketch maps revealed a significant main effect of *route number* ($F(1, 14) = 45.51$, $p < .001$). If segment limits are defined by changes in direction, there are eight segments in Route 1 and ten in Route 2. The sketch maps accurately reflected this difference, with an overall average of 7.46 for the first route, and 9.33 for the second. Taking the analysis one step further, the data were transformed from raw numbers of segments into absolute deviations from the actual number.

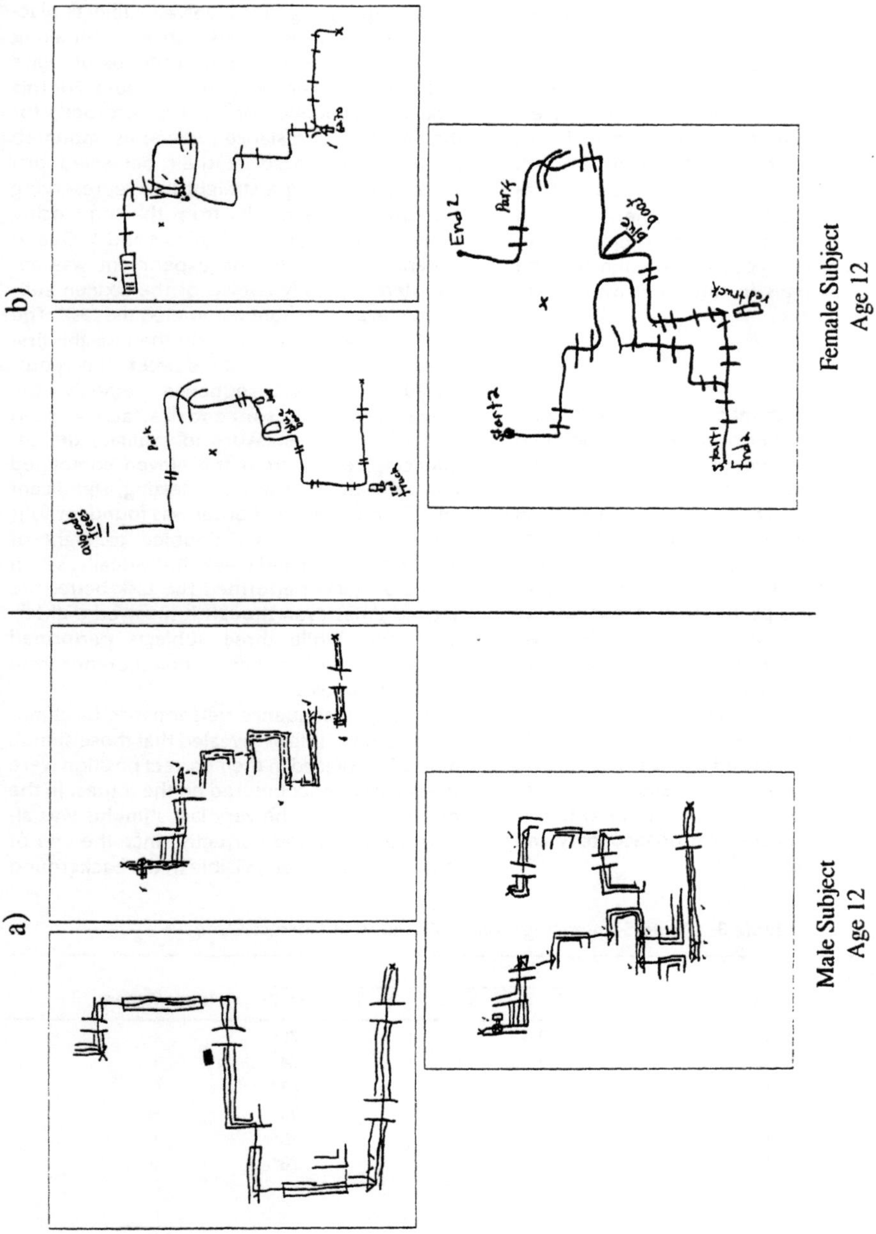

Figure 5. Examples of sketches of individual routes after fifth trial by a 12-year-old boy and a 12-year-old girl respectively.

Here there was no difference between routes, as each was represented equally accurately. There was, however, a significant main effect of *trial* ($F(2, 28) = 6.65$, $p < .01$), showing improvement in accuracy over time. On the first trial, the number of segments on the routes was either over or underestimated by an average of 1.66 segments. By the third trial, this number was reduced to .94, and by the fifth trial, it reached .91. Thus, it appears that regardless of the variability in style and visual quality of the sketch maps, the subjects were, on the whole, very accurate in estimating the number of changes in direction involved on each of the routes.

Features

Finally, the sketch maps were analyzed with regard to the number of features represented along the route. The only significant result found in this analysis was due to the main effect of *trial* ($F(2, 28) = 3.59$, $p < .05$). With more experience, a greater number of features was represented: 2.90, 4.56, and 4.59 for Trials 1, 3, and 5 respectively. In general the number of distinct features portrayed on the maps was quite small. This finding may reflect the difficulty of recall and graphic reproduction, but it also may indicate that relatively few major landmarks are necessary for successful route navigation. The efficient storage of task-oriented spatial knowledge thus may involve recall level information of a small set of key landmarks, supplemented by the ability to recognize a much broader set of secondary and tertiary landmarks.

Sequencing

The sequencing task involved subjects placing in correct order, from start to finish along a route, a series of mounted photos of views or scenes experienced along the route. For this task, only sequence order was recorded; for the next task of distance judgments, mounted photos were placed at their perceived distances apart along a straight line representing each route. The results from the sequencing experiment are given in Tables 3 and 4. Due to material constraints, the experiment was administered to only twelve of the sixteen subjects, eleven of whom completed the task. The twelfth subject refused to do the task the first week, but then the second week she reproduced the correct sequence perfectly—the only subject to do so. Kendall's Tau was used as a descriptive measure to evaluate the sequencing results from the eleven completed tasks. On the first week of testing, significant (at .05) congruence of order was found for four subjects; this number doubled to eight of eleven by the second week. Individually, seven of the subjects performed the task better the second time, even though it involved a different route, while three subjects performed worse; one subject showed no difference from week 1 to week 2.

Looking at sequence performance by stimulus location (Table 4) revealed that those stimuli most often placed in their correct position were the first ones encountered on the routes. In the case of Route 2, the very last stimulus was almost always placed correctly since the end of the route was, in fact, visible in the background

Table 3. Route Sequencing: Kendall's Tau with Correct Order

Subject Id.	Sex	Condition	Week 1 Rt. 1	Week 1 Rt. 2	Week 2 Rt. 1	Week 2 Rt. 2	Improvement
01	M	1	.20			.76*	+
02	M	1	.20			.24	+
10	F	1	.82*			.82*	0
11	F	1	.73*			.36	−
12	F	1	.47			.76*	+
13	F	1	.29			.70*	+
05	M	2		.67*	.51*		−
06	M	2		.48*	.38		−
07	M	2		.12	.91*		+
15	F	2		.42	.56*		+
16	F	2		.27	.69*		+

*Significant at .05

Table 4. Route Sequencing Results by Location

Route 1			Route 2		
Ordered stimuli	No. times in correct position	$\Sigma\|O_c - Os\|$*	Ordered stimuli	No. times in correct position	$\Sigma\|O_c - Os\|$*
1	6	14	1	8	21
2	6	17	2	5	15
3	5	23	3	1	26
4	4	15	4	3	24
5	4	15	5	3	22
6	3	18	6	1	22
7	1	37	7	1	22
8	2	10	8	2	24
9	1	13	9	4	17
10	2	18	10	1	40
			11	4	25
			12	10	2

*O_c = correct order; O_s = subjective order.

Table 5. Distance Judgments: Percentage Correct

	Week 1			Week 2			
Subject Id.	Trial 3	Trial 4	Trial 5	Trial 3	Trial 4	Trial 5	Combined
1	38	33	35	46	44	46	61
2	58	48	60	38	54	60	46
3	44	46	50	48	48	44	61
4	58	52	65*	60	54	58	61
5	38	48	65*	35	48	50	43
6	44	46	44	54	54	56	61
7	46	53	48	60	56	65*	64
8	42	40	40	56	65*	73*	43
9	58	54	48	46	50	38	54
10	46	60	58	52	44	58	43
11	52	46	56	56	58	48	50
12	46	48	58	44	50	44	57
13	46	58	54	48	33	46	54
14	58	69*	69*	48	69*	69*	57
15	44	56	63	48	40	42	50
16	52	52	58	40	46	50	50

*Significant at .05

of the photo. In general, the worst performance was associated with stimuli on the mid-to-latter segments of the routes. The results of the sequencing experiment suggest that the ordering of stimuli is considerably more difficult than either actual navigation or simple landmark recognition. Clearly, successful completion of the task implies the development of relational knowledge, at least at the level of one-dimensional topological associations. At the end of two weeks of experience in a novel environment, this type of knowledge was acquired by the majority of subjects tested, but not to the same degree as were navigation and recognition abilities.

Distance Judgments

The data from the distance judgment experiment were scored in terms of the percent correct responses in each segment location category. But since it was impossible, with the chosen referent locations, to create a sufficient number of task items in the inclusion violation category, only three classes were used. Overall results, combining classes, are summarized by *trial* in Table 5. Using the binomial probability distribution to evaluate significance, *none of the subjects was able to perform the distance judgment task at a level significantly better than chance on the first attempt*. By the last attempt (Trial 5) three out of the sixteen subjects performed significantly better than chance on the Route 1 task, and another three had significant results on Route 2; only one subject scored significantly high on both tasks. It is interesting that the five subjects with the best performance were all in Condition II—the group of subjects that scored significantly better with respect to the orientation data derived from the sketch maps.

The distance judgment data were then analyzed using a 2 × 2 × 3 × 3 (*condition* × *route number* × *trial* × *difficulty*) ANOVA design. The analysis revealed a significant main effect for *trial* ($F(2, 28) = 9.12$, $p < .001$). A marginally significant main effect for *difficulty* and a significant three-way interaction between *condition*, *trial*, and *difficulty* showed no consistent ordering of performance over the different tasks. The results of this task suggest that the *level of spatial knowledge necessary to make consistently accurate distance judgments was, by and large, not attained with the limited route learning experience given the subjects*. Two important implications arise from these results. The first is that *successful route navigation is not necessarily dependent on* the ability to make metrically correct distance estimations; and second, *the comparison of these results to the sketch map orientation data suggests that such distance judgment ability may be related to the acquisition of a frame of reference within which distance relations may be more accurately represented.*

Discussion

It is relatively easy to observe and measure actual route learning behaviors, and the navigation tasks of this study appeared to be quickly and completely mastered. It is much more challenging to measure the knowledge components that are presumed to develop as a consequence of such navigation behavior. Nevertheless, the evidence is strong in support of the notion that a key element of spatial knowledge acquisition is the identification of particular locations whether they be choice points or landmarks. The features represented on the sketch maps, although few in number, indicated that information about the environment was stored in a manner that facilitates recall of certain salient loci. From the scene recognition results, it was evident that many more landmarks were recognized when visually encountered, and that images of views along the direction of travel—most similar in composition and viewpoint to those seen during field navigation—were easier to identify than were more simple scenes of individual house plots. In addition, there was further evidence supporting earlier studies (e.g., Allen 1981; Doherty and Pellegrino 1985; Golledge et al. 1985; Gale et al. 1990) that the highest sensitivity for scenes is associated with intersections where navigation decisions and actions must be taken.

The sketch maps, in conjunction with the results of the sequencing experiment, suggest that landmark information is not disjoint, but can be integrated into route representations. The routes were, for the most part, accurately depicted with respect to the number of directional changes and segments each contained. Moreover, by the end of the second week of experience, a majority of those tested were able to place scenes of decision points in reasonably correct order—indicating that spatial relations, at least at an ordinal level, were successfully stored and accessed. There was clear evidence from the route navigation results that experience gained on one route facilitated procedural learning on the second route. Presumably, accumulated navigation experience over interlocking routes should eventually lead to more sophisticated knowledge of spatial relations, but the processes involved in the acquisition, storage, and retrieval of relational information are not yet well understood.

The failure in this field study, with multiple route exposures and navigation trials, to produce even rudimentary distance-judgment performance stands in contrast to research by Allen (1981) that indicated that much briefer exposures to route information produced the necessary integration of metric information from learned routes. This contrast may be understood in the light of key differences between our research and that of Allen. First, in this study, subjects experienced the route through actual field experience, whereas Allen's subjects experienced the route through a series of slides. Second, the duration of route experience within a single exposure trial was much briefer in Allen's research. Third, Allen's route was through a much more diversified urban setting, whereas our route was through a more homogeneous suburban area. On the basis of these experimental differences and contrasting results, at least two major hypotheses can be advanced regarding key factors affecting spatial information integration. One such hypothesis is that such integration is more difficult when the key locations or landmarks are experienced with a greater temporal separation along with a high density of intervening information. Another hypothesis is that the nature of the environment, particularly its distinctiveness and heterogeneity, significantly contributes to the ease of route segment assignment and localization and thus systematically affects performance on metric tasks such as distance judgment.

The implications of our results for various spatial problem solving tasks such as navigation and wayfinding, as well as for cognitive mapping tasks generally, are significant. For occupations requiring constant exposure to unfamiliar environments (e.g., parcel delivery systems, emergency vehicle dispatch, inner-city bicycle delivery messengers), for independent blind travelers trying to explore new environments, for the child learning his or her way home from a new school, it would seem that successful route learning can take place in a small number of trials. In conjunction with route learning, landmark recognition and choice point definition are included as part of the learning process, but route learning does not by itself appear to produce the type of spatial knowledge that is required to understand layout and configuration. Interpoint distance estimates can be confused if segment sequencing is violated. Choice points can be recognized often using a priming cue, but this is usually unidirectional and cannot be used for route retrace or learning in the reverse direction. Compiling a cognitive map with configurational properties on the basis of what is learned by experiencing a route may introduce significant distortions and errors, and may account for much of the spatial variance and inaccuracy previously found in many cognitive mapping studies. The question of how best to integrate the information obtained from learning separate routes into a configurational whole remains unanswered, but our results indicate that this process of integration is not simple and justifies considerable research. And, if our understanding of space and spatial properties is primarily derived from travel experiences, how well can we expect individuals to understand even the simplest spatial processes such as nearest neighbor, distance decay, intervening opportunity, regionalization, and hierarchical orderings? Needless to say, these are all fundamental concepts for understanding things such as maps and diagrams and interpreting those maps and diagrams in ways to facilitate human activities. One may even surmise that even low-level spatial understanding is not as common a feature as we usually presume it to be. Thus, it might make little sense to undertake activities such as making geographic information systems accessible to naive users. We might also be laying the foundations for a solid argument in favor of the need for a discipline of experts (i.e., geographers) who can understand low and high-level, basic and derived, spatial properties, and can develop effective ways for teaching spatial reasoning and spatial understanding to populations at large. And finally, in the theoretical domain, we ask the question: Is the capability of this age group to perform seqencing and distancing tasks a result of development or something related to the process of spatial reasoning and route learning as a facet of that reasoning process? In other words, are some of the poor results we achieved, particularly in the distance judgments tasks, simply a result of the fact that preteenage children are yet to experience entry into a concrete or abstract stage of reasoning—stages which, according to Piaget, occur towards the entrance and exit (respectively) of the teenage years. Ongoing research by our group is currently investigating this particular hypothesis.

Concluding Remarks

As in our other studies (Gale et al. 1990), these results indicate that having a certain amount of declarative knowledge does *not* imply that the procedures to operate on that knowledge exist. Such development is dependent on both the learning experience and the environment in which learning takes place. It is obvious that the term "route knowledge" can become ambiguous if it is found that subjects can learn routes without having clearly defined procedures for operating on a declarative knowledge base. In the context of a major goal of this paper (i.e., to assess whether or not information learned on different routes can be integrated to provide configurational understanding), it appears that knowledge can be acquired about specific routes while procedures for integrating that knowledge into a configurational understanding may be absent. This finding suggests that developing a configurational or survey level understanding may not be a simple consequence of combining declarative and procedural knowledge systems, but might involve something more. Precisely what additional information or skills this involves is unclear at this time, but the question is certainly worthy of further investigation. We can hypothesize that access to frames of reference, either by active field participation in an environmental learning task or by providing frames of reference via map or other visual or tactile devices, may be an important initial step in the processes of reasoning about space, understanding spatial relations, and in developing configural understanding.

A general finding of this study is that the orientation, sequencing, and topological arrangement of components of a route can be developed parsimoniously. Being able to navigate a given route apparently does not require extensive knowledge about all the scenes or plots along it. Specific places where choices must be made become important organizing features. Intervening route segments appear to be chunked and regarded as more or less homogeneous. Our distance judgment task proved to be extremely difficult when segment inclusion features were violated, as in those instances where a near comparison to a given referent node was in an adjoining segment, while the further referent was in the same segment as the referent node.

Obviously environmental learning and the cognitive mappings that develop from this learning involve a highly selective process, and much information can be ignored in specific task situations such as those of navigating over preset routes. Parsimonious selection of route information might lead to errors in behavorial performance on tasks such as sketch mapping, distance estimation, orientation, or even sequencing, because of selective elimination of scene, view, or plot information along the route. To understand the wayfinding process, whether in a familiar or unfamiliar environment, the reasons for selecting critical bits of information must be known. This raises questions that helped initiate behavioral research in geography, such as: What processes underlie the selection of certain environmental features and the elimination of others? Why do some cues develop greater degrees of familiarity or significance than others? Is the objective or goal of a decision maker the most critical factor in selecting information from the environment, or do task-specific instructions heighten selected aspects of the environmental learning process? Can spatial behavior be understood better and predicted if we understand the spatial reasoning behind the overt behavior?

Our research offers one further observation: that environmental knowledge structures do not consist *solely* of objects, features, or facts (i.e., declarative knowledge) and sets of procedures that allow connections among them, but *also require some superordinate processes that encourage understanding of associations, relations, patterns, and configurational properties*. Thus, developing the capacity to integrate sets of features into patterns, even those as simple as the sequence of features along a route, must be an integral part of spatial knowledge acquisition. Our research joins the growing quantity of work in related disciplines in trying to define the nature of spatial knowledge, its component parts, and the processes by which such knowledge is acquired, stored and used. Solutions to these problems are coming slowly but inexorably as detailed controlled experiments using active and passive field and laboratory conditions become more a feature of geographic research.

Acknowledgments

Research reported was supported by NSF Grant #SES84-07160, NSF Grant #SES897-20597, and a 1987–88 fellowship from the John Simon Guggenheim Foundation.

Notes

1. The study was restricted to a single neighborhood because we had neither the personnel nor the resources to duplicate it elsewhere. We have made every attempt, however, to construct an experiment that could be duplicated in any neighborhood regardless of its complexity or simplicity. Although it would be most interesting to reproduce the experiment using a variety of physical, economic, and social neighborhood structures, such a task was beyond the scope of our resources and is beyond the scope of this paper. We would encourage replication of the experiment to investigate more thoroughly whether differences in the physical or social structure of neighborhoods have a greater impact on the route learning process.
2. Subjects were recruited from fifth and sixth graders at a local elementary school, which had provided subjects for our previous experiments (reviewed earlier in this paper). Since tasks required 2–3 hours per subject per day to complete, and scheduling difficulties frequently prevented us from completing more than one subject each morning and each afternoon, and since each trial had to be carried out under direct experimenter supervision (both in the field and in the mobile laboratory located at the end of routes), contact between supervisor and subjects was extremely labor-intensive and time-consuming. Sixteen subjects completing five trials required more than 150 hours of scheduling in the field. This exhausted the summer time period over which the experiment took place. Even so, there was a risk that environmental changes might occur during the data gathering period (e.g., a house being painted a new and outstanding color, construction work beginning, radical landscaping changes associated with specific houses, etc.). All such changes would have altered the study environment and might have invalidated the experiment. No such changes occurred, but it was clearly impossible to schedule more than sixteen subjects within this temporal constraint.
3. Before beginning the experimenter-guided trial, subjects were told that they would be undertaking multiple trials in both forward and reverse directions over two separate but partially overlapping routes in an unfamiliar residential neighborhood. They were told to learn the route and remember all the detail that they could about the area through which they traveled, for they would be questioned about the route itself and features along it in a debriefing session after each trial. During the trial route experience (i.e., experimenter-guided) the trial experimenter discouraged conversation but answered all questions. Subjects knew they would be drawing maps, identifying features, and answering questions about the features.

References

Acredolo, L. 1976. Frames of reference used by children for orientation in unfamiliar spaces. In *Environmental knowing*, ed. G. Moore and R. Golledge, pp. 165–72. Stroudsburg, PA: Downden, Hutchinson and Ross.

———. 1978. Development of spatial orientation in infancy. *Developmental Psychology* 14:224–34.

———. 1981. Small- and large-scale spatial concepts in infancy and childhood. In *Spatial representation and behavior across the life span: Theory and application*, ed. L. Liben, A. Patterson and N. Newcombe, pp. 63–81. New York: Academic Press.

———. 1987. Early development of spatial orientation in humans. In *Cognitive processes and spatial orientation in animal and man*, vol. 2: *Neurophysiology and developmental aspects*, ed. P. Ellen and C. Thinus-Blanc, pp. 185–201. NATO ASI Series, Series D: Behavioral and Social Sciences—No. 37. Dordrecht, the Netherlands: Martinus Nijhoff.

Aiello, J. 1987. Human spatial behavior. In *Handbook of environmental psychology*, vol. 1, ed. D. Stokols and I. Altman, pp. 389–504. New York: John Wiley & Sons.

Allen, G. L. 1981. A developmental perspective on the effects of subdividing macrospatial experience. *Journal of Experimental Psychology: Human Learning and Theory* 7:120–32.

———. 1985. Strengthening weak links in the study of the development of macrospatial cognition. In *The development of spatial cognition*, ed. R. Cohen, pp.301-21. Hillsdale, NJ: Lawrence Erlbaum.

———. 1987. Cognitive influences on the acquisition of route knowledge in children and adults. In *Cognitive processes and spatial orientation in animal and man*, vol. 2: *Neurophysiology and developmental aspects*, ed. P. Ellen and C. Thinus-Blanc, pp. 274–83. NATO ASI Series, Series D: Behavioral and Social Sciences—No. 37. Dordrecht, the Netherlands: Martinus Nijhoff.

———, and Kirasic, K. 1985. Effects of the cognitive organization of route knowledge on judgments of macrospatial distance. *Memory and Cognition* 13:218–27.

Anderson, J. R. 1976. Language, memory and thought. Hillsdale, NJ: Lawrence Erlbaum.

———. 1982. Acquisition of cognitive skill. *Psychological Review* 89:369–406.

Böök, A., and Gärling, T. 1980. Processing of information about location during locomotion: Effects of concurrent task and locomotion patterns. *Scandinavian Journal of Psychology* 21:185–92.

———, and ———. 1981. Maintenance of orientation during locomotion in unfamiliar environments. *Journal of Experimental Psychology: Human Perception and Performance* 7:995–1006.

Briggs, R. 1973a. On the relationship between cognitive and objective distance. In *Environmental design research*, vol. 2, ed. W. Preiser, p. 187. Stroudsburg, PA: Dowden, Hutchinson & Ross.

———. 1973b. Urban cognitive distance. In *Image and environment: Cognitive mapping and spatial behavior*, ed. R. Downs and D. Stea, pp. 361–88. Chicago: Aldine.

———. 1976. Methodologies for the measurement of cognitive distance. In *Environmental knowing*, ed. G. Moore and R. Golledge, pp. 325–34. Stroudsburg, PA: Dowden, Hutchinson and Ross.

Burroughs, W., and Sadalla, E. 1979. Asymmetries in distance cognition. *Geographical Analysis* 11:414–21.

Cadwallader, M. 1979. Problems in cognitive distance: Implications for cognitive mapping. *Environment and Behavior* 11:559–76.

Canter, D., and Tagg, S. 1975. Distance estimation in cities. *Environment and Behavior* 7:59–80.

Carey, S., and Diamond, R. 1977. From piecemeal to configurational representation of faces. *Science* 195:312–14.

Doherty, S. 1984. Developmental differences in cue recognition and spatial decision making. Ph.D. dissertation, Graduate School of Education, University of California, Santa Barbara.

———, and Pellegrino, J. W. 1985. *Developmental changes in neighborhood scene recognition* (Technical Report 8603). Santa Barbara: University of California, Center for the Study of Spatial Cognition and Performance.

———, and ———. 1986. Developmental changes in neighborhood scene recognition. *Children's Environments Quarterly* 2:38–43.

———; Gale, N.; Pellegrino, J. W.; and Golledge, R. G. 1989. Children's versus adult's knowledge of places and distances in a familiar neighborhood environment. *Children's Environment Quarterly* 6:65–71.

Evans, G., and Pezdek, K. 1980. Cognitive mapping: Knowledge of real world distance and location information. *Journal of Experimental Psychology: Human Learning and Memory* 6:13–24.

———, Marrero, D., and Butler, P. 1981. Environmental learning and cognitive mapping. *Environment and Behavior* 13:83–104.

———; Brennan, P.; Skorpanich, M.; and Held, D. 1984. The effects of pathway configuration, landmarks, and stress on environmental cognition. *Journal of Environmental Psychology* 4:323–35.

Freundschuh, S. 1989. *Can survey (map view) knowledge be acquired from procedural knowledge?* Paper presented at the annual meeting of the Association of American Geographers, Baltimore, MD.

Gale, N. 1985. Route learning by children in real and simulated environments. Ph.D. dissertation, Department of Geography, University of California, Santa Barbara.

———; Golledge, R. G.; Pellegrino, J.; Doherty, S. 1990. The acquisition and integration of neighborhood route knowledge. *Journal of Environmental Psychology* 10:3–25.

Gärling, T., and Böök, A. 1981. *The spatiotemporal sequencing of everyday activities: How people manage to find the shortest route to travel between places in their home town.* Manuscript, Department of Psychology, University of Umeå, Sweden.

———, and Gärling, E. 1987. *Distance minimizing in downtown pedestrian shopping.* Manuscript, Department of Psychology, University of Umeå, Sweden.

———, and Golledge, R. G. 1989. Environmental perception and cognition. In *Advances in environmental behavior and design*, vol. 2, ed. E. Zube and G. Moore, pp. 203–36. New York: Plenum Press.

Golledge, R. G. 1977a. Learning about an urban environment. In *Timing space and spacing time*, ed. N. Thrift, D. Parkes, and T. Carlstein, pp. 76–98. London: Aldine.

———. 1977b. Multidimensional analysis in the study of environmental behavior and environmental design. In *Human behavior and environment: Advances in theory and research*, vol. 2, ed. I. Altman and J. Wohlwill, pp. 1–42. New York: Plenum Press.

———. 1987. Environmental cognition. In *Handbook of environmental psychology*, ed. D. Stokols and I. Altman, pp. 131–74. New York: John Wiley and Sons.

———, and Timmermans, H. 1990. Applications of behavioural research on spatial problems I: Cognition. *Progress in Human Geography* 14:57–99.

———; Briggs, R., and Demko, D. 1969. The configuration of distances in intra-urban space. *Proceedings of the Association of American Geographers* 1:60–65.

———; Rayner, J. N.; and Parnicky, J. J. 1980. *The spatial competence of selected populations: The case of borderline retarded and socio-economically disadvantaged groups.* National Science Foundation Final Report. Ohio State University Research Foundation.

———; Pellegrino, J. W.; Gale, N.; and Doherty, S. 1988. *Integrating spatial knowledge.* Paper presented at the International Geographical Union meeting, Sydney, Australia.

———; ———; ———; and ———. 1991. Acquisition and integration of route knowledge. *Na-*

tional Geographical Journal of India 37(1–2):130–46.

———; **Smith, T. R.; Pellegrino, J. W.; Doherty, S.; and Marshall, S. P.** 1985. A conceptual model and empirical analysis of children's acquisition of spatial knowledge. *Journal of Environmental Psychology* 5:125–52.

Gopal, S.; Klatzky, R. L.; and Smith, T. R. 1989. NAVIGATOR: A psychologically based model of environmental learning through navigation. *Journal of Environmental Psychology* 9:309–31.

Grier, J. 1971. Nonparametric indexes for sensitivity and bias: Computing formulas. *Psychological Bulletin* 75:424–29.

Hart, R. A., and Moore, G. 1973. The development of spatial cognition: A review. In *Image and environment: Cognitive mapping and spatial behavior*, ed. R. Downs and D. Stea, pp. 246–88. Chicago: Aldine.

Heft, H., and Wohwill, J. 1987. Environmental cognition in children. In *Handbook of environmental psychology*, ed. D. Stokols and I. Altman, pp. 175–204. New York: John Wiley & Sons.

Herman, J. 1980. Children's cognitive maps of large scale spaces: Effects of exploration, direction, and repeated experience. *Journal of Experimental Child Psychology* 29:126–43.

Hirtle, S., and Hudson, J. 1991. Acquisition of spatial knowledge for routes. *Journal of Environmental Psychology* 11(4):335-45.

Kuipers, B. 1977. *Representing knowledge of large-scale space*, AI-TR-418. Cambridge: Artificial Intelligence Laboratory, Massachusetts Institute of Technology.

———. 1978. Modeling spatial knowledge. *Cognitive Science* 2:129–53.

———. 1985. *The map-learning CRITTER*, AI-TR85-17. Austin: University of Texas.

Lee, T. 1964. Psychology and living space. *Transactions of the Bartlett Society* 2:11–36.

Leiser, D., and Zilbershatz, A. 1989. The TRAVELLER: A computational model of spatial network learning. *Environment and Behavior* 21:435–63.

Liben, L. 1982. Children's large-scale spatial cognition: Is the measure the message? In *New directions for child development: Children's conceptions of spatial relationships*, ed. R. Cohen, pp. 51-64. San Francisco: Jossey-Bass.

———; **Patterson, A.; and Newcombe, N., ed.** 1981. *Spatial representation and behavior across the life span*. New York: Academic Press.

Lindberg, E., and Gärling, T. 1981. Acquisition of locational information about reference points during blindfolded and sighted locomotion: Effects of a concurrent task and locomotion paths. *Scandinavian Journal of Psychology* 22:101–08.

———, **and** ———. 1982. Acquisition of locational information about reference points during locomotion: The role of central information processing. *Scandinavian Journal of Psychology* 23:207–18.

Lloyd, R. 1989. Cognitive maps: Encoding and decoding information. *Annals of the Association of American Geographers* 79:101–24.

Lowry, R. 1973. A method for analyzing distance concepts of urban residents. In *Image and environment: Cognitive mapping and spatial behavior*, ed. R. Downs and D. Stea, pp. 338–60. Chicago: Aldine.

Maki, R. 1981. Categorization and distance effects with spatial linear orders. *Journal of Experimental Psychology: Human Learning and Memory* 7:15–32.

Moar, I., and Bower, G. 1983. Inconsistency in spatial knowledge. *Memory and Cognition* 11:107–13.

Montello, D. R. 1991. The measurement of cognitive distance: Methods and construct validity. *Journal of Environmental Psychology* 11:101–22.

Pellegrino, J. W.; Golledge, R. G.; and Gale, N. 1990. Integrating spatial knowledge: From routes to configurations. Paper presented at the Symposium on Environmental Cognition of the International Congress of Applied Psychology, Kyoto, Japan.

Piaget, J., and Inhelder, B. 1967. *The child's conception of space*. New York: Norton.

Potts, G. 1974. Storing and retrieving information about ordered relationships. *Journal of Experimental Psychology* 103:431–39.

Reitman, J., and Reuter, H. 1980. Organization revealed by recall orders and confirmed by pauses. *Cognitive Psychology* 12:554–81.

Sadalla, E. K., and Staplin, L. J. 1980a. An information storage model for distance cognition. *Environment and Behavior* 12:183–93.

———, **and** ———. 1980b. The perception of traversed distance: Intersections. *Environment and* Behavior 12:167–82.

———; **Burroughs, W. J.; and Staplin, L. J.** 1980. Reference points in spatial cognition. *Journal of Experimental Psychology: Human Learning and Memory* 6:516–28.

Siegel, A. W., and White, S. 1975. The development of spatial representation of large scale environments. In *Advances in child development and behavior*, ed. H. Reese, pp. 9–55. New York: Academic Press.

Smith, T. R.; Pellegrino, J. W.; and Golledge, R. G. 1982. Computational process modelling of spatial cognition and behavior. *Geographical Analysis* 14:305–25.

Tversky, B. 1981. Distortions in memory for maps. *Cognitive Psychology* 13:407–33.

Wohlwill, J. F. 1976. Searching for the environment in environmental congition research: A commentary on research strategy. In *Environmental knowing*, ed. G. T. Moore and R. G. Golledge, pp. 385–92. Stroudsburg, PA: Dowden, Hutchinson and Ross.

Zimring, C., and Gross, M. 1992. The environment in environmental cognition research. In *Environment, cognition, and action: An Integrated approach*, ed. T. Gärling and G. Evans, pp. 78-95. New York: Oxford University Press.

Submitted 12/90, revised 7/91, accepted 12/91.

[17]

Gender and Individual Access to Urban Opportunities: A Study Using Space–Time Measures*

Mei-Po Kwan
The Ohio State University

Conventional accessibility measures based on the notion of locational proximity ignore the role of complex travel behavior and space–time constraints in determining individual accessibility. As these factors are especially significant in women's everyday lives, all conventional accessibility measures suffer from an inherent "gender bias." This study conceptualizes individual accessibility as space–time feasibility and provides formulations of accessibility measures based on the space–time prism construct. Using a subsample of European Americans from a travel diary data set collected in Franklin County, Ohio, space–time accessibility measures are implemented with a network-based GIS method. Results of the study indicate that women have lower levels of individual access to urban opportunities when compared to men, although there is no difference in the types of opportunities and areas they can reach given their space–time constraints. Further, individual accessibility has no relationship with the length of the commute trip, suggesting that the journey to work may not be an appropriate measure of job access. **Key Words: accessibility, gender, space–time constraints, urban opportunities.**

Introduction

Recent debate on gender/ethnic differences in access to employment opportunities raises concerns about the difficulties in using the length of the commute trip as a measure of job access (McLafferty and Preston 1996; Wyly 1996). As either long or short work trips can be taken as indicators of spatial inequality in the access to employment (the former for inner-city minorities, the latter for suburban women employed in female-dominated occupations), it is recognized that "correct" interpretations of results depend on a contextualized understanding of the life situations of the individuals being studied. Other studies also suggest that the accessibility experiences of individuals in their everyday lives is much more complex than what can be represented by conventional measures of accessibility (Pickup 1985; Kwan 1998; Talen and Anselin 1998).

In this paper, I examine the problem of using conventional measures to evaluate personal accessibility. I argue that these measures ignore the situational complexities of activity-travel behavior and the role of space–time constraints in shaping the accessibility experience of individuals. As these factors bear significantly upon women's everyday lives, all conventional accessibility measures suffer from an inherent "gender bias" that has hitherto gone unnoticed. In view of these limitations of conventional accessibility measures, this study argues that defining accessibility as space–time feasibility instead of locational proximity would enable a better understanding of individuals' accessibility experiences. It provides formulations of alternative accessibility measures for studying gender differences in access to urban opportunities based on the construct of the space–time prism.

Despite the advantages of space–time accessibility measures, few studies have implemented them due to operational difficulties such as the geoprocessing capabilities required. To examine gender differences in individual accessibility, I use a network-based GIS method to operationalize space–time measures. Data for the study are derived from a travel diary data set collected in Franklin County, Ohio, a digital transportation network, and a land parcel geographic database of the study area. Results of the study, based on a subsample of European Americans, indicate that there are significant gender differences in individual access to urban opportunities for full-time employed women and men, although no difference was observed in the composition and types of opportunities available to these individuals. Further, individual accessibility has no

* This research was supported by grants from the National Science Foundation (grant #SBR-9529780) and the Committee on Urban Affairs, Ohio State University. Another NSF grant (#SBR-9512451) enabled the acquisition of the SGI Power Challenge server used for the computation in this study. I am grateful to Stuart Aitken and the anonymous reviewers for their helpful comments.

significant relationship with the length of the commute trip, suggesting that the journey to work may not be an appropriate indicator of job access.

Conventional Accessibility Measures

Conventional accessibility measures are based on three fundamental elements (Hanson and Schwab 1987). First, a reference location serves as the point from which access to one or more other locations is evaluated. The reference location most often used is the home location of an individual, or the zone where an individual's home is located when zone-based data are used. Second, a set of destinations in the urban environment is specified as the relevant opportunities for the measure to be enumerated. This set may include employment opportunities (for evaluating job access) or particular types of shops, services or facilities. Further, each opportunity may be weighted to reflect its importance or attractiveness. Third, the effect of the physical separation between the reference location and the set of urban opportunities on access is modeled by an impedance function, which represents the effect of distance decay on the attractiveness of the relevant opportunities.

Based on these three elements, various types of accessibility measure can be specified. Morris et al. (1979) provide a helpful typology of accessibility measures comprising two broad groups: relative and integral measures. Relative accessibility measures describe the degree of connection between two locations (Ingram 1971). They are expressed in terms of the presence or absence of a transport link, or the physical distance or travel time between two locations. Examples of this type of measure are commuting distances or times used in recent studies on the spatial mismatch between population subgroups and jobs (e.g., Ihlanfeldt 1993). Integral measures, on the other hand, represent the degree of interconnection between a particular reference location and all, or a set of, other locations in the study area. They have the general form:

$$A_i = \Sigma W_j f(d_{ij})$$

where A_i is the accessibility at location i, W_j is the weight representing the attractiveness of location j, d_{ij} is a measure of physical separation between i and j (in terms of travel time or distance), and $f(d_{ij})$ is the impedance function. When impedance is expressed in the form of a distance decay function similar to those found in gravity models, the access measure is a gravity-based measure (Handy and Niemeier 1997; Helling 1998; Levinson 1998). The most commonly used impedance functions are the inverse power function and the negative exponential function. In the case where an indicator function is used as the impedance function to exclude opportunities beyond a given distance limit, the measure is a cumulative-opportunity measure (Wachs and Kumagai 1973; Talen 1997). This measure indicates how many opportunities are accessible within a given travel time or distance from the reference location.

A further distinction could be made depending on whether an access index is enumerated and used as an indicator of physical or place accessibility (how easily a place can be reached or accessed by other places), or personal or individual accessibility (how easily a person can reach activity locations). As will be shown below, this distinction is important if the objective is to understand the accessibility experienced by individuals in their everyday lives, and to avoid the problem of ascribing the accessibility of a location or area (e.g., census tract) to a person at that location or in that area (Pirie 1979).

Limitations of Conventional Accessibility Measures

One major difficulty in using conventional measures to evaluate individual accessibility stems from the concept of accessibility they operationalize, namely, accessibility as locational proximity of opportunities with respect to a single reference location. There are several problems with this notion. First, as accessibility is measured with respect to a single reference location such as home or the workplace, all travel or activities (whether actual or potential) which contribute to personal accessibility are assumed to be based on this single origin. This amounts to assuming that all potential trips, which contribute to the accessibility of a person residing at a particular location, start from that single reference point. Such an assumption, however, deviates quite significantly from many characteristics of activity-travel behavior observed in recent research.

As past studies have shown, a substantial proportion of intra-urban travel consists of multi-purpose, multi-stop journeys, and potential stops at various locations may become more accessible by virtue of their proximity to sites other than home or the workplace (Hanson 1980; O'Kelly and Miller 1983; Kitamura et al. 1990; Golledge and Stimson 1997). A study by Richardson and Young (1982) demonstrates that, without considering the effect of linked trip or trip chaining behavior, conventional measures may considerably underestimate the accessibility to activities of non-central urban locations. Similarly, Arentze et al. (1994) conclude that methods taking multipurpose travel into account may lead to different evaluations of the differential accessibility in real-world situations. The complex linkages and interdependencies between activities in a person's daily life therefore render the evaluation of accessibility in terms of a single reference location problematic. An example of this is the multipurpose journey to work where the work trip is linked to travel to essential facilities. As observed in several studies, the availability of child-care facilities is often an important factor determining the accessibility of certain job locations for women with young children (Michelson 1985, 1988; Tivers 1985, 1988; Hanson and Pratt 1990; Bianco and Lawson 1996; England 1996a, 1996b; Myers-Jones and Brooker-Gross 1996; Gilbert 1998). In other words, access to jobs depends not only on where the jobs are, but also on the location of child care facilities which renders some job locations more feasible than others. Using the length of the commute trip as a relative measure of job access can therefore obscure the accessibility experience of individuals, especially women, in their everyday lives, since job location may be constrained by the location of essential facilities rather than the reverse.

A second problem of conventional accessibility measures is their ignorance of the role of individual time-budget and space–time constraints in determining personal accessibility. As every individual faces a particular set of space–time constraints imposed by the space–time fixity of obligatory activities in their everyday lives, personal accessibility is contextually constituted in the sequential unfolding of these activities (Pirie 1979). In other words, individual accessibility is determined not by how many opportunities are located close to the reference location, but by how many opportunities are within reach given the particularities of an individual's life situation and adaptive capacity (Dyck 1989, 1990; England 1993). Conventional access measures based on locational proximity of opportunities to a single reference point simply cannot reflect interpersonal differences associated with these contingencies of everyday life: they tend to reflect place accessibility rather than individual accessibility.

Such inability of conventional accessibility measures to reflect individual differences besides those captured by the impedance function leads to an inherent bias when these measures are used to examine gender differences in accessibility. As many studies observe, women's access to jobs and urban opportunities is determined more by their space–time constraints in everyday life than by factors such as travel mobility or relative location to opportunities (Palm 1981; Miller 1982, 1983; Hanson and Pratt 1990). Further, those constraints associated with socially ascribed gender roles tend to be spatially and temporally more binding for women than for men (Pickup 1984, 1985; Tivers 1985; Pratt and Hanson 1991; Johnston-Anumonwo 1992; Kwan 1999a, 1999b). These factors led Pickup (1985, 106) to conclude that

> Conventional "spatial" accessibility measures of access to shops or jobs were meaningless for women whose activity choices were continually facing additional time constraints from their gender role. For the analysis of men's job choices, such measures might have more adequacy.

Similarly, Kwan (1998) observes that the spatial pattern of individual accessibility for men has a stronger relationship with place accessibility than does that for women. To overcome the inherent gender insensitivity of conventional accessibility indices, space–time measures based on the time-geographic framework provide attractive alternatives.

Space–Time Accessibility Measures

All space–time measures of individual accessibility are based on Hägerstrand's (1970) time-geographic framework and the space–time prism construct. Lenntorp (1976) provides the earliest operational formulations of space–time accessibility measures based on the notion of an individual's reach, which is the physically accessible part of the environment given the individual's space-time constraints. In three-dimensional terms, this accessible portion is the space–time prism or potential path space (PPS). The projection of the prism onto planar geographic space delimits the reachable area by the individual and is called the potential path area (PPA). The volume of the space–time prism and the area delimited by the PPA were specified as accessibility measures by Lenntorp (1976). Another early formulation was provided by Burns (1979), who conceptualizes individual accessibility as the freedom to participate in different activities. Based on the concept of space–time autonomy, he formulates two accessibility measures using the space–time prism construct. One is based on the set of locations reachable by an individual, while the other is based on the set of reachable routes of the transportation network.

There are several extensions of the original formulations of space–time accessibility measures after Lenntorp and Burns. Landau et al.'s (1982) study identifies the feasible locations to be included in an individual's destination choice set based on the person's space–time constraints and the minimum amount of time required for an activity. Kwan and Hong (1998) incorporate the effect of spatial knowledge on the size and spatial configuration of the choice set based on an extension of Golledge et al.'s (1994) work on feasible opportunity sets. By extending the problem of destination choice from a particular activity to all non-fixed activities in a day, the size of the feasible opportunity set delimited in these studies can be formulated as space–time measures of individual accessibility. For instance, Villoria (1989) implements such a measure using geometric methods. To overcome the limitations of geometric implementation, Miller (1991) and Kwan (1998) develop network-based GIS methods for operationalizing space–time accessibility measures. Miller (1999) further extends space–time measures through integrating three notions of accessibility into a single analytical framework. He also develops computational procedures for enumerating the proposed measure within transportation networks.

Operationalizing Space–Time Measures: Concept and Implementation

What follows is an explanation of the concept behind the operational method used to enumerate space–time measures in this study. Every individual has a daily activity program consisting of a number of out-of-home activities. Among these activities, some are spatially and/or temporally fixed for the individual (e.g., workplace), while others can be undertaken at various locations or times of the day (e.g., grocery stores). The former type of activities are referred to as "fixed activities" which serve as "pegs" in the daily space–time trajectory of the individual (Cullen et al. 1972). Since the area in the urban environment within reach by an individual is determined by the space–time requirements associated with these fixed activities, the feasible region for an individual on a particular day can be derived based on the space–time coordinates of consecutive pairs of fixed activities. The procedures for deriving this region are described as follows.

For any given pair of consecutive fixed activities, a given amount of time is available for travel and participation in "flexible" activities between them. Based on the locations of the pair of fixed activities in question, a spatial search is performed on the transportation network to find all urban opportunities within reach given this specific time constraint. The area so identified is the PPA, which delimits an area containing all feasible routes and urban opportunities given the space–time constraints defined by the particular pair of fixed activities (Burns 1979; Miller 1991). If the individual has n out-of-home fixed activities to perform in a day, there will be $n+1$ pairs of consecutive fixed activities when the starting and ending home stops are regarded as fixed space–time pivots. This in turn means that $n+1$ distinct PPAs can be delimited. To obtain the set containing all feasible opportunities for the day, these individual PPAs are aggregated to form a daily potential path area (DPPA).

These procedures for deriving the DPPA are illustrated in Figure 1 using the actual travel diary data of a resident of Franklin County, Ohio. This person performed three out-of-home activities on the day: work, attending an event at a church, and having a meal at a restaurant before returning home. With the former two activities being fixed, three PPAs are identified based on three consecutive pairs of fixed locations along the person's space–time path (with the starting and ending home stops included as fixed locations). These three individual PPAs (in the middle of the diagram) are then aggregated to form the daily potential path area (DPPA). Two-dimensional representations of

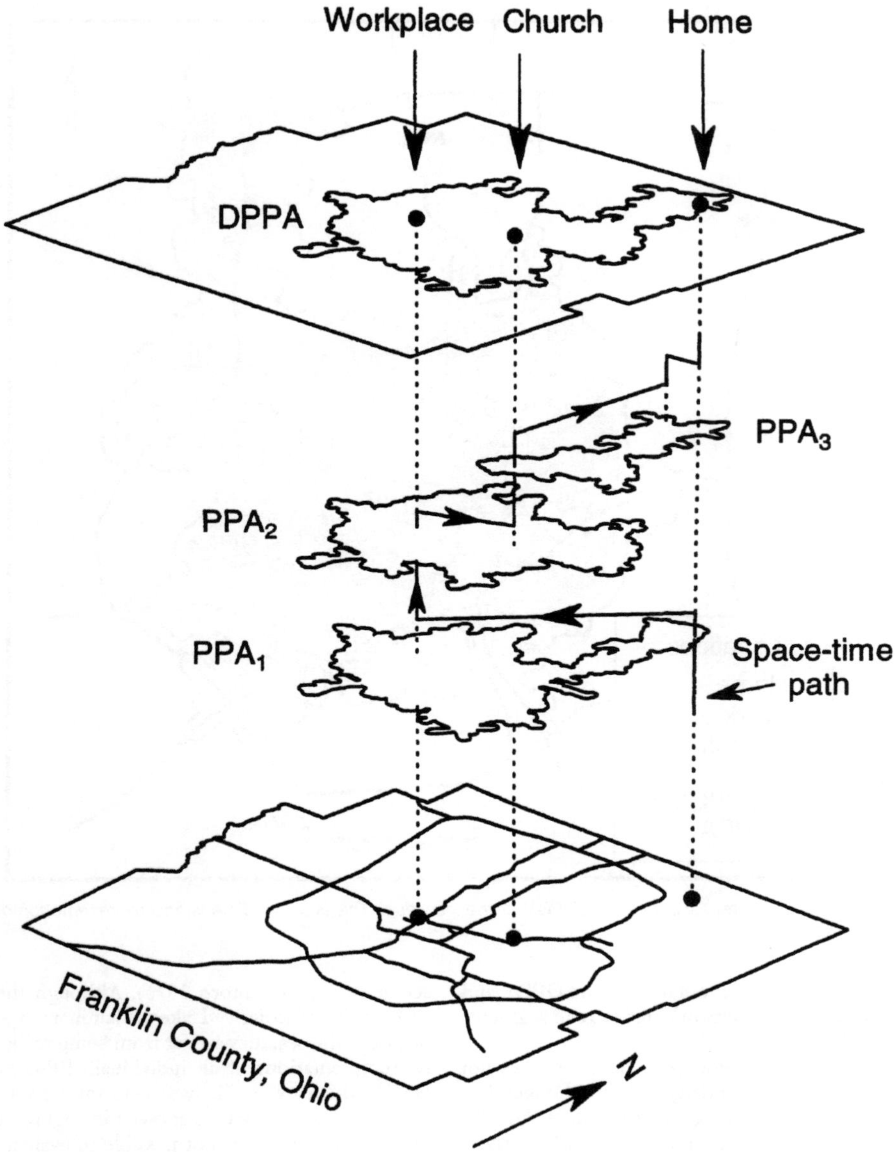

Figure 1: *Derivation of the daily potential path area (DPPA) using actual data of an individual in the subsample. While all locations are real activity sites, the timing and duration of these activities were modified to improve visual clarity.*

the daily potential path area (DPPA) derived and the network arcs it includes are shown in Figure 2. Based on this delimitation of the DPPA, three space–time measures of individual accessibility can be specified: (a) the number of opportunities included in the DPPA, (b) an area-weighted sum

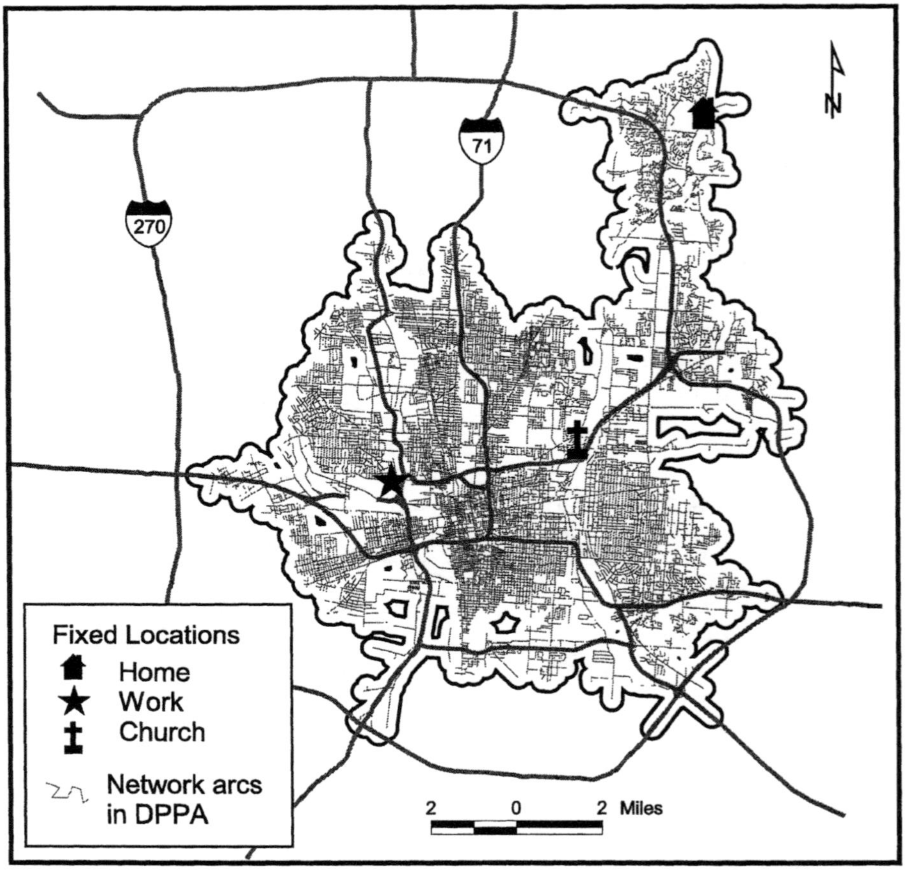

Figure 2: *Daily potential path areas (DPPA) of the individual. This is a two-dimensional representation of the DPPA in Figure 1.*

of all opportunities contained in the DPPA, and (c) length of the network arcs or road segments included in the DPPA.

To implement these space–time measures, not only detailed travel diary and time-budget data are needed. Geoprocessing capabilities are also required to handle a person's daily activity schedule (including the locations of all fixed activities), the complexities of a realistic travel environment (e.g., transport network geometry and variations in travel speed), and the spatial distribution of urban opportunities. Without these, past attempts to operationalize the prism construct had to resort to geometric methods based on the straight-line distance between fixed activities (e.g., Lenntorp 1976). Although the PPAs or DPPA so derived take the familiar shape of spatial ellipses, they are far from being realistic representations of an individual's PPAs or DPPA, which are not likely to be in any convenient geometric shapes (as showed in Figures 1 and 2). Further, it was not possible to evaluate the number and composition of opportunities included in an individual's DPPA without incorporating information about urban opportunities into a geographic database. With the spatial data handling capabilities of recent GIS and the development of network-based GIS methods (Miller 1991; Kwan and Hong 1998), meeting some of these requirements for computing space–time accessibility has become possible. This study uses the procedures formulated in Kwan (1998) for this purpose.

Study Area and Data Collection

The study area for this research is Franklin County, Ohio (Fig. 3). The county is located at the center of the seven-county Columbus Metropolitan Statistical Area (MSA). Its main urban area consists of the city of Columbus and several smaller cities. It is 542 square miles in area and had an estimated population of one million in 1995 (City of Columbus 1993). A two-day travel diary data set I collected in the study area in 1995 provides the individual activity-travel data for this study. Among the detailed activity-travel data collected, two specific items used in this paper are street addresses of all activity locations and the subjective space–time fixity ratings of all out-of-home activities (i.e. the difficulty in changing the location or time of an activity).

Two digital data sets provide additional information needed for this study. One of these is a detailed digital street network of Franklin County called Dynamap/2000. The network database contains 47,194 arcs and 36,343 nodes of Columbus streets and comes with comprehensive address ranges for geocoding locations. Activity locations collected in the travel diary survey are geocoded on this street network using ArcView GIS. The GIS procedures for enumerating the space–time measures are also implemented on this network after creating a realistic representation of the travel environment of the study area (e.g., variations in travel speed with road types). The home-work distances for all individuals in this study are measured in terms of network distance and travel time on the network.

Another source of data is a digital database of all land parcels in the study area maintained by the Franklin County Auditor's Office. Among the 34,442 non-residential parcels in the database, 10,727 parcels belonging to seven land-use categories are selected as the urban opportunities for this study.[1] These land-use types include shopping and retail facilities, restaurants, personal-business establishments such as banks, indoor entertainment (e.g., theaters), outdoor recreational activities, educational institutions (except higher education), and office buildings (vacant, agricultural, and industrial parcels are excluded). Since the average area of these parcels is 0.00379 square miles, they can be treated as point entities given the spatial scale of the study area. Subsequent analysis uses a point coverage of the centroids of these parcels.

The spatial pattern of these urban opportunities is shown in Figure 4 in the form of an opportunity density surface (outdoor recreation

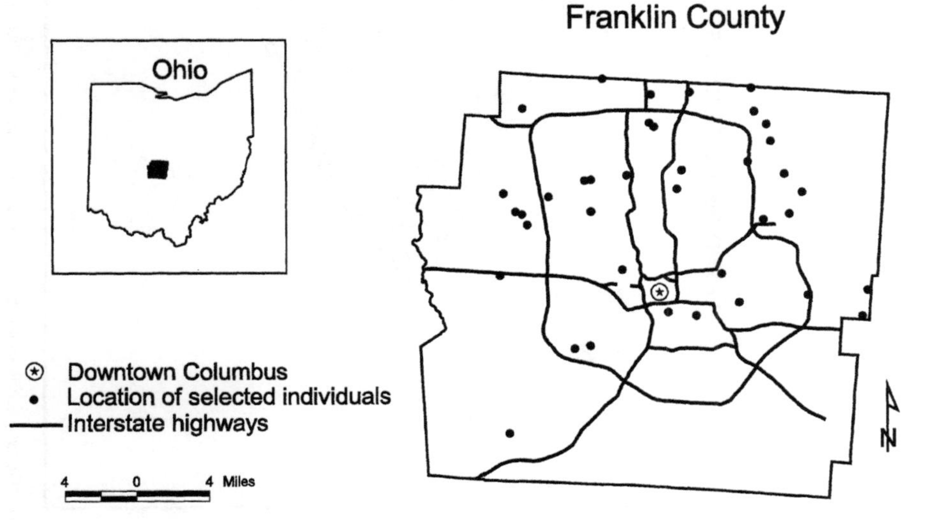

Figure 3: *The study area and home locations of the individuals in the subsample.*

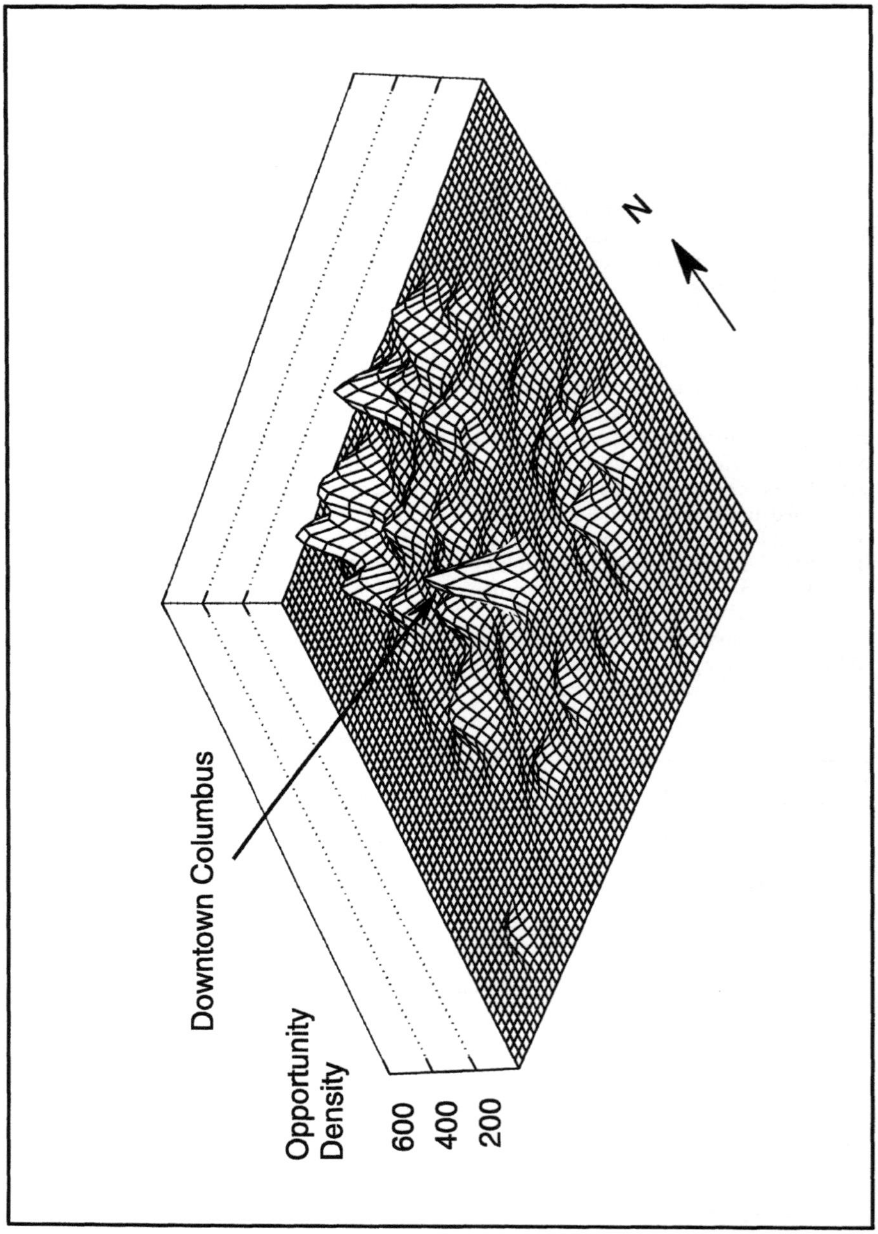

Figure 4: *Density surface of urban opportunities in Franklin County, Ohio.*

is excluded due to large areal extent). The surface was generated by covering the study area with 18,542 grid cells of 0.036 square miles, and using kernel estimation based on each parcel's weighted area and a search radius of one mile (Gatrell 1994). The surface indicates that, besides a peak in downtown Columbus (center of surface), urban opportunities are mainly concentrated along major transport arterials and their intersections. Several peaks are found in the northern part of the County along segments of the I-270 and Morse Road.

Analysis and Results

Despite the original attempt to obtain a representative random sample of the population of the study area, many gender/ethnic subgroups are underrepresented in the final sample. For instance, less than 2% of the respondents are African Americans, while that group accounts for 16% of the County's population in 1990. As many population subgroups in the sample are too small for comparative analysis (e.g., African Americans and part-time employed European American men), gender/ethnic groups with 15 or more employed individuals were first identified.[2] This led to a subsample of 56 full-time employed European Americans (27 women and 29 men) for the study. Understandably, this subsample is small in size and does not represent the population of the study area. When compared to the population of Franklin County, Ohio, individuals in the subsample have a higher level of access to automobiles (all regularly use their own cars) and a higher household income (median household income is about $75,000 a year in contrast to the County's average of $30,375 in 1990). 76% of the individuals in the subsample hold managerial and professional occupations, while only 30.5% of the population in the County hold such positions. Home locations of these individuals are spatially scattered in the study area and are largely outside the older urban core of Columbus (Fig. 3).

Results reported in this paper are thus specific to these highly mobile, upper-middle-class European Americans. Further, the focus of the paper on the simple dichotomy of women and men is largely the result of the limitations imposed by the small size and characteristics of the original sample. If a larger and more representative sample had been available, more refined groupings which take into account the effects of and interactions between gender, race, life-cycle stage, income, occupational status, and age could have been used in the analysis (e.g., Janelle and Goodchild 1983).[3] Although limited to examining individual accessibility in terms of the over-generalized categories of women and men, I do not presume or intend to argue that women's (and men's) diverse, complex and gendered experiences of everyday life can be reflected by using such a simplistic dichotomy (the problem of doing so is noted by Fuss 1989; McDowell 1993; Pratt and Hanson 1993; Staeheli and Lawson 1995; Kobayashi 1997). The terms "gender" and "gender differences" in this paper are used, and should be understood, in terms of the context of this study. Further, the effects of local geography on the results are not examined due to data limitations.

To examine gender differences in access to urban opportunities, three space–time accessibility measures were computed for each individual in the subsample. These measures are the number of opportunities in the daily potential path area (DPPA), the sum of the weighted area of these opportunities, and length of the network arcs included in the DPPA. The weighted area of a parcel equals the parcel's area (in acres) multiplied by a building-height factor based on the number of stories of the structure on the parcel. Further, the length of network arcs included in the DPPA is used as a surrogate for the area or size of the DPPA. Although GIS procedures can be used to find this area, computational intensity of the topological operations involved renders them infeasible. The value of all measures reported here is the average of the two survey days for each individual. The geoprocessing required for enumerating these space–time accessibility measures was performed in ARC/INFO GIS.

Gender Differences in Space–Time Accessibility

In terms of all three space–time accessibility measures, full-time employed women in the subsample have lower levels of access to urban opportunities when compared to full-time employed men (Table 1). Their daily potential path areas (DPPAs) are 64% smaller in area (in terms of the length of network arcs) and contain 44% less urban opportunities (in terms of their number and weighted area) than those of men.

Further, all gender differences in space–time accessibility are significant at $p < 0.10$, while many are significant at the $p < 0.05$ level as the ANOVA results in Table 1 indicate.

As these women and men are full-time workers and on average work 8.2 hours a day, they have about 6.8 hours a day for pursuing non-employment activities (assuming that each has 15 hours left after allowing for sleep and other essential maintenance activities). Since these individuals face similar time-budget constraints, gender differences in space–time access to urban opportunities observed here are largely the result of the constraints associated with the space–time fixity of the non-employment activities they need to perform in their everyday lives. In other words, women in the subsample experience lower level of accessibility because they face more space–time constraints in their non-work hours than men. As many of these constraints are gender-role-related, these results corroborate past observations that women's employment is not accompanied by major role shifts among members of the household (Hanson and Hanson 1981; Michelson 1985, 1988; Gregson and Lowe 1993; Aitken 1998). Instead, these studies found that domestic and childcare responsibilities are still primarily borne by women in dual-career households independent of their employment status or the birth of a child. This suggests that changing the share in gender-role-related responsibilities within a household between women and men may be an important step for increasing individual access to urban opportunities for women.

Despite the lower levels of space–time accessibility of women when compared to that of men, women's access to different types of urban opportunities in relative terms is similar to that of men. In terms of either the number or weighted area of opportunities, both women and men have virtually the same composition of various types of opportunities in their DPPAs (Table 2). Individual access to office buildings and shopping facilities is highest for both groups. Further, there is no major difference in the composition of opportunities in the DPPAs of these individuals when compared to the County's pattern (except a slightly higher percentage of office buildings and a slightly lower percentage of outdoor recreational facilities for both groups of individuals).

In addition to the lack of gender difference in the composition of opportunities in the DPPAs, there is also no significant difference in the average size of the parcels in the DPPAs of these individuals (Table 3). Since downtown Columbus is characterized by a large number of small land parcels, while areas outside the older urban core have more large land parcels in relative terms, the average size of land parcels included in a DPPA indicates the spatial tendency of the DPPA to a certain extent. If the average size of land parcels in a DPPA is smaller than that of the County's average, the DPPA includes more parcels in the older urban core than those in suburban areas. So this result suggests that there is no major difference in the spatial tendency of the DPPAs of the full-time employed women and men in the subsample. Further, since average

Table 1 *Space–Time Accessibility of Individuals in the Subsample*

	Full-time employed women		Full-time employed men		Franklin County	
	Number of opportunities	Weighted area of opportunities	Number of opportunities	Weighted area of opportunities	Number of opportunities	Weighted area of opportunities
Land-use types						
Shopping	546.5*	662.1*	798.6*	962.5*	3052	4020.9
Food	213.5	98.0*	313.9	147.5*	1195	689.8
Personal business	294.8*	217.7*	423.9*	316.3*	1611	1367.9
Entertainment	6.9	18.1	9.8	25.3	45	130.8
Recreation (outdoor)	118.6	1206.9	175.8	1750.6	1015	10338.6
Education	193.8	614.3*	274.7	891.1*	996	3946.0
Office buildings	542.5*	1070.2	777.6*	1449.1	2813	5502.1
All land-use types	1916.5*	3887.3	2774.4*	5542.4	10727	25996.1
Length of network arcs in DPPA (in miles)	1008.0*		1652.7*			

** Difference between women and men significant at $p < 0.05$ (all other differences are significant at $p < 0.10$)*
Sources: Travel diary survey conducted in Columbus, Ohio, 1995; Franklin County Auditor's Office.
Note: Values for both groups of individuals are group means for a particular land use type. Weighted areas are in acres.

Table 2 *Percentage of Various Types of Urban Opportunities in DPPA*

	Full-time employed women		Full-time employed men		Franklin County	
	Percent of number of opportunities	Percent of weighted area of opportunities	Percent of number of opportunities	Percent of weighted area of opportunities	Percent of number of opportunities	Percent of weighted area of opportunities
Land-use types						
Shopping	28.5	17.0	28.8	17.4	28.5	15.5
Food	11.1	2.5	11.3	2.7	11.1	2.7
Personal business	15.4	5.6	15.3	5.7	15.0	5.3
Entertainment	0.4	0.5	0.4	0.5	0.4	0.5
Recreation (outdoor)	6.2	31.0	6.3	31.6	9.5	39.8
Education	10.1	15.8	9.9	16.1	9.3	15.2
Office buildings	28.3	27.5	28.0	26.1	26.2	21.2
All land-use types	100%	100%	100%	100%	100%	100%

Sources: Travel diary survey conducted in Columbus, Ohio, 1995; Franklin County Auditor's Office.

Table 3 *Average Size of Land Parcels in DPPA*

	Full-time employed women (area in acres)	Full-time employed men (area in acres)	Franklin County (area in acres)
Land-use types			
Shopping	1.21	1.21	1.32
Food	0.46	0.47	0.58
Personal business	0.74	0.75	0.85
Entertainment	2.62	2.58	2.91
Recreation (outdoor)	10.18	9.96	10.10
Education	3.17	3.24	3.96
Office buildings	1.97	1.86	1.96
All land-use types	2.03	2.00	2.42

Sources: Travel diary survey conducted in Columbus, Ohio, 1995; Franklin County Auditor's Office.
Note: Values are group means for a particular land-use type.

sizes of the land-use parcels in the DPPAs of the two groups are slightly smaller than that of Franklin County, their DPPAs cover more areas in the old urban core than suburban areas.

The lack of difference in the composition and average parcel size of the opportunities included in the DPPAs of the women and men in the subsample indicates that there is no significant difference in the spatial tendency of their DPPAs in terms of their coverage of various parts of the study area. Underlying this similarity in the spatial tendency of the DPPAs, however, is a subtle gender difference in the density of opportunities included in the DPPAs (in terms of the number of opportunities and their weighted area per mile of network arcs in the DPPAs). For instance, opportunity density for the full-time employed women is 3.26 per mile of network arcs, while that for the full-time employed men is 2.59 (difference significant at $p < 0.05$). This difference may be due to the adaptive strategies adopted by the women in the subsample who, when facing more restrictive space–time constraints than men, attempt to negotiate their space–time paths through areas with a higher density of urban opportunities as a means of compensating the disadvantages of their smaller DPPAs. It supports the assertion of past studies that women are creative urban actors with transformative capacities who are actively engaged in redefining the meaning and use of urban space when trying to juggle their multiple roles (Dyck 1989, 1996; England 1993, 1996b). Although gender roles may impose additional space–time constraints on the daily lives of women, and the spatial structure of the urban environment may further disadvantage them relative to men (Saegert 1981; McDowell 1983; Hayden 1984; Weisman 1992), women are not passive victims of their circumstances but are active agents involved in reconstituting the conditions of their lives as they engage in creating their particular geographies in the routine practices of everyday life.

Space–Time Accessibility and the Journey to Work

Past research on the journey to work reveals that gender/ethnic differences in commuting are not only outcomes of the complex interaction between gender roles, occupational segregation, and travel mobility, but are also highly dependent on sociospatial context, including the residential and employment location of the population subgroups in question (e.g., see studies reviewed in Blumen 1994; Gilbert 1997; McLafferty and Preston 1997). While most of these studies observe that European-American women in general work closer to home than do European-American men, there are also cases where certain subgroups of European-American women travel as far as their male counterparts (England 1993; Hanson and Pratt 1994, 1995). Contrary to what these studies have observed, the full-time employed women in the subsample have longer commutes than men (both groups being European Americans) in terms of network distance between home and the workplace (9.8 miles for women and 7.2 miles for men) and travel time (19.9 minutes for women and 16.4 minutes for men) (differences significant at $p < 0.05$, Table 4). Given that these women and men have similar racial and socio-economic characteristics, come from the same set of affluent dual-earner households in suburban Columbus, and work at spatially scattered locations in the study area, the longer journey to work of the women in the subsample can be ascribed to their high occupational status when compared to men in the subsample (82% of women hold managerial and professional occupations as compared to 71% of men).

But what makes this observation particularly significant is the relationship between individual accessibility and journey to work. Since women in the subsample have lower levels of space–time access to urban opportunities (Table 1) and face more space–time constraints than men (Kwan 1999b), their longer journeys to work are contrary to the interpretation of past studies that long commutes for upper-middle class women reflect their less constrained life situations. If these European-American women can actually experience lower levels of access to urban opportunities even though they have longer journeys to work, the length of the commute trip may not be an appropriate indicator of how constrained their everyday lives are. Although this particular relationship between individual accessibility and the length of the work trip is specific to the subsample and may be different for other gender/ethnic subgroups and in other spatial contexts (as shown in Johnston-Anumonwo 1997; Preston and McLafferty 1993), it calls into question an often adopted view about this relationship: that the long commutes of upper-middle-class European-American women indicate that they are relatively free of constraints or oppressive life situations. While this, by extension, does not provide direct support for questioning the assertion that the short journeys to work of minority women indicate the availability of jobs near their homes, it does highlight the need to be more cautious about using the length of the work trip as a measure of job access (Kwan 1999c).

Table 4 *Correlations of the Home-Work Distance with the Number of Opportunities and Weighted Area in DPPA*

	Full-time employed women		Full-time employed men	
	Number of opportunities and home-work distance	Weighted area and home-work distance	Number of opportunities and home-work distance	Weighted area and home-work distance
Land-use types				
Shopping	-0.07	-0.08	-0.01	-0.05
Food	-0.05	-0.04	-0.01	-0.05
Personal business	-0.03	-0.05	-0.00	0.01
Entertainment	-0.06	-0.01	-0.06	-0.10
Recreation (outdoor)	0.22	-0.03	-0.10	-0.08
Education	-0.02	0.02	0.00	-0.06
Office buildings	-0.06	-0.08	0.01	-0.02
All land-use types	-0.04	-0.05	-0.01	-0.05
Home-work distance (in miles)	9.8*		7.2*	
Travel time to work (in minutes)	19.9*		16.4*	

** Difference between women and men significant at p < 0.05*
Source: Travel diary survey conducted in Columbus, Ohio, 1995.

In recent debate on gender/ethnic differences in access to employment opportunities, the length of the commute trip is often used as a measure of job access. In their study of the differences in job accessibility between African-American and European-American youths, Ihlanfeldt and Sjoquist (1990, 268) express this view by stating that, "If jobs are nearby, commuting time will be low. Conversely, if jobs cannot be found nearby, travel time will be high." However, data of the Columbus subsample indicate that the length of the commute trip (in terms of travel distance) has no relationship with individual access to urban opportunities (no correlation coefficient is larger than 0.25 or is significant at $p < 0.05$ in Table 4). This suggests that the use of work trip length as a measure of access to jobs, which is one type of urban opportunity, may be inadequate for understanding the access experience of individuals belonging to different gender/ethnic subgroups.[4] Commuting time can be low or high for a variety of reasons other than job access, and the often-assumed determinate relationship between them needs to be critically reexamined for each sociospatial context (Giuliano 1988; England 1993, 1994; Brun and Fagnani 1994; McLafferty and Preston 1996; Johnston-Anumonwo 1997; Immergluck 1998).

This suggests that situating the journey to work in the wider context of the totality of everyday life instead of treating it as an isolated problem of job access may be a more fruitful direction for future research. As England (1993, 237) argues, "journeys to work should be viewed as part of a much larger time-space budgeting problem." In other words, treating the journey to work as an outcome of a highly complex decision-making process involving elements of constraints, choice, and the transformative capacity of individuals is more desirable than using it as an indicator of job access (Burnett and Hanson 1982; McLafferty and Preston 1996). In view of this, more appropriate means for representing or measuring individual access to urban opportunities and jobs are needed for revealing how, and to what extent, individuals of different gender/ethnic subgroups are disadvantaged in their everyday lives (Johnston-Anumonwo 1997). As the space–time measures of individual accessibility used in this study are based on the notion of space–time feasibility instead of locational proximity, they have several advantages over conventional accessibility measures for incorporating the effect of the interaction between a person's constraints and the availability of urban opportunities in space–time.[5]

Discussion and Future Research Directions

Using space–time accessibility measures, this study suggests that access to urban opportunities for women in the subsample is significantly less than that of men. These gender differences in individual access to urban opportunities are largely due to differences in space–time constraints that results in different sizes of DPPAs. The study also indicates that there is no relationship between individual accessibility and the journey to work, which is often used as a measure of job access. This suggests that the length of the commute trip may not be an appropriate indicator of how much constraint an individual faces. Much future research is needed to decipher the complex relationships between access to urban opportunities (including jobs) and the length of the commute trip.

Although space–time measures have advantages over conventional accessibility measures, several qualifications are warranted at this point for the proper interpretation of these results. First, since the land parcel data set used in this study does not provide information about the opening hours of establishments, their effect on individual accessibility was not taken into account by either the operational measures or results presented in this paper. For instance, since women in the subsample undertook more out-of-home activities in the evening (when many facilities were closed) than men, the space–time measures as implemented in this study might have overestimated the number of opportunities they can access. Given this limitation and the fact that women in the subsample are still observed to have lower levels of accessibility when compared to men, the effect of this problem is an underestimation of the gender gap in accessibility rather than a change in the direction of gender differences. Since different operational methods for dealing with the interaction between individual space–time constraints and opening hours of facilities have been developed in past studies (e.g., Landau et al. 1982; Kwan and Hong 1998), future research needs to collect and incorporate data about opening hours into space–time formulations. This is especially important if the method is to be applied to other groups of individuals who engage in a variety of work-hour arrangements (e.g., rotating or night shifts) that affects both the time-budget and space–time constraints a person faces.

Second, findings of this study are based on a specific population subgroup which consists of affluent, employed European Americans from suburban Columbus, Ohio, who rely on their own cars for transportation. Given the particularity of this sociospatial context and sample, and the contextual nature of gender differences in commuting and individual accessibility, it should be emphasized that the paper's main contribution is methodological rather than empirical. I intend to highlight the point that relationships between space–time constraints, individual accessibility and journey to work are not as determinate as often assumed in the past and are therefore difficult to generalize. The paper also reveals some of the methodological problems involved in using work trip length as a measure of job access. It attempts to open new avenues for future exploration of these complex relationships. In light of Rose's (1997, 307) assertion that feminist research should not aim at producing knowledge that claims to have universal applicability, this paper does not intend to make broad generalizations or knowledge claims with applicability to other sociospatial contexts or gender/ethnic subgroups.

Further, since my purpose in this paper is to propose more gender-sensitive accessibility measures in comparison with conventional accessibility indices, the space–time measures are formulated largely in analytical-quantitative terms. Using this particular discursive form, the study suggests that it is possible to uncover the gender bias of a particular technique in its own terms (that is, in the language of quantitative methods) even without invoking ontological or epistemological arguments. Based on the premise that quantitative methods can still be useful for answering certain questions in feminist research (Lawson 1995; McLafferty 1995), results of this study suggest that future feminist research using quantitative methods needs to subject these methods to critical scrutiny before accepting them as appropriate for addressing a particular set of questions.

A cautionary note on the theoretical underpinnings and limitations of the space–time measures used in this study is also warranted. Hägerstrand's time-geographic framework, on which the space–time measures of this study are based, has been subject to several rounds of critiques (e.g., Giddens 1984; Gregory 1985; Harvey 1989). Dyck (1989) and England (1993) have also emphasized that women are not just victims of gender-role constraints, but are also active agents with transformative capacity. What is particularly relevant to this study, however, is the feminist critique of Rose (1993), who criticizes time-geography for its masculinist conception of space and the body, its repression of feminine subjectivity, and its erasure of the specificity and positionality of the master subject.[6] As Rose (1993) argues, time-geography (and thus the space–time measures used in this study) cannot take into account the effect of women's fear of violence or attack in public spaces on their mobility (as elaborated in Valentine [1989] and Pain [1991]). To reveal the richness and complexities of the lived experiences of women (and men), the space–time measures in this study therefore need to be complemented by a contextualized understanding furnished through ethnographic methods such as in-depth interviews or participant observation. Future research needs to consider how multiple methodologies can be used in this kind of study. ■

Notes

[1]Since this parcel data set does not provide information about the opening hours of establishments, their effect on individual accessibility is not taken into account by either the operational measures or results presented in this paper.

[2]Identification of population subgroups at this stage was based on gender, ethnicity and employment status (full-time versus part-time). Due to the group-size requirement of 15 or more and the small sample size, only the gender dimension remains as the major differentiating criterion between subgroups.

[3]The higher degrees of freedom provided by such a sample would also allow the use of log-linear and structural equation models as originally intended.

[4]Although space–time measures are used in this study to evaluate access to urban opportunities, some of the results in Table 4 allow a connection between urban opportunities and job opportunities. As most individuals in the subsample hold managerial and professional occupations, the weighted area of office buildings in the DPPA can be used as a surrogate for

job opportunities for this particular group of individuals. For other population subgroups, different land parcel categories may be specified to better represent the relevant job opportunities.

[5]Another fruitful direction is the application of spatial-statistical methods to small-area data (e.g., Talen 1997; Immergluck 1998).

[6]Gregory (1994), however, is rather positive about time-geography in his reassessment of Hägerstrand's framework while calling into question Rose's assertion about the inherent phallocentrism of the perspective (pp. 127–128). Laws (1997) indicates the possibility of a post-structuralist appropriation of the time-geographic perspective for understanding feminine spatiality based on theories of embodied female identities and corporeality. Hanson and Pratt's (1995) re-interpretation of their earlier observations in terms of feminist containment narratives based on the reciprocal constitution of feminine subjectivity and bodily comportment suggests a similar direction. Further, some of Rose's (1993) criticisms may also be addressed through extending Adams' (1995) time-geographic re-theorization of personal boundaries. Lastly, a variety of representational strategies may be used to compose rich pictorial narratives of women's (and men's) everyday experiences (Gregory 1994, 249–251). These approaches together suggest that a feminist re/deconstruction of time-geography, perhaps based on the work of Grosz (1990, 1994), Foucault (1977), and Young (1989), may provide new ways for understanding the spatio-temporal experiences of women's (and men's) everyday lives in their particular sociospatial contexts.

Literature Cited

Adams, Paul C. 1995. A reconsideration of personal boundaries in space–time. *Annals of the American Association of Geographers* 85(2):267–85.

Aitken, Stuart C. 1998. *Family Fantasies and Community Space*. New Brunswick, NJ: Rutgers University Press.

Arentze, Theo A., Aloys W.J. Borgers, and Harry J.P. Timmermans. 1994. Geographical information systems and the measurement of accessibility in the context of multipurpose travel: A new approach. *Geographical Systems* 1:87–102.

Bianco, Martha, and Catherine Lawson. 1996. Trip-chaining, childcare, and personal safety: Critical issues in women's travel behavior. In *Women's Travel Issues: Proceedings from the Second National Conference*, 122–143. Washington, DC: Federal Highway Administration.

Blumen, Orna. 1994. Gender differences in the journey to work. *Urban Geography* 15(3):223–45.

Brun, J., and J. Fagnani. 1994. Lifestyles and locational choices - trade-offs and compromises: A case study of middle-class couples living in the Ile-de-France region. *Urban Studies* 31:921–34.

Burnett, Patricia, and Susan Hanson. 1982. The analysis of travel as an example of complex human behavior in spatially-constrained situations: Definition and measurement issues. *Transportation Research* 16A(2):87–102.

Burns, Lawrence D. 1979. *Transportation, Temporal, and Spatial Components of Accessibility*. Lexington, MA: Lexington Books.

City of Columbus. 1993. *Growth Statement 1993*. Columbus, OH: Development Department, City of Columbus.

Cullen, Ian, Vida Godson, and Sandra Major. 1972. The structure of activity patterns. In *Patterns and Processes in Urban and Regional Systems*, ed. A.G. Wilson, 281–96. London: Pion.

Dyck, Isabel. 1989. Integrating home and wage workplace: Women's daily lives in a Canadian suburb. *Canadian Geographer* 33(4):329–41.

———. 1990. Space, time, and renegotiating motherhood: An exploration of the domestic workplace. *Environment and Planning D* 8:459–83.

———. 1996. Mother or worker? Women's support networks, local knowledge and informal child care strategies. In *Who Will Mind the Baby?*, ed. Kim England, 123–40. London: Routledge.

England, Kim. 1993. Suburban pink collar ghettos: The spatial entrapment of women? *Annals of the Association of American Geographers* 83:225–42.

———. 1994. Reply to Susan Hanson and Geraldine Pratt. *Annals of the American Association of Geographers* 84(3):502–4.

———. 1996a. Who will mind the baby? In *Who Will Mind the Baby?*, ed. Kim England, 3–19. London: Routledge.

———. 1996b. Mothers, wives, workers: The everyday lives of working mother. In *Who Will Mind the Baby?*, ed. Kim England, 109–22. London: Routledge.

Foucault, Michel. 1977. *Discipline and Punish: The Birth of the Prison*. Trans. by Alan Sheridan. New York: Vintage Books.

Fuss, Diana. 1989. *Essentially Speaking: Feminism, Nature and Difference*. New York: Routledge.

Gatrell, Anthony C. 1994. Density estimation and the visualization of point patterns. In *Visualization in Geographic Information Systems*, ed. H.M. Hearnshaw and D.J. Unwin, 65–75. New York: John Wiley and Sons.

Giddens, Anthony. 1984. *The Constitution of Society: Outline of the Theory of Structuration*. Berkeley, CA: University of California Press.

Gilbert, Melissa R. 1997. Feminism and difference in urban geography. *Urban Geography* 18(2):166–79.

———. 1998. "Race," space, and power: The survival strategies of working poor women. *Annals of the American Association of Geographers* 88(4):595–621.

Giuliano, Genevieve. 1988. Commentary: Women and employment. *Urban Geography* 9(2):203–8.

Golledge, Reginald G., and Robert J. Stimson. 1997. *Spatial Behavior: A Geographic Perspective*. New York: Guilford.

Golledge, Reginald G., Mei-Po Kwan, and Tommy Gärling. 1994. Computational process modeling of household travel decisions using a geographical information system. *Papers in Regional Science* 73(2):99–117.

Gregory, Derek. 1985. Suspended animation: The stasis of diffusion theory. In *Social Relations and Spatial Structures*, ed. Derek Gregory and John Urry, 296–336. London: Macmillan.

———. 1994. *Geographical Imaginations*. Cambridge, MA: Blackwell.

Gregson, Nicky, and Michelle Lowe. 1993. Renegotiating the domestic division of labour? A study of dual career households in north-east and south-east England. *Sociological Review* 41(3):475–505.

Grosz, Elizabeth. 1990. Inscriptions and body-maps: Representation and the corporeal. In *Feminine, Masculine and Representation*, ed. Terry Threadgold and Anne Cranny-Francis, 62–74. London: Allen and Unwin.

———. 1994. *Volatile Bodies: Toward a Corporeal Feminism*. Indianapolis: Indiana University Press.

Hägerstrand, Törsten. 1970. What about people in regional science? *Papers of the Regional Science Association* 24:7–21.

Handy, Susan L., and Debbie A. Niemeier. 1997. Measuring accessibility: An exploration of issues and alternatives. *Environment and Planning A* 29(7):1175–94.

Hanson, Susan. 1980. The importance of the multipurpose journey to work in urban travel behavior. *Transportation* 9:229–48.

Hanson, Susan, and Perry Hanson. 1981. The impact of married women's employment on household travel patterns: A Swedish example. *Transportation* 10:165–83.

Hanson, Susan, and Geraldine Pratt. 1990. Geographic perspectives on the occupational segregation of women. *National Geographic Research* 6(4):376–99.

———. 1994. On suburban pink collar ghettos: The spatial entrapment of women? by Kim England. *Annals of the American Association of Geographers* 84(3):500–4.

———. 1995. *Gender, Work, and Space*. London: Routledge.

Hanson, Susan, and Margo Schwab. 1987. Accessibility and intraurban travel. *Environment and Planning A* 19:735–48.

Harvey, David. 1989. *The Condition of Postmodernity*. Cambridge, MA: Basil Blackwell.

Hayden, Dolores. 1984. *Redesigning the American Dream: The Future of Housing, Work and Family Life*. New York: W.W. Norton.

Helling, Amy. 1998. Changing intra-metropolitan accessibility in the U.S.: Evidence from Atlanta. *Progress in Planning* 49(2):55–107.

Ihlanfeldt, Keith R. 1993. Intra-urban job accessibility and Hispanic youth employment rates. *Journal of Urban Economics* 33:254–71.

Ihlanfeldt, Keith R., and David L. Sjoquist. 1990. Job accessibility and racial differences in youth employment rates. *American Economic Review* 80:267–76.

Immergluck, Daniel. 1998. Job proximity and the urban employment problem: Do suitable nearby jobs improve neighborhood employment rates? *Urban Studies* 35(1):7–23.

Ingram, D.R. 1971. The concept of accessibility: A search for an operational form. *Regional Studies* 5, 101–7.

Janelle, Donald G., and Michael F. Goodchild. 1983. Space–time diaries and travel characteristics for different levels of respondent aggregation. *Environment and Planning A* 20(7):891–906.

Johnston-Anumonwo, Ibipo. 1992. The influence of household type on gender differences in work trip distance. *Professional Geographer* 44:161–69.

———. 1997. Race, gender, and constrained work trips in Buffalo, NY, 1990. *Professional Geographer* 49(3):306–17.

Kitamura, Ryuichi, Kazuo Nishii, and Konstadinos Goulias. 1990. Trip chaining behavior by central city commuters: A causal analysis of time-space constraints. In *Developments in Dynamic and Activity-Based Approaches to Travel Analysis*, ed. P. Jones, 145–70. Aldershot: Avebury.

Kobayashi, Audrey. 1997. The paradox of difference and diversity (or, why the threshold keeps moving). In *Thresholds in Feminist Geography: Difference, Methodology, Representation*, ed. John P. Jones III, Heidi J. Nast, and Susan M. Roberts, 3–9. New York: Rowman and Littlefield.

Kwan, Mei-Po. 1998. Space–time and integral measures of individual accessibility: A comparative analysis using a point-based framework. *Geographical Analysis* 30(3):191–217.

———. 1999a. Gender, the home-work link, and space–time patterns of non-employment activities. *Economic Geography*, in press.

———. 1999b. Gender differences in space–time constraints and their impact on activity patterns. Manuscript. Ohio State University.

———. 1999c. Measuring individual accessibility in everyday life: Methodological and operational issues. Manuscript. Ohio State University.

Kwan, Mei-Po, and Xiao-Dong Hong. 1998. Network-based constraints-oriented choice set formation using GIS. *Geographical Systems* 5:139–62.

Landau, Uzi, Joseph N. Prashker, and Bernard Alpern. 1982. Evaluation of activity constrained choice sets to shopping destination choice modelling. *Transportation Research A* 16:199–207.

Laws, Glenda. 1997. Women's life courses, spatial mobility, and state policies. In *Thresholds in Feminist Geography: Difference, Methodology, Representation*, ed. John P. Jones III, Heidi J. Nast, and Susan M. Roberts, 47–64. New York: Rowman and Littlefield.

Lawson, Victoria. 1995. The politics of difference: Examining the quantitative/qualitative dualism in post-structuralist feminist research. *Professional Geographer* 47(4):449–57.

Lenntorp, Bo. 1976. *Paths in Time-Space Environments: A Time Geographic Study of Movement Possibilities of Individuals*. Lund: Royal University of Lund, Department of Geography.

Levinson, David M. 1998. Accessibility and the journey to work. *Journal of Transport Geography* 6(1):11–21.

McDowell, Linda. 1983. Towards an understanding of the gender division of urban space. *Environment and Planning D* 1:59–72.

———. 1993. Space, place and gender relations: Part II. Identity, difference, feminist geometries and geographies. *Progress in Human Geography* 17(3):305–18.

McLafferty, Sara. 1995. Counting for women. *Professional Geographer* 47(4):436–42.

McLafferty, Sara, and Valerie Preston. 1996. Spatial mismatch and employment in a decade of restructuring. *Professional Geographer* 48(4):420–31.

———. 1997. Gender, race, and the determinants of commuting: New York in 1990. *Urban Geography* 18(3):192–212.

Michelson, William. 1985. *From Sun to Sun: Daily Obligations and Community Structure in the Lives of Employed Women and Their Families*. Totowa, NJ: Rowman and Allanheld.

———. 1988. Divergent convergence: The daily routines of employed spouses as a public affairs agenda. In *Life Spaces: Gender, Household, Employment*, ed. Caroline Andrew and Beth M. Milroy, 81–101. Vancouver: University of British Columbia Press.

Miller, Harvey J. 1991. Modelling accessibility using space–time prism concepts within geographic information systems. *International Journal of Geographical Information Systems* 5(3):287–301.

———. 1999. Measuring space–time accessibility benefits within transportation networks: Basic theory and computational procedures. *Geographical Analysis*, 31(1):1–26.

Miller, Roger. 1982. Household activity patterns in nineteenth-century suburbs: A time-geographic exploration. *Annals of the American Association of Geographers* 72:355–71.

———. 1983. The Hoover® in the garden: Middle-class women and suburbanization, 1850-1920. *Environment and Planning D* 1:73–87.

Morris, J.M., P.L. Dumble, and M.R. Wigan. 1979. Accessibility indicators for transport planning. *Transportation Research* 13A, 91–109.

Myers-Jones, Holly J., and Susan R. Brooker-Gross. 1996. The journey to child care in a rural American setting. In *Who Will Mind the Baby?*, ed. Kim England, 77–92. London: Routledge.

O'Kelly, Morton E., and Eric J. Miller. 1983. Characteristics of multistop multipurpose travel: An empirical study of trip length. *Transportation Research Record* 976:33–9.

Pain, Rachel. 1991. Space, sexual violence, and social control: Integrating geographical and feminist analysis of women's fear of crime. *Progress in Human Geography* 15(4):415–31.

Palm, Risa. 1981. Women in nonmetropolitan areas: A time-budget survey. *Environment and Planning A* 13(3):373–78.

Pickup, Laurie. 1984. Women's gender-role and its influence on their travel behavior. *Built Environment* 10:61–8.

———. 1985. Women's travel need in a period of rising female employment. In *Transportation and Mobility in an Era of Transition*, ed. G.R.M. Jansen, P. Nijkamp, and C.J. Ruijgrok, 97–113. Amsterdam: North Holland.

Pirie, Gordon H. 1979. Measuring accessibility: A review and proposal. *Environment and Planning A* 11:299–312.

Pratt, Geraldine, and Susan Hanson. 1991. Time, space, and the occupational segregation of women: A critique of human capital theory. *Geoforum* 22(2):149–57.

———. 1993. Women and work across the life course: Moving beyond essentialism. In *Full Circles: Geographies of Women over the Life Course*, ed. Cindi Katz and Janice Monk, 27–54. New York: Routledge.

Preston, Valerie, and Sara McLafferty. 1993. Gender differences in commuting at suburban and central locations. *Canadian Journal of Regional Science* 16(2):237–59.

Richardson, A.J., and W. Young. 1982. A measure of linked-trip accessibility. *Transportation Planning and Technology* 7:73–82.

Rose, Gillian. 1993. *Feminism and Geography: The Limits of Geographical Knowledge*. Minneapolis, MN: University of Minnesota Press.

———. 1997. Situating knowledges: Positionality, reflexivities and other tactics. *Progress in Human Geography* 21(3):305–20.

Saegert, Susan. 1981. Masculine cities and feminine suburbs: Polarized ideas, contradictory realities. In *Women and the American City*, ed. Catherine R. Stimpson, Elsa Dixler, Martha J. Nelson, and Kathryn B. Yatrakis, 93–108. Chicago: University of Chicago Press.

Staeheli, Lynn A., and Victoria A. Lawson. 1995. Feminism, praxis, and human geography. *Geographical Analysis* 27(4):321–38.

Talen, Emily. 1997. The social equity of urban service distribution: An exploration of park access in Pueblo, Colorado, and Macon, Georgia. *Urban Geography* 18(6):521–41.

Talen, Emily, and Luc Anselin. 1998. Assessing spatial equity: An evaluation of measures of accessibility to public playgrounds. *Environment and Planning A* 30:595–613.

Tivers, Jacqueline. 1985. *Women Attached: The Daily Lives of Women with Young Children*. London: Croom Helm.

———. 1988. Women with young children: Constraints on activities in the urban environment. In *Women in Cities: Gender and the Urban Environment*, ed. Jo Little, Linda Peake, and Pat Richardson, 84–97. New York: New York University Press.

Valentine, Gill. 1989. The geography of women's fear. *Area* 21:385–90.

Villoria, Olegario G. Jr. 1989. An Operational Measure of Individual Accessibility for Use in the Study of Travel-Activity Patterns. Ph.D. diss., Graduate School of the Ohio State University.

Wachs, Martin, and T. Gordon Kumagai. 1973. Physical accessibility as a social indicator. *Socio-economic Planning Science* 7:437–56.

Weisman, Leslie K. 1992. *Discrimination by Design: A Feminist Critique of the Man-Made Environment*. Urbana, IL: University of Illinois Press.

Wyly, Elvin K. 1996. Race, gender, and spatial segmentation in the Twin Cities. *Professional Geographer* 48(4):431–44.

Young, Iris M. 1989. Throwing like a girl: A phenomenology of feminine body comportment, motility, and spatiality. In *The Thinking Muse: Feminism and Modern French Philosophy*, ed. Jeffner Allen and Iris M. Young, 51–70. Bloomington, IN: Indiana University Press.

MEI-PO KWAN (Ph.D., University of California, Santa Barbara) is Assistant Professor in the Department of Geography, The Ohio State University, Columbus, OH 43210-1361. E-mail: kwan.8@osu.edu. Her research interests include gender/ethnic issues in transportation geography, GIS, spatial analysis, spatial cognition, and travel behavior.

[18]

GENDER, RACE, AND COMMUTING AMONG SERVICE SECTOR WORKERS*

Sara McLafferty
Hunter College, CUNY

Valerie Preston
York University

The generality that women work closer to home and have shorter commuting times than men needs to be assessed for racial groups. Statistical analysis of commuting times for a large sample of service workers in the New York metropolitan area shows that black and hispanic women commute as far as their male counterparts and their commuting times far exceed those of white men and women. Workplace factors, such as income, occupation, and job accessibility, are important in explaining these findings. **Key Words: gender, race, commuting.**

Recent articles on differences in commuting time and distance between men and women in advanced industrialized societies establish that women work substantially closer to home and have shorter commuting times than men. The consistency of these findings prompted a search for explanations, stimulating investigations into the gender division of labor in the home and workplace and the spatial "entrapment" of women (Hanson and Pratt 1988; Institute of British Geographers 1984; Nelson 1986). Most studies, however, focused on women as a homogeneous group, without considering divisions among women based on race and ethnicity. The length, time, and direction of commuting vary significantly among racial and ethnic groups, and should affect the gender differential in work trip length (Davies and Albaum 1972; Kain 1968). The combined effects of gender and race put minority women at a disadvantage in their access to employment opportunities. Do the well-known differences in commuting between men and women persist when we disaggregate the working population by race? Does a common gender effect cut across racial groups, or is the effect of gender on commuting contingent on the worker's membership in a racial group?

We report here the results of analyses employing a very large sample of service sector workers in the New York City metropolitan area. Within this sample we investigate whether gender differences in commuting exist for the two main minority groups in the New York area, blacks and hispanics, and we analyze the influence of class, as described by income and occupation, on gender and racial differences in commuting time. Our results indicate that the well-established disparity in commuting between men and women differs significantly among racial groups. Black and hispanic women commute as far as black and hispanic men, and their commuting times far exceed those of white males and females. These results persist even after controlling for income, occupation, and industry of employment. By revealing significant disparities between

* We thank Ellen Hamilton and Danny Kahn for their research assistance and Susan Hanson and the reviewers for their comments on an earlier draft of the manuscript.

white and minority women, this study shows how gender and race interact to affect urban commuting flows and local labor markets. At a more general level, these findings support Sanders' (1990, 229) recent contention that "gender studies have fallen victim to the myth of 'universal womanhood' and have not been sensitive to the experiences and contributions of women from various races and classes." It is necessary to move beyond a homogeneous treatment of women to address how minority womens' social and spatial contexts influence spatial behavior.

The first section of this paper is a brief review of the literature on gender differences in commuting and the postulated links between commuting and the economic and social status of white and minority women. The second section describes the study area and data set utilized in the research. The third section presents the results of the statistical analysis of differences in commuting times among gender and racial groups, and their implications for gender-based analyses of commuting patterns and local labor markets.

Background

A large body of research in the past two decades indicates that in advanced industrialized societies, women work significantly closer to home and have shorter commuting times than men (Hanson and Johnston 1985; Howe and O'Connor 1982). The degree of disparity depends on whether distance or time is used in measuring work-trip length. Each measure has certain advantages. Distance is an obvious indicator of geographic proximity to employment and is invariant from day to day. In contrast, commuting time provides a more direct measure of the "cost" incurred in commuting relative to individuals' time constraints, but it can fluctuate significantly depending on traffic congestion and other factors. Studies that use distance as a measure show unequivocally that women work closer to home than men and thus have more localized labor markets (Fagnani 1983; Howe and O'Connor 1982; Madden 1981). Hanson and Johnston (1985), for example, found that women worked more than two miles closer to home on the average than men.

Analyses of commuting time are generally consistent with those of commuting distance. Women typically spend less time in commuting than men, though the difference is proportionally less than that for commuting distance (Gordon et al. 1989; Hanson and Johnston 1985). In a study of commuting patterns in suburban Toronto, womens' average commuting time was even slightly longer than men's, although the commuting distance was shorter (Rutherford and Wekerle 1988). The authors explained this discrepancy as a result of public transit use: women were more likely to use public transportation in getting to work, and travel times are typically longer for public transit users. Despite the slight differences in conclusions based on analysis of distance or time, the weight of evidence confirms that women work closer to home than men.

Why do sharp differences exist in male and female commuting patterns? Many studies have analyzed and reviewed this question (Hanson and Johnston 1985; Wekerle and Rutherford 1989). Explanations of gender differences in commuting initially focused on women's household and childcare responsibilities and patriarchal relations within the home (Madden and White 1980). Compared to men, women spend more time in child rearing and household work; they are less willing and able to devote time toward commuting. Though intuitively and theoretically appealing, this explanation does not hold up well under empirical scrutiny. Hanson and Johnston (1985), for example, found that the length of womens' work trips did not vary with the number and ages of children in the household. Similarly, studies by Madden (1981) and Gordon et al. (1989) concluded that the presence of children accounted for very little of the gender differential in commuting. Thus, while household responsibilities continue to affect women's labor force participation (Bergmann 1986), the effects on the length of women's work trips are unclear at this time.

An alternative explanation emphasizes the conditions and spatial distributions of female employment. Men and women hold very different class positions in the work force, positions that clearly influence commuting decisions. Despite women's increasing representation in high-status fields such as law and medicine, women still are much more likely than men to hold low-wage, part-time jobs that provide little payoff for commuting long distances (Bergmann 1986). Multivariate analyses indicate that women's lower wage rates are among the most important determinants of their shorter commuting distances (Hecht 1974; Madden 1981). "If women had the same job tenure, work hours and, most importantly, the same wages as their male counterparts in the household, their work trips would no longer be shorter, in fact they would be longer" (Madden 1981, 191). Not only do women earn less than men, but the economic return for each additional mile of commuting is significantly less for women than for men. Rutherford and Wekerle (1988) found that the increase in female incomes with additional commuting distance was just one-third that of males, giving women little incentive to travel long distances. Such studies suggest that women's shorter work trips are an economically rational response to the lack of well-paid, full-time job opportunities.

In addition to their lower wages, women make up a disproportionate share of the work force in low-status sales, clerical, and service occupations (Bergmann 1986). The effects of this occupational segmentation on commuting are complex. For the entire work force, work-trip length relates to occupational class: higher status, better paid managers and professionals commute significantly longer distances than blue collar and service workers (Gordon et al. 1989; Villeneuve and Rose 1988). Women's greater representation in clerical and service occupations should thus be associated with shorter commuting distances. The empirical evidence on occupational differences is mixed, however. Hanson and Johnston (1985) found no occupational differences in commuting distances for women in Baltimore. Elsewhere, occupation has complicated effects on women's commuting. For example, in a recent national study, female blue-collar workers unexpectedly reported the shortest trip distances, whereas female technicians reported the longest (Gordon et al. 1989). In the face of this contradictory evidence, several authors suggested that while women's concentration in lower-status occupations influences commuting, occupational effects are mediated by women's lower earnings and the spatial distribution of women's employment (Hanson and Pratt 1988; Wekerle and Rutherford 1989).

Researchers have only recently begun to analyze the intraurban spatial distribution of women's employment and its implications for commuting and labor force participation. Hanson and Pratt (1988) demonstrated that the so-called "female" occupations are unevenly distributed over space and the spatial pattern of "female" jobs differs greatly from that of "male" jobs (see also Blumen and Kellerman 1990). "Gender-based labor market segmentation is visible on the urban landscape at a very fine spatial scale" (Hanson and Pratt 1988, 198). The reasons for the spatial clustering of female jobs are complex. Some firms, especially back office and clerical firms, intentionally choose suburban locations that are accessible to a captive female labor force (Nelson 1986). On the other hand, firms employing large numbers of women may cluster for other reasons, such as agglomeration economies or lower land or transportation costs. The implications of the spatial clustering of female employment for women's commuting are poorly understood. Do women work in such clusters because they provide a convenient source of employment? Are the clusters accessible to all women or only to women with certain educational or social characteristics? Furthermore, although some types of "female" employment are spatially clustered, it is likely that other types are spatially dispersed. Women make up a significant share of the work force in retailing, health care, education, and other central place-type activities, and these tend to be spatially dispersed (Blumen and Kellerman 1990). The tendency for women to work in these industries could account in part for women's shorter commuting times. These recent studies imply that the spatial structure of opportunities affects women's commuting times and distances, just as it affects all types of spatial interaction. But further research is needed to investigate these effects in detail and tease out the relationships between the location of employment opportunities and women's commuting patterns.

This brief review of the literature indicates that the gender differential in commuting is rooted in women's labor market experiences and access to employment, as well as responsibilities in the home. Explanations of commuting must include the economic and social contexts in which women live and work. To the extent that these contexts vary among women of different racial and ethnic groups, commuting patterns should also vary.

The economic and labor force characteristics of white and minority women differ significantly in the United States and these in turn should lead to differences in commuting. Historically, black women have been much more likely to participate in the paid labor force than white women, although the gap has narrowed in recent years (Jones 1986). Earnings of black and hispanic women are significantly less than those of white men and also less than those of black and hispanic men and white women (Bergmann 1986; Cotton 1989; Foner 1980; US Bureau of Census 1988).

Black and hispanic women also differ significantly from white women in occupational status. Although both black and white women are often employed in service industries, black women within those industries tend to occupy lower-status occupations that offer limited opportunity for advancement (Jones 1985; Stafford 1985). Black and hispanic women are less likely to work in technical, administrative, and managerial occupations and are overrepresented in service, private household, and operative positions (Stolz 1985). Minority women often work for white women in domestic and office work (Jones 1985). Even within the same occupational categories, minority women often face greater barriers to career advancement than white women, reinforcing the concentration of minority women in low-status occupations. Hartmann (1988), for example, found that minority women workers were less likely to be promoted than white women. Minority women thus suffer the double burdens of gender and racial bias. Wage discrimination and poorer access to employment result in minority women's concentration in low-wage, low-status occupations.

Geographical access to employment opportunities also distinguishes white and minority women. Residential segregation of racial groups means that minority women and men often live in areas distant from and poorly connected to major centers of employment growth (Clark and Whiteman 1983; Davies and Albaum 1972; Goddard 1982). Kain (1968) called this phenomenon the "spatial mismatch" between jobs and available labor. Despite debate about the effects of the spatial mismatch on employment patterns and labor force participation among black men and women, most authors agree that the spatial mismatch is an important problem for the urban poor (Ellwood 1986; Hughes 1987; Leonard 1987). The spatial mismatch is clearly important to the study of gender and racial differences in commuting. Insofar as minority women reside in areas lacking job opportunities, their work trips should be systematically longer than those of white women. The implications of the spatial mismatch for differences between minority men and women in work-trip length are less obvious, however. Are minority women more sensitive to distance than minority men in choosing where to work, or does the lack of accessible job opportunities force both genders to travel long distances to work?

Minority women also differ from white women in their household characteristics and responsibilities. Minority women are much more likely to be single parents, facing enormous responsibilities for childcare and financial support (Simms and Malveaux 1986). Although some single parents are not in the paid labor force and receive welfare and AFDC payments, many work outside the home as sole breadwinners (Kodras 1986; Stolz 1985). The effects of single parenthood on commuting are complex and poorly understood. Conventional wisdom suggests that women who are single parents should work closer to home than other women, because of their greater household responsibilities and tighter time constraints. On the other hand, as sole wage-earners, single parents may be forced to endure long work trips in order to find decent-paying jobs (Stolz 1985).

The differences between minority and white women in wages, occupation, household responsibilities, and spatial access to employment indicate that the economic and social experiences of women vary significantly by race. These in turn should affect the length of commuting flows and the gender-based disparities in journey-to-work. The purpose of our research is to determine if gender-based differences exist for blacks and hispanics and to assess the effects of various economic characteristics—income, occupation, and industry of employment—on the time that men and women spend travelling to work. Because of the size and complexity of the problem, we only consider the association between economic factors and gender and racial differences in commuting time. Social factors, like single parenthood, are not addressed.

Data and Methods

The study area consisted of the 24 counties that make up the New York Consolidated Metropolitan Statistical Area (CMSA). With a population of 16.1 million in 1980, the New York CMSA was the largest urban settlement in the United States and contained the nation's largest and most diverse labor force (Gottmann 1962). New York City was the focal point of the region, which extended into southern Connecticut and western New Jersey. The region included several major cities, such as Newark and Patterson, NJ, that were important employment centers in their own right.

This analysis concentrates on people employed in the service sector of the economy. The service sector is broadly defined to include distributive and corporate services, health, education, government, retailing, and many others. These industries account for 80% of the region's paid labor force and are exceptionally important to the region's economic base (Port Authority of New York–New Jersey 1986). By focusing on service employment, we investigate the bulk of the region's work force and the major area of employment growth. Moreover, the service sector is especially important in providing employment for women and minorities (Christopherson 1989; Daniels 1989; Sheets et al. 1987).

To simplify the discussion, we group the service industries into six broad categories. The typology is based on previous classifications that combined service industries according to market characteristics and mode of provision (Sheets et al. 1987; Stanback et al. 1981). The first group, distributive services, involves the distribution of goods and information and includes transportation, communications, and wholesale trade. Advanced corporate services, which include finance, insurance, legal, and business services, provide capital, marketing, and ancillary services for businesses and corporations. This sector has been the focal point of New York's recent economic growth (Port Authority of New York–New Jersey 1986). The nonprofit services group consists of health, education, cultural, and social services provided typically by nonprofit organizations. The government sector comprises all public administration. Personal services include a wide range of activities, such as repair, recreation, and household services. Finally, retail services consist of private firms involved in distributing goods to consumers.

The various types of service industries differ greatly in their employment characteristics and these have important implications for commuting patterns. Wages are lowest in personal and retail services and relatively high in government and distributive services (Stanback et al. 1981). At the same time, female and minority workers make up a large share of the work force in nonprofit and personal services, but a relatively small share in distributive services (Bergmann 1986). Advanced corporate services, on the other hand, have a bifurcated labor force, with jobs concentrated in well-paid managerial positions and low-paid pink collar occupations (Stanback et al. 1981). Given that work trip length varies greatly with income, occupation, and other employment characteristics, we expect to find distinct differences in commuting among the six types of service industries.

Data for this study come from the journey-to-work subsample of the 1980 Census Public Use Microdata Sample (PUMS) (US Bureau of Census 1980). For a one percent sample of the population, the PUMS data set contains detailed individual- and household-level records describing labor force participation, occupation, industry of employment, places of work and residence, commuting time, income, and demographic characteristics. The information is organized by place of residence, so that employment in each service industry can be linked to the residential pattern of employees.

The PUMS data set presents three problems, however. First, the level of geographic detail is limited and varies within the region. Places of work and residence are identified only by county in New York state and by subcounty group in New Jersey and Connecticut. These areas vary greatly in population size and geographic extent and are not small enough to identify homogeneous social areas. Second, the data are subject to all the problems of the 1980 Census, such as undercount and sampling error. Finally, because the data are from 1980, they do not describe fully the most recent developments in service employment (Warf 1988).

From the PUMS data set, we extracted information on all persons living in the region who worked in the service sector, approximately 130,000 people. We excluded approximately 600 of those persons who listed their place of work as a distant state, such as Alaska or Illinois. The reported commuting times obviously made little sense for them.

Analysis of racial differences stressed the contrasts between the two major racial/ethnic minority groups—blacks and hispanics—and the remainder of the population. Blacks were identified based on the census's "race" variable, which reports an individual's self-identification as a member of a racial group. Identifying hispanics was more problematic, because hispanics comprise an ethnic minority rather than a racial group. We defined as hispanic those persons who either identified their race as "Spanish" or stated they were of hispanic origin. By this definition, more than 10,000 persons in the sample were classified as hispanic. The remainder of the sample, which comprises over 100,000 persons, consists mainly of white, non-hispanic workers, and is referred to as the "white" group. This group is ethnically and racially diverse, however, and it includes several thousand people of Asian descent, as well as numerous "white" ethnic groups. The small sample sizes for these other racial/ethnic minorities precluded analysis of their separate commuting flows.

To clarify the analysis of occupational effects, we combined occupational groups into four categories (Table 1). Blue-collar occupations include the skilled, semiskilled, and unskilled jobs of manual workers, craftspersons, operators, and laborers, who generally enjoy stable employment, relatively high wages, and fringe benefits, including some protection in the face of closures and layoffs.

Pink collar occupations are clerical, service, and sales positions that require a range of skills. Employment may be unstable, often seasonal or part time, wages and benefits are often low, and opportunities for advancement are often limited (Bluestone and Harrison 1986). White collar occupations are skilled occupations that typically require training beyond high school. This category cuts across industries and includes teachers, librarians, fire fighters, and health technicians. Managerial occupations involve the management and control of economic activities. Self-employed managers and proprietors are grouped with salaried administrators because both types of occupations benefit directly from a firm's economic growth and exert direct control over working conditions.

Analysis of variance (ANOVA) was used to assess the extent of gender and racial differences in commuting. In all cases the dependent variable is commuting time, a self-reported measure of the usual time (in minutes) spent commuting to work one way. The values reported are clearly estimates, as commuting times can vary substantially from trip to trip depending on congestion, transit reliability, and many other factors. We assume that errors in commuting times do not vary systematically from group to group. Unfortunately, we were unable to examine commuting distance, because the PUMS file does not include such information. The gross areal units used in coding place of work and place of residence also precluded estimates of commuting distance. Therefore, our conclusions are limited to the analysis of commuting time.

In addition to assessing variations in commuting times among gender and racial groups, we analyzed the extent to which such variations resulted from differences in occupation, income, and industry of employment. These employment characteristics are primary determinants of economic divisions among white and minority women, and they are known to affect commuting. By emphasizing economic factors, the analyses shed light on the influence of economic divisions on gender and racial differences in commuting flows.

TABLE 1
EXAMPLES OF OCCUPATIONS IN EACH OCCUPATIONAL CATEGORY

Category	Occupations
Blue collar	Transport workers, craftsmen, laborers, operators, foremen
Pink collar	Clerical, sales, service workers
White collar	Nurses, teachers, social workers, technicians
Managerial	Self-employed, administrators, doctors, lawyers

Results

The comparison of men's and women's commuting times indicates that women work significantly closer to home than men (Table 2). Women's commuting times average 5.6 minutes less than men's (significant at $P = .05$). Median commuting times confirm this pattern: the median commuting time for women is 20 minutes, compared with 30 minutes for men. Only 6.7% of the female workers sampled travel more than one hour to work, as opposed to 10.5% of male workers. These results confirm many past studies' findings that women's work trips are significantly shorter than those of men.

Commuting times also differ significantly and predictably among racial groups. Whites spend the least time travelling to work and blacks the most (Table 2). The average commuting time for hispanics is less than that for blacks, but still significantly greater than that for whites. Median commuting times follow the same general pattern: 20 minutes for whites and 30 minutes for blacks and hispanics. These findings support several past studies that demonstrate minorities' poorer geographical access to employment (Davies and Albaum 1972; Leonard 1987).

Occupation and industry of employment also affect average commuting times. Managerial workers have the longest commuting times among occupational groups (Table 2). This result is not sur-

TABLE 2
MEAN COMMUTING TIME BY OCCUPATION, GENDER, AND RACE

Factor		Commuting time (min) Mean	Standard deviation	n
	Male	33.9	24.3	69,998
Gender	Female	28.3	22.5	60,675
		$F = 1941.0^{a}$		
Race	White	29.9	23.5	102,182
	Black	37.2	23.7	18,117
	Hispanic	34.9	23.0	10,110
		$F = 849.7^{a}$		
Occupation	Blue collar	30.2	22.4	18,956
	Pink collar	30.1	23.3	59,007
	White collar	30.3	25.3	25,208
	Managerial	35.7	25.3	27,502
		$F = 459.0^{a}$		

[a] Significantly different from zero (at $P = 0.05$).

prising given their typically higher incomes and job status (Scott 1988; Wheeler 1967). On the other hand, average commuting times are almost identical for blue-, pink- and white-collar workers. The main distinction among occupational groups is between managerial and other workers.

More significant than the occupational differences are those based on industry of employment. Workers in distributive and advanced corporate services have the longest average commuting times and workers in retail, nonprofit, and personal services the shortest (Table 3). Commuting times in the former are a full 10 minutes more than those in the latter. These differences relate in part to the underlying spatial patterns of employment in the industries. Jobs in both distributive and advanced corporate services are spatially concentrated in the metropolitan region. Almost one-half of all jobs in the advanced corporate sector are located in Manhattan. Jobs in distributive services are slightly more dispersed, but still clustered in Manhattan and the surrounding counties in New York and New Jersey. Not surprisingly, these concentrated patterns of employment are often associated with longer work trips as employers must draw upon a wider spatial labor market (O'Connor 1980). In contrast, retail and nonprofit services, as classic low-order central place activities, are more spatially dispersed and correspond closely to the distribution of residential population. This dispersion leads, in general, to shorter commuting times.

TABLE 3
MEAN COMMUTING TIME BY INDUSTRY[a]

Industry	Commuting time (min) Mean	Standard deviation	n
Distributive	35.7	24.1	25,475
Advanced corporate	38.5	25.7	30,643
Nonprofit	26.2	21.1	32,546
Government	33.6	24.2	8531
Personal	28.9	22.6	8628
Retail	24.7	20.5	24,850

[a] $F = 1480.8$, significantly different from zero (at $P = 0.05$).

To investigate the separate and combined effects of gender, race, occupation, income, and industry on commuting times, we computed a series of multiple analysis of variance (ANOVA) models, first for all industries and then for each industry separately. In each model, gender, race, and occupation served as independent variables and income was introduced as a covariate. The F statistics for the main and interaction effects indicate

TABLE 4
MULTIPLE ANOVA RESULTS FOR ENTIRE SAMPLE

		Parameter value	F statistic
Main effects:			
Gender			30.53
	Male	0.78	
	Female	−0.78	
Race			524.64
	White	−4.70	
	Black	3.18	
	Hispanic	1.52	
Occupation			35.47
	Blue collar	−1.55	
	Pink collar	1.45	
	White collar	−0.94	
	Managerial	1.04	
Covariate:			
Income		0.00035	53.70[a]
Interaction effects:			
Gender × race			50.60
Gender × occupation			8.34
Race × occupation			17.05
Gender × race × occupation			ns[b]

[a]t statistic.
[b]Not statistically significant at $P = 0.05$.

the significance of the independent variables, alone and in combination, in affecting commuting times. In addition to the ANOVAs, linear models were computed to evaluate the effects of the independent variables on commuting times. In the linear models, the independent variables (except for income) represent categories of race, gender, and occupation, and the parameter values represent the difference between that category's mean commuting time and the grand mean, controlling for other factors in the model. For example, the parameter value for the category "men" shows the deviation of men's average commuting times from the grand mean after controlling for income, race, and occupation. As a continuous variable, income serves as a covariate in the model. Its parameter is analogous to a regression coefficient, and a t statistic indicates the significance of the parameter. Because of the very large sample size, many of the F and t statistics were statistically significant (at $P = .05$); we therefore focus attention on the relative magnitudes of the statistics as well as their significance.

For all industries combined, income and race have by far the largest impacts on commuting time, as shown by their F statistics (Table 4). In contrast, the effects of gender and occupation, though statistically significant, are not nearly as large as those of income and race. The parameter values (Table 4) show the nature and direction of the effect of each variable on differences in commuting times. Income is positively associated with commuting time, confirming that higher-income workers spend more time travelling to work. We also find that commuting times for blacks and hispanics significantly exceed those for whites. Holding constant income and occupation, blacks spend almost eight minutes more on average than whites in commuting (Table 4). Differences based on gender are smaller than those based on race. The parameter values for the gender variables indicate that after controlling for income, race, and occupation, women's commuting times differ

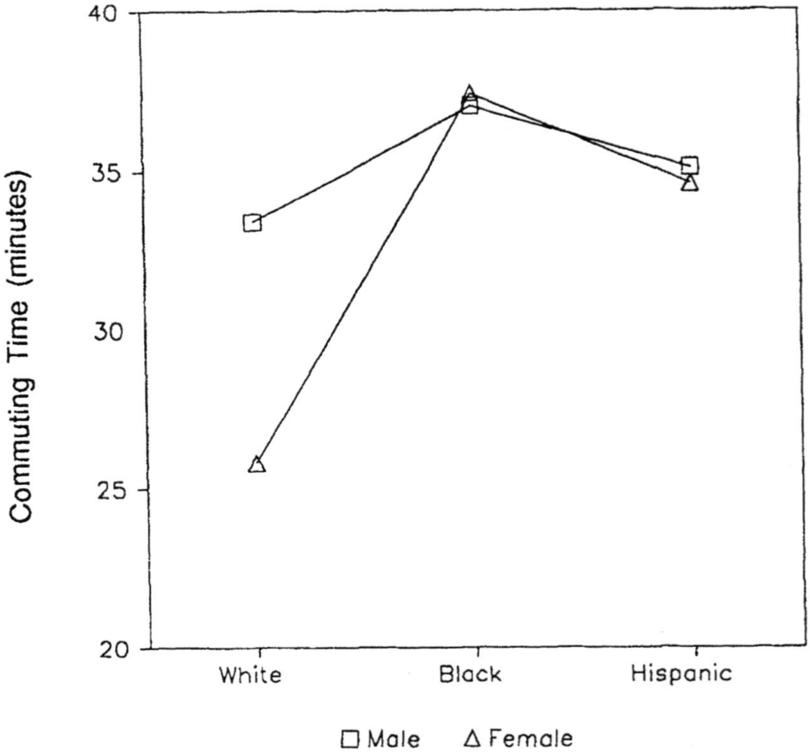

Figure 1. Average commuting time by gender and race.

from men's by only 1.56 minutes—hardly a substantial difference.

Complicating the interpretation of the main effects is the fact that all of the two-way interactions are statistically significant. Especially important is the interaction between gender and race, which has an *F* statistic second only to that of race alone. To examine in detail the interaction between gender and race, we plotted the average commuting times for various combinations of the two variables (Fig. 1). The plot reveals large differences in commuting times for white men and women, but no gender difference for blacks and hispanics. Black and hispanic women commute as long as their male counterparts, and on average, their commuting times far exceed those of white men and women. Thus, we observe no gender differential in commuting for blacks and hispanics; the large and significant difference in men's and women's commuting times so often noted in the literature exists only for the white workers in our sample.

Disaggregating the data by industry confirms the importance of the interaction between gender and race. The average commuting times for each type of service industry show significant gender differences only for whites (Table 5). Black women's commuting times are consistently similar to those for black men. In fact, in five of the six industries, black women commute longer on average than black men, although the difference in times is relatively small. Particularly

TABLE 5
MEAN COMMUTING TIMES BY GENDER, RACE, AND INDUSTRY OF EMPLOYMENT

Industry		Race		
		White	Black	Hispanic
Distributive	Men	36.3	38.2	35.8
	Women	31.5	40.3	37.6
Advanced corporate	Men	41.7	40.7	39.2
	Women	33.6	42.9	41.1
Nonprofit	Men	27.3	34.5	32.2
	Women	21.7	34.0	32.0
Government	Men	34.1	37.6	38.7
	Women	28.0	39.0	35.9
Personal	Men	27.6	34.9	34.2
	Women	22.8	40.0	32.4
Retail	Men	25.6	34.3	32.3
	Women	20.1	35.1	28.7

noteworthy is the exceptionally long commuting time for black women in personal service industries. This finding could reflect the concentration of black women in domestic jobs that require long trips to distant, high-income areas. The results for hispanic women reveal a similar pattern, except that in two industries, government and retailing, hispanic women spend less time commuting than hispanic men. In these two industries, hispanic women's shorter work trips may be associated with their greater concentration in low-wage, pink-collar jobs. However, the weight of evidence shows that minority women commute as far as minority men.

To what extent do these important findings reflect differences in the economic status and class position of white and minority men and women? The difference in earnings and occupational status between minority men and women is much less than that between white men and women, because of the concentration of minority men in low-wage, low-status jobs (Stafford 1985). Insofar as economic factors affect commuting times, such differences in economic status could account for the relative equality of minority men's and women's commuting times. To examine this hypothesis, separate multiple ANOVA models were computed for each type of service industry, incorporating the effects of income and occupation along with gender and race. If the observed differences in commuting are due to differences in economic status, then the direct effect of gender and interaction effect of gender and race should diminish after controlling for income, occupation, and industry.

The *F* statistics for these models reveal that in every industry, race and income have the largest effects on commuting time (Table 6). The nature and direction of effects conform to those described earlier: higher incomes are associated with longer commuting times and, holding income constant, black and hispanic workers spend significantly more time commuting than white workers. The effects of gender diminish when we examine spe-

TABLE 6
F STATISTICS FOR MULTIPLE ANOVA BY INDUSTRY

Effect	Industry					
	Distributive	Advanced corporate	Nonprofit	Government	Personal	Retail
Gender	ns[a]	5.2	7.7	ns	ns	ns
Race	50.5	17.5	157.9	7.7	48.0	25.1
Occupation	14.0	14.3	ns	ns	ns	ns
Gender × race	19.0	9.4	8.3	ns	ns	ns
Race × occupation	ns	2.1	13.2	2.4	ns	3.3
Gender × occupation	ns	ns	ns	ns	ns	4.7
Gender × race × occupation	3.8	ns	ns	ns	ns	ns
Income (*t* statistic)	19.7	22.5	9.1	12.6	10.4	25.6

[a]ns = Not statistically significant at $P = 0.05$.

cific industries. Gender has a statistically significant effect (at $P = .05$) on commuting time in only two of six industries—advanced corporate and nonprofit services. And even in these two industries the difference in commuting time between men and women is small (less than two minutes) *after* controlling for income and occupation. Much of the difference in commuting times between men and women results from differences in income and the industries and occupations in which they are employed.

The interaction variable for gender and race also decreases in significance after controlling for income, occupation, and industry of employment (Table 6). In three of the six industry-specific models, the interaction effect is *not* statistically significant. For government, personal, and retail services, differences in income account for the observed differences in commuting times among gender and racial groups. In these predominantly consumer service industries, neither gender nor occupation has a significant *F* statistic: income and race are by far the most significant determinants of commuting time. The significance of the race and income variables, and lack of significance of the gender–race interaction variable, could reflect the lesser availability of retail and personal services in minority neighborhoods. The relative scarcity of locally available jobs may require black and hispanic workers to commute farther, on average, to obtain employment in these industries. For the other three industries—distributive, advanced corporate, and nonprofit services—differences in commuting times persist after controlling for income and occupation, and the gender-race interaction variable remains statistically significant. Two of these industries, distributive and advanced corporate, are producer service industries able to locate certain functions, such as back offices, to take advantage of a particular labor force. The persistence of the gender and race interaction may result from firms' efforts to locate certain activities near a white, female labor force. This locational strategy would lead to lower than expected travel times for white women, after controlling for income and occupation.

Conclusion

Gender and race produce fundamental divisions within modern society, and the interactions between them give rise to distinctly different spatial relations and spatial structures. Gender relations vary not only among cultures, but also among racial groups, depending on the historical, economic, and cultural circumstances of the groups (Epstein 1988). The intersection between gender and race is critically important in understanding women's commuting patterns and commuting flows. Although white women have significantly shorter commuting times than white men, no such gender gap emerges for the minority population. Commuting times for black and hispanic women equal those for black and hispanic men and far exceed those for both white men and women. There is no evidence that minority women are more sensitive to commuting time than minority men in seeking employment. The conditions under which minorities live and work create long commuting times for both men and women.

The lack of gender differences in commuting for minorities derives from two economic factors: industry of employment and class, as described by income and occupation. These factors account for a large proportion of the variance in commuting times for both white and minority workers. Controlling for income and industry reduces the significance of the interaction between gender and race to the point where it becomes statistically insignificant in three of six industries. In these industries, the similarity in minority men's and women's commuting times is due in part to the similarity in their social class: they differ much less in income and occupational status than white men and women.

This study confirms the importance of economic factors in explaining white women's shorter commuting times. After controlling for income, industry, and occupation, the difference in men's and women's commuting times diminishes and often becomes statistically insignificant. Although we did not compare the importance of economic factors relative to household factors, such as number of children and marital status, if the latter were important, the gender gap should remain after controlling for differences in economic status.

Despite their overall significance, economic factors do not explain why minority workers have significantly longer commuting times than white workers. To the contrary, minority workers' concentration in low-wage, low-status jobs should lead to shorter commuting times, not longer ones. Several factors may explain minority workers' longer commuting times. Minority workers are more likely to be "transit captives," reliant on slow public transportation to get to work (Hanson and Schwab 1988; Wekerle and Rutherford 1988). Minority workers' long travel times may result from the cumulative effects of decades of racial discrimination in job and housing markets. Many of the minority workers in our sample live in inner city areas that lack service-sector job opportunities. To find suitable employment, they must endure long work trips. Underfunded public policies to redress this problem have met with limited success, and most have targeted manufacturing firms rather than services (Hughes 1987). Consequently, the spatial mismatch remains as evident today as it was in the 1960s.

Many recent studies discuss the spatial entrapment of suburban women, with their short commuting times and restricted labor markets (Nelson 1986). But this research points to a more insidious form of spatial entrapment: that of minority women who spend a great deal of time commuting to low-wage, low-status jobs. Although this finding is not entirely new, it demands much more research attention. At the aggregate level, statistical studies are needed to confirm these findings in other study areas and industries and to examine the effects of, for example, single parenthood and public transit dependence on minority women's commuting times and labor force participation. At the same time, disaggregate studies of minority women's experiences in the home and workplace are urgently needed. What kinds of spatial constraints do minority women face in finding and keeping well-paid jobs? How do minority women cope with household and child-care responsibilities, while commuting long distances? Such studies will help to unravel the complex forces underlying the spatial entrapment of minority women.

Literature Cited

Bergmann, B. 1986. *The Economic Emergence of Women*. New York: Basic Books.

Bluestone, B., and B. Harrison. 1986. *The Great American Job Machine: The Proliferation of Low-wage Employment in the U.S. Economy*. A study prepared for the Joint Economic Committee, US Congress, Washington, DC.

Blumen, O., and A. Kellerman. 1990. Gender differences in commuting distance, residence, and employment location: Metropolitan Haifa, 1972 and 1983. *The Professional Geographer* 42:54–71.

Christopherson, S. 1989. Flexibility in the U.S. service economy and the emerging spatial division of labour. *Transactions of the Institute of British Geographers* 14:131–43.

Clark, G., and J. Whiteman. 1983. Why poor people do not move: Job search behavior and disequilibrium in local labor markets. *Environment and Planning A* 15:85–104.

Cotton, J. 1989. Opening the gap: The decline in black economic indicators in the 1980s. *Social Science Quarterly* 70:803–19.

Daniels, P. 1989. Some perspectives on the geography of services. *Progress in Human Geography* 23:427–37.

Davies, C. S., and M. Albaum. 1972. The mobility problems of the poor in Indianapolis. *Antipode* 1:67–87.

Ellwood, D. 1986. The spatial mismatch hypothesis: Are there teenage jobs missing in the ghetto? In *The Black Youth Employment Crisis*, ed. R. Freeman and H. Holzer, 147–90. Chicago: University of Chicago Press.

Epstein, C. 1988. *Deceptive Distinctions: Sex, Gender, and the Social Order*. New York: Russell Sage Foundation.

Fagnani, J. 1983. Women's commuting patterns

in the Paris region. *Tijdschrift voor Economische en Sociale Geografie* 74:12–24.

Foner, P. 1980. *Women and the American Labor Movement*. New York: Free Press.

Goddard, H. 1982. Recent evidence on black employment and metropolitan decentralization. *Growth and Change* 13:32–39.

Gordon, P., A. Kumar, and H. Richardson. 1989. Gender differences in metropolitan travel behavior. *Regional Studies* 23:499–510.

Gottmann, J. 1962. *Megalopolis: The Urbanized Northeastern Seaboard of the U.S.* New York: Twentieth Century Fund.

Hanson, S., and I. Johnston. 1985. Gender differences in work-trip length: Explanations and implications. *Urban Geography* 6:193–219.

Hanson, S., and G. Pratt. 1988. Spatial dimensions of the gender division of labor in a local labor market. *Urban Geography* 9:180–202.

Hanson, S., and M. Schwab. 1988. Describing disaggregate flows: Individual and household activity patterns. In *The Geography of Urban Transportation*, ed. S. Hanson, 154–78. New York: Guilford Press.

Hartmann, H. 1988. Internal labor markets and gender: A case study of promotion. In *Gender in the Workplace*, ed. C. Brown and J. Pechman, 59–105. Washington, DC: Brookings Institution.

Hecht, A. 1974. The journey-to-work distance in relation to the socio-economic characteristics of workers. *Canadian Geographer* 18:367–78.

Howe, A., and K. O'Connor. 1982. Travel to work and labor force participation of men and women in an Australian metropolitan area. *The Professional Geographer* 34:50–64.

Hughes, M. 1987. Moving up and moving out: Confusing ends and means about ghetto dispersal. *Urban Studies* 24:503–17.

Institute of British Geographers, Women and Geography Study Group. 1984. *Geography and Gender: An Introduction to Feminist Geography*. London: Hutchinson.

Jones, B. 1986. Black women and labor force participation. An analysis of sluggish growth rates. In *Slipping Through the Cracks: The Status of Black Women*, ed. M. Simms and J. Malveaux, 11–31. New Brunswick, NJ: Transaction Books.

Jones, J. 1985. *Labor of Love: Labor of Sorrow*. New York: Basic Books.

Kain, J. 1968. Housing segregation, negro employment, and metropolitan decentralization. *Quarterly Journal of Economics* 82:175–97.

Kodras, J. 1986. Labor market and policy constraints on the work disincentive effects of welfare. *Annals of the Association of American Geographers* 76:228–46.

Leonard, J. 1987. The interaction of residential segregation and employment discrimination. *Journal of Urban Economics* 21:323–46.

Madden, J. 1981. Why women work closer to home. *Urban Studies* 18:181–94.

Madden, J., and M. White. 1980. Spatial implications of increases in the female labor force: A theoretical and empirical synthesis. *Land Economics* 56:432–46.

Nelson, K. 1986. Female labor supply characteristics and the suburbanization of low-wage office work. In *Production, Work and Territory*, ed. A. Scott and M. Storper, 149–71. London: Allen and Unwin.

O'Connor, K. 1980. The analysis of journey to work patterns in human geography. *Progress in Human Geography* 14:475–99.

Port Authority of New York and New Jersey. 1986. *The Regional Economy: Review 1985, Outlook, 1986*. New York: Port Authority of New York and New Jersey.

Rutherford, B., and G. Wekerle. 1988. Captive rider, captive labor: Spatial constraints on women's employment. *Urban Geography* 9: 173–93.

Sanders, R. 1990. Integrating race and ethnicity into geographic gender studies. *The Professional Geographer* 42:228–30.

Scott, A. 1988. *Metropolis: From the Division of Labor to Urban Form*. Berkeley: University of California Press.

Sheets, R., S. Nord, and J. Phelps 1987. *The Impact of Service Industries on Underemployment in Metropolitan Areas*. Lexington, MA: D. C. Heath.

Simms, M., and J. Malveaux, eds. 1986. *Slipping Through the Cracks: The Status of Black Women*. New Brunswick, NJ: Transaction Books.

Stafford, W. 1985. *Closed Labor Markets: Underrepresentation of Blacks, Hispanics and Women in New York City's Core Industries*. New York: Community Service Society of New York.

Stanback, T., P. Bearse, T. Noyelle, and R. Karasek. 1981. *Services: The New Economy*. Totowa, NJ: Rowman and Allanheld.

Stolz, B. 1985. *Still Struggling: America's Low-Income Working Women Confront the 80s*. Lexington, MA: D. C. Heath.

US Bureau of Census. 1980. *Census of Population and Housing, 1980: Public Use Microdata Samples Technical Documentation*, Washington, DC: US Government Printing Office.

US Bureau of Census. 1988. *Statistical Abstract of the United States: 1988*. Washington, DC: US Government Printing Office.

Villeneuve, P., and D. Rose. 1988. Gender, separation of employment in metropolitan Montreal, 1971–1981. *Urban Geography* 9:153–79.

Warf, B. 1988. The New York region's renais-

sance: Examining the causes and effects. *Economic Development Commentary* 12:13–17.
Wekerle, G., and B. Rutherford. 1989. The mobility of capital and the immobility of female labor: Responses to economic restructuring. In *The Power of Geography*, ed. J. Wolch and M. Dear, 139–72. London: Allen and Unwin.
Wheeler, J. 1967. Occupational status and work-trips: A minimum distance approach. *Social Forces* 45:508–15.

SARA MCLAFFERTY is Associate Professor of Geography at Hunter College, New York, NY 10021. Her interests are medical geography, urban restructuring, and locational analysis. VALERIE PRESTON is Associate Professor of Geography and Director of the Institute for Social Research at York University, North York, Ontario, Canada M3J 1P3. Her interests are urban and behavioral geography and gender issues.

Part III
Interregional Transport

[19]

A geographer's analysis of hub-and-spoke networks[1]

Morton E O'Kelly

Department of Geography, The Ohio State University, Columbus, OH 43210, USA

Hubs, as discussed in this paper, are *special nodes* that are part of a *network*, located in such a way as to facilitate connectivity between interacting places. Hubs are examined from the spatial organization viewpoint: that is, the linkages, hinterlands, and hierarchies formed by hub-and-spoke networks are described. Features of the hub-and-spoke system that make them different from basic facility location problems are emphasized. Special attention is paid to the contrasts between air passenger and air express freight applications. The paper discusses various broad types of models that are appropriate for network analysis. The paper includes a simulation exercise with a hypothetical network. The data are for interactions between 100 city pairs in the US, and the characteristic features of the hub network under three alternatives scenarios are developed. Although there is not a perfect correspondence between any of these models and the 'reality' of actual air freight and air passenger nets, a rudimentary matching is suggested. The single hub allocation model would be an especially inconvenient network for passengers, but might be ideal for regional freight or communications systems. The multiple assignment model has much less passenger inconvenience. While the network appears to be ideal for the passenger system, it seems to make a miscalculation about the nature of flow economies of scale. In the final model flows must be deliberately routed to make up economical bundles, and the incentives are stacked in favor of large flows.

Keywords: hub-and-spoke networks, spatial organization, numerical examples

Introduction: what's so fascinating about hubs?

Hubs come in an incredible variety of sizes and functions: from a $150 Ethernet hub in a computer supply store to a multimillion dollar freight sorting facility with employees, loading docks, trucks, planes, logistical tracking systems and 'worlds' of their own. Hubs in all the examples to be discussed in this paper are *special nodes* that are part of a *network*, located in such a way as to facilitate connectivity between interacting places. The interactions might be between computers on a campus network, or national phone calls, or transactions between widely scattered bank machines, cash registers, and banks. In most cases, hub network design involves a decision about *where* to place the hub (a location decision) and *how to route the traffic* that is to flow between the origins and destinations over the resulting network. [A detailed review and classification of the approaches to this problem have been presented recently by O'Kelly and Miller (1994) and in further work published by Campbell (1994), Aykin (1995a,b) and others.]

Hub-and-spoke networks, for geographers interested in transportation, are like check-out lines for a queuing theorist; i.e. they are a tangible physical manifestation of a process. Hubs are *geographical* in that they serve a specific regional area and they often confer benefits on the region in which they are located. Hubs are usually a catalyst for agglomeration and scale economies. Hubs, because of their direct connection to many spoke cities, are ideal accessible places at which to gather, and from which to distribute, material. Hubs require circuitous routing and a greater individual travel time between spoke cities than would be the case in a directly connected system (Taaffe *et al*, 1996; Middendorf *et al*, 1995): yet they are widely used in the overall system of spatial organization. Hubs have been blamed for high fares; and, at the same time, credited with increasing connections between places. They pose

[1]Presented at the AAG, as The Fleming Lecture in Transportation Geography, Fort Worth, TX, 3 April 1997.

a very rich set of fundamental questions for geographers, and some of these will be reviewed in this paper. For geographers interested in spatial organization, hubs are a challenge, and yet they offer reassurance that there is something keenly interesting about the way in which spatial interaction is organized.

O'Kelly and Miller (1994) devised a classification scheme which places the hub-and-spoke network into one of eight classes, called Protocols A, B, …, H. These cases are formed by answering the following three questions: are the hubs fully interconnected; are connections allowed to several of the hubs from each node; and are direct node-to-node flows allowed to by-pass the hubs in the network? Recently, advances have taken place in many aspects of hub-and-spoke research. These advances allow a more comprehensive analysis of the distinguishing features of hub systems than was possible in the early 1990s. I refer the interested reader to the recent special issue of *Location Science* (October 1996) which contains research articles on hub-and-spoke networks. One aim of this paper is to elaborate on, and to extend, the classification by O'Kelly and Miller (1994). It is interesting to consider the functional characteristics of the various sectoral applications of hub-and-spoke systems, with a view to determining the appropriate network design: in other words, are there certain types of hub networks that are better suited to specific modes? To make this question more concrete: would we expect the same hub-and-spoke analysis to be equally applicable to air passenger, air freight, telecommunications, and trucking? Of course not; but in making that emphatic statement we are calling on intuitive insights that we, as geographers, have about the nature of these various networks, and in this paper, I hope to explicate some of the bases for our collective intuition.

In this paper, the intent is not to review algorithmic and mathematical issues in hub-and-spoke network design, but rather to focus on inherently geographical questions that arise in the course of understanding and modeling hub location. I propose to examine the air passenger and the air freight sectors, with specific reference to the features of these systems that lead us to expect different hub-and-spoke outcomes. The particular features of these systems which I want to review are: location, routing, rates, and agglomeration. The remainder of the paper is organized as follows: in the next section, the features of the hub-and-spoke system that are interesting from a basic economic geography viewpoint are surveyed; then, examples of air passenger and air freight networks are given, with special emphasis on the building blocks of spatial organization (i.e. location of nodes, routes and rates along linkages, agglomeration, and patterns of hinterlands and hierarchies). Finally, the implications of different routing models are explained in the case of a sample network. The paper ends with a brief concluding section.

Features of the hub-and-spoke system that make it difficult to model

Hubs pose some basic theoretical challenges for the analyst because they require a re-thinking of fundamental assumptions about location, routing, transportation rates, and agglomeration. The following four subsections give a few examples of the contrasts between hubs and the conventional locational and network analysis in terms of the main elements of spatial organization (nodes, links, hinterlands, and hierarchies).

Location of nodes

First, let us add to the O'Kelly and Miller (1994) classification scheme a dimension that was previously used in mainstream locational analysis, but whose application to the hub-and-spoke problem has been overlooked until now: distinguish between the concept of delivery systems, and user attracting systems. In *delivery systems*, the same decision maker positions the facilities and determines the rules of allocation to the centers. A classic example would be a fire department that decides where to place fire stations and also has a predetermined emergency response plan. In such models it makes sense for the entire problem to be treated as a unified single goal optimization task. In *user attracting* systems, in contrast, the facility is located by one agent, but the allocation decisions are decentralized, and the planner has to make some reasonable guesses as to how the public will make use of the facilities. How does this distinction help with the location of hubs in network analysis? In the air freight hub-and-spoke case, we are certainly considering a delivery system, because the operator decides where to place the sorting centers, and has complete control over the rules for routing packages between these centers. Of course the service must attract demand, but it is fair to assume that the attraction of the service for the end user is not a function of the routing of the packages. (Ultimately, the cost of operating the system, which is a function of location, will translate into the price charged for the service, and this will affect the demand for the air express delivery system.) Since air freight systems may organize themselves without worrying about the travel itinerary of the individual packages, they may exploit the hub-and-spoke method to a great degree, and therefore, many express package delivery companies have similar locational incentives and imperatives. This observation goes a long way to explaining the clustering of air freight hubs in the same locations: we have a relatively dense cluster of hubs in the Ohio Valley, in an area that is demonstrably a low cost solution to the hub-and-spoke system design.

For air passenger traffic, on the other hand, the role of consumer behavior and the inconvenience of making intermediate stops cannot be ignored. In the air passenger system, the location of hubs and the routing of aircraft is under the control of a single decision maker (see Wollmer, 1990; Daskin and Panayotopoulos, 1989), but the impact of these decisions on air passenger acceptance of the carrier and the level of service/demand must be considered. The critical difference between the air passenger system (a user attracting system) and the delivery systems mentioned above, is this tension between price, routing and demand. Weidner (1995) has stated that the more efficient the network from the operator's point of view, the less convenient that network would be for passengers. Thus there is great scope in the air passenger system for competition, and the need for air passenger systems to locate their hubs with a view to much more than aggregate system travel time: they must also worry about aggregate demand, and the level of diversion of demand to other systems from key market pairs if the service falls below an acceptable level (see related research by Dobson and Lederer, 1993). While certain origin–destination (O–D) pairs are adequately accommodated by the hub-and-spoke system, others are not, and this provides the opportunity for competitive services to emerge. A key ingredient in a competitor's ability to offer a preferable alternative level of service is the position of *that* carrier's hubs vis-à-vis the particular market. Consider three regions: if the origin/destination pairs from region 1 to region 2 and from 2 to 3 are well served, but pairs from region 1 to 3 are poorly served, then the possibility of an inter-regional carrier to take the 'slack' arises.

Linkages: routes and rates

Routes. In many conventional logistics problems, we expect to see the *shortest path* emerge as an ideal candidate for shipments between origins and destinations. In hub-and-spoke systems, however, determination of the optimal routing for any particular origin–destination pair is a complex question, which is sensitive to the *allowable* connections between nodes and hubs. If we assume that spoke cities are allowed to connect to all the hubs (a situation seen in some air passenger networks), then it is reasonable to assume that the flow from an origin to a destination will proceed in a very logical fashion: go from the origin city to the hub which gives the best connection service to the destination city, and then complete the transaction by proceeding to the destination. In this simple case, the hub-and-spoke system operates as a series of shortest path problems. But there are many ways in which this intuitive first guess about routing might break down: (1) the network may be organized so as to economize on the number of spoke-to-hub connections, so that the origin city cannot simply select its choice of hub; (2) there may be capacity limitations on the preferred shortest route, forcing some interactions to be re-routed; and (3) the network may operate under some economies of scale, density or scope, giving incentives to the operator to route the flows away from their myopically preferred alternative. In *delivery systems*, network planners may unquestionably direct flows along paths that are optimal for the system, with the lowest cost for the entire network being achieved by indirect routing and the amalgamation of flows (O'Kelly and Bryan, 1997). The best level of service in a delivery system, therefore, does not necessarily require shortest paths between origins and destinations, as traffic is deliberately detoured to form bulk shipments. In a *user attracting system* the network planner would like to have a highly efficient flow system, but users would resent the indirect routes, opening the possibility for other carriers to fill gaps in the market (see the discussion of Southwest Airlines below). In the case of air passenger transportation systems, the ability of the carriers to exploit the hub-and-spoke concept is tempered by the need to recognize the impact of the network design on revenue as well as costs.

Rates. A major element of the hub-and-spoke network design is the presence of inter-hub connectors. A key question is the comparative cost of routing flow, indirectly, over these connectors. In models of network flow, the analysis of so-called 'system optimal' or 'user equilibrium' networks is typically driven by a cost function that is an *increasing* function of the congestion on a link. In hub networks, we actually want to encourage controlled congestion, as greater load factors lower per unit costs for aggregate interactions. There are many sources of economies of scale in hub networks, ranging from the efficient bundling of all flows from a source onto a single flight to a hub, to centralized ground support and maintenance shops at a location where the majority of an airline's equipment can be expected to make a stop. The development of high capacity connectors between hubs has been thought of as the major source of scale effects. To continue the varied examples: in air passenger and air freight systems the inter-hub flights could be serviced with larger aircraft, and in communications, the backbone connections between major hubs are often T3 lines or better.

Hinterlands and service regions

In the typical location model, such as the median location problem and the Weber problem, it is usually safe to assume that service will be provided to the user from the nearest facility. For example, in a system of distribution centers, warehouses are expected to have non-overlapping service areas or 'hinterlands', whereby the goods are shipped to cities from the closest available source. In turn, this fact can be exploited to great effect in the design of solution methods for typical location models. Somewhat surprisingly, this *nearest center* property is not appropriate for the service areas of hubs! An immediate implication is that the non-overlapping hinterland is *not* an expected property of hub service regions. In the case of hub networks, we seek the facility locations that provide the best means of completing a number of transactions between the many origins and their desired destinations. A few moments reflection on the nature of the trip from origin to destination will convince us that the 'nearest is best' rule does not apply in a straightforward way to hubs. Examples of *origins* include an automatic teller machine (ATM); a point of embarkation on a journey or flight; or an office manager sending off a shipment by FedEx. These interactions are completed when an account is credited or debited at a bank, a passenger reaches their destination city, or when the package reaches its intended recipient. The entity demanding service is therefore an *interacting pair*, and so hub placement should be made not just to minimize distance from an origin to a hub, but rather it should be done in a way that facilitates interactions between pairs of places (see also Ray *et al*, 1989).

In locational terms, then, it makes no sense to place hubs to facilitate these interactions with regard to just the origin or the destination: we must consider the routing of the transaction via one or more intermediate sorting centers. This observation is especially true for time sensitive delivery services. In such cases, the length of the trip from home to hub to finish point, within an available time window, is a critical factor in judging a system's feasibility. In each of the three examples, the central hub function is performed by a facility: in the case of the ATM customer transaction the hub is a *bank clearing house*; in the case of the air passenger, the hub is an *airport*; and in the case of the express package, the hub is a *sorting center*.

The key to contrasting hub-and-spoke location models with simple location-allocation models, is that while in basic location theory, the unit of demand is represented by a point, line, or area, the entity demanding service in a hub-and-spoke system is a *pair* of places. At one end of the ordered pair is the origin of the transaction and the second half of the transaction is to be completed at a destination.

Hierarchy

Finally, in this brief survey of the ingredients of the spatial organization of hub-and-spoke networks, we come to the topic of hierarchy. The hub-and-spoke system is at least a two-level system and is often analyzed as such in the telecommunications literature (Balakrishnan *et al*, 1995; Gavish, 1992). In the air transport sector, Chou (1990) and Kuby and Gray (1993) have discussed examples of systems which allow hierarchical levels of service to emerge naturally from the extent of connection provided at a node. Less directly, hierarchies are evident in the way that some nodes and hubs become the focal points for agglomerating economic activity.

Another way to see the role of agglomeration as being different in the case of the hub-and-spoke system is to observe that the conventional definition of agglomeration economies reflects the reduction in the costs of operation (or increased revenues) that accrue to a business as a result of its location near to other firms in similar or dissimilar industries. If the industries are *similar* we call these localization economies, and if the industries are *dissimilar* we call them urbanization economies. In the case of the savings generated by a hub-and-spoke system, the classification is a little bit more cloudy: in freight, many firms from a wide range of industries can benefit from location near to a hub, largely from cost savings. Certainly the ability to manage the supply chain is a key to cost reduction, but, at the same time, top quality customer service ultimately leads to greater demand and revenue.

In the air passenger sector, the kinds of agglomeration effects that accrue to a hub are related to conferences, meetings, and convention business. A hub increases a city's ability to attract conventions and business meetings. Also the opportunity to call a one day fly-in business meeting is enhanced if the meeting is located at a hub airport.

In the next section, I would like to focus on the issues which arise in selected applications of hub-and-spoke networks. The generic concern is for the provision of transportation infrastructure to facilitate interaction between places.

Air passenger examples

Some of the most interesting recent research from a geographical perspective has been concerned with the effects of 'fortress' hubs on the domination of local air markets, and the implication of these hubs for passenger fares. Consider the domination of certain nodes by major airlines. The following example was discussed recently in Zhang (1996) and Borenstein (1992): Atlanta is Delta's major hub; United Airlines offers a non-stop service to Atlanta only from its *major hubs* in Denver, Chicago O'Hare, and Washington Dulles. The point here is that if a city becomes a hub for a major carrier, it makes it difficult for another non-hub carrier to match the level of service from that hub, and essentially removes this city from consideration for entry as a hub by another airline. This argument is basically sound, but has been subjected to intense analysis by Hansen (1988) and Hansen (1990). He determined that: (a) there are a few selected hub cities; and (b) the domination of a hub market by one (or a small number) of competitors is the outcome of a competitive game played between major airlines. There has been a lively debate on the economics of hubbing: for example, Hansen and Kanafani (1989) were unable to find significant cost savings due to hub-and-spoke operations, and so fell back on 'structural' reasons for airlines to operate hubs; in other words, the ability to control the local traffic from a city and the possibility of creating local monopolistic conditions are attractive. The pre-emption of entry by an established hub operation is so intense that the ability of the non-hub carrier to compete for the demand that is serviced by that hub is weakened. Tretheway and Oum (1992), on the other hand, confirm the existence of economies of traffic density (declining cost per passenger as the number of passengers per station increases). Evidence for economies of scale in the airline industry is reviewed by Antoniou (1991).

Other practices which enable the successful operation of a hub-and-spoke system include the ability to offer many connections, over the course of a day, to passengers routed through various intermediate stops/hubs. The fact that aircraft are full, or close to full, with a blend of passengers with various *elasticities of demand* means that the airline can engage in very sophisticated demand management and pricing schemes, effectively micro-managing the yield from the contents of the flight based on the passengers' ability and willingness to pay.

To make the examples a bit more concrete, consider the case of United Airlines, with hubs in Denver (DEN), Chicago O'Hare (ORD) and Washington Dulles (IAD). A passenger flying from Columbus (CMH) to Boston (BOS) on United has route choices which include transit stops at Dulles or ORD. A passenger on United flying to San Diego (SAN) from Columbus would have route choices (depending on the time of day) via DEN or ORD. This observation is consistent with the multiple allocation model, which will be illustrated further in the final section. Notice that the same pair of trips (CMH to BOS and CMH to SAN) on American Airlines would also use hub stops: in this case the most likely routing on American to BOS is through O'Hare or JFK, but the most likely routing to SAN is either through O'Hare or Dallas/Fort Worth. The point here is that for both airlines the connection of their spoke cities (in this case CMH) to a variety of hubs allows the carrier to offer multiple alternative times, with reasonably convenient trips for an air market pair that, on its own, would hardly suffice to support direct service. The interesting side bar to this is that the CMH to BOS market is served by one carrier for which this is a direct connection: America West offers three direct flights, departing at 7:10 a.m., 9:30 a.m., and 1:40 p.m., and their fare for the 27–30 March 1997 round trip, if booked by 25 March, was quoted as $458. Not bad considering that the cheapest indirect flight on United is around $844, although that is a 'no penalty' fare. This example clearly shows the *connectivity* advantage that a city has if it is a hub for an airline (as CMH is for America West), compared to networks in which it is only a spoke city (United and American Airlines). The implications in terms of price depend on the degree of domination of the market, with higher fares from cities which are fortress hubs for a major carrier (see, for example, Borenstein, 1989).

If the plane flying into a hub from a spoke city is filled with passengers with a variety of destinations who will fly onwards from the hub, how should the airline price or cost the spoke-to-hub service? Should they take into account the fact that, by delivering a group of passengers to the hub, and providing an enhanced load factor from flights out of the hub to many destinations, the outbound flights will have improved profitability? There are obviously more questions than can be answered in one research paper: they require a full equilibrium analysis beyond the scope of this particular work. The interested reader is referred to recent theoretical analysis of networks by, among others, Brown (1991), Brueckner *et al* (1992), Skorin-Kapov (1995), and Sorenson (1990).

Air passenger networks: linkages

Recently, a fascinating advertisement appeared in the *New York Times* for Northwest Air (*New York Times*, Tuesday 4 February 1997): it uses the term 'hubs and spokes' in bold lettering and shows a stylized diagram of the Northwest hub-and-spoke system. The interesting thing about the stylized diagram is that the hubs are shown as a set of interlocking cogged wheels, thereby emphasizing the connection between hubs, while de-emphasizing the need for spokes to reach hubs and the need for changes between aircraft at the hub locations. It is as if the airline wants to create an illusion that the hub-and-spoke system requires very few transfers and connections. Yet connectivity is what hubs are all about, as is painfully obvious as soon as this connectivity breaks down. On a recent stormy Friday (18 February 1997) in the Southeastern US, the air traffic system was placed under a lot of stress (not to mention the passengers)! The delay to inbound flights to the US Airways hub in Charlotte reverberated through the late afternoon and early evening as flights were canceled or delayed beyond the departure time of the next flight from the gate. Equipment could not be moved to gates because of the backlog. The information boards in the departure area could not keep track of all the delays. Attendants reassured passengers that, although the gate might say Rochester (NY), the flight was actually going to Cleveland; and, of course, the nature of the way hub-and-spoke systems bundle flows ensured that as soon as a single flight is delayed or canceled, passengers scramble to get bookings on other flights, which, in turn, are already packed with other delayed passengers.

As two contrasting examples, consider US Airways and Southwest Airlines. The former is heavily invested in hub-and-spoke systems, while the latter is noted for serving selected high demand city pairs with point-to-point service. US Airways in early 1997 has over 500 daily departures (including Express) from Pittsburgh, PA, and almost as many (481 including Express) from the Charlotte hub that was mentioned above. A hub-and-spoke system with this degree of concentration has implications for the fare structure out of the dominated city (Kling *et al*, 1991; Brueckner *et al*, 1992) and results in an intense complex array of interchanges between gates during successive waves of flights peaking in the evening rush hour. Many hub airports have the kind of bustling big city crowds that are not seen at comparable densities anywhere else during the day in many US cities. Certainly this kind of intense crowd creates a very good market for business, and several modern hub facilities including Denver and Charlotte have plaza-like food courts which rival major shopping centers for selection and quality (though not price!). It is these kinds of economic realities that give hub-and-spoke networks a very mixed set of costs and benefits for late 20th century passengers.

While the hub-and-spoke system has become a much studied feature of the air transport system and is unlikely to be dismantled any time soon, there *are* alternative means of connection. There are many markets that have sufficient demand to warrant frequent service, and the emerging new technology of smaller jets give airlines the flexibility to service these markets. The best example of a point-to-point service is the case of Southwest Airlines. Quoting from the Southwest homepage fact sheet:

> The airline began service June 18, 1971 with flights to Houston, Dallas, and San Antonio. Southwest has grown to become the fifth largest U.S. airline (in terms of domestic Customers carried). Southwest has 23 years of consistent profitability, and was the only major carrier in 1990, 1991, and 1992 to make net and operating profits. Southwest became a major airline in 1989 when it exceeded the billion dollar revenue mark. Southwest is the U.S.'s only major shorthaul, low-fare, high-frequency, point-to-point carrier.

Southwest has earned the 'Triple Crown' for its 1992, 1993, 1994, and 1995 performance. This consists of best baggage handling, fewest customer complaints, and best on-time performance, according to statistics published in Department of Transportation (DOT) Air Travel Consumer Reports.

The contrast between point-to-point and hub-and-spoke systems is stark. The fact that there are successful examples of each kind of system indicates that there are multiple solutions to the design of air passenger networks. Point-to-point systems have an advantage in short-haul market pairs with a dense level of demand, and work well provided the carrier does not try to offer service between all pairs of places. Hub-and-spoke systems in contrast seem to be the ideal solution (from the carrier's viewpoint) whenever the flows can be channeled through convenient switching points (such as Chicago or Atlanta). The hub-and-spoke system leaves potential markets for others to fill in, especially when a region has a low level of service from a hub carrier.

Freight

As much as air passengers love to hate the hub-and-spoke system, the emotional reaction of air packages can be safely ignored. The air freight system is free to make extensive use of vast sorting centers, strategically located to make the inter-regional transfer of materials as efficient and as time sensitive as possible. Although an air passenger has to worry about making a connection by running quickly to a gate, packages on the other hand are easily sorted and bustled through miles of conveyor belts in the middle of the night.

There is a continuum in terms of time sensitivity for express packages. In the case of canceled checks, time is literally money, and a very clever hub-and-spoke system based in Columbus (US Check) operates a high tech delivery system to make sure that items reach clearing banks by very exacting deadlines. This is fine-tuned to such a degree that the company can justify expenditure on specific routes, with specific high speed and level of service. A recent company newsletter mentions an example of tailoring service to deadlines; quoting from their newsletter: 'We believe that there may be considerable interest in the Utica 07:30 deadline and, given enough interest (read "money"!) we may be able to create a Lear routing to satisfy that deadline.'

Of course not all air express has the same degree of time sensitivity. Packages and plans can be delivered by the next business day, or the second business day. Apart from the marketing advantages that emerge from segmenting the service, the express package delivery system has the ability to use cheaper modes for connections that are less time sensitive. Trucking packages to a sorting center is always going to be cheaper than flying them there, provided that the truck connection can reach the hub in good time. A recent paper by O'Kelly (1998) builds on the earlier analysis by O'Kelly and Lao (1991) to determine the optimal hub connection strategy. What is the best way to connect cities? The trade-off between the need to sort and the need to ship in economical bundles is a key to understanding the hub-and-spoke arrangement. The conflicts are between sort times, volumes, and time sensitivity.

Stockpiles of important parts, supplies, and medicines may be held at hub locations in preparation for rapid distribution to the spokes. Specific examples include Frankfurt (Germany) which is a major military hub. Emery, with a domestic superhub in Dayton (Ohio), is an example of a company that provides integrated logistical services to their customers. For example, quoting from their homepage:

> Emery's logistics unit provides customers an international base of multi-modal transportation and logistics services to

help them streamline operations and improve customer service. Some of the coordinated single-source services provided by Emery Global Logistics include scheduled or chartered air, ocean, and truck transport; shipment monitoring and expediting; warehousing, inventory management and order fulfillment; customs clearance; management reports; and single-source invoicing. (3/24/97)

Emery Global Logistics manages 'over 20 logistics centers around the world, including multi-user regional distribution facilities in strategic locations,' including Dayton, Singapore, and Brussels.

According to information obtained from the homepage of Emery (3/24/97), companies served by Emery's logistics unit include General Motors — India; 3COM; Motorola Land Mobile Products Sector; Dade International, a provider of *in vitro* diagnostic products; Flexlink; IOMEGA; Liebert; Compaq; GTE/Codetel; Durametal; and Dennison Hydraulics.

Numerical analysis of different allocation systems

Much of the preceding part of the paper has dealt with general contrasts between air freight and air passenger systems. Now, the analysis proceeds with some numerical examples. The basic premise is that the air passenger side can be best represented as a multiple allocation hub model, where each origin and destination is potentially linked to a variety of hub cities, and therefore the passenger has an option to pick a convenient hub through which to make a transfer. In other words, it is possible for the passenger to go from origin to destination through hub 'A' or 'B' depending on the relative locations of the destinations. This arrangement of course invites the possibility that a niche carrier (such as Southwest Airlines) can profitably operate in competition with hub-and-spoke airlines, and indeed that the hub-and-spoke airline will choose to service some O–D pairs directly (see Aykin, 1995b for further discussion). The identification of the multiple allocation hub model with passenger air networks is therefore a simplification of reality.

For a new entrant to the air passenger network, the problem would be to optimize the service provided to passengers, by identifying those links where interaction is poorly served. The difficult part of this problem that needs further research is to measure the sensitivity of demand to the level of service provided by the network. A more partial approach, which looks at some of the problem, but not all, is to ask how the allocation of the nodes to hubs responds to the 'rules' and to the location of major centers. For air freight, on the other hand, the opportunity to maximize load factors, regardless of routing, gives the carrier every incentive to divert flow towards major hubs, and to make allocation decisions depend on the ability of flows to capture scale economies.

The general idea in this part of the paper is to show that the construction of a pure single hub allocation model would result in an efficient system, but one with great inconvenience for passengers. In contrast, the multiple hub allocation model has better convenience, but dilutes the inter-hub flows to such an extent that the original premise for hubbing (scale economies in the hub-to-hub links) is greatly weakened. Finally, a newer model, one where flows are assigned a cost on the basis of scale economies, which must be earned, is illustrated, and is suggested as a good prototype for freight networks, and perhaps for future policy related studies.

These points of contrast are illustrated for a large hypothetical case: a 100 node interaction system representing sample flows in 1970 between major US cities. Approximately 16.5 million origin to destination interactions are represented in this data set. The older data have been used extensively in other spatial interaction model exercises and so are used to provide continuity in the examples, and to emphasize the academic as opposed to the applied aspects of the methods and analysis. There is nothing to stop the analysis being re-created with company data if these methods were to be used for an actual case study.

The purpose of this exercise is simply to examine logical applications of flow models to some reasonable sets of hub locations. A heuristic method called *tabu search* was used to generate five sets of hub locations, and each set has four selected hubs (Skorin-Kapov and Skorin-Kapov, 1994). These sets of hub locations will be used as the basis for calculating the performance of all three models. The sets of hub locations used are listed in *Table 1*. Then, using the locational patterns shown in *Table 1*, I solve three different versions of the hub allocation model. The first uses a SINGLE hub allocation rule, the second uses a MULTIPLE hub allocation rule, and the third, called FLOWLOC, uses a method that gives a discount for larger flow.

In solving the model in three different ways, each version is designed to produce a network that is a stylized example of a different kind of case: [A] *Single hub allocation* which economizes on node-to-hub links and ignores inconvenience of routing; [B] *Multiple hub allocation* which is similar to the assumption that matches the passenger system and where the travelers select their own preferred path. While the flows may not actually warrant a discount, I am going to assess the costs of the inter-hub links at a lower level, presumably because lower cost, high capacity equipment can be used there. [C] The final case is a *flow*

Table 1 Sets of hub locations used

		Hub locations			
Hub set	**α**	**Eastern**	**Midwestern**	**Southern**	**Western**
1	0.1	NYC	Chicago	Birmingham	Los Angeles
2	0.2, 0.3	NYC	Chicago	Birmingham	Las Vegas
3	0.4, 0.6	Allentown	Chicago	Birmingham	Las Vegas
4	0.8	Harrisburg	Cincinnati	Tulsa	Las Vegas
5	1.0	Pittsburgh	Chicago	Tulsa	Las Vegas

responsive model which gives a smaller per unit cost to larger flows.

The aim of this portion of the paper is to examine some readily comprehensible features of the hub-and-spoke system, as illustrated by some idealized examples. The key points of comparison are: allowable connections and routing (allocation rules); inter-hub flow volume; complexity of the analytical steps needed; and last, but not least, the geographer's interpretation of the results, pointing to appropriate examples where the ideal or mathematically optimal solution (while fitting within the constraints and parameters of the system) might be especially difficult to rationalize or understand. The analysis is presented without any mathematical formulae, but for each of the major types of analysis that are presented, reference is made to recent formulations, where a more or less standard set of terms, variables, and equations have been formulated in a common mathematical model (see, for example, Campbell, 1994, 1996). The FLOWLOC formulation is detailed in O'Kelly and Bryan (1997). Space does not permit a complete formulation here, but one important feature of the model should be mentioned: the sample data set used has symmetric flows between places, and this symmetry is exploited to keep the number of variables in the model as small as possible. Obviously, network design is sensitive to changes in the structure of the flows. In some cases, such as air freight, the data would be very likely asymmetric. The formulation is quite general, however, and is not dependent on the assumption of symmetry.

Throughout the following discussion, I will refer to an inter-hub discount factor 'α', which can be between zero and one. The idea of this effect is to apply a lower transportation rate to the inter-hub links than to the spokes. If α is large (i.e. close to 1), then the inter-hub links do not enjoy any special discount, and they have the same rate as spokes; in this case the calculations basically assess the costs of interaction between hubs at the same rate as those between spokes and hubs. On the other hand, a value of α approaching zero is the other extreme case and assumes that the inter-hub portion of the flow costs are deeply discounted. The role of α as a representation of the inter-hub discount factor has been discussed in detail in many of the operational models in the literature (O'Kelly *et al*, 1996). Recent interest has also focused on the creation of a cost per unit of flow that is more closely representative of how much flow is actually allocated to the links. The problem with the α method for inter-hub costs is that it applies to all flows between hubs regardless of the volume. In fact this discount would apply to *any* inter-hub flow, and does not realistically require an amalgamation or bundling of flows. In the FLOWLOC model, the inter-hub discount is determined by the amount of flow actually utilizing the inter-hub links and is determined separately for each inter-hub link. In this way, agglomeration of flows on the inter-hub links is explicitly encouraged since the discount increases as the amount of flow traveling across the inter-hub link increases.

Single allocation

Allowable connections and routing (allocation rules) in the case of the single allocation model are limited to one direct link from each origin to a single hub. The result is a very economical network in terms of connections. In the case of the 100 node problem with four hubs, there are 96 non-hub cities connected directly to one hub each, plus of course the fully interconnected inter-hub links. The level of inter-hub volume is expected to be high in the case of this network, because all the nodal flow is channeled to the hubs and they have to make use of the inter-hub links to get from one regional system to another. An example for the 100 node problem is shown in *Figure 1* which illustrates the connections from origins to hubs in Allentown, Chicago, Las Vegas, and Birmingham; this happens to be the best known (from the tabu heuristic) location pattern for $\alpha = 0.6$. The inter-hub flow volumes created by assigning all the flow to the appropriate shortest path between each origin and destination are shown by line width. Examination of the map reveals several two stop trips, and a lot of inconvenient links: for example Seattle, Portland and Spokane are allocated to a hub in Las Vegas and must interact with each other via that hub. The inter-hub link between the East coast and the Southeast (in the example, Allentown to Birmingham) is very heavily loaded, as is the main street connection from Allentown to Chicago. This network has some especially easy to calculate properties, once the nodal allocations are determined, as all the origin to destination flows are connected by an obvious shortest path. The routing from each place to every destination is a *spanning tree* routed at the origin. An example for Dallas to some selected cities in the Midwest and East is shown in *Figure 2*. For a given location and connectivity pattern, the routes are fixed. If the inter-hub discount applies equally to all inter-hub links, then the routing stays the same even if the discount is varied. The location and allocation stages of this model are difficult to solve, however, and a powerful optimization program (CPLEX) was used to confirm the optimality of the allocation variables. *Table 2* reports the results of the calculation of the network cost for each of the five hub sets, for seven different values of the inter-hub discount. Reading down any column of this table shows the values of α for which this is the 'winning' set of hub locations in boldface type. For example, hub set three turns out to be the low cost alternative *if* α is 0.4 or 0.6. Reading across the rows, we see that for a fixed level of α, the five different hub sets have costs which vary quite dramatically: thus if $\alpha = 0.2$, we could operate hub set 2 for 666.98 but operating hub set 5 with the same

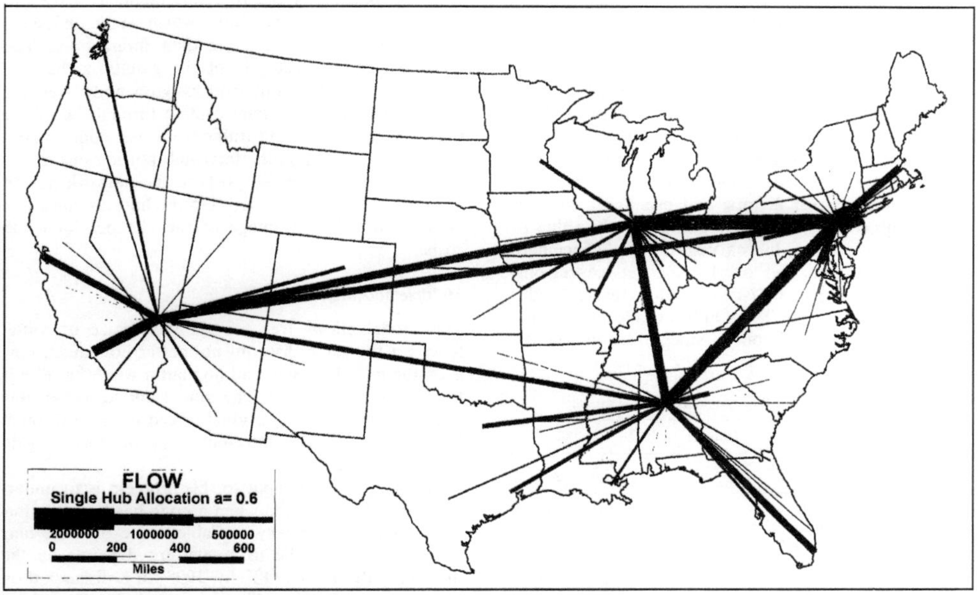

Figure 1 Single hub allocation model, with $\alpha = 0.6$. Lines are proportional to flow.

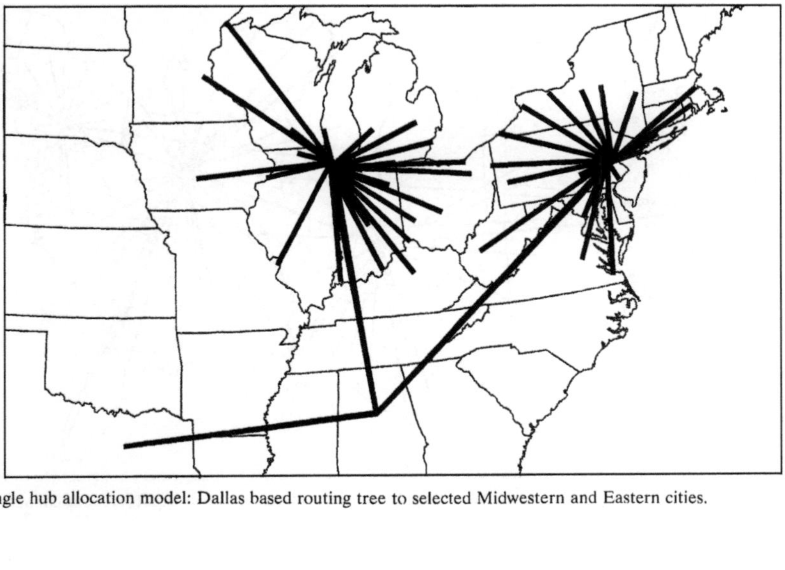

Figure 2 Single hub allocation model: Dallas based routing tree to selected Midwestern and Eastern cities.

discount factor would cost almost 100 units more (763.95). Previous research (O'Kelly, 1987) has shown that, as α approaches zero, the model effectively acts as a set of medians (the inter-hub costs are washed out completely when $\alpha = 0$). In this case the nearest center allocation is recovered, and the hinterlands of the hubs are non-overlapping as shown in the example in *Figure 1*. The assigned regional service areas for another hub location set are shown in *Figure 3*. As we move to higher inter-hub cost factors, and examine a hub set (this is hub set 4, which was optimal for the value of $\alpha = 0.8$) with locations in Las Vegas, Tulsa, Cincinnati, and Harrisburg, we get overlapping service regions. Notice, for example, in *Figure 3*, that several nodes which are close to the Tulsa hub are allocated to the hub in Cincinnati. Without a steep discount on the inter-hub portion of the trip, this model is forced to assign the nodes to the hubs which make the most sense in terms of completing their interactions. The geographical interpretation of the results is that the model has to decide where to assign a node; this is a delicate balancing problem: routing through hub A or B effectively commits all traffic from that node to go a particular path. When the inter-hub portion of the trip costs less than the spoke portion, the network has an incentive to connect the nodes to the hubs as quickly as possible, to take advantage of the cheaper inter-hub costs.

Table 2 Average cost per unit flow: single assignment model

	Hub set				
α	**1**	**2**	**3**	**4**	**5**
1.0	1223.54	1197.29	1181.08	1157.05	**1153.66**
0.8	1084.83	1066.56	1055.68	**1049.33**	1057.54
0.6	946.11	933.39	**927.85**	941.39	960.55
0.4	807.40	800.22	**800.03**	833.01	862.60
0.3	738.04	**733.63**	736.11	778.68	813.40
0.2	668.62	**666.98**	672.20	723.94	763.95
0.1	**597.89**	600.26	608.25	669.00	714.38

Multiple allocation

This model allows free play in the choice of route: flows get the same discount no matter how much flow is on the inter-hub link, and no matter who/what else is flowing on that linkage. The model mimics reasonably well the way that an individual forced to route through hubs would go. The inter-hub discount factor again plays a key role.

The geographer's input to this problem is to understand the fact that even when a node has several alternative routes technically available, some of these may not be used, and the usage pattern depends on the discount pattern (see O'Kelly, 1998 for a discussion on hub allocation). When there is no discount, and we allow nodes to connect to all hubs, they will of course

Figure 3 Single hub allocation model, with $\alpha = 0.8$. Allocations to four hubs illustrate non-nearest center asssignment.

do so, yielding a comparatively large number of links (96 non-hub nodes connected to all four hubs, plus the inter-hub connection). Of course, since almost every flow has an opportunity to by-pass the inter-hub links, there are very small flows on the inter-hub connections, and the incentive to build the hubs in the first place as a conduit of flows onto dense inter-hub corridors is lost (see Aykin, 1995a,b). As soon as an incentive to route flows between hubs is introduced, the grip of the triangle inequality is loosened, and some trips will be routed via the seemingly longer path from i to k to m to j, using the longer (but cheaper) k to m portion of the network. The analysis can still use an all or nothing assignment rule, which operates to load up the flow on the shortest path between each origin and destination. Those links with zero flow, or an uneconomically small flow, can be eliminated. An example of four hubs serving 100 places is shown in *Figure 4*. In drawing this figure, an inter-hub discount factor of $\alpha = 0.6$ was used.

The results of optimization for the five hub sets, with the seven different values of inter-hub discount are shown in *Table 3*. Generally, using the multiple hub allocation models, only hub sets 1, 2 and 3 need to be considered because they always beat hub sets 4 and 5.

The multiple allocation model also produces a different routing tree, and a different hinterland diagram from that seen in the single assignment model. In *Figure 5*, for example, consider the Dallas connections in contrast to the single allocation example shown previously. It is now clear that Dallas makes use of two connections to hubs: Chicago and Birmingham. Interestingly, it does not have any direct connection to the third hub in Allentown. Because of the wide variety of allowable connections from each origin, the inter-hub linkages are relatively underused, and in the particular case of the multiple hub system shown in *Figure 4*, no inter-hub link has more than 500 000 units of flow. Finally, the extent of the overlap between service areas can be seen, when the links to each hub are colored separately: thus we see in *Figure 6* that the regional service areas around each major hub overlap.

Flow related costs

An example that matches the freight case is explored next: this is the case where the nodes are allowed to link to as many hubs as make sense, but the costs are assessed on a sliding scale that gives an incentive to bundle materials together on high volume links. I use four different cost functions, with the five different locational schemes, which results in 20 runs (*Table 4*). The five different ranges in each cost function are:

(1) 100–250 000 units of flow;
(2) 250 000–500 000;
(3) 500 000–750 000;
(4) 750 000–2 000 000;
(5) > 2 000 000.

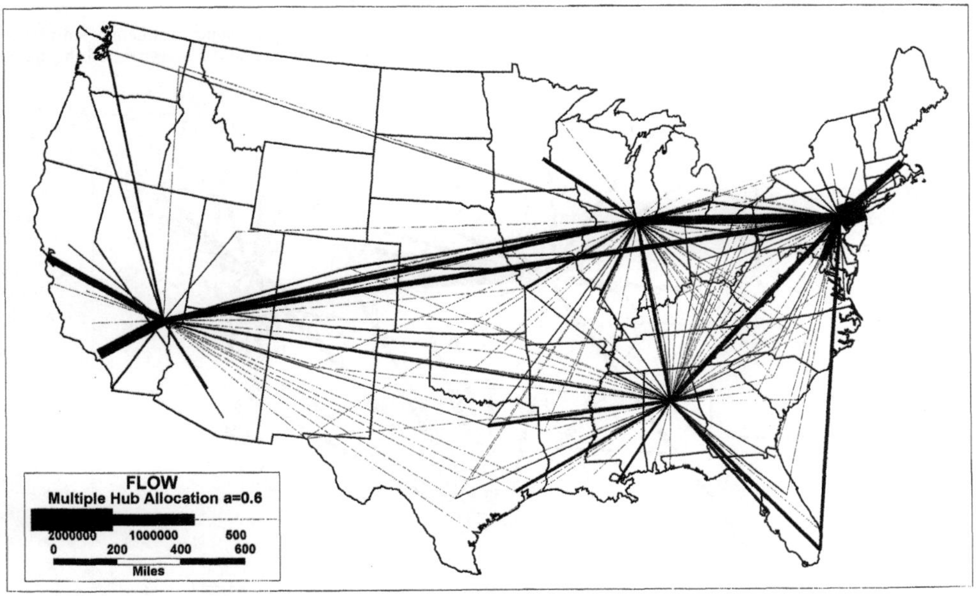

Figure 4 Multiple hub allocation model, with $\alpha = 0.6$. Lines are proportional to flow.

So if the total inter-hub flow on a certain link is 100000 units, it will receive the discount associated with that flow — the smallest allowable. If another inter-hub link carries 2.5 million units of flow, it will be at the fifth level and will receive the largest available discount.

The results of this analysis show that out of the five hub sets considered, the optimal one is hub set 3. The level of inter-hub flows on this linkage pattern is quite heavy compared to the multiple allocation model. The results of the allocation of the flows using the economies of scale approach are shown in *Figure 7*. Note now that the fact that flows have an incentive to bundle together makes the routes emphasize the Allentown-to-Chicago and Chicago-to-Las Vegas connections. In order for these flows to reach a size sufficient to ensure that the discount is earned, there inevitably has to be some re-routing of flows away from their myopically preferred shortest path. The best way to see this is to consider the Dallas based tree, in contrast to the example seen above for the multiple allocation models (refer now to *Figure 8* and compare it to *Figure 5*). In the case shown in *Figure 8*, the routes are very similar to the multiple allocation model, with two significant differences. First, Dallas to Washington, DC in FLOWLOC is routed via Birmingham only, but in the multiple allocation model the flow is routed via Birmingham and Allentown. The reason for this difference is that in the FLOWLOC model, the Birmingham to Allentown inter-hub link is rarely utilized and so no major discount is earned. In this case then it makes no sense to route the Dallas to Washington, DC flow over an undiscounted inter-hub link. This exemplifies the flaw in the multiple allocation model: the Birmingham to Allentown inter-hub link gets a cheaper rate simply by being designated an inter-hub connection, and does not have to earn this lower cost by virtue of its combined flow. The second point of contrast is much more significant in terms of the number of routes affected: while the Eastern cities would prefer to be routed via the Birmingham to Allentown inter-hub link (as shown in the multiple allocation solution), notice that in the FLOWLOC model they are forced to utilize the Chicago–Allentown inter-hub link, even though it is more expensive for the individual O–D pairs. These flows are deliberately routed over a longer path so that they may contribute to the bundle of flows between Allentown and Chicago, a bundle that eventually ends up paying substantially less per unit of flow as a result of this routing.

Table 3 Air passenger: multiple assignment model

	Hub set				
α	1	2	3	4	5
1.0	961.13	956.71	**956.36**	972.58	1001.81
0.8	924.38	916.56	**916.15**	938.99	962.77
0.6	862.52	856.32	**855.67**	879.58	905.81
0.4	775.09	**772.19**	772.65	801.78	835.08
0.3	**719.75**	720.42	723.32	759.55	796.30
0.2	**658.57**	662.91	668.28	714.95	756.35
0.1	**594.93**	599.68	607.19	667.19	712.98

Conclusions

This paper has reviewed some distinctive features of hub-and-spoke networks, and has contrasted the spatial

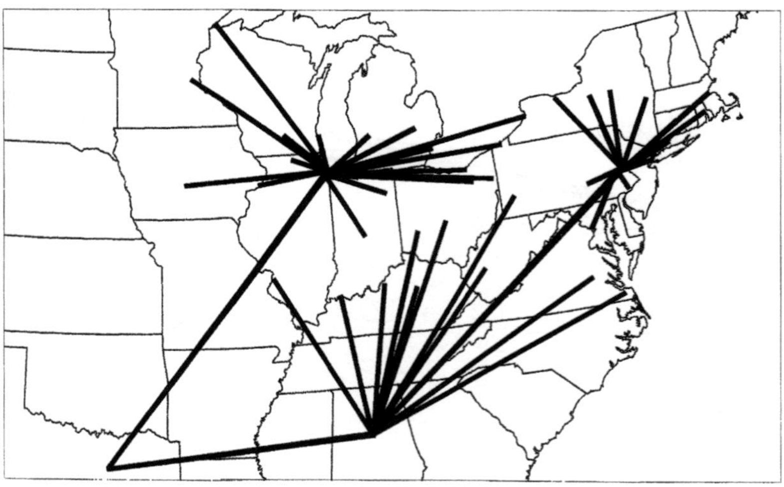

Figure 5 Multiple hub allocation model: Dallas based routing tree to selected Midwestern and Eastern cities.

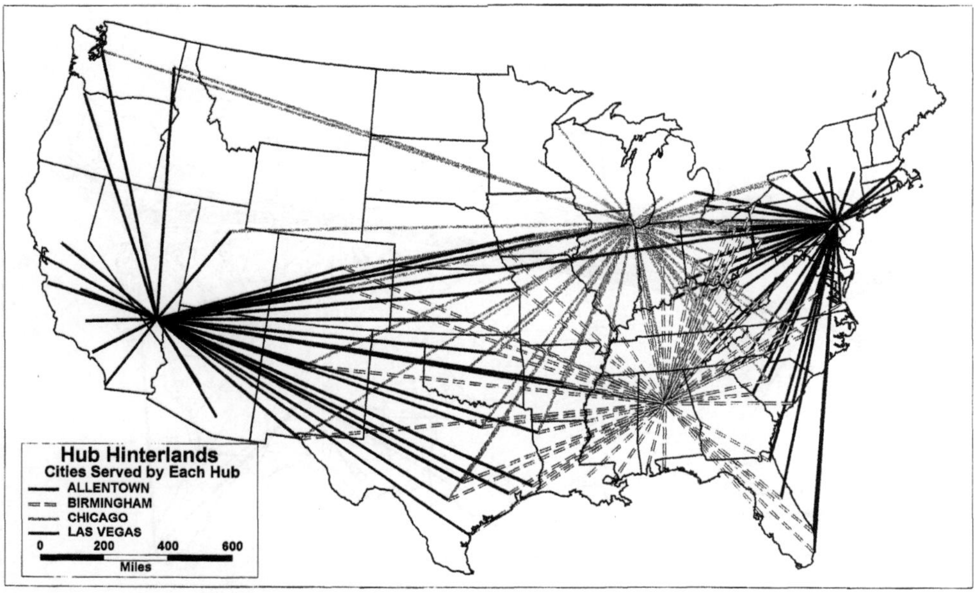

Figure 6 Multiple hub allocation service areas around the four hubs.

organization of such nets to more conventional problems in locational analysis. So far the literature has presented a large number of strong analyses of abstract models, and to get to a more robust application of these models to real network systems, more research is needed. Certainly to get to more realistic applications, we will need to consider multiple objectives, non-systematic (stochastic) changes in flows, making sure to design and build networks that do not lock in on flows which will inevitably change over time. Network designs that are capable of handling multiple objectives as well as uncertainty must be worked out. The data analysis examples in this paper were performed with interactions that reflect the urban system from almost 30 years ago, and clearly there should not be any attempt made to extract from these results prescriptions for actual contemporary networks. The interactions, though, could easily be plugged in for any appropriate urban system, for any date, and the results would then be more useful to an applied decision analysis. The paper's simulation exercise has shown quite clearly the merits of various prototypical hub-and-spoke networks: the single hub allocation model minimizes the number of links, at the expense of relatively inconvenient routing between city pairs. This would be an especially inconvenient network for passengers, but might be ideal for regional freight or communications systems.

Table 4 Freight

	Hub set				
Cost function	**1**	**2**	**3**	**4**	**5**
A	959.50	954.63	**954.31**	972.09	999.62
B	956.92	948.89	**947.94**	966.45	993.65
C	951.23	937.90	**936.33**	950.98	981.62
D	941.55	920.89	**917.59**	931.22	964.50

The multiple assignment model has much less passenger inconvenience. If the inter-hub links are not discounted, the model will use its ability to route to many hubs to create a set of complete linkages between every node and every hub, thereby by-passing the inter-hub links. If the inter-hub links are given a substantial discount, then some nodes route via the hubs, but probably not enough of the flow goes through the hubs to warrant the discount that is offered. Thus while the network appears to be ideal for the passenger system, it seems to make a serious miscalculation in the nature of flow economies of scale. It is the last simulated model, FLOWLOC, that has gone a long way towards improving this situation. In this latest model, discussed fully elsewhere (O'Kelly and Bryan, 1997) the flows must be deliberately routed to make up economical bundles, and the incentives are stacked in favor of large flows. The resultant pattern certainly requires the ability on the part of the network operator to route and direct traffic, but if that ability is

Figure 7 Flow related costs. Lines are proportional to flow.

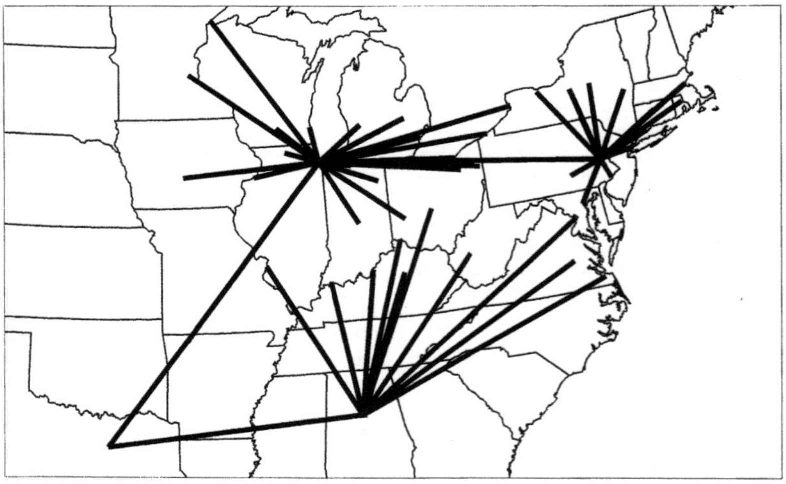

Figure 8 FLOWLOC: Dallas based routing tree to selected Midwestern and Eastern cities.

granted, the resulting network is an elegant compromise between sparse connections and reasonably direct routing.

Note

The following references were useful in peparing the conference version of this paper, and may be valuable sources for further details for some of the general comments made in this paper: Berry *et al* (1996); Butler and Huston (1990); Fleming (1991); Ghobrial (1983); Hall (1989, 1996); Helme and Magnanti (1989); Jaillet *et al* (1994, 1996); Murphy and Hall (1995); Oum *et al* (1995); Skorin-Kapov *et al* (1996); Taha and Taylor (1994); Taylor *et al* (1995).

Acknowledgements

I thank the Department of Geography at the University of Washington for their invitation to present this paper. I am grateful to Debbie Bryan for excellent research assistance. The data in the numerical examples were kindly provided by Stewart Fotheringham. The maps for this paper were produced using TransCAD 3.0.

References

Antoniou, A. (1991) Economies of scale in the airline industry: the evidence revisited. *Logistics and Transportation Review* **27**, 159–184.

Aykin, T. (1995a) The hub location and routing problem. *European Journal of Operational Research* **83**, 200–219.

Aykin, T. (1995b) Networking policies for hub-and-spoke systems with application to the air transportation system. *Transportation Science* **29**, 201–221.

Balakrishnan, A., Magnanti, T. L. and Wong, R. T. (1995) A decomposition algorithm for local access telecommunications network expansion planning. *Operations Research* **43**, 58–76.

Berry, S. T., Carnall, M. and Spiller, P. (1996) Airline hubs: costs, markups and the implications of customer heterogeneity. NBER Working Paper No. 5561.

Borenstein, S. (1989) Hubs and high fares: dominance and market power in the US airline industry. *RAND Journal of Economics* **20**, 344–365.

Borenstein, S. (1992) The evolution of US airline competition. *Journal of Economic Perspectives* **6**, 45–74.

Brown, J. H. (1991) An economic model of airline hubbing-and-spoking. *The Logistics and Transportation Review* **27**, 225–239.

Brueckner, J. K., Dyer, N. and Spiller, P. (1992) Fare determination in airline hub-and-spoke networks. *RAND Journal of Economics* **23**, 309–333.

Butler, R. V. and Huston, J. H. (1990) Airline service to non-hub airports ten years after deregulation. *The Logistics and Transportation Review* **26**, 3–16.

Campbell, J. F. (1994) A survey of network hub location. *Studies in Locational Analysis* **6**, 31–49.

Campbell, J. F. (1996) Hub location and the *p*-hub median problem. *Operations Research* **44**, 923–935.

Chou, Y.-H. (1990) The hierarchical-hub model for airline networks. *Transportation Planning and Technology* **14**, 243–258.

Daskin, M. S. and Panayotopoulos, N. D. (1989) A Lagrangian relaxation approach to assigning aircraft to routes in hub-and-spoke networks. *Transportation Science* **23**, 91–99.

Dobson, G. and Lederer, P. J. (1993) Airline scheduling and routing in a hub-and-spoke system. *Transportation Science* **27**, 281–297.

Fleming, D. K. (1991) Competition in the US airline industry. *Transportation Quarterly* **45**, 181–219.

Gavish, B. (1992) Topological design of computer communication networks — the overall design problem. *European Journal of Operational Research* **58**, 149–172.

Ghobrial, A. A. (1983) Analysis of the air network structure: the hubbing phenomenon. Ph.D. dissertation, University of California, Berkeley.

Hall, R. W. (1989) Configuration of an overnight package air network. *Transportation Research* **23A**, 139–149.

Hall, R. W. (1996) Pickup and delivery systems for overnight carriers. *Transportation Research* **30A**, 173–187.

Hansen, M. (1988) A model of airline hub competition. Ph.D. dissertation, University of California, Berkeley.

Hansen, M. (1990) Airline competition in a hub-dominated environment: an application of noncooperative game theory. *Transportation Research* **24B**, 27–43.

Hansen, M. and Kanafani, A. (1989) Hubbing and airline costs. *Journal of Transportation Engineering* **115**, 581–590.

Helme, M. P. and Magnanti, T. L. (1989) Designing satellite communication networks by 0–1 quadratic programming. *Networks* **19**, 427–450.

Jaillet, P., Yu, G. and Song, G. (1994) Network design and hub facility location. Paper presented at the ORSA/TIMS Meeting, Detroit, MI.

Jaillet, P., Song, G. and Yu, G. (1996) Airline network design and hub location problems. *Location Science* **4**, 195–212.

Kling, J. A., Grimm, C. M. and Corsi, T. M. (1991) Hub-dominated airports: an empirical assessment of challenger strategies. *The Logistics and Transportation Review* **27**, 203–223.

Kuby, M. J. and Gray, R. G. (1993) The hub network design problem with stopovers and feeders: the case of Federal Express. *Transportation Research* **27A**, 1–12.

Middendorf, D., Bronzini, M. S., Peterson, B. E., Liu, C. and Chin, S.-M. (1995) Estimation and validation of mode distances for the 1993 Commodity Flow Survey. *Proceedings of the 37th Annual Transportation Research Forum*, pp. 456–473. Transportation Research Forum, Virginia.

Murphy, P. R. and Hall, P. K. (1995) The relative importance of cost and service in freight transportation choice before and after deregulation: an update. *Transportation Journal* **35**, 30–38.

O'Kelly, M. E. (1987) A quadratic integer program for the location of interacting hub facilities. *European Journal of Operational Research* **32**, 393–404.

O'Kelly, M. E. (1998) On the allocation of a subset of nodes to a mini-hub in a package delivery network. *Papers in Regional Science* **77**, 77–99.

O'Kelly, M. E. and Lao, Y. (1991) Mode choice in a hub-and-spoke network: a zero-one linear programming approach. *Geographical Analysis* **23**, 283–297.

O'Kelly, M. E. and Miller, H. J. (1994) The hub network design problem: a review and synthesis. *Journal of Transport Geography* **2**, 31–40.

O'Kelly, M. E., Bryan, D., Skorin-Kapov, D. and Skorin-Kapov, J. (1996) Hub network design with single and multiple allocation: a computational study. *Location Science* **4**, 125–138.

O'Kelly, M. E. and Bryan, D. (1997) Hub location with flow economies of scale. *Transportation Research B* (forthcoming).

Oum, T. H., Zhang, A. and Zhang, Y. (1995) Airline network rivalry. *Canadian Journal of Economics* **28**, 836–857.

Ray, J. J., Solanki, R. and Southworth, F. (1989) The location of flow-support facilities on an airline network. Paper presented at the Regional Science Association Conference, Santa Barbara, CA.

Skorin-Kapov, D. (1995) Cost allocation in hub networks. Unpublished manuscript.

Skorin-Kapov, D. and Skorin-Kapov, J. (1994) On tabu search for the location of interacting hub facilities. *European Journal of Operational Research* **73**, 502–509.

Skorin-Kapov, D., Skorin-Kapov, J. and O'Kelly, M. E. (1996) Tight linear programming relaxations of uncapacitated *p*-hub median problems. *European Journal of Operational Research* **94**, 582–593.

Sorenson, N. (1990) Airline competitive strategy: a spatial perspective. Ph.D. dissertation, University of Washington.

Taaffe, E. J., Gauthier, H. L. and O'Kelly, M. E. (1996) *Geography of Transportation*, Second Edition. Prentice Hall, Saddle River, NJ.

Taha, T. T. and Taylor, G. D. (1994) An integrated modeling framework for evaluating hub-and-spoke networks in truckload trucking. *The Logistics and Transportation Review* **30**, 141–166.

Taylor, G. D., Harit, S., English, J. R. and Whicker, G. (1995) Hub-and-spoke networks in truckload trucking: configuration, testing and operational concerns. *The Logistics and Transportation Review* **31**, 209–237.

Tretheway, M. W. and Oum, T. H. (1992) *Airline economics: foundations for strategy and policy*. Center for Transportation Studies, University of British Columbia, Vancouver.

Weidner, T. J. (1995) Hub choice equilibrium in the US airnet. Paper presented at the Regional Science Association International Meeting, Cincinnati, OH.

Wollmer, R. D. (1990) An airline tail routing algorithm for periodic schedules. *Networks* **20**, 49–54.

Zhang, A. (1996) An analysis of fortress hubs in airline networks. *Journal of Transport Economics and Policy* **30**, 293–307.

[20]

INTERMODAL TRANSPORTATION IN NORTH AMERICA AND THE DEVELOPMENT OF INLAND LOAD CENTERS*

Brian Slack

Concordia University

This paper analyzes the changes that are transforming intermodal transportation in the United States and Canada. Traffic is being concentrated at a relatively small number of inland load centers that serve as regional truck distribution points, and are linked by high volume dedicated train services. Both theoretical and practical implications of the changes are examined, and a range of research questions are presented. **Key Words: intermodal transportation, load centers, containerization.**

In the last six years radical changes have occurred in intermodal transportation in North America. Developments are affecting many facets of the transportation industry. One of the most striking features has been the emergence of load centers where intermodal traffic is assembled and distributed. The geographical concentration of intermodal shipments is an evolving process with far reaching consequences for regional economies, the modal split, transportation industry structure, and spatial patterns of freight flows.

This paper focuses on the evolving character of intermodal transportation in North America and the emergence of traffic hubs. The transformation of old piggyback operations and their replacement by modern intermodal systems have forced the transportation industry to restructure. Traffic concentration is occurring and this is making it necessary to evaluate the viability of established networks. In particular, reassessments of the location, size, and function of intermodal terminals are being undertaken.

This paper analyzes the changes that are occurring and compares the evolving networks with a widely recognized general model of network evolution. The results suggest that intermodal transportation, as an advanced system, requires the existing network model to be extended beyond its "final" stage. The study goes on to relate the emergence of load centers in intermodal networks with traffic hubs in other modal systems. Finally, the paper examines the geographical implications of the important changes in intermodal transportation and explores the research directions geographers should take.

Recent Developments in Intermodal Transportation

The introduction of trailer on flat car (TOFC) or piggyback by North American railroads during the 1950s was an early example of intermodalism (Mahoney 1985). The system was adopted because it was thought that it would enhance the railways' competitive position vis-a-vis the trucking industry for the transport of domestic freight. Most major North American railway companies invested heavily in flat cars and terminal facilities during the 1950s. Although Schwind (1967) saw TOFC providing a streamlined rail service, intermodal transfers were accomplished by circus ramps that were built in most rail centers in order to provide maximum market coverage. In North America as a whole over 2500 piggyback terminals were established (Archambault 1987).

Piggyback grew rapidly, providing one of the few growth markets for the rail-

---

* The author acknowledges the financial support of Transport Canada in this project. The views expressed are those of the author, and do not necessarily reflect those of Transport Canada. I am particularly grateful for the helpful comments of an anonymous reviewer and the editors.

roads. In 1977 piggyback traffic represented 7% of all US rail revenues, and TOFC tonnage had increased by 40% since 1969 (Mahoney 1985). The overall effectiveness of piggyback was much less than these figures would suggest, however. Mahoney (1985) showed that the system handled less than 1% of all domestic intercity freight movements in the United States. Because it was introduced to compete with the trucking industry, truckers were extremely reluctant to cooperate. Further problems arose because of the decisions made to locate ramps wherever rail service was possible. This location decision was a direct challenge to truckers in local markets, the markets they are best able to serve. Long service delays also resulted, since it took days to assemble and deliver the individual flat cars and their trailers from one ramp to another, even when the distances involved were not great (Traffic Management 1985). Piggyback thus earned a reputation for slowness and inefficiency, a perception amongst shippers that has proved to be difficult to change. The railroads' own experience with TOFC was also not very satisfactory, since it produced unacceptably small profit margins.

In the 1960s, containers began to transform maritime transport (Hayuth 1985, 1987). As an intermodal system, they subsequently revolutionized land-water cargo flows. Containerization was an innovation, however, that US railway companies largely ignored. They regarded it as a maritime technology, and were reluctant to make the required capital investments. Their position must be seen in light of the limited success of piggyback, which caused the railways to look upon yet another intermodal development with skepticism. The result was that as many as 80–85% of all containers landed in US ports were hauled by truck.

The situation was somewhat different in Canada, where the two major railway companies adopted containerization from the beginning (Archambault 1987). They made important investments in marine terminals and rolling stock, and after 1974 began to offer intermodal tariffs in conjunction with the maritime conferences. The addition of large volumes of this new container on flat car (COFC) traffic to the existing TOFC business encouraged both Canadian National (CN) and Canadian Pacific (CP) to operate combined TOFC and COFC unit trains along the major transportation corridors. Significantly, US railways were prevented by regulatory restrictions from copying many of the Canadian initiatives.

The Staggers Act of 1980, and the more recent Shipping Act (1984) have substantially transformed the regulatory environment in the United States (Hayuth 1987). Intermodal tariffs can now be filed, and intermodal mergers are permissible. Deregulation has been combined with important technological developments, double stacking in particular, to create great opportunities for intermodalism in the United States. Although the developments that have occurred on land involve rail transport to a considerable extent, they have been initiated primarily by the shipping lines (Archambault 1987; Hayuth 1987). Sea-Land was the innovator of double stack rail service in 1981, but it was the decision of American President Lines (APL) to begin double stack service from Los Angeles to Chicago in 1984 that may be credited with revolutionizing the intermodal scene in the United States. The shipping line provided the stack cars and containers hauled by Union Pacific (UP) locomotives.

The scale and pace of subsequent developments have been remarkable. By 1987, just three years after APL's initiative, there were 122 double stack train services per week in the United States (Fig. 1). APD (the domestic arm of APL) alone accounted for 36 of these services. APD owns 41% of all stack cars in the United States, with Sea-Land being the second most important proprietor with 16% (Containerisation International 1987). The primary involvement of the shipping lines in intermodal transport on land has caused some observers to see these recent changes as evidence of a further stage in

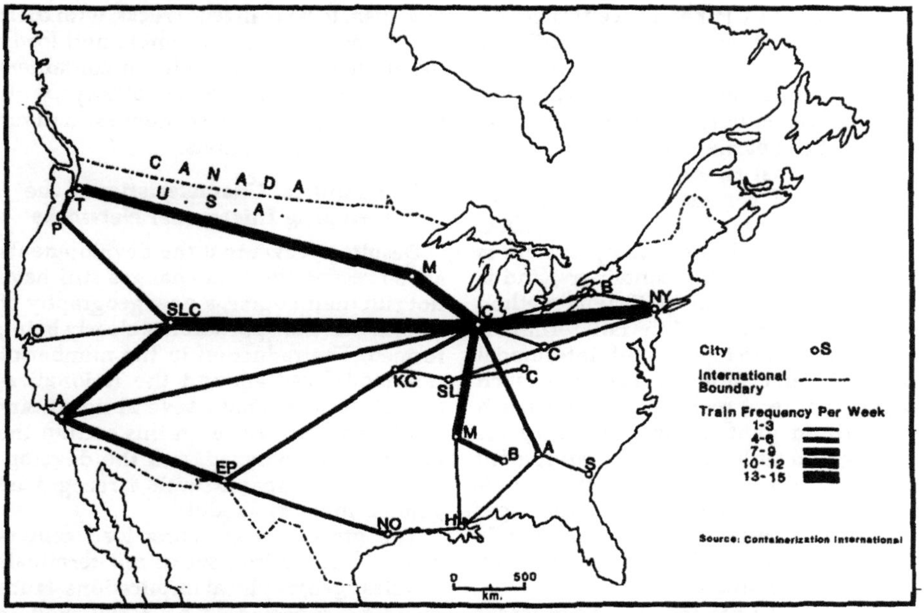

Figure 1. Frequency of US double stack trains, 1987.

the maritime container revolution, which is now being extended to include landward legs (Archambault 1987; Hayuth 1987). These developments go much further than the transformation of terrestrial flows of maritime containers, however, since their impacts are no longer confined to impex (import-export) traffic. The imbalanced flow of containerized double stack traffic in the western United States has resulted in APD offering attractive rates for domestic westbound freight in maritime containers. Domestic traffic in containers represents a new and growing business. Two hundred thousand of APD's total 350,000 loads represent domestic freight. This traffic is being converted largely from piggyback (Containerisation International 1987).

The Development of Load Centers

The introduction of containers, initially exclusively maritime, more recently domestic as well, has necessitated a reevaluation of rail intermodal operations. The main element that distinguishes the modern systems from earlier piggyback configurations is a marked reduction in the number of terminals. Over the last few years there has been a trend towards network rationalization and a concentration of traffic at a relatively small number traffic hubs. Archambault (1987) estimated that at their maximum there were 2500 piggyback ramps in North America. Even as recently as 1978 there were 1176 intermodal centers in the United States alone (Down and Wise 1986). Eight years later, the number of intermodal terminals with mechanized equipment totaled 175 (Official Intermodal Guide 1987). These remaining terminals are load centers, where unit loads are assembled and dispersed.

Although the pattern of concentration varies among the different companies, many of the major railroads in North America have gone through major streamlining. For example, CN has reduced its number of intermodal centers from 80 to six; CP has concentrated its business at eight centers; Santa Fe (SF) has reduced its intermodal terminals from 100 to 28; and, Burlington Northern (BN) has consolidated its business at 22 terminals instead of the earlier 140 rail ramps.

The principal factor behind the emergence of load centers has been the remarkable growth of the railways' COFC business. The old circus ramp system is incapable of handling containers, and is a relatively slow and inefficient method of loading trailers. With rail terminals being required to handle both intermodal forms of traffic, side lift machines (such as piggypackers) or top lift cranes (such as overhead gantries) are necessary for loading and unloading. This equipment is expensive, representing capital investments of up to one million dollars per gantry. As a result the railways are able only to allocate such lift devices to a small number of terminals.

The evolving pattern of rail services is another factor behind the rationalization process. The establishment of double stack services in the United States and dedicated trains in Canada has resulted in intermodal services being offered in a few high density corridors. Only those markets capable of generating sufficient volumes of business to form unit train loads along these corridors are being selected as load centers.

A further contributing factor has been the line mergers that have taken place in the United States since 1976. Many of the newly merged companies encountered overlapping services and redundancies in their combined operations, and have been forced subsequently to streamline their networks (Archambault 1987).

System reorganization is also producing a radically different form of local network structure. Instead of the dense market coverage of the old piggyback ramps, a hub and spoke configuration is emerging. Each load center (hub) serves a regional market using truck pickup and delivery (spokes), and is linked by dedicated rail services to the other hubs. This system permits the economies of intermodalism to be realized. Trucks, with their flexibility and speed, assemble unit loads (whether traditional trailer or container) at the hub, from where the railways, with their long line haul economies, deliver them to distant markets.

Geographical Characteristics of the Developing Intermodal Networks

Despite the fact that the developments are so recent and that changes still have not run their course, a new geography of intermodal transportation is clearly being formed. The reduction in the number of intermodal centers and the rationalization of networks have several significant spatial characteristics. In this section the changes are analyzed, and the developments are compared with a recognized general network model.

The process of selecting load centers from a large initial set of rail terminals has clear geographical implications. Hubs could be expected to be established in those market centers which generate sufficient traffic. A certain critical size threshold may exist, and load centers will be established in all major markets that lie above this critical level.

This relationship does not exist in many cases because load centers operate in geographical space. The hub and spoke configuration of todays' intermodal systems implies a spatial constraint. To combine the advantages of rail with trucking requires tributary areas to be designated for exclusive truck service. The size of these zones is determined by the cost differentials between the modes. The structure of rail and truck freight rates is a complex issue, but it is widely recognized that truck rates are in general competitive in a radius of 400–700 km around a market area (Wright 1973). Trucking may compete in some instances up to 1000 km, but on average 500 km may be taken as a median radius. Load centers should, therefore, be established up to 1000 kms apart, and within their hinterlands markets should be large enough to generate sufficient volumes of business to warrant dedicated rail service. Such a service would involve a minimum of 74,000 loads per year (Slack 1988).

Inevitably this theoretical arrangement has to be modified by other limiting factors. The decisions regarding the location of inland load centers are made primarily by the railroads. Their choices are constrained by the actual alignments of their networks. Many cities which meet the criteria of market size and regional accessibility may in fact be served by a number of competing lines. This competition may result in the market shares of individual lines being suboptimal for the establishment of a load center. But no railroad will willingly relinquish its presence in such markets. Competition then forces individual railroads to maintain intermodal hubs, but their market share may not be capable of sustaining a sufficient volume of business to warrant complete unit train service. The result is that in practice many hubs handle less than complete trains. They may contribute a few cars that may be added to a dedicated train as it passes through that center, or may be joined to other segments and blocked at another location. Thus the minimum threshold size of a inland load center tends to be much smaller than 74,000 loads per year.

A more serious constraint is the amount of lift capacity available in load centers (Down and Wise 1986). A wide range of terminal equipment is available to handle double stack, COFC, and TOFC traffic. The least expensive type of lift device is a sideloader, which if operated to its maximum capacity, is capable of handling approximately 40,000 loads annually. Lift devices are essential, constituting the minimum equipment requirement of any intermodal terminal. Their capacities become important determinants of threshold sizes of load centers. However, it is unrealistic to expect that the maximum capacity of a sideloader should become the minimal threshold capacity of a load center. Equipment utilization is never optimal given the imperfections in terminal operations, layout, and maintenance. In addition, the flow of freight through a terminal usually exhibits temporal variations. The number of working spots and storage capacities frequently constrain terminal capacities still further. Investment decisions regarding lift equipment are made with the expected growth of traffic in mind, resulting in lower actual threshold values being accepted. If a 50% level of lift capacity is taken as the minimal acceptable operational activity, a throughput of 20,000 loads would be the absolute minimum threshold volume for a load center.

Theoretically, therefore, load centers should be established to command a market area with a radius of 500 km and be capable of generating at least 20,000 intermodal loads per year. Actual hub performance cannot be evaluated against these theoretical estimates because intermodal rail traffic data are market sensitive and not published. Limited evidence suggests that most railway companies still operate many of their modernized terminals at below these minimum thresholds. In the case of one US railroad, which is developing a hub configuration, 27% of its hubs handled less than 20,000 loads in 1987 and over half handled fewer than 40,000 loads.

Individual company networks can, however, be compared with the hypothesized spatial configuration of load centers. The extent to which modern intermodal concepts have been adopted is reflected in the density of COFC/TOFC terminals. In the following paragraphs the systems of three railroads are analyzed by comparing their actual networks with the hypothesized market areas of 500 kms radii.

Conrail's present intermodal network has changed little from the old piggyback system (Fig. 2). Although the company has moved to establish some double stack services, and is in the process of upgrading some of its terminals, the overall picture is one of dense local coverage, with little rationalization so far.

BN's intermodal network is more streamlined. In 1982 BN ran a highly successful trial test of a hub system in Minnesota. Subsequently the railroad has concentrated its intermodal business at 23

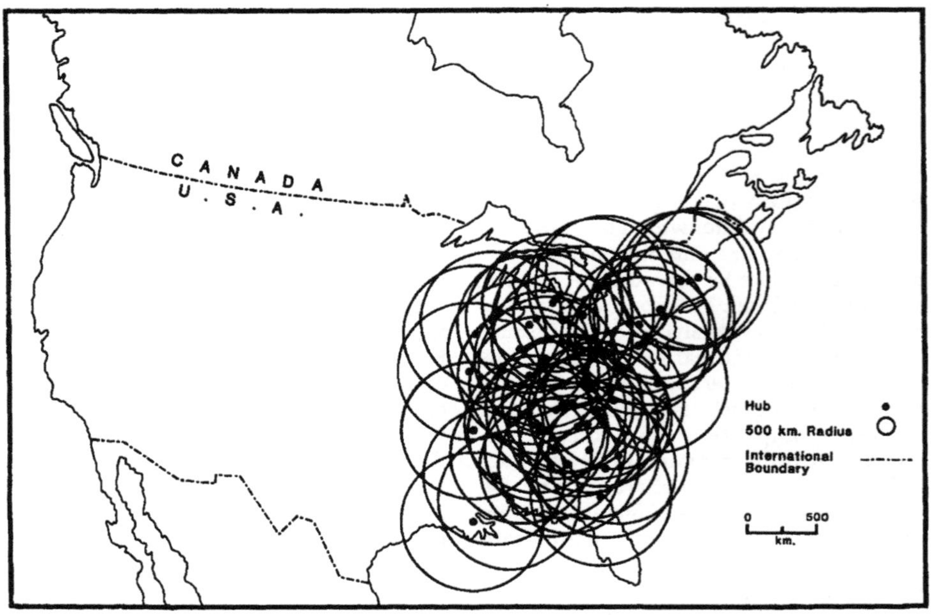

Figure 2. Conrail's intermodal hubs.

hubs (Fig. 3), each capable of handling an intermodal train without switching. The present load centers represent a sharp reduction from the former 140 rail terminals. BN has deliberately sought to replace rail delivery by truck transfer within a 200 mile radius of each hub (Janes Railways 1988). The immediate goal of the company is to convert to a 100% dedicated intermodal train service. The present network represents a marked rationalization of operations, but there are still a number of spatial redundancies in the northwest and in the central Mississippi valley (Fig. 3).

The present network of CN represents an advanced stage of intermodal operations. CN maintains over 100 "intermodal points" scattered across Canada. Over the last few years, intermodal traffic has been concentrated at six Canadian and two US hubs. The load centers provide basic coverage of most of southern Canada, including the major markets (Fig. 4).

The simplicity of the CN network reflects the commitment made by the company to intermodal transport over a longer time frame than its US counterparts. The networks of BN and other western railroads appear to be moving in the same direction, and there are signs that other US lines, such as CSX and Norfolk Southern, are beginning to rationalize their intermodal operations. The changes that have already been put in place by US railroads may be expected to continue over the next few years, and dramatically different intermodal networks will be established.

The developments may be captured in a simple intermodal network model (Fig. 5). Phase one of the process represents the situation of a railroad achieving maximum market coverage. Its intermodal facilities are located throughout the service area, and are linked by a dense network of lines. Intermodal traffic is a small segment of the total business activity and is integrated in overall freight operations. Phase 2 is reached when growth in in-

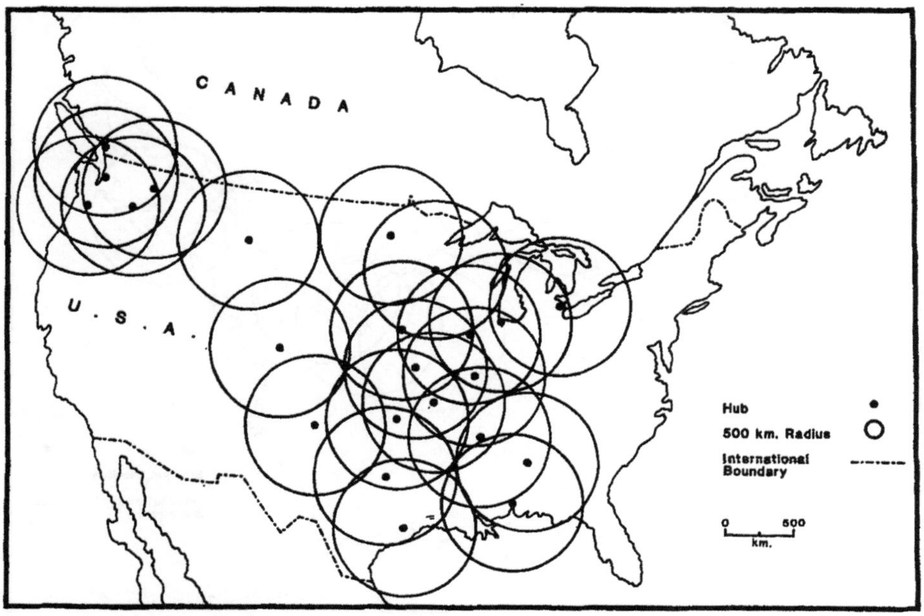

Figure 3. Burlington Northern's intermodal hubs.

termodal transport enables the railway company to offer specialized TOFC and COFC service between certain major traffic centers. Based on empirical evidence, the stimulus for this reorganization is provided by maritime clients seeking to extend mini-bridge or land bridge container services. The railway company begins to exploit the economies of intermodalism, necessitating some rationalization of its railnet and the mechanization of certain intermodal yards. This stage is one that characterizes most US railroads today.

Phase 3 represents the condition where the entire system has been reformed and a distinct intermodal network is put in place. The network of hubs recognizes the role of trucking in local and regional pick-up and delivery of both COFC and TOFC traffic. The hubs act as load centers, and they are connected by dedicated intermodal rail services. The CN system is perhaps the best example of this network at present.

The sequence described above may be compared with a well-known model of transport network evolution (Taaffe et al. 1963). The early stage of dense market coverage of piggyback services corresponds to the penultimate stage in the general model, that of "complete interconnection." The latest intermodal networks bear some similarities to the final stage of general model, that of the "emergence of high priority linkages." Intermodal networks appear to be comparable with the general transport network model at its latermost stages, therefore. Intermodal transport is an advanced system. It comes into being only after a basic transport infrastructure is established. The earliest stages of the network development model from "scattered ports" to "interconnection" (Taaffe et al. 1963), are unlikely to involve developed intermodal systems.

There is one main difference between intermodal networks and the general model. In the final stage of the model,

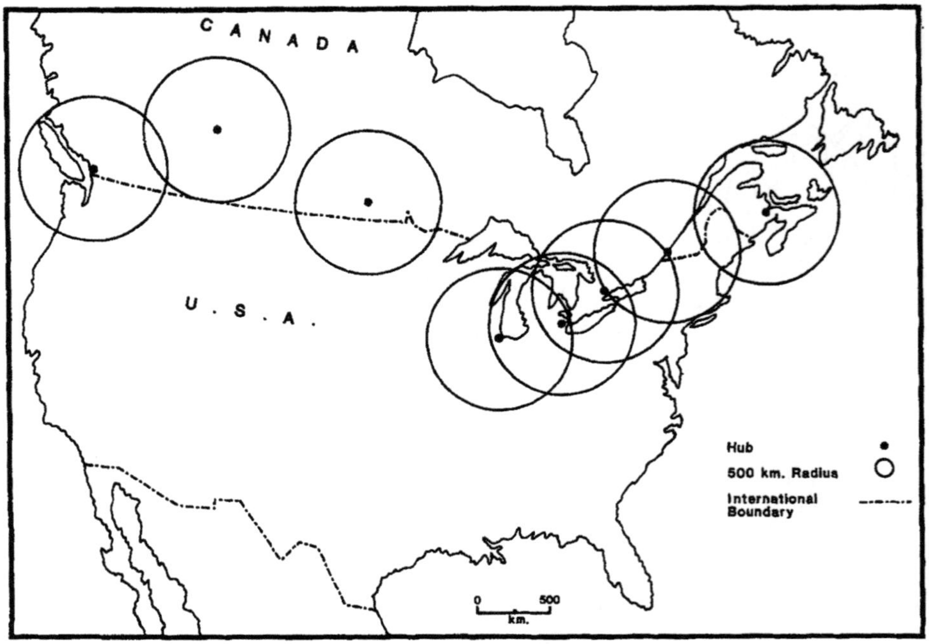

Figure 4. Canadian National's intermodal hubs.

"high priority linkages," the terminals off the main routes are maintained. In modern intermodal networks these nonhub centers no longer engage directly in TOFC or COFC traffic, and are redundant. The disappearance of these terminals from intermodal networks suggests that a fully developed intermodal system goes beyond the stages encompassed by the general network development model. One more stage needs to be added to the Taaffe-Morrill-Gould model to reflect the realities of intermodal systems. This additional stage would exclude all former terminals that no longer serve as hubs.

Inland Load Centers and the Hub Phenomenon

The concept of load centering is well established in many fields of transportation. In ocean transport, containerization has led to the concentration of general cargo at a relatively small number of ports. Although there is some debate as to whether or not the degree of concentration is growing (Hayuth 1988), there is little doubt that container traffic is spatially agglomerated. Economies of scale are frequently cited as the major cause of container load centering. The selection of a very small number of ports on each maritime range permits containers to be concentrated in sufficient volumes to allow optimal use of container handling equipment and large cellular ships. An analogous process is occurring in air transport (although this is not for the most part involving intermodal activities). Airline networks are becoming increasingly nodal with the establishment of traffic hubs (Grove and O'Kelly 1986). For example, the network of American Airlines is focused on Dallas-Fort Worth, and United Airlines shows a concentration of flights in and out of Chicago and Denver. The result is that an increasing number of passengers at airports are in transit, being forced to change flights at the airline hubs.

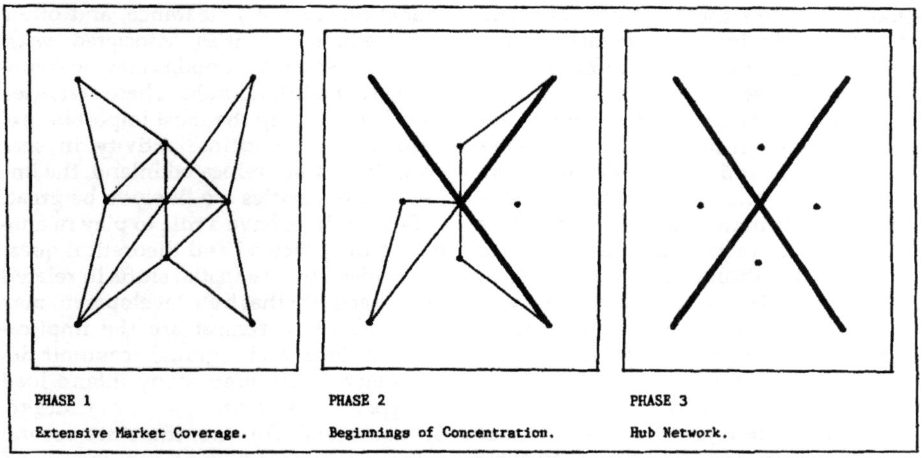

Figure 5. Model sequence of hub network development.

At Atlanta, for example, over 70% of passengers using the airport are in transit (Kanafani and Ghobrial 1985). Economies of scale are again cited for the development of air traffic hubs (Kanafani and Ghobrial 1985). The concentration of passengers at hubs permits the airlines to employ the new generation of fuel efficient aircraft, while offering the public at the same time a higher level of flight frequency than is possible with networks characterized by city pair linkages.

Intermodal rail traffic appears to be following the trend already established in the other systems. This paper has already detailed the emergence of new railnets and hubs, and it is clear that the principal forces behind their development are economies of scale and technology. There are some differences between the evolving rail hub systems and the networks of the other modes, however. These differences need to be outlined because they will lead inevitably to a different set of problems and impacts.

In the marine and air systems load centering is being instituted by the carriers. The shipping consortia and the airlines are making the important decisions regarding the choice of load centers. The terminals themselves are usually publicly owned and the authorities are rarely, if ever, involved in the load centering decisions. On the other hand, many of the costs arising out of hub selection are borne by these terminal authorities. Ports have to invest to attract the shipping lines, and because of the intense rivalry on each maritime range, the port administrations are frequently pushed into making investment decisions that may be entirely speculative. Existing load centers may become congested, forcing ports to expand and upgrade their infrastructures, while nonhub ports, trying to make themselves more attractive, may have to invest in facilities that may not be immediately financially sustainable. Similarly, the costs of congestion in airport hubs are borne in part by the authority. No scale economies are realized by the airports themselves (Kanafani and Ghobrial 1985). Average costs per passenger do not decline with increased traffic volume. For these reasons, some of the most critical problems in transportation today involve airport and port congestion at the major load centers. Ports are striving to secure adequate transfer areas and rail access, and many airports are facing crises in terminal congestion and air safety. The costs and benefits of load centering are not borne equally between the terminals and the carriers, producing the imbalance between capacity and actual traffic that exists in many load centers.

The intermodal load centers are somewhat different. Although the balance between capacity and actual traffic will never be perfect, the fact that the railway companies act both as terminal operators and long haul carriers means that there can be a more rational approach to system design. The selection of hub centers and the appropriate investments can be made with far greater certainty. Obviously market conditions represent a major unknown, but the hub operation will not have to face the crisis that is sometimes faced by a port or an airport when their users and carriers relocate to a competing city after the terminal has made a large capital investment.

Evolving Intermodal Systems: Questions and Implications

The changes that are being wrought presently in intermodal transportation are considerable. Although several impacts may not yet be fully discernable, a number of effects are quite clear. In this section the most important implications of the changes in intermodal transportation are discussed. The goal is to identify the impacts, and to outline a number of questions with significant geographical content that require attention.

Load centers are concentrating the economic advantages formerly dispersed over a wider geographical area. This concentration will very likely have negative effects on cities and regions that may lose direct access to rail corridors. The effects will be measured not only in the loss of direct transport-related employment, but also in terms of comparative accessibility and attractiveness for industries with high transport locational requirements. Even more important may be the effects of inland hubs on seaport cities. Some workers have speculated that a major few inland load centers will usurp many of the functions traditionally performed in seaports (Hayuth 1984, 1985). Warehousing, consolidation, customs clearance, and other intermediate activities associated with major transportation nodes may be transferred to the inland hubs. These activities have been among the most important externalities of maritime activity in seaports. If they are relocated inland, the impacts on port cities are likely to be great.

Geographers have a role to play in analyzing the practical and theoretical questions raised by the spatial shifts in related service activity that hub development may bring about. Foremost are the implications for local and regional economic development. Although many inland load centers are major cities, such as Chicago, St. Louis, and Toronto, there are several hubs, such as Spokane, Billings, and Moncton, that have been selected for their regional nodality. Will the hub function serve as a stimulus to economic growth? Will there be a shift in the geographic location of intermediate services away from ports? To what extent will load centers act as growth poles? The whole process of network rationalization raises important questions about relative accessibility and regional development.

Despite the potentially significant economic impacts of intermodal system development, it is a process that has hardly any public input. The selection and promotion of the intermodal hubs is a process that involves the railways only. If geographers can measure the spatial shifts and their economic consequences, we may be able to alert various levels of government to the kinds of changes that are occurring, and bring the process of network change further into the public domain.

One of the imponderables at this time is the future of piggyback. On both sides of the border several companies are proclaiming the cost and operational advantages of containers for domestic traffic. APD and CP have made significant investments in containers rather than trailers. They argue that containers can be better integrated with impex traffic at terminals, and can lead to significant economies in the yards because they can be stacked on top of each other. CP claims a 10–20% cost saving of containers over road trailers (Containerisation International 1986). Furthermore, CP's use of non-standard container sizes gives customers greater flexibility (Down and Wise 1986).

Proponents of domestic containers see them as not only competing effectively with piggyback, but also giving intermodal traffic an opportunity to capture a larger share of the long distance intercity traffic, which in the United States represents nearly half a billion loads per year (Smith 1986). At present only 1% of this traffic is handled by the railways intermodally. Some experts believe domestic containers may permit the railways to capture 5% of the market, which, although still a small share, would represent a massive increase in the volume of traffic (Containerisation International 1987).

The future of domestic containerization involves questions relating to the modal split and the future of long haul trucking. Will the market share of the railroads increase? What will be the effect on rates and the spatial ranges of the different modes? Undoubtedly the trucking industry will not sit back in face of the new competition. The longer term effects of the competition are likely to be complex. Geographers have made a number of important contributions concerning the effects of containerization on maritime transport (Hayuth 1987, 1988). The question arises how comparable are the conversions of maritime and domestic traffic to containers? Will the changes that have occurred in shipping activity and port competition be reproduced on land with the emergence of domestic containers?

One of the most striking impacts of intermodal transport development is the realignment of traffic flows. The hub and spike configuration of intermodal networks is resulting in new patterns of freight movements. As has been mentioned earlier, traffic is being concentrated at a relatively small number of hubs, which serve as local and regional distribution centers.

In addition, major interregional shifts in intermodal traffic are occurring. One example involves the diversion of US impex containers via the port of Montreal. Linked by an efficient rail system to the Midwest, the port of Montreal has been able to compete most effectively with US East Coast ports, primarily New York and Baltimore, for the region's containerized trade with Europe. Nearly 50% of Montreal's container traffic, which has doubled over the last four years, involves "captured" US trade. There are many contributing factors, but the most important element is the ability of the shipping lines on the St. Lawrence gateway to offer through bills of lading by a fully intermodal service to the Midwest (Archambault 1987).

A further, more recent, example involves the considerable eastward expansion of the trade hinterlands of US West Coast ports. The double stack rail services operated by US and Japanese shipping companies have significantly restructured trans-Pacific container traffic flows. Their services out of West Coast ports have brought the mini-landbridge concept into reality (Hayuth 1988; Mahoney 1985). Pacific Rim trade originating even in eastern North America may now be quickly and cheaply routed by rail to a Pacific port. East Coast ports are thus facing losses of Far East trade; the removal of Japanese container services from the port of Saint John has already created one Canadian casualty. The latter has suffered because Canadian railways are feeding southern Ontario containerized traffic through the Chicago hub center to US West Coast terminals (Slack 1988).

These shifts in regional flows require detailed analysis. How big is the diversion of trade from east to west coasts? Which ports are likely to be most affected? What have been the actual cost reductions for shippers in the Midwest? Has this change had any impact on the import-export traffic of the Midwest? In addition to these practical matters, theoretical and conceptual questions are raised by the shifts in flows. The long established notion of discrete port hinterlands has to be reevaluated. This has been one of the most enduring concepts in transportation geography, but the development of mini-bridge and land bridge services from West Coast ports has destroyed notions of natural tributary areas for individual ports.

Conclusion

This paper has drawn attention to the restructuring presently underway in intermodal transportation in North America. Regardless of how far the developments go, a number of fundamental changes already have clearly taken place, each with important spatial consequences that require much further measurement and analysis. The impacts are likely to be differentially severe. The benefits and losses will not be equally distributed across the continent. The questions raised in this study of contemporary intermodalism in North America should be added to the research agenda of transportation geographers.

Literature Cited

Archambault, M. 1987. *Developments in Intermodal Transport in North America.* Report TP8848E. Ottawa: Transport Canada.

Containerisation International 1986. CN and CP seeking differing intermodal solutions. April: 66–71.

Containerisation International 1987. Waiting for the off. . . . April:53–55.

Down, J. W., and D. H. Wise. 1986. Domestic containerization: Overview of terminal design and operating issues. In *Facing the Challenge: The Intermodal Terminal of the Future* (State of the Art Report 4), ed. E. W. Kaplan, 116–22. New Orleans: Transportation Research Board.

Grove, P. G., and M. E. O'Kelly. 1986. Hub networks and simulated scheduled delay. *Papers and Proceedings of the Regional Science Association* 59:103–19.

Hayuth, Y. 1984. Port development in light of intermodal transport. In *New Challenges for Shipping and Ports,* 126–33. Haifa: Israel Shipping and Aviation Research Institute.

Hayuth, Y. 1985. Seaports: The challenge of technological and functional changes. In *Ocean Yearbook 5,* ed. E. Borgese and N. C. Ginsburg, 79–101. Chicago: University of Chicago Press.

Hayuth, Y. 1987. *Intermodality: Concept and Practice.* Essex, UK: Lloyd's of London Press.

Hayuth, Y. 1988. Rationalization and deconcentration of the US container port system. *The Professional Geographer* 40:279–88.

Janes Railways 1987–88. 1988. London: Janes Publishing Co.

Kanafani, A., and A. Ghobrial. 1985. Airline hubbing—some implications for airport economics. *Transportation Research* 19A:15–28.

Mahoney, J. H. 1985. *Intermodal Freight Transport.* Westport, CT: Eno Foundation.

Official Intermodal Guide. 1987. Washington: National Railway Publication Co.

Schwind, P. J. 1967. The geography of railroad piggyback operations. *Traffic Quarterly* 21:237–48.

Slack, B. 1988. *Locational Determinants of Inland Load Centres.* Ottawa: Transport Canada.

Smith, D. 1986. Domestic containerization: How big can it get? *Transportation Research Forum* 27:289–95.

Taaffe, E., R. Morrill, and P. Gould. 1963. Transport expansion in underdeveloped countries. *The Geographical Review* 53:503–29.

Traffic Management. 1985. Piggyback shipments: What's the upturn in service? September:58–63.

Wright, W. B. 1973. Maximizing the rail-truck link. *Distribution Worldwide* (April):55–56.

BRIAN SLACK (Ph.D. McGill) is Professor of Geography at Concordia University, Montreal. His research interests include port development, intermodal transportation, and port service industries.

[21]

Air cargo services and the electronics industry in Southeast Asia

Thomas R. Leinbach and John T. Bowen, Jr.***

Abstract

Air cargo services are among the most important producer services for manufacturers with internationalized production networks. In the context of supply chain management strategies, such services allow firms to respond more effectively to competitive forces in global markets. Our research aims to relate to and deepen the theoretical literature on advanced producer services and global production networks by examining the demand for air cargo services in the electronics industry in Southeast Asia, an industry and a region in which air cargo services are particularly significant. Using an air cargo intensity index for each of over 120 firms from Singapore, Penang, Kuala Lumpur, and Manila, we show that the diversity of air cargo usage is related to several aspects of firm structure and operation. Our research suggests that product type, internationalization, localization, product cycle (obsolescence), and other factors such as a firm's material management strategy are especially critical. New directions for research that will build upon these findings are noted.

Keywords: producer services, air cargo, electronics industry, Southeast Asia, global production networks, supply chain management, logistics
JEL classifications: L93, R41, L63, O14
Date submitted: 14 October 2002 **Date accepted:** 2 June 2003

1. Air cargo services in the global economy

The thousands of routes operated by the world's airlines have become vital arteries of trade. Between 1980 and 2000, the volume of international air freight traffic, measured in freight tonne-kilometers, grew fivefold (*ICAO Journal*, 2001). With its growing importance, air cargo has taken a prominent place among the services from which states, regions, and firms can derive competitive advantage in global markets. Air cargo has become the principal mode of international transport for a wide variety of (especially knowledge-intensive) goods. The range of 'air-eligible' goods continues to widen as more firms look to the speed and precision of their supply chain management strategies as sources of competitive advantage. At the same time, air cargo services facilitate the internationalization of production networks at new scales, better enabling firms to leverage the assets of more distant regions.

* Department of Geography, University of Kentucky, Lexington, KY 40506–0027, USA.
email <leinbach@uky.edu>
** Department of Geography and Urban Planning, University of Wisconsin, Oshkosh, Oshkosh, WI 54901–8642, USA.
email <bowenj@uwosh.edu>

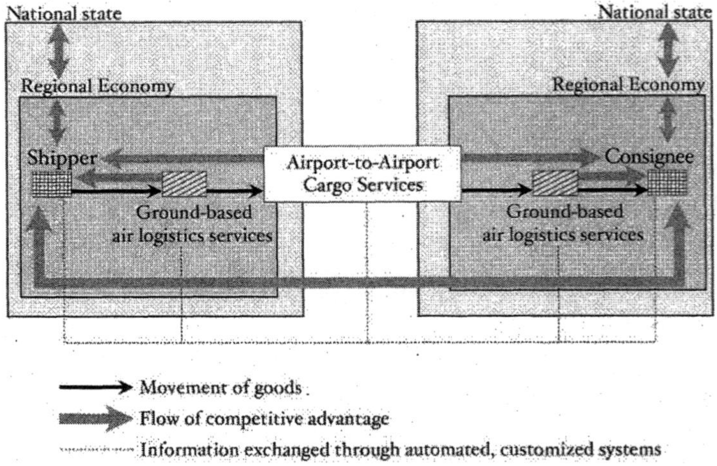

Figure 1. Situating air cargo services.

We examine the use of air cargo services in Southeast Asia, a region that is distant from the world's principal economies and important as an electronics export platform. Air cargo services are vital, of course, to the incorporation of peripheral regions like Southeast Asia into the international economy, but because the quality and capacity of air cargo services (including ground-based air logistics services performed in airport-adjacent logistics parks) vary spatially, such services can be a source of competitive advantage to firms and to the regional and national economies in which those firms are embedded (Figure 1). In turn, the quality and capacity of air cargo services are dependent, in part, on the policies of national states and the importance of air-eligible products (e.g. especially light-weight, high-value electronics) within regional economies. In this context, we argue that air cargo services can play a critical role as an advanced producer service that 'holds down' global production networks in particular places, a theme we explore in detail below.

Air cargo services include both the airport-to-airport carriage of goods performed by airlines as well as related air logistics services performed on the ground, usually by freight forwarders. Air logistics include, for instance, truck transport from an airport to a client firm's manufacturing plant but also highly customized services that use information technology and airport-adjacent third-party[1] warehouses to accelerate the distribution of finished goods and minimize the inventory of component parts. A growing number of manufacturers rely on such services to secure an advantage in international competition. As suggested in Figure 1, the competitive advantage enjoyed by air cargo services firms (e.g. the degree to which they are 'world class' or based in a city with unusual centrality in global airline networks) can be the basis for competitive advantage in a user firm.

In this paper, we seek to understand variation in the intensity of air cargo services usage among electronics manufacturing firms in Malaysia, the Philippines, and Singapore. These firms use such services both as shippers (when exporting components or

1 In the context of logistics services, third-party refers to a firm that carries out logistics services like warehousing and distribution on behalf of either the shipper (first party) or consignee (second party).

finished products) and as consignees (when importing components and raw materials). We relate the insights drawn from the analyses of over 120 interviews with firms, both locally and foreign-owned, to the existing literatures concerning global production networks and advanced producer services. We argue that foremost among the major factors influencing firms' air cargo services usage are the degree to which production has been internationalized, the nature of the good being produced, the importance of speed in a firm's supply chain (and distribution chain), and the degree to which a firm is a logistics decision-maker or decision-taker in the production networks in which the firm is positioned.

While air cargo still accounts for only a relatively small share of world trade (particularly when compared with the physical volume of sea freight), its importance has risen rapidly in the past two decades. The expansion of air cargo has been driven by several interconnected factors. First, the development of faster, larger, longer-range aircraft has contributed to a long-term downward trend in real air freight rates of about 3% per year, rendering air cargo services more affordable for more goods (Boeing Commercial Airplane Group, 2000). The introduction of the Boeing 747–400F, especially in the Pacific Basin, during the past decade was very important in this regard; and the launch of the ultra-large Airbus A380 freighter in 2008 will perpetuate the same trend. Second, the liberalization of air freight services in many markets has permitted an expansion of services and an intensification of competition, reinforcing the decline in air freight rates. Third, the internationalization of economic activity and concomitant disaggregation of production networks have increased the significance of air cargo services not only to firms but also to regions and nations. Fourth, the proliferation of just-in-time material management practices has helped to redefine air cargo for many firms from an emergency recourse to a regular feature of supply chain management strategies.

The term supply chain management denotes the ever-increasing emphasis on managing the external relations of production and the control of resource flows from source to consumer (Mentzer, 2001). The purposes are: first, to aim for improvement in logistics performance including greater reliability, smoother flow through the chain, and more efficient connections between the various links in the chain and second, to realize the lowest possible cost for the chain as a whole. The nature of the supply chain and, we would argue, its incorporation of air cargo services depend on several factors. These are the complexity of the product, the number of available suppliers, and the distances over which the supply chain is stretched (Lambert, 2001, pp.103–105). Supply chain management is raised here as a key factor in the explanation of the demand for air cargo service (Sturgeon, 2001). Porter (1990) argued that the competitive advantage of firms rests on product cost and/or differentiation. Air cargo services, as part of a supply and value chain management strategy, can contribute to both product cost and/or differentiation: through inventory minimization in the case of the former and speed-to-market in the case of the latter (see Gallacher, 1998 and Meersman and Van de Voorde, 2001, on time-based competition).

Supply chains embedded in the Pacific Rim are particularly reliant on air cargo services. Though air cargo has grown rapidly in markets worldwide, its expansion in Asia has been stunning. Air cargo on routes to, from, and within Asia-Pacific grew at an average annual rate of 9.8% from 1980 to 2000 (Boeing, 2002), faster than for any other region. By 2001 nearly half of the world's freighter aircraft capacity was deployed on Asia-Pacific routes (Bowen, 2004). Until the onset of the Asian financial crisis, cargo growth on routes *within* Asia exceeded those between Asia and the rest of the world. Intra-Asian cargo flows manifest the increasingly complex relationships among Asian economies (Poon, 1997). In the computer industry, for instance, Singapore, Taiwan, Malaysia, and Thailand occupy different niches with a consequent flow of components among them. Much of this traffic is intra-firm, reflecting the empowerment of the transnational corporation as manager of the new international division of labor (Yeung, 2001). The increasingly complex overlay of manufacturing and services in these markets exemplifies the concurrence of industrialization and tertiarization in some developing economies, while the most advanced economy in the region (Singapore) better fits the conjunction of deindustrialization and tertiarization in the developed world (Williams, 1997, p.18).

2. Air cargo services as advanced producer services

The stature of air cargo services in the contemporary international economy illustrates several features of services more generally. First, services facilitate economic transactions and play a vital role in development (Daniels and Moulaert, 1991; Daniels, 1993; Begg, 1994; Illeris, 1996). Second, services may be used to secure competitive advantage in the global economy (Porter, 1990, 1996; Dicken, 1998). Third, national economies are becoming less insular and more interdependent, a process that is inextricably linked to the presence of 'enabling technologies'—transportation, communication, and organizational innovations that aid in the internationalization process (Dicken, 1992, pp.106–107). Together these changes are manifest in the marked increase in production, consumption, and trade of services, especially producer services (Wood, 2002).

One aim is to situate our work in the theoretical literature on advanced producer services (Daniels and Moulaert, 1991). In this regard we seek to show the degree to which prevailing explanations of demand for producer services apply to air cargo services (e.g. Goe, 1990, 1991; O'Farrell, 1993; O'Farrell et al., 1993; Beyers and Lindahl, 1996). At the outset it is worth noting that air cargo services have received little attention in the broad economics (e.g. Button, 1990; Fennes, 1997; Button and Owens, 1999) and geographical literatures (e.g. Raguraman, 1997; Rimmer, 1997; Loughlin, 1998) and virtually none in the producer services literature (Nusbaumer, 1987; O'Connor and Hutton, 1998; Ho, 1998). The neglect of air cargo services, along with other transport services, by producer services researchers is rooted in the perception that transport services lack the sophistication (e.g. skill requirements, specialization) of 'advanced producer services' (Harrington, 1995) and consequently display a 'more even spatial distribution' (Williams, 1997, p.34), thereby minimizing their impact on urban and regional development. At least in the case of air cargo, this perception is contradicted by the development, especially in the 1990s, of specialized air cargo and ancillary logistics services whose supply and sophistication vary across the hierarchy of world cities in a fashion that is similar to other producer services.

Our exploration of air cargo services is intended to respond to the call for research on services and the new economy (Beyers, 2002; Wood, 2002). In this regard, our research draws upon several themes in the advanced producer services (APS) literature (e.g. Daniels and Moulaert, 1991, and special issues of *Growth and Change: A Journal of Urban and Regional Policy*, Vol. 22, No. 4, 1991 and *Papers in Regional Science*, Vol. 75, No. 3, 1996). Most critical to our work are the factors affecting demand. In deepening our understanding of the demand for producer services, we must distinguish among those factors related to the structure of firms, industries and regional economies. For example what is the importance of scale of both production and corporate organization? Scale may be neither a necessary nor sufficient condition for a greater and more diversified demand, as the actual structure, development phase, and strategy of the firm further complicate the picture. In particular, firms covering several stages of the production process (vertically integrated) may demand a wider range of services. The type of industry and product (as well as its life-cycle stage) also affect demand. The demand for producer services by a firm manufacturing standardized products for mass markets will differ from one offering customized products for a specialized market. At the same time, the innovative content of products and processes will likely affect the nature of demand (Martinelli, 1991, p.25).

A separate, important theme of the producer services literature has been the performance and competitiveness of business service firms. This aspect of the literature is critical because the competitiveness of a business service firm can be a source of competitive advantage for a user firm, especially if the latter is the focal (or dominant, organizing) firm in a production network. For example, one study examined the effectiveness of producer services in promoting the efficiency of manufacturing enterprises in regional economies (O'Farrell and Hitchens, 1990). That research attempted to develop methodologies to test the effectiveness of producer services in promoting the efficiency of manufacturing firms in peripheral areas by measuring the utility of service inputs and the quality of producer service firm outputs. Still another study has pointed to the role of producer service outsourcing in the innovative performance of manufacturing firms (MacPherson, 1997).

Another related line of inquiry concerns the externalization of services functions by manufacturing firms. Conventional wisdom suggests that the growth of producer services has been attributable to cost-driven factors and vertical disintegration processes. Yet research in this field shows that cost-driven externalization is not the most important force underlying growth. Rather the need for specialized knowledge combined with a variety of other cost, quasi-cost, and non-cost driven forces are critical (Beyers and Lindahl, 1996). Evidence from still another study suggests that the expansion of demand is the primary cause of increasing business service output and not restructuring strategies as predicted by a flexible firm model (O'Farrell et al., 1993).

The externalization of services may seem taken for granted with respect to transportation services since they have been so seldom performed by manufacturers themselves (in contrast, for example, to in-house legal services). But in fact the trend towards externalization has affected the air cargo services industry, too. Both the search for specialized knowledge and the desire to reduce costs have been important drivers in the rapid growth of third-party logistics that are increasingly provided in conjunction with conventional transport services. Since the mid-1990s, freight forwarders and other firms have begun to offer more customized logistics services to manufacturers enabling the latter to minimize inventories, more tightly time production and delivery schedules, and reduce transit times. Such services are most advanced in the United States where third-party contract logistics grew faster than gross domestic product for eight consecutive years to 2002 (Armbruster, 2003); but they are also growing rapidly in Asia (*South China Morning Post*, 2002). The quality and availability of such services varies widely (Bowen and Leinbach, 2004).

O'Connor and Hutton (1998) examine the applicability of advanced producer services themes more generally to the Asia Pacific region. In a special issue of *Asia Pacific Viewpoint*, Daniels (1998) explores the idea that in this region the relationships between economic growth, industrial transformation and the rise of producer services appears to be different from those found in North America and Europe (e.g. the role of ethnicity in shaping business linkages) and such services are less prominent than expected. K. C. Ho (1998), in the same issue, sketches the locational dynamics of regional functions in the Asia Pacific region, emphasizing how generic factors such as proximity to company affiliates and market access apply. Among his findings are that industry-specific dynamics for air delivery and online information services relate to location and competitive advantage. Our research casts further light on the demand for producer services in this dynamic region.

3. Air cargo services in global production networks

Producer services, including air cargo services, have not only facilitated the internationalization of economic activity but have also fostered the development of more sophisticated production relations. Especially important has been the development of more sophisticated supply chain management practices. In fact one of the most significant paradigm shifts in modern business management is that individual businesses no longer compete as solely autonomous entities, but rather as supply chains (Lambert, 2001, p.99). A remarkable shift in corporate strategy and operational activity has occurred over the last decade with the externalization of production rendering many corporations heavily reliant on external resources (Meersman and van de Voorde, 2001, p.69). A result is that a significant proportion of competitive advantage rests with the management of these external resources (Hall and Braithwaite, 2001, p.81). Dicken and Thrift (1992) have pointed up that the production chain is a complex, dynamic system and note that 'the inter-firm structure of large corporations is better represented as a network than a hierarchy'. But increasingly we have learned that even small and medium enterprises reveal a large degree of network organization to enhance flexibility, delivery and cost competitiveness (Hall and Braithwaite, 2001, p.94).

These developments are manifest most in the emergence of the global production network, a term used to describe 'the globally organized nexus of interconnected functions and operations through which goods and services are produced and distributed' (Coe et al., 2003, p.18). In the context of contemporary internationalization, regional development is contingent upon the 'coupling' of the strategic needs of 'trans-local' firms and region-specific assets (e.g. low-cost labor) through a global production network. The term network refers to 'both a governance structure and a process through which disparate actors and organizations are connected in a coherent manner for mutual benefits and synergies' (Yeung, 2000, p.302). Relational networks have been propounded as 'the foundational unit of analysis for our understanding of the global economy, not individuals, firms or nation states' (Dicken et al., 2001). Such networks unfold across multiple scales and are powerfully shaped by the state.

How do producer services, in general, and air cargo services (and related logistics services), more specifically, factor in global production networks (Figure 2)? First, the cost, capacity, and reliability of producer services help to define the scale over which such networks are realized. Despite the fascination with globalization, much of the internationalization of economic activity has not been global in scale but has instead been concentrated within Europe, within North America, and within East Asia (Poon, 1997; Yeung, 2001; Lai and Yeung, 2003). There are a number of reasons for this pattern but one is the continued high cost (in terms of money, time, and managerial complexity) of maintaining truly global supply and distribution chains. The development and elaboration of advanced producer services mitigates those constraints. In particular,

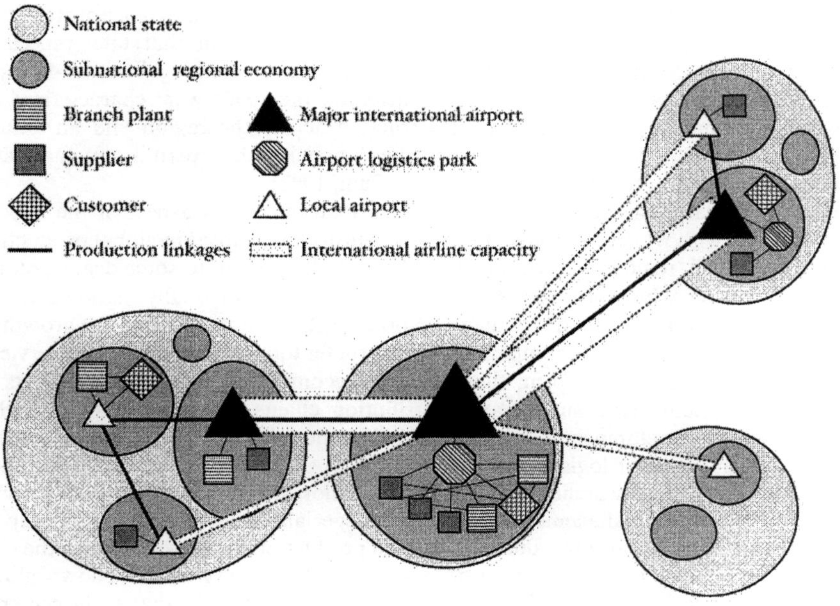

Figure 2. Air cargo services and global production networks.

the continued decline in real air freight rates and the expansion of air freight networks facilitate the rescaling of production networks.

Second, beyond their role as an enabling mechanism for internationalization, producer services play a more specific role in funneling the extension of global production networks along certain corridors. The geography of internationalization has created an 'archipelago economy' (Veltz, 1996, cited in Coe et al., 2003) in which places well integrated into the global economy are surrounded by less privileged places that are not. Air cargo services, like other dimensions of the distributive infrastructure (Martinelli, 1991) of the international economy, reflect and reinforce that geography. Firms evaluate the viability of a new node in a global production network based in part on the cost and availability of transport and communications infrastructure tying that node to others already in the network. In Asia-Pacific, for instance, the density of global production networks and the density of transport networks (represented by the overlap of production linkages and high capacity corridors in Figure 2) augment one another.

Third, producer services are, like the service sector more generally, still subject to pervasive state regulation, illustrating the refusal of the state to 'whither away' as an influence upon the global economy (Smith et al., 2002). On the contrary, the territorially specific policies of the nation-state create bounded spaces within which global production networks are articulated 'in highly differentiated ways' (Coe et al., 2003). The state's influence on air cargo services and, through those services, its influence on the structure of global production networks is manifold. States regulate airline competition to varying degrees (thereby affecting the quality, quantity, and costs of services provided), play an important (though diminishing) role in infrastructure provision, and implement labor policies that affect the cost of air cargo services. And yet, the importance of attracting investment by trans-local firms in global production networks means that states, especially small states, are compelled to pursue similar strategies in creating business-friendly environments. In Asia-Pacific aviation, for instance, the 1990s were characterized by massive investments in new airport infrastructure across the region and an almost region-wide shift towards liberal 'Open Skies' policies that permit much greater freedom of operation to international airlines (Oum, 1998).

Fourth, air cargo services, broadly defined, are an important aspect of the region-specific assets with which the needs of trans-local firms are coupled in global production networks. Indeed, regional development has become dependent, to some degree, on the effectiveness with which producer services expertise interacts globally, nationally, and regionally to support productive demand (Wood, 2002). As noted earlier, the growth of air cargo volumes has been accompanied by the proliferation of related logistics services; in their most sophisticated form, these services become highly customized efforts to restructure a client-firm's supply and distribution chains. Such services are highly uneven in their development across the global economy (represented by the uneven distribution and size of logistics parks in Figure 2). In Southeast Asia, for instance, Singapore hosts large warehouses from which major air freight forwarders provide region-wide logistics management services. Such specialized services enable Singapore to better enhance and capture value in the global production networks in which the city-state is incorporated. In this respect air transport linkages and services 'hold down' global production networks (particularly in the vicinity of major hubs like that at the center of Figure 2) and may unleash, or conversely constrain, regional potential.

Fifth, global production networks are governance structures through which power is effected; air cargo services and related logistics services can be a conduit through which such power is exercised. For instance, an important development in air cargo services has been the development of electronic tracking that permits an unprecedented degree of precision for firms monitoring the movement of components. Many major manufacturers have developed rigorous standards that supplier firms must meet; focal firms then use the information from electronic tracking to evaluate the suppliers' performance. Consequently, production linkages, such as those shown in Figure 2, are characterized by reciprocal flows of goods and information, with both types of traffic moving at unprecedented speed and in unprecedented volumes.

An alternative way of viewing air cargo services as a conduit for the governance of a global production network is to distinguish between push supply chains and pull supply chains (Hall and Braithwaite, 2001, p.71). In the former, costs are transmitted up the chain with little control over the cost structure of the entire chain. In contrast, a pull supply chain operates on the principle that the supply chain must be able to deliver a product to market at an affordable level. In a global production network organized around a pull supply chain, the focal firm often orchestrates the logistics management strategies of other firms in the network.

Sixth and finally, air cargo services facilitate the realization of relational rents in global production networks (Kaplinsky, 1998; Khan and Jomo, 2000). Relational rents are abnormally high profits secured through a firm's superior relationships with other firms or with institutions of the state (e.g. cronyism). Rent is partially contingent upon access, and because air cargo services shape patterns of access within a global production network, such services can engender relational rents. Links among firms that can foster an overwhelming competitive advantage (i.e. the ability to secure rent) such as the much faster acquisition of spare parts or raw materials and components, may depend upon the availability of sophisticated producer services such as air cargo. Much as a family connection in an important government regulatory agency might enable a firm to achieve an artificially high profit margin, so too an unrivaled connection to a supplier through a network of air cargo services might enable a firm to also realize relational rent.

In sum, air cargo services are, for a growing number of firms and a growing variety of products, an integral part of supply chain management which is in turn a system through which goods, services, money, and power flows among the actors linked by a global production network. Our principal interest is in better understanding variation in the demand for air cargo services among electronics manufacturers in Southeast Asia. All of these firms are incorporated into global production networks, most into multiple networks. Almost all are under great pressure to accelerate their supply and distribution chains. And yet, as shown below, their reliance on air cargo services varies widely. We model that variation as the function of product, firm, and place characteristics.

4. Data gathering

A basic survey instrument was constructed using a series of categorical and ranking questions. In addition several open-ended questions were included to generate discussion by respondents of key factors linking a firm's use of air cargo services and the firm's competitive advantage (Schoenberger, 1991; Fowler, 1993; Markusen, 1994). The instrument was field tested in Singapore prior to beginning formal interviews. Data was gathered from a total of 126 firms in three markets with the following breakdown: Penang (41 firms), Singapore (38), Manila (24), and Kuala Lumpur (23) (see Figure 3). Firms were selected through a random sample stratified by market location. Business directories for each market were used to generate lists of electronics firms. Firms were then contacted by telephone to arrange survey and interview schedules.

For the purposes of this study, firms were classified using definitions modified from the 1997 North American Industry Classification System (NAICS).[2] Five firm classes were used:

1. Precision Equipment ($n = 22$)
2. Wires and Cables ($n = 21$)
3. Consumer Electronics ($n = 20$)
4. Computers and Peripherals ($n = 28$)
5. Semiconductors ($n = 35$)

These classifications are subdivided into manufacturing, assembly, and sales and service or headquarters firms. The majority of firms surveyed are involved in manufacturing activity. These firms produce components on-site, which distinguish them from assembly operations, which do not.

The survey instrument was administered either on-site at each firm in a face-to-face situation, or via fax/email when scheduling for face-to-face administration was not possible. Nearly all of the responses to the survey instrument were obtained by administering the questionnaire on-site at each firm in a face-to-face situation. Of the 126 firms interviewed, only nine preferred to respond by fax/email. The survey instrument comprised four sections of which Section 1 contained questions regarding firm type, employee characteristics, firm ownership, and geographical location of operations.

2 See http://www.census.gov/epcd/naics/NDEF334.HTM for definitions.

Figure 3. Study region.

Table 1. Sample firm characteristics by place

	Singapore	Penang	Kuala Lumpur	Manila
Sample size	38	41	23	24
Percent manufacturing/assembly	63.2	90.2	82.6	95.8
Percent sales/service/regional HDQ	36.8	9.8	17.4	4.2
Percent locally owned	32.4	11.4	13.6	15.8
Percent knowledge intensive	50.0	17.1	17.4	33.3
Mean air cargo intensity index	5.39	6.02	3.16	6.34
Mean cycle time (days)	33.9	18.1	27.6	8.9
Mean DISTANCE* (kilometers)	6,649	7,873	6,180	5,946

*See Appendix for variable definition.

Section 2 sought data on principal logistics for a firm's primary product moving by air cargo. This section requested data on the geography of source materials, transport mode, product path from firm to customer, and final product destination. A sub-section dealt with the use of special logistics services used by firms (e.g. overnight delivery; third party logistics). Section 3 contained questions regarding a firm's interaction with the air cargo industry including use of freight forwarding services and criteria for the selection of air cargo carriers. Finally, Section 4 included several ranking and open-ended questions regarding air cargo services and firm competitive advantage.

5. Characterizing the sample of air cargo user firms

Though Singapore, Malaysia, and the Philippines share a dependence upon the electronics industry and were selected as study settings for that reason, they occupy different places in the global electronics industry. Those differences are apparent in some of the basic characteristics of the 126 sample firms. First, firms classified as manufacturers or assemblers dominate the samples in Penang, Kuala Lumpur, and Manila but represent only about 60% of the sample in Singapore (Table 1). A large share of the Singapore firms comprises sales, service, and/or regional headquarters reflecting the shift in the city-state towards a postindustrial economy.

It is important to note that the large share of non-manufacturing firms does not mean that air cargo services are becoming less important to the economy of Singapore. On the contrary, most of the non-manufacturers interviewed there specifically cited the availability of air cargo services as a significant factor in the decision to locate functions in Singapore. For example, one joint Japanese-American firm in the sample, AVX-Kyocera, had shifted the manufacture of passive components (e.g. capacitors and resistors) out of Singapore to nearby Batam, Indonesia and Penang in the 1990s but transformed its Singapore operation into a large global distribution center. The distribution of products by the firm is critically dependent on the ample air cargo capacity from Singapore as well as Singapore's status as a free trade zone (in which it is less expensive to hold stocks).

Second, the four sample sites were also distinguished by the types of electronics firms that were interviewed (Figure 4). The 126 sample firms were categorized into five groups based on their principal product: computers and peripherals (28 firms), consumer electronics (20), precision equipment (22), semiconductors (35), and wires and cables (21). It should be noted that not all firms were assigned easily to one of these categories. To improve the accuracy of assignment, several individuals with expertise in the electronics industry were consulted.

Perhaps most strikingly, the proportion of sample firms classified as semiconductor firms differed markedly across the four cities. Singapore, which was a key player in the semiconductor industry in the 1960s and 1970s, has now lost much of this business, particularly the lower value-added assembly operations. It remains, however, an important center in the fabrication of valuable silicon wafers from which the semiconductor chips are cut. Malaysia occupies an intermediate position in which both high-end (wafer fabrication) and low-end (cutting and processing of chips) functions are undertaken. And the Philippines' semiconductor industry remains concentrated in labor-intensive operations.

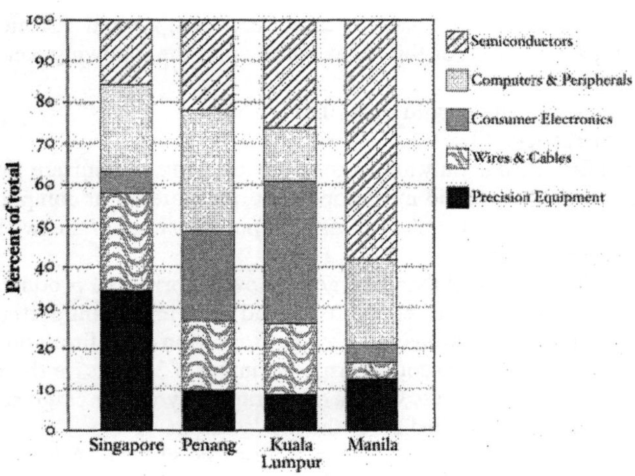

Figure 4. Product class by place.

Firms in other product categories balanced the smaller proportion of semiconductor firms in Malaysia and Singapore. In particular, a large number of firms in both Penang and Kuala Lumpur produce consumer electronics; examples include walkie-talkies and color television sets. In Singapore, precision equipment firms comprised a disproportionately large share of the sample; examples include the manufacturers of specialized alternators and motors.

Finally, once more reflecting differences in the level of development across the four sample markets, Singapore's sample was distinguished by the higher level of local ownership and a higher proportion of knowledge-intensive operations (Table 1). Nearly a third of the Singapore firms were locally owned. In contrast, the same proportion was below 15% in the other three sampled markets. And half of the Singapore respondents described their workforces as mainly knowledge-intensive. Conversely, more than 65% of Philippine firms and 80% of Malaysian firms described their workforces as mainly labor-intensive.

6. Analysing the intensity of air cargo usage

As a comprehensive measure of the intensity with which firms use air cargo services, an index was formulated based on firms' responses to several questions. The index was constructed by assigning points to firms that had certain characteristics or exhibited certain behaviors. The number of points earned for each characteristic or behavior is necessarily somewhat arbitrary, but we have drawn on our knowledge of the industry and that of our interviewees in developing an index. For example, the disparity in the number of points a firm earned for using express air cargo services on a regular basis versus only in emergencies was based on the difference in costs associated with express air cargo services versus conventional air cargo services. After evaluating alternative ways of formulating the index, we are satisfied that the approach presented below offers the best approach for estimating an air cargo index that is both reliable and valid.

The index incorporates elements measuring:

1. The degree to which a firm used air cargo services in general. A firm was assigned points (in parentheses) for this index based upon the following conditions:
   on a regular basis (3.0)
   to meet peak season demand only (0.5)
   only in emergencies (0.25)
2. Whether a firm used mainly air cargo to move the most important raw material or component and the second most important raw material or component used to make its principal product. A firm earned 1 point for each of the two inputs moved mainly by air freight.
3. Whether a firm used mainly air cargo to move its principal product to its most important destination market and its second most important destination market with each index increment weighted by the proportion of the firm's output that went to each destination. A firm could earn a maximum of 2 points for this item if each of the two destination markets were served principally via air cargo services and if those two markets absorbed all of the firm's output.
4. The degree to which a firm used express air cargo services. A firm was assigned points (in brackets) for this index based upon the following conditions:
   on a regular basis for most or all products (5.0)

on a regular basis for some products (2.5)
to meet peak season demand only (1.0)
only in emergencies (0.5)
no express (0)

The measure was constructed by adding the four components with the appropriate weights being assigned to individual elements noted above. The resulting air cargo intensity index ranges between 0 and 12. For example, AVX-Kyocera in Singapore has an index value of 6.3 because it uses general air cargo services on a regular basis (3.0 points), it uses mainly air cargo for the import of both of its two main raw materials (tantalum and ceramic capacitors) (2.0 points), it delivered goods to its second most important destination market (South Korea which absorbed about 40% of the firm's output) by air (0.8 points), and uses express for emergency situations (0.5 points). The mean value of the index for the 126 sample firms was 5.4. The index is somewhat bimodal, with one peak near zero and a second, stronger peak just above the overall sample mean.

The intensity index showed a marked variation across the five product categories (Figure 5). Unsurprisingly, the highest mean index was recorded by firms grouped in the semiconductor category (7.6) and the lowest by firms in the consumer electronics category (2.5). Semiconductor firms produce goods that have an extremely high value-to-weight or value-to-volume ratio and therefore a high capacity to bear the costs of air transport. More importantly, semiconductors have very short product life cycles and semiconductor manufacturers are under great pressure to shorten cycle times—that is the elapsed time between a new order being placed and the order being filled at the customer's premises. Conversely, consumer electronics (e.g. computers, printers, refrigerator ice-makers) are bulkier; less valuable; and less sensitive to corrosion at sea and are characterized by somewhat longer product obsolescence cycles. Moreover, the long experience of the semiconductor industry in the use of air cargo services to integrate global production networks is reflected in the greater sophistication of semiconductor firms in the use of such services and in the willingness of many such firms to accept higher transportation costs to secure speed-to-market advantages.

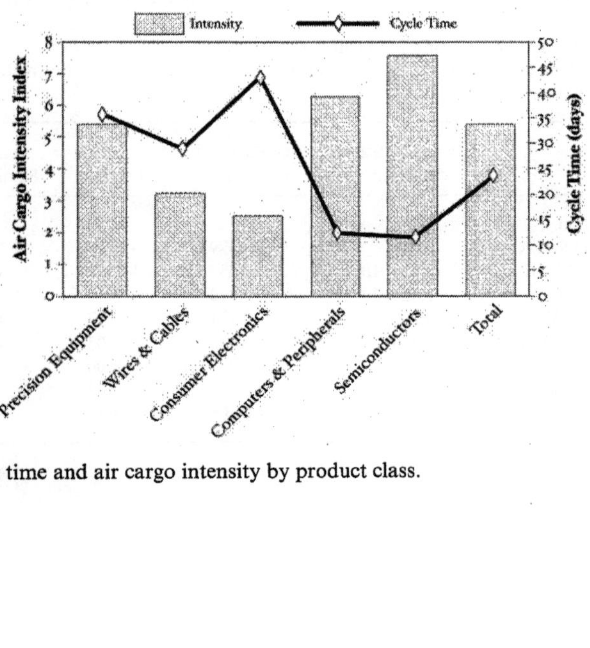

Figure 5. Cycle time and air cargo intensity by product class.

The air cargo intensity index also varied by location, with Manila recording the highest average value and Kuala Lumpur the lowest (Table 1). The variation across the four markets can be largely explained by the aforementioned differences in the proportion of different product categories (e.g. the preponderance of semiconductor firms in Manila and consumer electronics manufacturers in Kuala Lumpur).

7. Analysing the use of air cargo services

In attempting to learn more about the variability of the use of air cargo services and its correlates we examined a variety of plausible firm characteristics represented by eight independent variables (variables are listed in the Appendix). In the initial stage of ordinary least squares modeling, the most significant variable in the analysis was DISTANCE (Table 2) with which the response term was significantly and positively related. DISTANCE is a ratio scale variable measuring the distance associated with both the backward (major inputs) and forward (major destination markets) linkages for a firm's principal product. Greater distances are indicative of greater internationalization in which the effort to leverage uneven factor endowments in dispersed economies is often predicated on the regular use of air cargo services.

The importance of this variable in a specific context is revealed by the example of Seagate Technology, the world's largest manufacturer of disk drives. The Seagate plant at Penang has been operating there for 12 years and currently employs approximately 4,000 workers. The principal product is a 'slider' (read-write device). The main input material is the wafer, which is shipped to the Penang manufacturing plant 'just-in-time' from the US and Ireland. The finished slider is then shipped to both Bangkok and Seoul where it is mounted on a head gimbal or arm of the disk drive. Subsequently, the loaded gimbals move forward in the production chain from those nodes to computer assembly operations (principally Dell and Acer) across Asia. The distances over which this global production and distribution network is spread and the necessity for speed mandate the use of air cargo services.

The other significant variable in the initial model was CYCLTIME, which was negatively related to air cargo services intensity. CYCLTIME, a ratio scale variable measuring the number of days from the time an order is received until the goods are

Table 2. Explaining variation in air cargo intensity
Dependent variable: air cargo intensity index; $n = 106$ (incomplete information was available for 20 firms)

	Unstandardized coefficient	Standardized coefficient	*t*-score	Significance
Constant	4.406	2.72	0.008	
CUSTOMER	−0.701	−0.099	−1.14	0.256
CYCLTIME	-3.02×10^{-2}	−0.234	−2.61	0.011
DISTANCE	3.02×10^{-4}	0.310	3.41	0.001
EMPLOYEE	0.200	0.171	1.17	0.244
KNOWLABR	0.384	0.095	1.05	0.296
MATSRC	−0.332	−0.087	−1.00	0.318
MCTIME	−0.117	−0.090	−0.98	0.329
PRODDEST	−0.739	−0.100	−1.15	0.254

Adj. R-square 0.255.

delivered to customer, is a critical dimension of a firm's logistics management strategy. Short cycle times, often only several days in length, are achieved by trading off the high cost of air cargo (and even express air cargo) for the competitive advantage gained through faster deliveries to customers. Firms with faster cycle times were much more likely to use air cargo services and to use them intensively. For instance, among firms that reported they used air cargo services on a regular basis, mean cycle time was 17.1 days while the mean cycle time for firms that used such services only in emergencies was 46.6 days. In other words, incorporation into a global production network was insufficient to engender reliance on air cargo services. The nature of the product and the manner in which it is produced mediate the influence of spatial disaggregation of production.

Semiconductor firms had shorter cycle times than others and that helps to explain the aforementioned variation in air cargo intensities by product class (Figure 5). The differences in cycle times by product give rise to differences in cycle time by place (Table 1). Interestingly, Singapore firms had the highest mean cycle time. This result is, at first glance, counter-intuitive but makes sense insofar as Singapore has moved towards a post-industrial economy. Within that economy, manufacturing has shifted towards highly specialized, very high value-added operations for which the pressure to deliver finished goods rapidly to a customer has eased. Conversely, many of the semiconductor manufacturers that dominate the electronics industry in Penang and Manila are under intense competitive pressure. Time is a key dimension along which that competition is waged. And yet for firms in each of these markets, despite their varied position along an industrializing-postindustrial spectrum, the quality of and capacity of air cargo services are important aspects of the region-specific assets that help to hold down global production networks.

Another case study further illustrates the importance of cycle times. Intel, a US-based firm, the world's leading manufacturer of microprocessors, has operations throughout Southeast Asia. Its Manila and Penang branch plants were included in the sample for this study, and both had very high values for the air cargo intensity index (12.0 and 11.6 respectively) but differing cycle times (4 days and 14 days, respectively). Both branches indicated that express air cargo is used for most shipments. With the economic slowdown in the semiconductor industry, Intel in Manila has begun to de-emphasize express services but there are significant constraints on the extent to which it can do so. Indeed the trend towards e-commerce and smaller, door-to-door shipments are likely to increase Intel's reliance on complex air cargo services (not simply express). In these cases, air cargo services are customized and akin to the custom-based production emphasized by Nilsson (1996, p.8). Custom-based manufacturers cater to the needs of 'sophisticated and demanding buyers', much as air cargo service firms increasingly offer specialized, customized services to sophisticated and demanding manufacturers.

The remaining variables (described briefly in the Appendix) were statistically insignificant. This outcome and the sizeable unexplained variation fueled our effort to develop a richer model more consonant with the complexity of firm behavior. Our examination of the data for nonlinearity among variables did not yield any insights, but as described in the next section, the incorporation of several interaction terms in the model was helpful.

7.1. The introduction of interaction effects

In addition to the additive effects of our initial variables, the multiplicative effects produced by certain interaction effects were incorporated into the model (see variable descriptions in the Appendix). In this fashion, we examine how certain variables mute or enhance or otherwise mediate the effect of others. We were particularly interested in the manner in which the size of firms and the knowledge intensity of their workforces filter or catalyse other variables. For example, the variable CYCL*EMPL is the product of cycle time and the number of local employees. This variable assesses the degree to which the impact of cycle times on air cargo intensity differs between large and small firms. As noted above, cycle time is significantly and negatively related to air cargo intensity. The interaction variable CYCL*EMPL can be used to measure the degree to which the local size of a firm magnifies or minimizes that effect.

The use of interaction terms like those described above may introduce the problem of collinearity (Aiken and West, 1991). To minimize collinearity, we 'centered' each interaction term (Draper and Smith, 1998). That is, each variable was expressed in terms of each value's deviation from the variable mean (e.g. each firm's value for CYCL*EMPL is the product of its deviation score of CYCLTIME and EMPLOYEE). The use of centered terms largely eliminates the problem of collinearity and its deleterious impact on parameter estimates.

In the enhanced model, DISTANCE and CYCLTIME remain the most significant variables. One more main effect variable emerged as significant in the second model: the binary variable indicating whether the customer has the greatest influence over the choice of air cargo services. Somewhat surprisingly, the sign on the variable is negative suggesting that in firms where the customer drives logistics ('pull' supply chains), the intensity of air cargo usage is somewhat lower. Firms that identified the customer as the greatest influence in air cargo usage tend to be firms whose position in relationships with customers is subordinate and that subordination is likely related to the performance of relatively low value added functions. Of the firms that gave the answer 'the customer', 36% were assemblers specializing in low value added operations; in contrast, assemblers represented only 27% of the firms where the customer was *not* the key decision-maker.

Although the knowledge versus labor intensity of a firm's workforce and employment size did not emerge as directly significant, both were implicated in the interaction-effect terms that were significant in the second model. First, the variable CYCL*KNOW was significantly and positively related to the air cargo intensity index. CYCL*KNOW is a product of CYCLTIME and an ordinal variable KNOWLABR indicating whether a firm's local workforce was mainly labor-intensive or knowledge-intensive. The significance of this interaction effect variable result can be interpreted as indicating that for firms with knowledge-intensive workforces, longer cycle times were nevertheless associated with high air cargo intensity indices. For example, Advantest, a Japanese-owned firm with a substantial sales and distribution facility in Singapore, has a very long cycle time but is highly dependent on air cargo services. The firm manufactures extremely sensitive automatic test equipment for use in the semiconductor industry. The sophisticated nature of that equipment is manifest in the degree to which the Advantest's workforce is knowledge-intensive. The firm has a well-articulated logistics strategy within which air cargo services play a decisive role. Advantest's products may be ordered four to six months in advance of delivery, but once produced the test equipment is so valuable and so sensitive that no mode other than air is suitable for long-distance transport.

The intensity of air cargo services usage is transmitted through the global production networks of which Advantest is a part from semiconductor manufacturers (for whom the intensity of time-based competition is pervasive) to this equipment manufacturer. It is worth noting that the nature of a firm's logistics management strategy is just one of several forms of firm practice that are transmitted along such networks. Far more than simply flows of goods, global production networks are also manifest in the flow of money, information, ideas, and practices. The late 1990s Asian financial crisis, for instance, was preceded by the cascade of bullish capacity expansion from one firm to another through the networks that linked them which gave way, when the crisis began, to the flow of mediating strategies from firm to firm (Lai and Yeung, 2003).

Another interaction-effect term, CYCL*DIST, has a simpler interpretation. The negative coefficient indicates that, while long production network distances are strongly associated with air cargo services, that association is muted when cycle times are long (and therefore delivery is often less urgent). For instance, Xerox's Penang branch plant had a moderate air cargo intensity index (5.5) despite the fact that the distances across which its raw materials and finished products move were vast. Xerox manufacturers solid ink printers in Penang. The principal components are sourced in the USA, Japan, and Germany; and the main destination markets for the finished goods are the USA, the Netherlands, and Japan. The attenuation of Xerox's production linkages is somewhat offset by the more relaxed cycle times (30 days). One result is that while Xerox relies on air freight for the delivery of its most important components, the completed printers are shipped mainly by sea freight.

Finally, DIST*EMPL was significantly and negatively related to the air cargo intensity index, indicating that for large firms the impact of distance was muted. This outcome again contradicts our expectation that large firms would be more likely to use air cargo services. At least in some parts of the electronics sector, large manufacturers may be leading a movement away from air cargo services and their high cost. An important example is Sony whose three plants in the region manufacture 12 different products related to hi-fi component systems, personal audio goods, and data storage devices. In Penang, the firm employs over 10,000 people and was the largest employer in all of our samples in the region. Sony's intensity index of 4.6 is well under the average of 6.0 for all firms in Penang. Surprisingly, integrated circuits are shipped by sea from China and Japan to Sony-Penang, depressing the Penang operation's air cargo intensity index. Sony has found that the goods arrive quickly enough if they are sea freighted from East Asia directly to Singapore's sophisticated cargo handling port and then by truck to Penang.

7.2. Place and industry influences

The addition of the interaction effects variables improved the explanatory power of the model to 44%. Analysis of outliers from this intermediate model pointed to the influence of industry and place-specific factors. Among the positive outliers are a large number of semiconductor firms (44% of the firms with a positive residual exceeding one standard deviation but only 28% of the sample firms). One of the most important influences upon the use of air cargo is the value per weight ratio of a good. Because firms were either unwilling or unable to provide this information, a variable directly measuring value to weight could not be incorporated in the analysis. Yet clearly, the very high value of semiconductors renders air cargo, even express air cargo, a relatively small expense. Moreover, because logistics represents an especially important source of competitive advantage for semiconductor manufacturers, firms in this industry are compelled towards a degree of conformity in their intensive use of air cargo services.

Among the negative outliers are a disproportionate number of Malaysian firms (67% of firms with a negative residual exceeding one standard deviation but only 49% of

Table 3. Enhanced model of air cargo intensity
Dependent variable: air cargo intensity index; = 107 (incomplete information was available for 19 firms)

	Unstandardized coefficient	Standardized coefficient	*t*-score	Significance
Constant	5.321		7.08	0.000
CUSTOMER	−1.244	−0.176	−2.38	0.019
CYCLTIME	-4.90×10^{-2}	−0.378	−4.54	0.000
DISTANCE	2.94×10^{-4}	0.301	3.72	0.000
MALAYSIA	−0.962	−0.139	−1.80	0.074
SEMICON	1.504	0.200	2.64	0.010
CYCL*DIST	-1.14×10^{-5}	−0.244	−3.00	0.003
CYCL*KNOW	3.22×10^{-2}	0.261	3.33	0.001
DIST*EMPL	-1.39×10^{-4}	−0.270	−3.29	0.001

Adj. *R*-square 0.472.

the sample firms). One factor behind the preponderance of Malaysian firms is localization. A large number of firms in the Kuala Lumpur region either sourced their principal components in Malaysia and Singapore or sent a large share of their output to these two ground-linked markets. Localization, which attends the development of increasingly dense intra- and interfirm linkages within a regional economy (Coe et al., 2003), is an important counterpoint to the increased dependence of manufacturing firms on air cargo services (and other long-range transport modes). A number of firms interviewed in the summer of 2001 indicated that they had pressured their suppliers to move closer or that they were under pressure to move towards their customers.

The importance of semiconductor firms as positive outliers and Malaysian firms as negative outliers suggest that the explanatory power of the analysis could be improved by adding two binary variables to the terms described earlier. The addition of SEMICON, a variable indicating whether or not the firm belonged to the semiconductor industry, and MALAYSIA, a variable indicating whether or not the firm was located in Malaysia, to the model improved its explanatory power to 47%. In fact, both variables proved significant in the final model, though MALAYSIA only weakly so (Table 3).

8. Conclusions

Our research demonstrates the variation in the articulation of air cargo services within the global production networks associated with the electronics industry in Southeast Asia. Because of the increasingly sophisticated nature and variety of these services we argue that they should be treated collectively as an advanced producer service. Much of the literature on advanced producer services has focused on the supply of, rather than the demand for, such services. Our study is among the first to examine the demand for a producer service across several economies at different levels of industrialization and to situate advanced producer services within global production networks.

Much of the variation in air cargo services usage is related to product characteristics that go beyond simply the value-to-weight ratio. For example, the high air cargo intensity among semiconductor firms is driven not only by the high value of their lightweight products but also rapid product cycles and the greater risk of damage associated with sea freight. Customer preference and management policy are also often key in explaining the demand for air cargo. While, as noted above, specific products, such as semiconductors, are much more likely to use air transport than are goods defined as consumer electronics, even within the latter category there is also considerable variation. The production location with respect to major markets, seasonality, and the opportunities for local sourcing of inputs are reasons for the varying behavior. Just as important as types of products as a discriminating variable for air cargo intensity is the form of production. For example if a firm relies on and utilizes specific inventory control procedures (e.g. just-in-time), air cargo may be more critical and used more heavily.

One of the most significant findings of the research to date is the extent to which air cargo usage is associated with the degree to which a firm has internationalized, not only its production sites and final markets but also its material procurement sites. A measure of distance associated with forward and backward production linkages emerged as the most important explanatory factor of air cargo use. A second important variable was cycle time, an indication of a firm's material management strategy. In the enhanced model, the role of the customer as an influence on logistics choices and the interaction among several other variables pointed to the complexity of air cargo services usage. The interaction terms, in particular, highlighted the role of firm size and production process (and its associated knowledge versus labor intensity) as forces that mediate the impact of cycle times and distance. Our research thus suggests that product type, internationalization, localization, product cycle (obsolescence), and other factors such as a firm's material management strategy are especially critical.

Above all, our research points to the importance of drawing together research on advanced producer services and global production networks. Producer services make possible the increasingly global scale of production networks and shape the spatial configuration of those networks. Some producer services, including those provided by the air cargo industry, are elements of the distributive infrastructure over which production networks are elaborated. Producer services are also crucial among the region-specific assets with which the needs of trans-local firms are coupled in global production networks. The development of the semiconductor industry in the Metro Manila region of the Philippines, for instance, has been contingent in part on the nature of the air cargo services available in that market.

In turn, the development of global production networks, with all of their concomitant complexity, has spurred the development of more sophisticated producer services. In the air cargo industry, for instance, conventional airport-to-airport transport services are now complemented by and integrated with knowledge-intensive logistics services that enable user firms to more effectively manage the complexity of international operations. In other producer services industries, too, the emergence of more highly specialized services has accompanied the globalization of manufacturing. Moreover, the internationalization of production means that demand for and use of sophisticated services has also been internationalized, and that is likely to erode the spatial differences in the availability of such services. To revisit the example of the semiconductor industry in the Philippines, for instance, that industry's requirements for complex air logistics compel third-party logistics firms to offer more sophisticated services in the Manila market, narrowing the gap between the quality of such services in the Philippines and in the region's services hub, Singapore.

In addition we also demonstrate the attachment to the externalization of services functions by manufacturing firms. Our findings offer a somewhat different perspective on the point which suggests that cost driven externalization is not the most important force underlying growth in demand in producer services (Beyers and Lindahl, 1996). In the air cargo context, the need for specialized services does indeed drive demand. The use of varied air cargo services in that regard can be likened to the use of other producer services to separate a firm from its competitors (Lindahl and Beyers, 1999).

Lastly, this work points toward several promising avenues of inquiry. First, although we have not tested the flexible firm model which predicts that restructuring strategies cause the expansion of producer service demand (O'Farrell et al., 1993), the relationship between the demand for air cargo services and user firms' expansion and contraction under varying circumstances is worth exploring. Second, externally provided supply chain management 'solutions' have an increasingly important impact on the way firms use air cargo, particularly in developing economies where the sophistication of logistics management has tended to be low in the past. Increasingly, the most lucrative business function of multinational freight forwarders is the development of specialized supply chain systems for large customers in which goals other than simple cost minimization are paramount (interview with Wilson, 2002). An ongoing interest focuses on improving our understanding of the manner in which forwarders design and implement these solutions and how they affect the operations of the forwarders' customers. Third, although this research has analysed the use of air cargo services in general, there is a subset of such services (including, for instance, time-definite express air cargo), which fit more clearly under the rubric of advanced producer services. We intend to examine how the use of advanced air cargo services differs from that of air cargo services in general.

Acknowledgements

We wish to gratefully acknowledge funding provided by the National Science Foundation, Geography and Regional Science Program for Air Cargo Services and Competitive Advantage in Industrializing Economies under BCS 0078734 (TRL) and BCS 0078621 (JTB). We also gratefully acknowledge assistance from our counterparts in Malaysia: Professor Morshidi Sirat, Dr Hassan Naziri Khalid, and Mohamad Haron Harashid, School of Humanities, Universiti Sains Malaysia, Penang and in the Philippines; Darlene Gutierrez, Daniel Mabazza, and Jonas Gaffud, Department of Geography, University of the Philippines, Diliman, Quezon City as well as support during the fieldwork from the Institute of Southeast Asian Studies, Singapore. Richard Gilbreath and the Cartographic Laboratory at the University of Kentucky prepared the figures. Josh Lepawsky was a valuable graduate research assistant who aided in carrying out interviews, questionnaire design and data analysis in Singapore and Malaysia. He also presented an earlier version of this paper at the 2002 Association of American Geographers Meeting in Los Angeles. Brian Zacho provided assistance in Singapore and the Philippines.

References

Aiken, L. S., West, S. G. (1991) *Multiple Regression: Testing and Interpreting Interactions*. Newbury Park: Sage.

Armbruster, W. (2003) US 3PL market grows 7%. *Journal of Commerce* (online edition), 2 April.

Begg, I. (1994) The service sector in regional development. *Regional Studies*, 27(8): 817–825.

Beyers, W. B. (2002) Services and the new economy: elements of a research agenda. *Journal of Economic Geography*, 2: 1–29.

Beyers, W. B., Lindahl, D. P. (1996) Explaining the demand for producer services: Is cost-driven externalization the major factor? *Papers in Regional Science*, 75(3): 351–374.

Boeing Commercial Airplane Group (2000) *World Air Cargo Forecast 2000/2001*. Available on-line at www.boeing.com/commercial/cargo/index.html.

Bowen, J. (2004) The geography of freighter aircraft operations in the Pacific Basin. *Journal of Transport Geography*, 12(1): 1–11.

Bowen, J., Leinbach, T. R. (2004) Market Concentration in the Logistics Industry and the Global Provision of Advanced Air Freight Services. *Tijdschrift voor Economische en Sociale Geografie*, 95(2): forthcoming.

Button, K. (1990) *Airline Deregulation: An International Perspective*. London: David Fulton.

Button, K., Owens, C. A. (1991) Transport and information systems: A case study of EDI deployment by the air cargo industry. *International Journal of Transport Economics*, 26(1): 3–21.

Coe, N. M., Hess, M., Yeung, H. W. C., Dicken, P., Henderson, J. (2003) 'Globalizing' regional development: a global production networks perspective. Paper presented at the 99th Annual Meeting of the Association of American Geographers, New Orleans, 5 March.

Daniels, P.W. (1993) *Service Industries in the World Economy*. Oxford: Blackwell Publishers.

Daniels, P.W. (1998) Economic development and producer services growth: the APEC experience *Asia Pacific Viewpoint*, 39: 145–159.

Daniels, P.W., Moulaert, F. (eds) (1991) *The Changing Geography of Advanced Producer Services: Theoretical and Empirical Perspectives*. London: Belhaven.

Dempsey, P.S., O'Connor, K. (1997). Air traffic congestion and infrastructure development in the Pacific Asia region. In C. Findlay, C. L. Sien and K. Singh (eds) *Asia Pacific Air Transport*. Singapore: Institute of Southeast Asian Studies, 23–47.

Dicken, P. (1992) *Global Shift: The Internationalization of Economic Activity*. London: Paul Chapman.

Dicken, P. (1998) *Global Shift: Transforming the World Economy*. New York: Guilford.

Dicken, P., Kelly, P. F., Olds, K., Yeung, H. W. C. (2001). Chains and networks, territories and scales: towards a relational framework for analysing the global economy. *Global Networks*, 1: 89–123.

Dicken, P., Thrift, N. (1992) The organization of production and the production of organization: why business enterprises matter in the study of geographical industrialization. *Transactions Institute of British Geographers*, New Series 17: 279–291.

Draper, N., Smith, H. (1998) *Applied Regression Analysis*. New York: Wiley.

Fennes, R. (1997) *International Air Cargo Transport Services: Economic Regulation and Policy*. Leiden: University of Leiden.

Fowler, F. J. (1993) *Survey Research Methods*. Newbury Park: Sage.

Gallacher, J. (1998) Cargo: Chasing the value chain. *Airline Business*, 14(11): 52–55.

Goe, W. R. (1990) Producer services, trade, and the social division of labor. *Regional Studies*, 24: 327–342.

Goe, W. R. (1991) The growth of producer service industries: sorting through the externalization debate. *Growth and Change*, 22: 118–141.

Hall, D., Braithwaite, A. (2001) The development of thinking in supply chain and logistics management. In A. M. Brewer, K. J. Button and D. A Hensher (eds) *Handbook of Logistics and Supply Chain Management*. Amsterdam: Pergamon, 81–98.

Harrington, J. W. (1995) Producer services research in US regional studies. *Professional Geographer*, 47 (1): 87–96.

Ho, K. C. (1998) Corporate regional functions in Asia Pacific. *Asia Pacific Viewpoint*, 39: 179–191.

ICAO Journal (2001) *Annual Civil Aviation Report*, 56 (6), July/August: 13.

Illeris, S. (1996) *The Service Economy: A Geographical Approach*. Chichester: John Wiley.

Kaplinsky, R. (1998) Globalization, industrialization and sustainable growth: the pursuit of the nth rent. IDS Discussion Paper 3, Institute of Development Studies, University of Sussex.

Khan, M. H., Jomo K. S. (2000) *Rents, Rent Seeking and Economic Development*. Cambridge: CUP.

Lai, P. Y. K., Yeung, H. W. C. (2003) Contesting the state: discourses of the Asian economic crisis and mediating strategies of electronics firms in Singapore. *Environment and Planning A*, 35: 463–488.

Lambert, D. M. (2001) The supply chain management and logistics controversy. In A. M. Brewer, K. J. Button and D. A. Hensher (eds) *Handbook of Logistics and Supply Chain Management*. Amsterdam: Pergamon, 99–126.

Lindahl, D. P., Beyers, W. B. (1999) The creation of competitive advantage by producer service establishments. *Economic Geography*, 75(1): 1–20.

Loughlin, M. (1998) Overseas air cargo service: airborne export producing industries and US cities, 1980–1995. PhD dissertation, University of Minnesota.

MacPherson, A. (1997) The role of producer service outsourcing in the innovation performance of New York State manufacturing firms. *Annals, Association of American Geographers*, 87(1): 52–71.

Markusen, A. (1994) Studying regions by studying firms. *Professional Geographer*, 46(4): 477–490.

Martinelli, F. (1991) A demand-oriented approach to understanding producer services. In P.W. Daniels and F. Moulaert (eds) *The Changing Geography of Advanced Producer Services: Theoretical and Empirical Perspectives*, 15–29. London: Belhaven.

Meersman, H., Van de Voorde, E. (2001) International logistics: a continuous search for competitiveness. In A. M. Brewer, K. J. Button and D. A. Hensher (eds) *Handbook of Logistics and Supply Chain Management*. Amsterdam: Pergamon, 61–77.

Mentzer, J. T. (2001) *Supply Chain Management*. Thousand Oaks: Sage.

Nilsson, J. (1996) Introduction: The internationalization process. In J. Nilsson, P. Dicken and J. Peck (eds) *The Internationalization Process: European Firms in Global Competition*, 1–12. London: Paul Chapman.

Nilsson, J., Dicken, P., Peck, J. (eds) (1996) *The Internationalization Process: European Firms in Global Competition*. London: Paul Chapman.

Nusbaumer, J. (1987) *Services in the Global Market*. Boston: Kluwer.

O'Connor, K., Hutton, T. A. (1998) Producer services in the Asia Pacific region: an overview of the issues. *Asia Pacific Viewpoint*, 39: 139–143.

O'Farrell, P. (1993) The competitiveness of business services firms in peripheral regions: An international comparison between Scotland and Nova Scotia. *Environment and Planning A*, 25: 1627–1648.

O'Farrell, P., Hitchens, D. (1990) Producer services and regional development: key conceptual issues of taxonomy and quality measurement. *Regional Studies*, 24: 163–171.

O'Farrell, P., Moffat, L., Hitchens, D. (1993) Manufacturing demand for business services in a core and peripheral region: Does flexible production imply vertical disintegration of business services? *Regional Studies*, 27: 385–400.

Oum, T. H., (1998) Overview of regulatory changes in international air transport and Asian strategies towards the US open skies initiative. *Journal of Air Transport Management*, 4: 127–134.

Poon, J. (1997) The cosmopolitanization of trade regions: global trends and implications, 1965–1990. *Economic Geography*, 73: 390–404.

Porter, M. (1996) Competitive advantage, agglomeration economies, and regional policy. *International Regional Science Review*, 19(1): 85–94.

Porter, M. (1990) *The Competitive Advantage of Nations*. New York: The Free Press.

Raguraman K. (1997) International air cargo hubbing: the case of Singapore. *Asia Pacific Viewpoint*, 38(1): 55–74.

Rimmer, P. J. (1997). Trans-Pacific Oceanic Economy Revisited. *Tijdschrift voor Economische en Sociale Geographie*, 88(5): 439–456.

Schoenberger, E. (1991) The corporate interview as a research method in economic geography. *Professional Geographer*, 44: 180–189.

Smith, A., Rainnie, R., Dunford, M., Hardy, J., Hudson, R., Sadler, D. (2002) Networks of value, commodities, and regions: reworking divisions of labour in macro-regional economies. *Progress in Human Geography*, 26(1): 41–63.

South China Morning Post (2002) Flexibility draws producers to third parties. 24 May, Supplement, 7.

Sturgeon, T. (2001) How do we define value chains and production networks? *IDS Bulletin*, 32(3): 9–18.

Veltz, P. (1996) *Mondialisation, Villes et Territoires: L'Economie d'Archipel*. Paris: PUF.

Williams, C. (1997) *Consumer Services and Economic Development*. London: Routledge.

Wilson, D. (2002). Vice President Supply Chain Solutions, Asia Pacific Division, FedEx. Interviewed by Bowen and Leinbach in Singapore, 10 July.

Wood, P. (2002) Services and the 'New Economy': an elaboration. *Journal of Economic Geography*, 2: 109–114.

Yeung, H. W. C. (2000) Organizing 'the firm' in industrial geography I: networks, institutions, and regional development. *Progress in Human Geography*, 24(2): 301–315.
Yeung, H. W. C. (2001) Organising regional production networks in Southeast Asia: implications for production fragmentation, trade, and rules of origin. *Journal of Economic Geography*, 1: 299–321.

Appendix

Main-effects and interaction-effects variables

1. CUSTOMER	a binary variable reporting whether an interviewee answered 'the customer' in response to the question, 'Who has the most influence over the choice of air cargo services?' Firms that identify the customer as the key force are likely to be under the influence of 'pull logistics' (described in the text).
2. CYCLTIME	a ratio scale variable measuring the number of days that elapse between the receipt of a new order and its fulfillment.
3. DISTANCE	a ratio scale variable summing the mean distance from the two main raw material or component sources to the firm AND the mean distance to the firm's two most important destination markets.
4. EMPLOYEE	an ordinal variable classifying the size of the firm in terms of local employment.
5. KNOWLABR	an ordinal variable indicating whether a firm's local workforce is best described as labor-intensive (1), balanced (2), or knowledge-intensive (3).
6. MCTIME	an ordinal variable indicating the rank given to 'time' among several factors influencing forward modal choice.
7. MATSRC	a binary variable indicating whether the major raw materials and/or components for a firm's principal product come from the same firm OR from other firms. This variable indicates whether the firm's backward logistics are predominantly intra-firm.
8. PRODDES	a binary variable indicating whether a firm ships mainly to other branch plants and subsidiaries of the same firm OR to other independent firms. This variable indicates whether the firm's forward logistics are predominantly intra-firm.
9. CYCL*DIST	the product of cycle times and the distance associated with a firm's production network, this variable shows the way in which the effect of distance may be conditioned by the level of cycle time.
10. CYCL*EMPL	the product of cycle times and the number of employees, this variable assesses the degree to which the impact of cycle times on air cargo intensity differs between large and small firms.
11. CYCL*KNOW	the product of CYCLTIME and KNOWLABR, this variable assesses the degree to which the influence of cycle times on air cargo services intensity is conditioned by the nature of a firm's workforce.
12. DIST*EMPL	the product of the number of employees and production network distance, this variable is intended to reflect the degree to which the effect of distance is conditioned by the size of a firm.
13. DIST*KNOW	the product of DISTANCE and KNOWLABR, this variable examines the degree to which the nature of a firm's workforce mutes or amplifies the influence of the distance in a firm's production network upon air cargo intensity.
14. EMPL*KNOW	the product of EMPLOYEE and KNOWLABR, this variable combines two variables that were insignificant in the initial stage of the analysis. Because we had expected both variables to be strongly significant, we incorporate this interaction variable in the second stage of the analysis to help separate out effects that might cloud the importance of either or both of these variables.
15. MALAYSIA	a binary variable indicating whether or not a firm is located in Malaysia.
16. SEMICON	a binary variable indicating whether or not a firm is a semiconductor manufacturer.

Part IV
Policy Issues

[22]

Reconsidering Social Equity in Public Transit

Mark Garrett and Brian Taylor

Over the course of this century, public transit systems in the U.S. have lost most of the market share of metropolitan travel to private vehicles. The two principal markets that remain for public transit systems are downtown commuters and transit dependents — people who are too young, too old, too poor, or physically unable to drive. Despite the fact that transit dependents are the steadiest customers for most public transit systems, transit policy has tended to focus on recapturing lost markets through expanded suburban bus, express bus, and fixed rail systems. Such efforts have collectively proven expensive and only marginally effective. At the same time, comparatively less attention and fewer resources tend to be devoted to improving well-patronized transit service in low-income, central-city areas serving a high proportion of transit dependents. This paper explores this issue through an examination of both the evolving demographics of public transit ridership, and the reasons for shifts in transit policies toward attracting automobile users onto buses and trains. We conclude that the growing dissonance between the quality of service provided to inner-city residents who depend on local buses and the level of public resources being spent to attract new transit riders is both economically inefficient and socially inequitable. In light of this, we propose that transportation planners concerned with social justice (and economic efficiency) should re-examine current public transit policies and plans.

Introduction

Public transit in the United States has become first and foremost a social service. Despite broad public support for mass transit, the automobile is the mode of choice for the vast majority of travelers. Eighty-six percent of all trips nationally are made by automobile (U.S. DOT, 1999). Rising personal income, the greater availability of automobiles, low fuel prices, and substantial public investment in metropolitan street and freeway systems have combined to reduce the general demand for public transit. Still, many people without regular access to automobiles depend on public transit as their main mode of transportation. For these "transit dependents" the continued availability of public mass transit is vital for access to jobs, schooling, medical care, and other necessities of life.

Over the past few decades, the proportion of low-income transit riders has been rising as more well-to-do travelers have shifted to automobiles. Outside of a few dense city centers like New York or San Francisco, the majority of local transit riders are poor. Most of these transit users live in the inner-city and many are members of minority groups, while so-called "choice" riders — those with regular access to private vehicles — are more likely to be in the suburbs and are predominantly white. Under public pressure to help address traffic congestion and air pollution problems in metropolitan areas, transit operators across the country are expected to provide services that will be attractive to automobile users, especially single-occupant commuters who tend to have higher incomes and far more travel options than transit dependents. However, the increased emphasis on commuter-oriented express bus and rail service is increasingly at odds with the growing inner-city ridership base of transit, who lack adequate access to private transportation due to age, income, or disability. The resulting inattention to many inner-city bus services raises troubling questions about how current public transit policies affect poor and minority urban residents.

This paper examines the growing tension transit planners face between meeting the strong demand for transit services by predominately low-income and minority inner-city residents on the one hand, and accommodating the political interests and desires of a more mobile, dispersed, and largely white, suburban-based electorate on the other. We argue that a number of exogenous and endogenous factors have contributed to a socially inequitable provision of public transit. In the next section, we provide a brief theoretical context for our discussion. In the following sections, we analyze the changing demographics of transit ridership and the current trends in U.S. transit policy. We conclude with a discussion of the role that politics has played in the increasing lack of connection between the needs of transit dependents for adequate, affordable local bus service and the policy response, which favors shifting resources to serving suburban commuters.

Transit Equity

The allocation of transit services between rich and poor, whites and people of color, suburbanites and inner-city residents, is not happenstance, but is directly connected to social and economic processes that have produced the current racial and economic polarization between suburbs and central cities. Mainstream planning has paid insufficient attention to the redistribution of economic and political power that is at least partly responsible for these patterns of uneven urban development. The tradition of equity planning, on the other hand, has been centrally concerned with reducing such urban inequalities.

Norman Krumholz (1982:163) has eloquently defined equity planning as an effort to provide more "choices to those...residents who have few, if any choices." In his tenure as Planning Director for the City of Cleveland, Krumholz formulated his notion of equity planning to counteract what he perceived to be the inherent unfairness and exploitative nature of the urban development process, a process that excluded the poor from the suburbs and concentrated them in declining inner-city areas. A key factor in the process of isolating the poor was the lack of adequate public transportation. Related to this was the government's policy during this era of massive public investment in urban freeways that helped to empty out central cities of middle- and upper-income residents.

Over the years, planners influenced by the ideas of equity planning have fought highway construction projects and urban renewal schemes that would have further displaced or disrupted low-income communities. Equity planners have also worked to improve public transit service for those who depend on it for access to jobs, shopping, school, and other services. In some cases, they have opposed expensive rail transit projects serving wealthier, suburban commuters at the expense of inner-city bus riders. For example, during the 1970s, city planners in Cleveland fought against costly city proposals to extend commuter rail lines and to construct a downtown people-mover system to serve the business community. They argued instead for lower bus fares and expanded bus service for transit dependent persons on the grounds that new fixed rail systems would not increase accessibility, but would draw resources away from suitable bus services (Krumholz and Forester, 1990).[1]

Nevertheless, planners in government agencies have too often tended to overlook the uneven distribution of public investment and public services in urban regions and their consequences for the lives of affected residents. Lately, though, some transit planners and others concerned with social equity have begun to address how regional political arrangements have led to allocations of public transit resources that have done little to increase transportation

[1]Krumholz and his fellow planners were successful in negotiating a deal providing for 1) a fare reduction to 25 cents for three years, 2) free non-peak fares for seniors and handicapped persons and half fares at peak hours, 3) improved neighborhood service, and 4) demand-responsive transit service for elderly and handicapped individuals.

choices for low-income residents. In a particularly interesting recent development, advocates for transit dependents have turned to the courts to confront transit policies that disadvantage poor and minority transit riders. By raising objections to fixed rail projects and agency transit fare policies, recent civil rights litigation against several major U.S. regional transit authorities represents in many ways a continuation of the work begun two decades ago by the Ohio planners in a new guise.[2]

Transit equity issues go well beyond disputes over particular projects, however, raising fundamental questions about the provision of urban transportation services. The policy-driven shift in population, particularly among middle-income whites, away from central cities and toward suburbs and outlying areas, has altered the historic ridership base for transit. Today, transit riders are, on average, much poorer than the general population, with disproportionate numbers of elderly and minority passengers. Current federal and state transit subsidy policies have generally not been consistent with these demographic shifts in urban transit use, but have tended to support suburban and downtown commuter services, including radial rail transit networks, in an effort to attract more discretionary commuters out of their automobiles. While this trend in funding priorities may have improved the range of options available to suburban commuters, the shift in emphasis toward serving suburban travelers and, in many cases, the resulting inattention to local bus service has diminished accessibility for inner-city residents, particularly to employment opportunities.

This issue of job accessibility has particular salience for the current debate over welfare reform, since nearly half (42%) of all trips on public transit are work-related (Hu and Young, 1993).[3] Although central cities contain only 20 percent of all workers, they still account for 69 percent of all transit use. In contrast, suburbs have half of all workers but generate only 29 percent of transit trips (Pisarski, 1996). Some public transit proponents argue, however, that commuter-oriented bus and rail transit systems are needed to

[2]See Committee for a Better North Philadelphia v. Southeastern Pennsylvania Transportation Authority (SEPTA), 1990 U.S. Dist. Lexis 10895 (E.D. Pa 1990), 935 F.2d 1280 (3rd Cir. 1991); New York Urban League, Inc. v. Metropolitan Transportation Authority, et al., 95 Civ. 9001 (RPP); reversed New York Urban League, Inc. v. The State of New York, Metropolitan Transportation Authority, et al., 71 F.3d 1031 (2nd Cir. 1995); Labor/Community Strategy Center, et al., v. Los Angeles Metropolitan Transportation Authority, et al., Case No. CV 94-5936 TJH.

[3]By comparison, only about 23 percent of private automobile trips are for work.

provide transit-dependent inner-city residents better access to suburban employment. While opportunities clearly exist to better link central cities and suburbs with public transit, the role of these so-called "reverse commutes" in metropolitan travel should not be overstated. Very long commutes are the exception, not the rule -- especially for low-income workers who must balance the time and expense of commuting against the wages from a given job. Further, most commutes are within suburbs or central cities, and not between them. A minority of commutes are from suburbs to central cities, and even a smaller share are reverse commutes to the suburbs (Pisarski, 1996). And while it is true that many *new* jobs are being created in the suburbs, the majority of job *opportunities* for low-income workers are still located in central cities (Shen, 1998). This is because most job openings are created by a worker vacating an existing position, and not through the creation of a new position.

Fixed-route transit systems work best at connecting dense suburban residential concentrations to dense central areas. They are far less effective in connecting inner-city residents to dispersed suburban employment sites, especially without time-consuming transfers. In a study of low-wage job access by mode in Los Angeles, Ong and Blumenberg (1999) find that the number of low wage jobs that can be accessed in a 30-minute trip by transit is 77.1 percent lower than by automobile in the central city neighborhood of Pico-Union. It is 97.1 percent lower in the low-income suburb of Watts.

The enormous employment access advantage of automobiles helps to explain why, in 1990, over 60 percent of the workers living in poverty households drove to work alone (Pisarski, 1992). It also explains why so many reverse-commute transit programs lose riders to automobiles when low-wage reverse commuters buy cars (Ong, 1996; Rosenbloom, 1992). Reverse commute transit programs can play a role in increasing job access for low-income central city residents. However, improving the quality of heavily patronized local transit service and reducing fares for short and off-peak trips would clearly do more to connect workers without cars to urban employment opportunities (Wachs and Taylor, 1998).

The incongruence between transit ridership patterns and subsidy policies has both social and spatial consequences that can potentially reinforce existing patterns of racial, ethnic, and economic segregation. Poor or mediocre public transit service in areas with high proportions of transit dependents exacerbates problems of social and economic isolation. From the standpoint of equity planning, this serves only to decrease choices for those who already have limited transportation options.

The Changing Demographics of Transit

Transit use peaked in the U.S. during World War II at over 23 billion trips annually but declined quickly thereafter. While transit use in general has remained fairly constant since the early 1970s, at about 7½ -8 billion annual trips, the proportion of all trips made by transit, particularly buses, has been decreasing, due to the increase in privately owned vehicle (POV) travel. In 1969, 7.8 percent of all unlinked trips were made by transit. In 1983, transit made up 2.3 percent of all trips, but declined to 1.8 percent by 1990 (Vincent, Keyes, and Reed, 1994).[4] By 1995, only 1.7 percent of all trips made were by transit (U.S. DOT, 1999).

Commute trips exhibit a similar pattern. Transit makes up a larger proportion of commute trips than of overall travel. Between 1980 and 1990 the number of commuters grew by nearly 19 million. However, the number of daily public transit riders fell slightly to about 5.9 million (Pisarski, 1996). As a result, the transit share of all commuter trips declined from 6.3 percent to 5.3 percent (Vincent, Keyes, and Reed, 1994). More importantly, there have also been changes in ridership demographics within public transit services, specifically, the distribution of riders across different modes of transit (bus, subway, and commuter rail) have become increasingly segregated both economically and racially.

The growing dichotomy in transit services can be seen in recent statistics on modal shifts within public transit from buses and subways to commuter rail. Between 1977 and 1995, the number of all transit trips rose from 7.28 billion to 7.76 billion. However, the number of bus trips declined somewhat from 4.94 billion to 4.84 billion and the number of heavy rail trips fell from 2.14 billion to 2.03 billion. As a result, the proportion of transit trips made by bus declined from 67 percent to 63 percent and those by subway fell from 29 percent to 26 percent. On the other hand, the number of trips by light rail and commuter rail have been increasing over the same period. Light rail trips rose from 103 million to 251 million annually, an increase from 3.3 percent to 4.4 percent. Between 1980 and 1995 annual commuter rail trips increased from 280 million to 344 million, going from 1.4 percent to 3.2 percent of all transit trips (APTA, 1999).

[4]In large urban areas with rail transit service the share of transit trips declined from 8.8 percent to 5.2 percent over the period.

Transit use varies significantly by income, gender, race, and ethnicity. For example, 57 percent of bus transit riders in Los Angeles earn under $15,000 a year compared to only 20 percent of all county residents. Of these riders, nearly 83 percent are nonwhite and most are female (MTA, 1991-1993). The typical Southern California commuter rail rider, by contrast, is a white male earning $65,000 with a monthly parking subsidy from his employer and ready access to alternative transportation (Rubin, 1994). Nationwide, Hispanic and especially African-American workers have much higher rates of transit usage than non-Hispanic whites, and these differences are particularly pronounced for bus and subway use. Among suburban workers nationwide, whites use commuter railroads slightly more than blacks or Hispanics (Pisarski, 1996). As shown in Figure 1, even as the overall share of transit trips has declined, the proportion of transit riders who are minority has been increasing. In 1977, about 20 percent of all rail transit and bus riders were nonwhite compared to about 14 percent of those traveling by private vehicle. By 1995, minorities made about two-thirds of all bus trips, compared to 60 percent of subway and commuter rail patrons. In contrast, during the same period the percentage of auto trips made by minorities rose slightly to 24 percent.

Figure 1. Unlinked Trips by Race and Travel Mode (1977-1995)

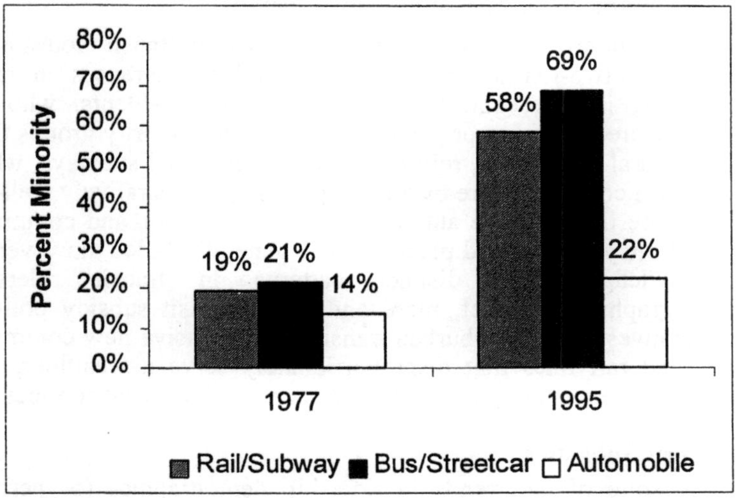

Source: National Personal Transportation Surveys 1977 & 1995.

There has also been a shift in transit mode used between those low-income "captive" riders who take local buses and wealthier, "discretionary" riders who use more express buses and commuter rail services. By 1995, the median household income of an urban bus passenger was below $20,000, compared to over $40,000 for commuter rail patrons and over $45,000 for drivers of private vehicles (U.S. DOT, 1999). Studies have shown that bus ridership declines with rising income, but the use of streetcars, subways and commuter railroads tends to increase with higher income (Pisarski, 1996).

Finally, there is a spatial dimension to the changes occurring in transit use. Public transit service is concentrated in the oldest, largest, and most densely developed American cities. Nearly 60 percent of transit passengers nationwide are served by the ten largest big city transit systems, and the remaining 40 percent by the other 5,000 plus systems (Taylor and McCullough, 1998). While overall transit use has declined slightly since the 1980s, the drop in the number of transit riders has been greatest in central cities, though ridership losses were proportionately greater in suburbs. Use of buses, streetcars, and subways is highest in central cities, while commuter railroads account for a higher percentage of all suburban trips (Pisarski, 1996). These shifting patterns of transit use mirror the growing economic and racial disparities in urban areas since central city residents tend to be poorer, mostly minority, and more transit dependent than suburbanites.

To summarize, the demographic shifts within transit modes have created a two-tier system characterized by differences in race, ethnicity, income, and location. Inner-city residents, who on average are much poorer and more often from minority groups than the general population, rely far more on buses and subways, while suburban commuters are by and large white, comparatively well-off and more likely to use automobiles, express buses, and commuter rail. Policy makers and planners have generally failed, however, to acknowledge these distinct patterns in transit ridership demographics. In fact, more and more, transit subsidy policies favor investment in suburban transit and expensive new commuter bus and rail lines that disproportionately serve a wealthier, less transit-dependent population than do central city transit services.

Transit Subsidies

In spite of the trends in ridership demographics (or perhaps because of them) transit systems around the U.S. have devoted substantial resources in recent years to building and operating commuter-oriented bus and rail services in an attempt to appease more affluent constituencies and lure middle-class riders back from automobiles. A number of factors — growing traffic congestion, public ambivalence toward further metropolitan highway construction, and heightened environmental awareness — have all contributed to a political base of support for this type of public transit.

In a 1981 study of public transit ridership, Pucher et al. (1981) found that the poor, the elderly, minorities, and women comprised a much higher percentage of bus ridership than of subway ridership, and a higher percentage of subway ridership than of commuter rail ridership. Moreover, the transit modes most used by these groups were the least subsidized modes:

> The average per-passenger operating subsidy to commuter rail in the United States is almost three times as great as that to bus service. Differences in capital subsidies by mode are even more to the disadvantage of bus riders (Pucher, Hendrickson, and McNeil 1981:481).

These patterns have only grown more pronounced in the intervening years. During the past two decades, over a dozen new rail transit systems have been constructed, mostly in lower-density, more auto-oriented cities like Miami, Portland, Sacramento, and San Jose. These costly new systems have required substantial public subsidy and have tended to attract far fewer new riders than expected (Pickrell, 1983; Pickrell, 1992).

All public transit service in the U.S. is heavily subsidized. Operating funds and capital expenditures over and above farebox revenues must be obtained from local, state, and federal sources. In 1997, transit providers nationally received about $7.7 billion in capital funds from various sources. The federal government provided about 54 percent from general revenues and the Mass Transit Account of the Highway Trust Fund, which is mainly funded through federal motor fuel taxes. State sources accounted for about 13 percent and local sources combined for another 11 percent. The remainder came from taxes levied by transit agencies and other directly generated sources (APTA, 1999).

Capital spending is skewed toward rail development and away from bus investment. As shown in Table 1, capital spending on rail transit amounted to over 60 percent of total 1997 expenditures even though bus usage is much higher than other transit modes.

Spending on commuter rail was considerably higher compared to patronage.[5]

Table 1. U.S. Passenger Trips, Capital and Operating Expenditures by Mode, 1997

	Passenger Trips	Capital Expenses	Operating Expenses
Bus	60.7%	30.0%	58.0%
Heavy Rail	28.4%	30.6%	18.3%
Light Rail	3.1%	10.7%	2.5%
Commuter Rail	4.2%	23.6%	12.0%
Other	3.6%	5.1%	9.2%

Source: APTA 1998 Transit Fact Book

Operating subsidies are provided mainly by state and local sources. Over half of the $18.9 billion in operating funds received by U.S. transit providers in 1997 came from farebox revenues and other directly generated sources. State and local sources accounted for about 21 percent each while the federal government supplied only about $578.1 million, or 3 percent (APTA, 1999.). Table 1 also shows that operating expenditures more closely matched patronage levels though spending on commuter rail was still proportionately higher.

Federal transit grants for 1997 totaled $4.38 billion of which $2.15 billion (49%) was allotted to formula grants and $1.9 billion (43%) for discretionary capital investment grants (APTA, 1999).[6] Under the Intermodal Surface Transportation and Efficiency Act (ISTEA) of 1991, the formula allocations were weighted heavily toward new construction; for fiscal year (FY) 1997, budget appropriations provided $1.58 billion for urbanized area capital

[5]Though the capital expenditure data are not amortized to reflect annualized capital expenditures, they do reflect the clear emphasis in transit subsidy policy toward rail capital investments.

[6]Federal capital grant approvals amounted to $4.05 billion for FY 97, of which $1.77 (43.8%) went to capital investment programs and $2.07 billion (51.2%) to formula grants. Of these funds, 39.1% went for buses, while 37.1% went to rail modernization and another 22.8% to new starts (APTA 1999).

projects but only $400 million for operating expenses. With passage of the Transportation Equity Act for the 21st Century (TEA-21),[7] areas over 200,000 population are no longer eligible for operating assistance. However, some preventative maintenance expenses can be funded through capital grants.[8]

Of the funds made available under the Urbanized Area Formula program (Section 5307), the federal government provides only a 50 percent match of operating costs but 80 percent of the net project cost of new capital projects.[9] The remaining funds must be supplied by state and local sources. This formula encourages local operators to cover a higher proportion of operating expenses from system revenues in order to be in position to leverage larger amounts of federal dollars for capital projects. In addition to encouraging capital expenditures, federal law favors expenditures for rail projects over expanding bus service. Slightly more than 90 percent of the funds available under Section 5307 are reserved for urbanized areas over 200,000 in population.[10] Of that share, approximately one-third is apportioned according to the amount of fixed guideway service provided by the transit operator and the remaining two-thirds based on bus service, despite the fact that approximately 95 percent of all transit service is provided by buses.[11]

Thus federal transit subsidies favor expanding service area coverage, over increasing ridership. Nearly 60 percent of the formula funds allocated for fixed guideway systems is apportioned according to the number of miles covered by vehicles in service

---

[7]Public Law 105-178, as amended by title IX of Public Law 105-206 [hereafter TEA-21"].

[8]TEA-21, section 3007 (amending 49 U.S.C. section 5307(b)). Areas under 200,000 population are eligible to receive operating assistance grants without prior limitations. Id. section 3027(b) (repealing 49 U.S.C. section 5336(d)).

[9]49 U.S.C. section 5307(e). A "net project cost" means the cost of a project that reasonably cannot be financed from revenues. 49 U.S.C. section 5302(a)(8).

[10]49 U.S.C. section 5336(a)(2). A total of 9.32% of the budgeted funds are available to areas with a population of less than 200,000 and are distributed through state governors. The funds are apportioned based 50% on population and 50% on population density weighted by population. Id. section (a)(1).

[11]49 U.S.C. section 5336(b)&(c). TEA-21 generally continues these funding formulas. Urbanized areas formula grants now receive 91.23% of the allocation while non-urbanized areas receive 6.37%. Grants for individuals with disabilities receive the remaining 2.4%. 49 U.S.C. section 5338(a)(2)(C)(iii). Over the operative period of TEA-21 (FY98-FY03), authorizations for Urbanized Area Formula grants (guaranteed and non-guaranteed) total $18.033 billion (FTA, 1999).

while close to 40 percent is allocated based merely on total track mileage.[12] Less than 5 percent is allotted on the basis of how many passengers are actually carried and this is weighted both by distance traveled, and by per-passenger operating costs.[13] In short, systems that cover larger areas and run more cars receive larger shares of federal subsidy almost irrespective of the actual number of patrons served. The same is true for the two-thirds portion allocated on the basis of existing bus service. Only 9.2 percent of the amount available is apportioned based on the number of bus passenger miles traveled weighted by operating costs.[14] Over 90 percent of these funds are distributed to individual urbanized areas by a formula that is weighted as follows: 50 percent for miles of bus service, 25 percent for population, and 25 percent for population density.[15]

Discretionary funding also favors capital intensive programs. Of the $1.9 billion in budget appropriations for FY 97, the fixed guideway rail modernization and new starts programs each received $760 million. Bus and bus-related projects were allocated only $380 million in the FTA budget. Under TEA-21, rail modernization and new starts receive 40 percent each from guaranteed funding and bus capital projects only 20 percent.[16]

In addition to federal funding mechanisms, the states also contribute significantly to highway and transit finance. The State of California, for example, supplies funds to transit but does not allow funds collected from sales taxes in one county to be expended in another county. Within counties (with one exception), state law distributes transit funds based on the service area population only, not ridership. Since larger, more densely populated areas have a higher percentage of transit riders, the allocation favors smaller areas with low levels of transit ridership. The combined effect of these federal and state policies is that areas with low population, low density, and a large number of service miles receive a proportionately higher amount of transit funding per passenger than

---

[12] 49 U.S.C. section 5336(b)(2)(A).

[13] 49 U.S.C. section 5336(b)(2)(B).

[14] 49 U.S.C. section 5336(c)(2).

[15] 49 U.S.C. section 5336(c)(1). Of the 90.8 percent, 73.39% is apportioned to urbanized areas with a population of at least 1,000,000 and 26.61% to urbanized areas with a population of between 200,000 and 999,999.

[16] 49 U.S.C. section 5309. Over the operative period of TEA-21 (FY98-FY03), authorizations for rail modernization total $6.592 billion, new starts total $8.182 billion and $3.546 billion goes for bus and bus-related projects (FTA, 1999).

areas with higher population and densities. As a consequence, suburban systems tend to spend far more per transit rider than central city areas and generally can afford to operate newer buses over longer routes with fewer passengers (Taylor, 1991).

With respect to rail development, this shift in policy emphasis has been quite dramatic. Between 1983 and 1994, bus service nationwide increased 10.7 percent; during this same period subway and elevated rail transit service increased 28.8 percent, commuter rail service increased 31.6 percent, and light rail (streetcar) service increased 108.1 percent (TCRP, 1997). In 1993, buses carried over twice as many passengers (5.4 billion) as all rail transit modes combined (2.6 billion) (APTA, 1998), but total expenditures on bus and rail transit (most of which came from government subsidies) were approximately equal ($10.1 billion) (Price Waterhouse 1998).[17] Why the emphasis on subsidizing rail service over buses? A number of factors are at work.

The Politics of Public Transit

Given that both transit ridership and low-income transit dependents are concentrated on buses in central city areas, we might expect that transit providers would target more resources to improving central city bus service on both efficiency and equity grounds. But transit agencies, by and large, have shown more concern recently with attracting riders out of cars than with serving the needs of those who — due to age, poverty, or disabilities — must depend on public transit. The reasons for this include: (1) a public clamoring to reduce traffic congestion; (2) legal mandates to improve air quality; (3) inter- and intra-metropolitan competition for limited fiscal resources; and (4) a changing political landscape that makes it more difficult to implement redistributive social programs. As a result of these factors, transit planning and policy has been characterized by a shift in emphasis from local to commuter service, from bus operations to rail development, and from inner-city to suburban riders.

Given the overwhelming domination of private vehicle use in most metropolitan areas, even substantial increases in transit use would be unlikely to significantly reduce suburban traffic congestion. By the early 1980s, public transit captured only 3 percent of suburban journeys-to-work, and an even smaller share for other trip purposes (Downs, 1992). In 1990, public transit's share of overall metropolitan commuting (both central cities and suburbs) had declined to less than 5 percent of all journey-to-work

[17]Figures represent increased vehicle miles in revenue service.

trips (Rossetti and Eversole, 1993). Thus, even doubling the share of transit commuters would at best make only a small dent in reducing automobile travel (Downs, ibid.). Moreover, the share of metropolitan jobs located in central business districts is on a long declining trend, making radial, downtown-centered transit systems attractive to fewer and fewer commuters over time.

There is also no guarantee that the road space freed up by former auto travelers attracted onto buses and trains, will not be replaced by other travelers. This phenomenon, known as the "latent demand" for travel, is part of what Downs calls the "triple convergence" of drivers from (1) other routes, (2) other times, and (3) other modes on to newly uncongested roads (Downs, ibid.). If transit systems succeed in snaring a substantial share of former auto users, congestion could decline noticeably in the short-term. But other automobile travelers, who have chosen to avoid the previously congested routes will quickly be attracted onto the less crowded roads thereby diminishing the congestion benefit of higher transit use. The San Francisco-Oakland Bay Bridge corridor is an excellent example of this phenomenon. While the opening of the transbay tube under the bay attracted large numbers of former auto and bus travelers, the congestion reductions on the Bay Bridge following the addition of BART proved fleeting. Congestion levels quickly returned to, and then eventually exceeded, pre-BART levels. While it is possible that congestion today might be substantially worse had BART not been built, it is also unlikely that so many people would choose to work in downtown San Francisco and live in the far flung suburbs of Alameda and Contra Costa Counties to the east. Systems such as BART may make it easier to commute to the CBD without a car, but they also make it easier to live farther away and still work downtown (Webber, 1976). In fact, some argue that radial transit systems may increase congestion in some situations by encouraging downtown development and thereby attracting other commuters onto already congested highways (Downs, 1992).

Heightened public concern over air pollution has also focused attention on the role transit can play in reducing auto travel, thereby lowering exhaust emissions (Garrett and Wachs, 1996). Federal clean air and surface transportation legislation have been integrated in recent years to bring about reductions in motor vehicle emissions. Many air quality plans in federal "non-attainment" areas, including those for Los Angeles and the San Francisco Bay Area, call for reducing automobile use. Even though most air quality forecasts suggest that public transit will make very small contributions to air quality (Bae, 1993), transit systems are nonetheless charged with the task of attracting automobile drivers onto public transit on air quality grounds.[18] Transit policy is therefore geared to providing incentives to reduce the number of single occupant vehicle (SOV) trips. To compete with private automobiles, transit operators must offer drivers substantial incentives, since these automobile commuters tend to have higher incomes and more travel options than transit dependents. Providing high quality alternatives to the automobile typically entails expensive public investments in new fixed rail or express bus service that tends to raise the overall costs of transit service. This, in turn, can lead to pressure for fare increases or service reductions, both of which may lower ridership overall.

Not only is capital spending higher for suburban systems, but fare structures promote cross-subsidization of wealthier riders by poorer ones. Typically, higher-income persons are less sensitive to price changes than are lower-income people. With respect to transit fares, however, this relationship is just the opposite. With fewer available alternatives, low-income riders are *less* sensitive to fare increases than higher income riders who can often choose to drive rather than pay higher fares (Cervero, 1990). As a result, transit fares, on a per mile basis, tend to be lower on commuter and suburban transit systems than on central city bus systems in order to attract and retain discretionary commuters. In Los Angeles, for example, the base local fare on the central city system is $1.35, compared to $0.50 on the suburban Santa Monica system and $0.60 on the suburban Culver City system.

Beyond air quality and congestion concerns, large public works projects have always been popular with elected officials and voters,

---

[18]There is credible evidence that the air quality benefits of public transit are not measurably better than automobiles. Bae (1993) found in a study of transportation and land use measures in Los Angeles designed to achieve the air pollution emission reduction targets of the 1991 Air Quality Management Plan (AQMP), that measures aimed at reducing vehicle miles traveled would have only a modest impact on reducing air pollution and that more transit use is not necessary to achieve clean air objectives. According to regional planning forecasts cited by Bae, the then current $150 billion bus and rail improvement plan for the region would only achieve a 10 percent work trip transit share by 2010, not the 19.3% share outlined in the AQMP. All mode shift strategies combined (transit investments, ridesharing incentives, alternative work schedules and job-housing balance strategies) would account for only a 0.9 percent decrease in reactive organic gases (ROGs), a 2.0 percent decline in nitric oxides (NOx) and a 4.3 percent drop in carbon monoxide (CO) emissions.

and transit investments are no exception to this rule. Cutting ribbons to open new rail transit lines get elected officials and transit agencies media attention, reducing headways on existing bus service generally does not. Declining transit use also threatens transit agencies' political claims on public resources. As transit agencies have increasingly turned to local voters for financial support in recent years (TCRP, 1997), the focus on large transit capital projects has only heightened. When asked to approve county sales tax increases for transportation, California voters have shown a clear preference for major capital investments over increased funding for planning, operations, or maintenance (Zell 1989).

The general preference for large capital investments, concern over urban traffic congestion and air quality, and the lack of public consensus on what to do about these problems, has led many public officials to embrace rail transit as a clear and dramatic alternative to the automobile/highway system. Richmond (1991) has examined the popularity of rail transit among elected officials in Los Angeles and found that their support for rail transit is due more to positive, highly symbolic perceptions of trains than to any analyses or other direct evidence on the wisdom of rail transit investments.

The policy choice to favor new rail construction is reinforced by the overall spatial logic of federal and state regulations, which is to spread transit funds to voters on a roughly geographical basis rather than in accordance with transit use or need. And since the transit subsidy allocations are based on fixed characteristics such as population, density, and existing service, eligible areas do not need to compete directly for these funds. Therefore, each service area has an incentive to apply for and expend the full amount available regardless of any regional planning rationale to the contrary. The combination of federal funds for new rail starts, and dedicated local and state transportation funding programs, often produce politically powerful constituencies for rail development, even in situations where it fails to satisfy either the usual social equity or economic efficiency rationales. Rail is championed more frequently for its ability to stimulate local economic development than from any transit planning rationale (Richmond, 1991). Local business and civic interests that benefit from publicly-funded construction projects can be expected to lobby hard for a share of the funds made available.

Finally, it is important to recognize that transit dependents do not represent a strong constituency for improved bus service since fewer poor and minority persons are registered to vote, and are less likely to vote, compared to suburban residents. In addition, many urban transit users (especially in areas like Los Angeles) may also be new immigrants or undocumented persons and unable to vote (Meyers, 1996).

Voters who might support higher transit spending are increasingly located in newer, auto-oriented cities and suburbs. But since most transit riders have disproportionately low incomes, public spending on transit riders tends to redistribute tax revenues from wealthier to poorer individuals, and from suburbs to cities. In recent years, voters have clearly grown increasingly resistant to explicitly redistributive policies and programs, which does not argue for highlighting public transit's emerging role as a largely redistributive social service. Hence, transit operators often downplay this aspect of public transit subsidies in light of the declining popularity of explicitly redistributive fiscal policies, emphasizing instead its advantages in reducing traffic congestion, improving air quality, and stimulating economic development.

Transit providers thus have a strong political incentive to make transit service more attractive to suburban and discretionary riders in order to maintain broad public support for transit (Wachs, 1985; Wachs, 1989). At a policy level, this means providing wider service area coverage by shifting resources to new lines to capture additional riders. Consistent with the new suburban electoral majority, it also means focusing on improving the suburb to downtown work commute. In short, to secure popular, political, and financial support for their systems, transit operators and funding agencies must balance the demand for local service in high ridership central city areas, against the service preferences of suburban residents who tend to favor commuter transit systems. From an operational standpoint, these trends are particularly problematic since the total per-passenger subsidies needed to operate these new suburban lines are typically much higher than those for inner-city buses. While providing larger subsidies to certain lines or modes in an effort to attract new riders may make sense politically, such policies tend to decrease both efficiency *and* equity because low-income, central city riders are, on average, less costly to serve than suburban commuters. Research has consistently shown that the poor actually require lower subsidies per rider than do wealthier patrons (Hodge, 1995; Pucher, Hendrickson, and McNeil, 1981; Pucher, 1981; Pucher, 1983). Moreover, the small number of new riders brought onto the systems are often exceeded by the loss of existing ridership brought about by increased fares and the reduced quality of bus service (Rubin and Moore II, 1996). Declining revenues and increasing costs place even greater pressure on transit operators to either cut existing bus service or raise fares, further exacerbating these disparities.

Given all these factors it is not surprising that many transit systems have responded by directing their planning efforts toward expanding suburban commuter services over improving local operations and increasing rail service over buses, despite the shift in demand towards an increasingly poor ridership base. The combination of federal transportation funds for new rail projects and dedicated local and state funding programs have produced a natural political constituency for rail development, even in situations where it fails to satisfy either the usual social equity or economic efficiency rationales. The pressure to appeal to discretionary riders (who vote in larger numbers) over transit dependents (who do not) also favors capital intensive investments, such as rail transit. Such investments need heavy ridership to be cost effective, though fewer and fewer urban areas have sufficient residential and employment density to generate the required level of patronage. As we have noted, the result of this tension has been an increasing dichotomization of transit service and subsidies between those lines and systems serving more higher-income riders at substantial public subsidy on one hand, and those serving mostly poor, minority riders at substantially lower public subsidy on the other. Unfortunately, these implicit tradeoffs between transit dependents and discretionary users are rarely spelled out in the usual debates between bus and rail investment.

Conclusion

In summary, the dissonance between shifting ridership demographics and the policy response is a function of the diverging spatial logics shaping the demand for transit service on one hand, and guiding the public subsidization of transit service on the other. Indeed, while the transit demand is concentrated in high-density, low-income areas, subsidies favor lower-density, higher income areas. Since the majority of transit-dependent riders are poor and members of minority groups, the ongoing shifts in ridership patterns and the failure of transit authorities to respond to the growing disparity in service between transit-dependent and discretionary riders, have made transit planning a social justice issue. As a result, advocates for transit dependents throughout the nation have begun to challenge transit operators publicly, and even in court, over service policies that have a discriminatory impact on poor and minority communities.

The foregoing raises a number of normative questions regarding the value of public transit: how should fairness be defined in the context of public transit? Who is being served by the shift in transit investment to suburban services and, in some cities, from bus to rail? Should public transit policy strive for greater geographic *mobility*, regardless of the available alternative modes of transportation, or would it be preferable to improve *accessibility* for those with few private alternatives? How have transit planners responded to the changing spatial and social realities of cities and regions? Are current transit policies increasing or decreasing social equity? An important step in beginning to answer these questions is to clearly define the frame of reference for judging equity and fairness. Under our current system of public transit finance, equity is typically defined by comparing funding allocations among jurisdictions or agencies. Shifting the focus onto the distribution of subsidies for individual transit users or classes of transit users would significantly alter debates over transit equity by challenging the fairness of public transit service provision in the U.S. If indeed public transit is increasingly a social service for the poor and disadvantaged, then planners should begin to view the funding and deployment of public transit in a new light.

References

APTA (American Public Transit Association). 1998. *Transit Fact Book February 1997*. Washington, D.C.: American Public Transit Association.

APTA (American Public Transit Association). 1999. Web page [accessed 1 April 1999]. Available at www.apta.com.

Bae, Chang-Hee Christine. 1993. Air Quality and Travel Behavior: Untying the Knot. *Journal of the American Planning Association* 59, no. 1: 65-74.

Cervero, Robert. 1990. Transit Pricing Research: A Review and Synthesis. *Transportation* 17: 117-39.

Downs, Anthony. 1992. *Stuck in Traffic: Coping with Peak-Hour Traffic Congestion*. Washington, D.C.: The Brookings Institute.

FTA, (Federal Transit Administration). 1999. Web page, [accessed 1 April 1999]. Available at www.fta.dot.gov.

Garrett, Mark, and Martin Wachs. 1996. *Transportation Planning on Trial: The Clean Air Act and Travel Forecasting*. Thousand Oaks, CA: Sage Publications, Inc.

Hodge, David C. 1995. My Fair Share: Equity Issues in Urban Transportation. in *The Geography of Urban Transportation, 2nd ed.*, ed. Susan Hanson, 359-75. Guilford Press.

Hu, Patricia S., and Jennifer Young. 1993. *1990 NPTS Databook: Nationwide Personal Transportation Survey, Vol I.*, Federal Transit Administration, Washington, D. C..

Krumholz, Norman. 1982. A Retrospective View of Equity Planning: Cleveland 1969-1979. *Journal of the American Planning Association* 48, no. 2: 163-74.

Krumholz, Norman, and John Forester. 1990. *Making Equity Planning Work.* Philadelphia: Temple University Press.

Meyers, Dowell. 1996. Changes over Time in Transportation Mode for the Journey to Work: Aging and Immigration Effects. Paper presented at the *Conference on Decennial Census Data for Transportation Planning sponsored by the Transportation Research Board, Federal Highway Administration, Federal Transit Administration, and the Bureau of Transportation Studies.*

MTA (Metropolitan Transit Authority, Los Angeles). 1991-1993. *Origin-Destination Passenger Survey.*

Ong, Paul. 1996. Work and Automobile Ownership Among Welfare Recipients. *Social Work Research* 20, no. 4.

Ong, Paul, and Evelyn Blumenberg. 1999. *Measuring the Role of Transportation in Facilitating the Welfare-to-Work Transition*, Working paper, Department of Urban Planning, UCLA.

Pickrell, Don. 1992. A Desire Named Streetcar: Fantasy and Fact in Rail Transit Planning. *Journal of the American Planning Association* 58, no. 2: 158-76.

Pickrell, Donald. 1983. *The Causes of Rising Transit Operator Deficits*, Report No. DOT-I-83-47. U.S. Department of Transportation, Washington, D.C.

Pisarski, Alan. 1992. *Travel Behavior Issues in the 90s*, Office of Highway Information Management, Federal Highway Administration, Washington D.C.

Pisarski, Alan E. 1996. *Commuting in America II: The Second National Report on Commuting Patterns and Trends*, Eno Transportation Foundation, Inc., Landstowne, Va.

Price Waterhouse. 1998. *Funding Strategies for Public Transportation, Transit Cooperative Research Program (TCRP) Report 31*, Volume 1, National Academy Press.

Pucher, John. 1981. Equity in Transit Finance: Distribution of Transit Subsidy Benefits and Costs Among Income Classes. *Journal of the American Planning Association* 47, no. 4: 387-407.

___________. 1983. Impacts of Subsidies on the Costs of Urban Public Transport. *Journal of Transportation Economics and Policy* May: 155-76.

Pucher, John, Chris Hendrickson, and Sue McNeil. 1981. Socioeconomic Characteristics of Transit Riders: Some Recent Evidence. *Traffic Quarterly* 35, no. 3: 461-83.

Richmond, Jonathan E. D. 1991. "Transport of Delight: The Mythical Conception of Rail Transit in Los Angeles." Ph.D. Dissertation, MIT.

Rosenbloom, Sandra. 1992. *Reverse Commute Transportation: Emerging Provider Roles*, U.S. Department of Transportation, University and Training Program, Washington, D.C.

Rossetti, Michael A., and Barbara S. Eversole. 1993. *Journey to Work Trends in the United States and Its Major Metropolitan Areas, 1960-1990*, U.S. Department of Transportation, Federal Highway Administration, Washington D.C.

Rubin, Tom. 1994. *A Look at the Los Angeles Metropolitan Transportation Authority*, Metropolitan Transit Authority, Los Angeles.

Rubin, Tom, and James E. Moore II. 1996. *Why Rail Will Fail: An Analysis of the LA Metropolitan Transportation Authority's Long-Range Plan*. Los Angeles: Reason Public Policy Institute.

Shen, Qin. 1998. Location Characteristics of Inner-city Neighborhoods and Employment Accessibility of Low-wage Workers. *Environment and Planning B: Planning and Design* 25: 345-65.

Taylor, Brian. 1991. Unjust Equity: An Examination of California's Transportation Development Act. *Transportation Research Record* 1297: 85-92.

Taylor, Brian D., and William S. McCullough. 1998. Lost Riders. *Access* 13: 26-31.

TCRP. 1997. *Funding Strategies for Public Transportation*, Transportation Research Board, Washington, D.C.

U.S. DOT, (Department of Transportation). 1999. "1995 National Personal Transportation Survey." Web page, [accessed 1 April 1999]. Available at www.bts.gov/ntda/npts.

Vincent, Mary Jane, Mary Anne Keyes, and Marshall Reed. 1994. *NPTS Urban Travel Patterns: 1990 Nationwide Personal Transportation Survey*, U.S. DOT Federal Transit Administration, Washington, D. C.

Wachs, Martin. 1985. The Politicization of Transit Subsidy Policy in America. in *Transportation and Mobility in an Era of Transition.* eds G. Jansen, P. Nijbamp, and C. Riujgrok, 353-66. New York: North Holland.

_____________. 1989. American Transit Subsidy Policy: In Need of Reform. *Science* 244: 1545-49.

Wachs, Martin, and Brian D. Taylor. 1998. Can Transportation Strategies Help Meet the Welfare Challenge? *Journal of the American Planning Association* 64, no. 1: 15-21.

Webber, Melvin. 1976. The BART Experience — What Have We Learned? *Public Interest* 45: 79-108.

Zell, Eric S. 1989. A Vote Winning Combination: Contra Costa County Links Transportation Tax to Growth Management. *Urban Land* 48, no. 6: 6-10.

[23]

LAND USE POLICY AND TRANSPORTATION: WHY WE WON'T GET THERE FROM HERE

Genevieve Giuliano
School of Policy, Planning and Development
University of Southern California

The purpose of this paper is to consider the effectiveness of land use policy as an instrument for reducing environmental and other external costs associated with ownership and use of the private automobile. Emphasis is placed on the long run, since land use change is a slow process, and consequently can potentially have significant effects only in the long run. I will argue that land use change is driven by factors over which we have little policy control, and that current trends of decentralization will continue in the future. Although the link between urban form and travel behavior may be significant, it is highly unlikely that policy actions could shift urban form to patterns associated with less private vehicle travel. The paper begins by presenting some information on international trends in travel and land use patterns. Then I discuss explanatory factors associated with these trends. The final part of the paper addresses the future, and considers the potential of land use policies in the context of long run trends.

URBAN TRAVEL TRENDS

Urban travel trends are easily summarized. Car ownership and use is increasing, total travel is increasing, and both public transit use and non-motorized modes are decreasing.

Car Ownership and Use

Throughout the developed world, people own more private vehicles, use them more frequently, drive more miles, and are more likely to drive alone than ever before. The world's motor vehicle fleet has grown immensely over the past two decades. The total number of cars and trucks increased from 246 million in 1970 to 617 million in 1993, with most of the growth occurring outside the United States, as illustrated in Figure 1 (U.S. Department of Transportation, 1996). Average annual growth rates in the motor vehicle fleet over this period are 2.6 percent for the United States, 4.4 percent for other OECD countries, and 6.5 percent for non-OECD countries.

Patterns of vehicle ownership are further illustrated in Table 1, which gives average annual growth rates for car registrations in selected countries, grouped by level of per capita income and weighted by population. The low and low-middle income countries have the lowest car ownership rates, but the highest growth rates. These numbers suggest that absent severe policy intervention, the world car fleet will grow enormously in the coming decades as developing countries achieve higher levels of per capita income. It bears noting that China has the lowest 1992 car ownership rate (car per population ratio of 0.00162), even though the vehicle fleet increased by more than a factor of ten between 1970 and 1992. Another increase of this magnitude or greater is quite possible in the coming decade. At the opposite end of the spectrum, the United States continues to have the highest car ownership rate (car per population ratio of 0.6), but it had the slowest growth rate (2.2 percent) during this period, suggesting that car ownership in the United States may finally be reaching saturation.

TABLE 1 Growth in Car Ownership, by Country Per Capita Income Category, 1970–1992

	Annual Growth Rate 1970 - 1992, percent		
	Cars	Population	Cars/Pop. 1,992
Low Income Economies *(Examples: India, China, Nigeria)*	9.4	2.3	0.0034
Lower Middle Income Economies *(Examples: Peru, Thailand, Turkey)*	9.6	2.5	0.0350
Upper Middle Income Economies *(Examples: Mexico, South Korea, Brazil)*	7.2	2.4	0.0860
High Income Economies *(Examples: U.S., Japan, Germany)*	3.3	0.9	0.4760

Source: USDOT (1996), p. 219.

Car ownership is significantly related to per capita income. Figure 2 plots car ownership per 1000 population against the natural log of 1992 GDP in U.S. dollars for several European countries (east and west), the United States, Canada and Japan. The graph suggests that as economic well-being improves in lower income countries, car ownership will increase. The graph also shows that the greatest dispersion of car ownership rates is found among the higher income countries, with the United States at one extreme and Denmark and Japan at the other. In addition to per capita income, differences in car ownership and use across countries are attributed to population density, the density of cars relative to land area or road supply, and car ownership and fuel costs. High population density and limited land area may promote implementation of auto restraint policies to reduce congestion and other negative effects associated with auto travel in densely developed areas. Pucher (1988) associates the generally lower levels of car ownership and use outside the United States to public policies that make car ownership and use more costly and less convenient. Despite these policies, however, car ownership continues to increase.

More car ownership means more car use; annual vehicle-kilometers traveled has increased at about the same rate as car ownership. To illustrate, Table 2 gives annual average VKT growth rates for the US, Japan, and selected European Conference (EC) countries. Figure 3 shows VKT growth for 6 European countries, 1970 to 1995. Total VKT nearly doubled over the period, and the greatest growth occurred in private vehicle travel; the private vehicle share increased from 79 percent in 1970 to 85 percent in 1995.

TABLE 2 Growth in Car Use, by Country 1970–1993

	Avg. Annual Growth Rate (%)	
Country	VKT	Cars
United States	2.7	2.2
France	3.2	3.0
West Germany	3.0	3.6
Great Britain	3.8	3.2
Japan	6.5	6.9

Source: USDOT (1996), p. 209.

Mode Shifts

Table 3 gives information on mode shares for urban areas in various countries. Care must be taken in making such comparisons, because data are collected differently, and mode and trip definitions may differ across countries and across years. Data for all trips are not available for urban areas in the United Kingdom; hence only data for London and for the journey to work for Manchester are presented. Because London is such a large metropolitan area, it is not representative of the general level of car use in other U.K. urban areas. In all countries, the trend of increasing car use is obvious, but the rate of increase varies greatly. In the United States, where car use was already very high in 1969, increases have been quite small. In contrast, large increases have occurred in the urban areas of Norway and West Germany.

Increased car use has come at the expense of both public transport and non-motorized travel, depending on the urban area. In Germany, the public transport share has remained quite stable, while the non-motorized share has decreased. In the other countries, both public transport and non-motorized shares decreased. Decreases in non-motorized trips suggest substitution of longer trips for short trips, as well as population shifts out of core city areas to less dense (and therefore less bike or pedestrian accessible) areas. Although much of the transportation public policy debate focuses on car vs public transport, the observed decline in non-motorized trips is probably far more consequential from an environmental perspective.

Explanatory Factors

In addition to rising affluence, major explanatory factors for these trends include changing demographics and household structure, labor force participation, and changing land patterns. Higher income implies higher value of time, making travel time relatively more important in travel choice decisions. As the value of time increases, faster modes will be preferred, hence the increase in private vehicle travel. Higher income also implies greater demand for goods and services, and therefore more total travel. The relationship of car use, distance traveled, and trip frequency with household income is extensively documented. (e.g. Hu and Young, 1993; Pisarski, 1996).

TABLE 3 Mode Share Trends, All Person Trips, Selected Urban Areas

London	1975-76	1985-86	1989-91
Car	41	44.3	47.8
Public Transport	20	17.3	17.0
Bike	3	2.8	1.7
Walk	35	35.0	32.7
Manchester [a]	1971	1981	1991
Car	32	50	64
Public Transport	39	24	16
Bike	2	2	2
Walk	21	19	16
Norwegian city regions	1970	1985	1990
Car	32	60	68
Public Transport	20	11	7
Walk & Bike	48	29	25
W. Germany urban areas	1972	1982	1992
Car	34	43	49
Public Transport	17	17	16
Bike	8	10	12
Walk	41	30	23
US urban areas	1969	1977	1990
Car	79.8	82.3	84.3
Public Transport	4.9	3.4	2.8
Bike	0.7	0.7	0.7
Walk	11.5	10.7	9.1

[a] Journey to work only
Source: Pucher and Lefevre (1996); Hervick, Tretvik and Ovstedal (1993); Brog and Erl (1995).

Household size has declined both in the US and in Europe for several decades. Average number of persons per U.S. household was 2.75 in 1980 and 2.63 in 1990. Household composition has also changed: the most rapid increase in household growth was among non-family households, e.g. persons living alone or with other non-family persons (Pisarski, 1996). Similar patterns prevail in Europe; among the "EURO12, household size declined from 2.8 in 1981 to 2.6 in 1991 (ECMT, 1995). Declining household size is attributed to declining fertility rates, rising divorce rates, breaking up of the extended family system, aging of the population, and growing economic independence of women and young people (Masser, Sviden and Wegener, 1992). As birthrates continue to decline, smaller household size should be observed in less developed countries as well.

Declining household size means more travel for personal or household needs. Regular household activities (food shopping, laundry and cleaning, home maintenance, social visits, etc.) are shared among fewer household members. In addition, non-family households are less likely to share resources; consequently we would expect such members to behave more like individuals living alone, hence generating more household trips.

In both the United States and European countries, observed increases in the labor force participation rate are mainly due to increased participation by women. Increased participation in the labor market by women has at least two significant effects on travel. First, more working women means more households with multiple workers. In the U.S., 70 percent of all working households had two or more workers in 1990 (Pisarski, 1996). Housing location choice decisions are more complex for households with multiple workers; all else equal, it is more difficult for such households to live close to work, given dispersed job locations. Although research shows that women travel shorter distances to work than men, it seems reasonable to attribute some of the observed increase in commute travel distance to the rise in multiple worker households.

Second, increased participation of women in the workforce has not been accompanied by any major changes in household responsibilities. All else equal, working women are subject to greater time pressure, and consequently attribute high value to the efficiency of driving alone. The value women place on driving alone is demonstrated in the United States by the higher likelihood of women driving alone than men when household income is controlled (Rosenbloom, 1995). Also, although United States women in 1990 still drove fewer annual VMT than men, the rate of increase in VMT since 1983 has been higher for women (Pisarski, 1992).

LAND USE TRENDS

The major trend in urban spatial patterns for several decades has been decentralization. Suburbanization of population and employment has been evident in the United States throughout the Twentieth century. Large scale population suburbanization was followed by large scale employment decentralization and by the emergence of major agglomerations outside the traditional downtown (e.g. Muller, 1995). More recently, decentralization has been accompanied by dispersion, with most growth occurring outside major centers.

Table 4 gives population growth rates for United States metropolitan areas with 1 million or more population, by decade, 1960 through 1990, using United States census data. In each decade, population growth was more rapid in suburban counties than in central counties. In 1960, central counties accounted for a majority of the metropolitan population, but by 1970 the majority shifted to suburban counties, The suburban county share continued to increase through 1990.

TABLE 4 Population Growth for U.S. Metro Areas with 1 Million or More Population, Central and Suburban Counties

Population Growth Rates, Percent			
Years	Total Area	Central County	Suburban Counties
1960 - 1970	18.50	10.20	27.35
1970 - 1980	7.78	2.82	12.35
1980 - 1990	11.81	9.22	13.79
Population Shares, Percent			
Year		Central County	Suburban Counties
1960		51.60	48.40
1970		47.99	52.01
1980		43.28	56.72
1990		42.27	57.73

Source: Rosetti and Eversole (1993)

Population decentralization has been accompanied by employment decentralization. Empirical evidence of this trend is extensive. For example, Gordon and Richardson (1996) calculated average annual employment growth rates for 54 U.S. metropolitan areas for 1976 - 1980 and 1980 - 1986. Areas were segmented into CBD, remainder of the central city, and the remaining metropolitan area excluding the central city. In all cases, growth rates were highest in the suburban county. Similar results were found using annual employment data by county (Gordon, Richardson and Yu, 1996).

A similar process of population and employment decentralization is also evident within most metropolitan areas in Europe, although from a very different starting point and with a wider degree of variability of experience. Indeed, decentralization has been documented in major metropolitan areas throughout the developed world. Table 5 gives population and employment changes for several metropolitan areas, for core city areas and their suburbs. In all but one case (Liverpool employment), population and employment grew faster (or declined slower) in the suburbs than in the core city. Note that the table includes metro areas in several different countries, and that the most recent series ends in 1985 (more recent data are not available). It is possible that more recent data would reveal an acceleration of these trends, given the effects of globalization and the shift to an information-based economy. More recent population data are available for selected cities. Some examples of central city population shares: Paris central city population share declined from 32 percent in 1968 to 23 percent in 1990; Zurich form 38 percent in 1970 to 29 percent in 1995; Amsterdam from 80 percent in 1970 to 66 percent in 1994. Only London has held approximately steady at 41 percent in 1971 to 38 percent in 1994.

TABLE 5 Population and Employment Decentralization in Selected European Cities Average Annual Percentage Change

	Population			Employment		
City	Years	Core City	Suburbs	Years	Core City	Suburbs
Antwerp	1970-81	-0.8	+1.2	1974-84	-0.7	+0.4
Copenhagen	1970-85	-1.5	+1.0	1970-83	-0.3	+3.2
Hamburg	1970-81	-0.8	+1.9	1961-83	-0.8	+1.9
Liverpool	1971-80	-1.6	-0.4	1978-84	-2.6	-3.1
Milan	1968-80	-0.6	+1.3	1971-81	-0.9	+1.9
Paris	1968-80	-1.1	+1.1	1975-82	-1.1	+0.9
Rotterdam	1970-80	-1.6	+2.2	1975-84	-1.1	+1.5

Source: Jansen (1993)

Land Use and Commuting Patterns

Decentralization of population and employment is reflected in commuting patterns. To summarize, the traditional commute to the center city is no longer the dominant commute flow. Commuting between suburban locations is now the major flow in the United States, and is the fastest growing commute flow in European metropolitan areas. Table 6 gives commute flow data for the United States, drawn again from U.S. census data. Since the census only began asking the work location in 1980, comparisons are available only for 1980 and 1990. The data are compiled by county, a local political jurisdiction that can include one or more cities. Central counties therefore encompass the central city of the metropolitan areas as well as adjacent cities and county areas. Central county therefore overstates the central city portion in nearly every case. Several observations are to be drawn from Table 6. First, central counties were the location of the greatest share of job destinations in both years, but the share declines. Conversely, the share of job destinations in suburban and exurban locations increases. Second, the suburban resident worker share increases. Third, the largest flow is central county to central county in 1980, but is suburban county to suburban county in 1990.

TABLE 6 Commute Flows in U.S. Metropolitan Areas, 1980 and 1990

1980: 31 Metro Areas				
	Place of Work			
Place of Residence	Central county	Suburban county	Outside area	Subtotal
Central county	41.90	2.70	0.83	45.43
Suburban county	12.14	40.90	1.53	54.57
Subtotal	54.03	43.60	2.36	100.00
1990: 39 Metro Areas				
	Place of Work			
Place of Residence	Central county	Suburban county	Outside area	Subtotal
Central county	38.05	3.57	0.83	42.44
Suburban county	11.68	43.52	2.34	57.55
Subtotal	49.73	47.09	3.17	100.00

Source: Computed from Rosetti and Eversole (1993).

Using more disaggregate data, (1996) allocates the *increase* in commute flows between 1980 and 1990 as follows: 58 percent suburb to suburb, 20 percent suburb to central city, 12 percent central city to suburb, and 10 percent city to city. Thus the suburb to suburb commute continues to be the fastest growing commute flow segment.

With more suburban job destinations and fewer central city job destinations comes more use of the private car. Table 7 gives mode share for U.S. journey to work trips by destination location category. Public transit still carries a significant portion of work trips to central city destinations. In contrast, more people walk or bike to suburban jobs than take transit, and the private vehicle accounts for 90 percent of all trips.

TABLE 7 Journey to Work Mode Choice, 1990, by Job Location

	Mode Share, percent				
Job location	Drive alone	Carpool	Pub. Transp.	Walk/Bike	Other[a]
Central City	68.2	13.4	11.0	4.7	2.9
Suburbs	77.5	12.9	2.0	3.5	3.3

[a] Includes work at home.
Source: Pisarski (1996), p. 84.

The same trend of dispersing commute flows is evident in the EC. Limited data makes possible only a few examples. Commute flow data for the Paris region, 1975 and 1982, reveal that the greatest decline occurred in the central city to central city flow, while the greatest increase occurred in outer suburb to inner suburb commutes. Other large increases took place in central city to outer suburbs, and inner suburbs to outer suburbs, implying a significant dispersion of travel flows and longer distance commutes, which in turn implies greater use of private vehicles (Jansen, 1993).

In Germany, the share of workers living and working in the same city declined from 72 percent in 1970 to 61 percent in 1988. The increase in commuting by car that occurred is the result of both longer distance commuting and generally increased demand for car travel. For those living and working in the same city, the increase in car use was at the expense of non-motorized modes. For those working in a different city, the shift was from public transit (Jansen, 1993).

WHAT HAPPENED?

Before these trends were clearly evident and documented in Europe and other developed countries, decentralization and the dominance of the private auto were perceived as a uniquely American (U.S.) phenomenon. Explanations centered on public policy, cultural preferences, land availability, and rapid economic growth. Public policy factors include:

- Tax and pricing policies favorable to car ownership and use
- The Federal Interstate Highway construction program and the Highway Trust Fund
- Federal tax and mortgage policies that support home-ownership and favor suburban residential development
- Political fragmentation and powerful local governments that allow suburbanites to escape urban social and fiscal problems

It is claimed that these policies supported deeper social and cultural values:

- The tradition of strong private property rights
- Historical preferences for single family home-ownership
- The suburban ideal
- Ethnic and racial conflicts

It was argued that economic growth occurred throughout the developed world during the post-war era, albeit from a different base, therefore purely economic factors were not a satisfactory explanation for American-style decentralization. In light of similar trends now evident outside the US, however, explanations for decentralization merit further consideration.

If both population and jobs are decentralizing, even in countries where central governments have far more control over land use, cars are more costly to purchase and operate, public transit service is more extensive, and highways do not enjoy earmarked funding sources, then perhaps economic forces—rising per capita incomes and economic restructuring — play a more important role.

Rising Incomes

Rising per capita income increases demand for all sorts of consumer goods, including housing. Therefore, preferences for single family homes may not be so uniquely American after all. A 1985 survey conducted in West Germany provides a small piece of supporting evidence. When respondents were asked about their housing preferences; 59 percent chose single family detached house, 18 percent chose row house, and the remainder chose apartments and condominiums. At the time of the survey, just 40 percent actually lived in detached or row houses (Masser et al, 1992). Other evidence comes from the growing number of households that choose private homes in the suburbs of the United Kingdom, Paris and Australia, *even when such moves reduce accessibility to jobs and other activities* (Cullinane, 1992; Burnley, Murphy and Jenner, 1997, Baccaine, 1997). As demand for housing increases, households are willing to travel more in order to obtain preferred neighborhoods, housing characteristics, etc.

U.S. patterns of shopping and retailing are also evident in other countries. The suburban shopping center, conveniently accessible only by car and typically offering free parking, can be found along expressways in the suburbs of London, Milan, Munich and Paris. The emergence of the suburban shopping mall in European metropolitan areas may be explained by many of the same factors as in the United States: population suburbanization and rising consumer demand creates a market; shoppers are attracted by (relatively) lower prices, more variety, and convenient (car) access. A Royal Commission study of changing shopping patterns observes that shopping has become a leisure activity, and people are less willing to patronize the closest shops. Rather, they are willing to travel further to obtain greater variety, better quality, etc. (Royal Commission on Environmental Pollution, 1995).

Job Decentralization

The process of job decentralization is also evident outside the US, as described earlier. The shift to a service and information-based economy, together with improvements in information and telecommunications technology (ICT), have made firms more "footloose", and the agglomeration benefits of central locations have become less important for many types of activities. Service activities require less fixed infrastructure than manufacturing, and so are more easily relocated. As the workforce suburbanizes, these firms follow, taking advantage of lower land costs while maintaining or increasing labor force access. Expecting that workers will commute by car, these firms provide free or almost free (to the user) parking, further encouraging auto commuting. Declining agglomeration benefits also imply that congestion and other costs of agglomeration will not be as easily offset, and thus will promote additional decentralization. Suburban location in the United States has the additional advantages of lower business fees and taxes, as well as lower crime rates. Finally, as decentralization continues, regional accessibility becomes more homogenous, and the relative advantage of central location declines. The value of central location (all else equal) therefore declines for both households and firms.

FUTURE TRENDS

I have argued that rising incomes and changing economic structure have played a key role in the land use and travel patterns we observe today. What about the future? Would it be possible to reverse these trends, and, over time, to foster a reconcentration of activities in metropolitan areas? There are really two questions here. First, what magnitude of change would be required to significantly reduce private vehicle use; and second, is such change feasible?

The Evidence

There is now an extensive literature on sustainable development, and on the expected benefits of compact cities, transit-oriented land use, and pedestrian friendly neighborhoods. Proponents of compact development argue that increasing development densities and providing high quality transit will promote shifts to transit and non-motorized modes, and reduce use of the private auto. These expectations are based on empirically observed cross-sectional correlations between development density and measures of car use (Newman and Kenworthy, 1989a; 1989b). There are many questions about the validity of these findings, such as whether the environmental benefits of less car use are offset by more congestion, whether the relationship is significant at densities that might possibly be achieved, or whether there is any causal validity on which to base policy decisions.

Downs (1992) conducted some simple simulations, and concludes that very large increases in density would result in very small reductions in average commuting distance. Schimek (1996) found the relationship between person travel and residential density to be significant, but of very small magnitude. Specifically, a 10 percent increase in density is associated with a 0.7 percent decrease in VMT. From all the evidence available, it appears that in order to realize significant reductions in car travel, large magnitude changes in development density would be required.

The potential effects of pedestrian-friendly or transit-oriented neighborhood design is more uncertain. Crane (1996) considered the effects of various network designs, and concludes that there are possibilities for increased travel as well as decreased travel. Empirical work that has attempted to link aspects of neighborhood design to transit use or walk trips has yielded very mixed results (Cervero and Gorham, 1995; Ewing, Haliyur and Page, 1994; Handy 1992, 1996; Hanson and Schwab, 1987, Kitamura, Mokhtarian and Laidet, 1997). While in some cases a relationship between transit use or non-motorized travel and neighborhood design is demonstrated, a relationship with auto use is not demonstrated. That is, the effect of pedestrian or transit accessible designs may be to induce additional trip making, rather than to shift the mode of existing trips.

Implementing Effective Land Use Policies

On the basis of the existing evidence, it is difficult to support the use of any land use policy as a means for achieving environmental objectives associated with private vehicle use. Nevertheless, let me now consider the second question: are land use changes of a magnitude sufficient to significantly reduce private vehicle use feasible?

First, designing pedestrian friendly neighborhoods is quite possible, and indeed is happening in several new planned communities. Typically these communities are located in suburban (or even exurban) locations, often far from major job centers and accessible exclusively via automobile. They have the architectural attributes of New Urbanism--front porches, narrow streets, a town square—but are otherwise rather conventional middle or upper class planned communities, highway accessible and with plenty of room for the family's two or three cars. These new communities may have many benefits, but less private vehicle use is not likely to be one of them.

The real policy question is, therefore, can metropolitan densities be increased to a level that would lead to significantly less private vehicle travel? As noted above, this would require substantial increases in densities from existing levels and a reversal of development trends that have been in progress for many decades. I do not think such increases in density can be achieved, and increases in density that might be achieved would have at best little effect on private vehicle travel for the following reasons.

1. Most firms have no economic incentive to locate in dense, high cost centers. Agglomeration benefits are declining for all the reasons discussed above. Regulation would therefore be required to shift the incentive structure, either by offering large subsidies to locate in core areas, or imposing additional costs on locations in non-core areas, or imposing outright restrictions on development in non-core areas. In the United States, central city revitalization efforts have had very limited success, despite the large subsidies involved. There are, of course, some major success stories of downtown revitalization, and some types of activities that still value core locations. However, these are not the representative experiences of such efforts (e.g. Teaford, 1990). Furthermore, the metropolitan areas where central city growth has occurred have experienced even greater growth outside the central city (Gordon and Richardson, 1996). If the history of revitalization efforts are any indication, incentives to draw firms to core areas would have to be large indeed. Efforts to limit development in suburban areas have a mixed history; some studies have shown that the primary effect of such policies has been to shift growth to other areas; others have identified restrictions on housing supply that drive up prices. Higher housing prices create incentives for workers to seek less costly housing in more remote areas. (Rosen and Katy, 1981; Gyourko, 1991; Knaap, 1985)

2. Globalization makes it increasingly difficult to impose controls on where firms locate. Firms may respond in many ways to changes in conditions. As the share of footloose activities increases, more firms will have great flexibility in location choices. Through distributed production methods, out-sourcing, and other new forms of economic organization, firms can exploit the advantages of specific regions throughout the world. They can likewise avoid the disadvantages of specific locations. Large firms have been able to use this flexibility to promote "bidding contests among local communities for their business, as for example occurred in the case of GM's Saturn plant location in Tennessee. Also, if the cost of doing business in one location increases, activities can be shifted to other locations within the firm's spatial network. Examples abound of these shifts. In the United States, several types of product assistance telephone services, formerly performed in-house in central or branch offices, are contracted out to telephone service firms located in small communities in the Southern United States. These locations were chosen because there was a supply of workers willing to work swing and graveyard shifts for relatively low wages. In the United Kingdom, British Air shifted its reservation processing from several sites (including London) to Bangkok, where labor costs are much cheaper. Location flexibility transcends local, state and even national boundaries, and this flexibility can only increase as ICT continues to improve, making it ever more difficult to control the location of business activities via land use regulation.

3. Most households have no incentive to locate in dense, high cost centers. Demand for housing is related to household income. As incomes rise, so does demand for housing services — more living space. We are now observing population shifts to suburban areas in many countries; households are choosing suburban locations to obtain more housing. In doing so, they are willingly giving up access to jobs, downtown amenities, etc. The American Dream of the single family home (and garage) is not uniquely American at all, but rather reflects widely held preferences that can be acted upon as household income rises. There are of course some households that prefer urban living (young single persons, affluent empty nesters), and these niche markets would likely support high-density policies. However, these are niche markets, not mass markets.

Single family structures are not an option, if density must be greatly increased. It is important to note here that I am not arguing that residential densities cannot be increased; simply reducing the number of zoning restrictions that exist in most communities would increase densities and have many other beneficial effects as well. Rather, the issue is one of increasing densities to levels sufficient to reduce private vehicle use.

It is clear that most U.S. households prefer lower density living environments. According to a 1997 Fannie Mae survey, for example, just 9 percent of respondents stated that they preferred living in a "large city. The top two reasons given for not living in a central city were "pace of life and "crowding, traffic congestion. Because lower density living environments are preferred, households, like firms, will use their flexibility to act on their preferences. In the United States, households have historically demonstrated high levels of mobility. As development regulations are imposed to achieve high density in urban areas, households will likely search for more preferable surroundings in non-urban areas. And just as ICT gives firms more flexibility, it also gives households more flexibility: for example, telecommuting makes long commutes less costly and computers make possible a growing variety of home-based businesses.

4. Density policies required to achieve reductions in private vehicle use have no political constituency. If most firms and households have preferences against high density development, it follows that there would be little political support for the policies required to achieve such development. In the United States, land use control is vested in local governments, which have historically responded to the preferences of constituents. Those preferences have resulted in extensive application of policies that exclude various activities or social classes, limit development density, etc., but very few applications of inclusive policies. Efforts to control land use at the regional level are rare (Oregon, Florida and New Jersey have regional land use policies), and their success in achieving regional or statewide objectives has yet to be determined.

Perhaps more significant for this discussion is the very rapid proliferation of self-governed communities. The local homeowners' association (HOA) is one of the fastest growing types of non-governmental associations (NGOs) in the United States. There were an estimated 150,000 HOAs in 1992 (Kennedy, 1995). These associations typically operate and maintain common facilities, as well as enforce association rules and restrictions, including land use codes. Their authority may cover local (private) streets and other infrastructure, parks and recreational facilities, and policing. In effect, HOAs are taking on and privatizing many traditional functions of the public sector. They make it possible for homeowners to not only purchase their preferred package of housing and associated services (and thereby also restrict their tax contributions), but also assure its maintenance. I view the homeowner association as a means for individual households to exert more control over their local environment. Although governments still have all the traditional powers, including land use control, it is becoming more difficult to enforce policies for which there is little consensus.

The situation is different outside the United States. In Europe, land use control generally resides at the state or national level, and some countries (for example, The Netherlands) have very strict policies to direct and concentrate development. In the United Kingdom, a number of planning policies have been established in recent years to foster location of major traffic generators in existing activity centers, to balance housing and jobs, and to limit the extent of new development. A study of the Oxford region concludes that these policies do affect travel patterns, but their effect is limited (Curtis, 1996). In light of the population and employment trends described earlier in this paper, this conclusion seems reasonable. Land use policy has possibly slowed down the decentralization process.

It could well be argued that this evidence clearly supports land use policy strategies, but we need to get back to the fundamental objective of significantly reducing private vehicle use. Incremental changes in mode shares or distance traveled are not sufficient to measurably reduce vehicle pollution. Even in Europe, there are signs of trouble. For example, the Netherlands' widely acclaimed residential development planning program has encountered difficulties in producing residential communities with high enough densities to support transit because of the lack of demand for high density housing (Maat, 1999). It also bears noting that despite these policy efforts, private vehicle use continues to increase.

5. **Density policies that could be implemented will be swamped by larger trends.** The trends I have described — decentralization of population and employment, rising income, and

The growing impact of ICT — overwhelm just about any land use policy option that could be considered reasonably politically feasible. In Schimek's (1996) study, a 10 percent increase in household income is associated with a 3 percent increase in VMT, all else equal, an effect more than four times as great as that estimated for density. What is more likely to happen within the next 20 or 30 years, a 10 percent increase in household income, or a 10 percent increase in metropolitan density?

A Digression

Another way of putting the issue of land use policy efficacy in perspective is to consider pricing policy. The standard economic response to questions of environmental externalities is efficient pricing, or pricing that reflects the full costs of consumption. What kind of pricing policies are required to substantially reduce private vehicle use? The best example we have is Singapore, where, in addition to congestion pricing, permits to own private vehicles must be purchased at auction (the Vehicle Quota Scheme, or VQS), and a variety of taxes and fees are added to the retail price of a new car. Based on 1997 fees, for example, a private car with a retail price of $10,000 would cost a total of about $49,000, of which about $19,000 is the VQS average bid price. The VQS was introduced in 1990, in response to rapid increases in car ownership despite the already existing taxes and fees (the 1980s were a period of rapid economic growth and rising household incomes). A recent study has estimated that the VQS has reduced car ownership by 7 to 11 percent, compared to what would have occurred without the VQS (Chin and Smith, 1997). Note that in this example, the VQS increases the purchase price by 63 percent. Demand is highly inelastic in Singapore due to the very high price of car ownership. In view of the very low price of car ownership in the US, the Singapore example is not directly transferable. Nevertheless, if it takes price increases of this magnitude to further restrict car ownership by a few percentage points *in a very densely developed country with excellent mass transit*, it is difficult to imagine what would be required to do the same in the US or in Europe.

CONCLUSIONS

There are many problems associated with continued decentralization and low-density development. There are also many problems associated with growing private vehicle use. Although recognition of these problems is increasing, policy-makers have enjoyed few successes in reversing either trend. The greatest success in addressing automobile externalities has been realized by regulating the car, rather than regulating the driver. I have shown in this paper that the trends of car use and decentralization are powerful. They are supported by changing economic structure and rising affluence, and there is no reason to believe that fundamental shifts away from these trends will occur in the future. Land use policies that attempt to reverse these trends will be difficult to implement, and will have little effect on overall travel patterns. There are many good reasons to advocate changes in land use policy. In the United States, certain population segments (poor and minority households) are systematically excluded from many suburban communities; this spatial segregation is associated with a host of social and economic problems. Zoning and other restrictions increase prices, making housing less affordable particularly for lower income households. More specifically, there are good reasons to encourage higher development density and better urban design. With higher densities, a greater mix of housing choices can be offered. Mixed use development provides more opportunities for social and other activities. Pedestrian friendly design may encourage more recreational walking and biking, and perhaps even a few walk trips to the local store. These policies, however, will not help much in solving the environmental externalities of the private vehicle.

ACKNOWLEDGEMENT

Comments by participants at a Portland State University seminar and by Peter Gordon are appreciated. All errors and omissions are the responsibility of the author.

REFERENCES

Baccaini, B. 1997. Commuting and Residential Strategies in the Ile-de-France: Individual Behavior and Spatial Constraints. *Environment and Planning A*, Vol. 29, pp. 1801-1829.

Brotchie, J., M. Batty, P. Hall, and P. Newton, eds. 1993. *Cities in Competition: The Emergence of Productive and Sustainable Cities for the 21st Century.* Longman Cheshire, Melbourne, Australia.

Brog, W., and E. Erl. 1996. Germany. In *Changing Daily Urban Mobility: Less or Differently?* Report of the Hundred and Second Round Table on Transport Economics, European Conference of Ministers of Transport, Paris, France.

Burnley, I., P. Murphy, and A. Jenner. 1997. Selecting Suburbia: Residential Relocation to Outer Sidney. *Urban Studies*, Vol. 34, No. 7, pp.1109-1127.

Calthorpe, P. 1993. *The Next American Metropolis: Ecology, Community, and the American Dream.* Princeton Architectural Press, New York.

Cervero, R., and R. Gorham. 1995. Commuting in Transit Versus Automobile Neighborhoods. *Journal of the American Planning Association.* Vol. 61, pp. 210-225.

Chin, A., and P. Smith. 1997. Automobile Ownership and Government Policy: The Economics of Singapore's Vehicle Quota Scheme. *Transportation Research A.* Vol. 31, No. 2, pp. 129-140.

Crane, R. 1996. Cars and Drivers in the New Suburbs: Linking Access to Travel in Neotraditional Planning. *Journal of the American Planning Association.* Vol. 62(1), pp. 51-65.

Cullinane, S. 1992. Attitudes Towards the Car in the U.K.: Some Implications for Policies on Congestion and the Environment. *Transportation Research A*, Vol. 26A, No. 4, pp. 291-301.

Curtis, C. 1996. Can Strategic Planning Contribute to a Reduction in Car-Based Travel? *Transport Policy.* Vol. 3, No. 2, pp. 55-65.

Downs, A. 1992. *Stuck in Traffic: Coping with Peak Hour Congestion.* Brookings Institution, Washington, D.C.

Duany, A., and E. Plater-Zyberk. 1994. The Neighborhood, the District, and the Corridor. In P. Katz, ed. *The New Urbanism: Toward an Architecture of Community.* McGraw-Hill, New York, pp.xvii-xx.

Dunn, J. 1981. *Miles To Go: European and American Transportation Policies.* MIT Press, Cambridge, Mass.

European Conference of Ministers of Transport. 1995. *Urban Travel and Sustainable Development.* Organisation for Economic Co-operation and Development, Paris, France.

Ewing, R., P. Haliyur, and G.W. Page. 1994. Getting Around a Traditional City, a Suburban Planned Unit Development, and Everything in Between. In *Transportation Research Record 1466*, Transportation Research Board, National Research Council, Washington, D.C., pp. 53-62.

Giuliano, G. 1998. Urban Travel Patterns. In B. Hoyle and R. Knowles, eds. *Modern Transport Geography*, 2nd edition.

Gordon, P., and H. Richardson. 1996. Employment Decentralization in U.S. Metropolitan Areas: Is Los Angeles an Outlier or the Norm? *Environment and Planning A*, Vol. 28, pp. 1727-1743.

Gordon, P., H. Richardson, and G. Yu. 1996. Settlement Patterns in the U.S.: Recent Evidence and Implications. Presented at the TRED Conference, Cambridge, Mass.

Gyourko, J. 1991. Impact Fees, Exclusionary Zoning and the Density of New Development. *Journal of Urban Economics*, Vol. 30, pp. 242-256.

Handy, S. 1992. Regional Versus Local Accessibility: Neo-Traditional Development and its Implications for Non-Work Travel, *Built Environment*, Vol. 18, pp. 253-267.

Handy, S. 1996. Urban Form and Pedestrian Choices: A Study of Austin Neighborhoods. Presented at the Meeting of the Transportation Research Board, Washington, D.C., January 1996.

Hanson, S., and M. Schwab. 1987. Accessibility and Intra-Urban Travel. *Environment and Planning A*, Vol. 19, pp. 735-748.

Hervick, A., T. Tretvik, and L. Ovstedal. 1993. Norway: Crossing Fjords and Mountains. In I. Salomon, P. Bovy, and J-P Orfeuil, eds. *A Billion Trips a Day: Tradition and Transition in European Travel Patterns.* Kluwer Academic Publishers, Dordrecht, The Netherlands.

Hu, P., and J. Young. 1993. *1990 NPTS Databook, Nationwide Personal Transportation Survey.* Report FHWA-PL-94-010A. Federal Highway Administration, Washington, D.C.

Jackson, K.T. 1985. *Crabgrass Frontier: The Suburbanization of the United States.* Oxford University Press, Oxford, United Kingdom.

Jansen, G. 1993. Commuting: Home Sprawl, Job Sprawl and Traffic Jams. In I. Salomon, P. Bovy, and J-P Orfeuil, eds. *A Billion Trips a Day: Tradition and Transition in European Travel Patterns.* Kluwer Academic Publishers, Dordrecht, The Netherlands.

Kennedy, D. 1995. Residential Associations as State Actors. *Yale Law Journal*, Dec. pp. 761-793.

Kitamura, R., P. Mokhatarian, and L. Laidet. 1997. A Micro-Analysis of Land Use and Travel in Five Neighborhoods in the San Francisco Bay Area. *Transportation,* Vol. 24, pp. 125-158.

Knaap, G. 1985. The Price Effects of Urban Growth Boundaries in Metropolitan Portland, Oregon. *Land Economics*, Vol. 61, pp. 28-35.

Maat, K. 1999. The Compact City: Conflicts of Interest Between Housing and Mobility Objectives, Presented at the ESF/NSF Conference on Social Change and Sustainable Transport, Berkeley, Calif.

Masser, I., O. Sviden, and M. Wegener. 1992. *The Geography of Europe's Futures*. Belhaven Press, London, United Kingdom.

Muller, P. 1995. Transportation and Urban Form: Stages in the Spatial Evolution of the American Metropolis. In S. Hanson, ed. *The Geography of Urban Transportation*, 2nd ed. The Guilford Press, New York.

Newman, P. and J. R. Kenworthy. 1989a. Gasoline Consumption and Cities: A Comparison of U.S. Cities with a Global Survey. *Journal of the American Planning Association*, Vol. 55, pp. 24-37.

Newman, P. and J. R. Kenworthy. 1989b. *Cities and Automobile Dependence: A Sourcebook.* Gower Technical, Brookfield, Vt.

Phang, S-Y, and N. Asher. 1997. Recent Developments in Singapore's Motor Vehicle Policies. *Journal of Transport Economics and Policy*, pp.211-220, May.

Pisarski, A. 1992. *Travel Behavior Issues in the '90s*. Federal Highway Administration, Washington, D.C.

Pisarski, A. 1996. *Commuting in America II.* Eno Foundation, Landsdowne, Va.

Pucher, J. 1988. Urban Travel Behavior as the Outcome of Public Policy. *Journal of the American Planning Association*. Vol. 54, No. 3, pp. 509-519.

Pucher, J., and C. Lefevre. 1996. *The Urban Transport Crisis in Europe and North America*. MacMillan Press Ltd., Houndmills, Basingstoke, Hampshire, United Kingdom.

Rosen, K., and L. Katz. 1981. Growth Management and Land Use Controls: The San Francisco Bay Area Experience. *AREUA Journal*, Vol. 9, pp. 321-344.

Rosenbloom, S. 1995. Travel by Women. In *NPTS Demographic Special Reports*. Office of Highway Information Management, Federal Highway Administration, Washington, D.C.

Rosetti, M. and B. Eversole. 1993. *Journey to Work Trends in the United States and its Major Metropolitan Areas, 1960–1990.* Federal Highway Administration, Washington, D.C.

Royal Commission on Environmental Pollution. 1995. *Transport and the Environment: 18th Report*. Oxford University Press, Oxford, United Kingdom.

Schimek, P. 1996. Household Motor Vehicle Ownership and Use: How Much Does Residential Density Matter?

Teaford, J. 1990. *The Rough Road to Renaissance*. Johns Hopkins University Press, Baltimore.

U.S. Department of Transportation, Bureau of Transportation Statistics. 1996. *Transportation Statistics Annual Report 1996*. Bureau of Transportation Statistics, Washington, D.C.

[24]

Telecommunications and Travel

The Case for Complementarity

Patricia L. Mokhtarian

Keywords
complementarity
information and communication technologies (ICT)
passenger travel
substitutability
telecommunication
transportation

Summary

This article examines the conceptual, theoretical, and empirical evidence with respect to the impact of telecommunications on travel. The primary focus is on passenger travel, but goods movement is addressed briefly. I argue that although direct, short-term studies focusing on a single application (such as telecommuting) have often found substitution effects, such studies are likely to miss the more subtle, indirect, and longer-term complementarity effects that are typically observed in more comprehensive analyses. Overall, substitution, complementarity, modification, and neutrality within and across communication modes are all happening simultaneously. The net outcome of these partially counteracting effects, if current trends continue, is likely to be faster growth in telecommunications than in travel, resulting in an increasing share of interactions falling to telecommunications, but with continued growth in travel in absolute terms. The empirical evidence to date is quite limited in its ability to assess the extent of true causality between telecommunications and travel, and more research is needed in that area. At this point, what we can say with confidence is that the empirical evidence for net complementarity is substantial, although not definitive, and the empirical evidence for net substitution appears to be virtually nonexistent.

Address correspondence to:
Patricia L. Mokhtarian
Department of Civil and Environmental Engineering and Institute of Transportation Studies
University of California
Davis, CA 95616, USA
plmokhtarian@ucdavis.edu
www.its.ucdavis.edu/telecom/

FORUM

Introduction

The potential of telecommunications to substitute for travel has long been appreciated. Indeed, such potential has often been not just a later realization, but an integral impetus behind the development of the technology. Early communication devices such as jungle tom-tom drums, trumpet alarms, smoke signals, and flashing lanterns were surely conceived precisely to replace the need for a physical messenger. The same cannot necessarily be said of the more recent (1876) invention of the telephone. Alexander Graham Bell's own initial vision of its uses seemed to be more along the lines of broadcast radio than personal communication, the president of Western Union dismissed it as an "electrical toy," and the chief engineer of the British Post Office in 1879 sniffed that the "superabundance of messengers, errand boys, and things of that kind" in Great Britain obviated the need for the telephone there (Dilts 1941, cited in de Sola Pool et al. 1977). It did not take long, however, for speculation to begin about the uses of the new technology to eliminate travel. Albertson (1980) referred to a letter to the editor of the *Times* published May 10, 1879, suggesting that the telephone could provide relief from travel for harried businessmen. The utopian science fiction of H. G. Wells ([1899, 1954] 1968) and E. M. Forster (1909) portrayed society taking part in teleconferencing on a large scale, in lieu of physical travel.

That was the speculation; what is the reality? Contributing to a retrospective on the 100 year anniversary of the invention of the telephone, Pierce (1977, 164–165) anticipated some of the arguments raised in this article:

> We have seen that telephony has grown steadily since its inception. What has this done to other modes of communication? Is telephony replacing travel? No. Very roughly, in recent years the number of telephone calls and the number of air miles flown have increased at about the same rate, and the number of car miles traveled [has increased] about half as fast. Undoubtedly, a telephone call sometimes substitutes for a trip, but more and faster communication tends to engender widespread associations and activities that result in trips.

In the additional quarter century since the telephone's centennial, new communication technologies and services have been introduced and adopted at an ever-accelerating pace: facsimile machines, teleconferencing, e-mail and the Internet, and mobile telephony, to name just a few major ones. With the increasing power, realism, flexibility, user friendliness, and ubiquity of these devices and services, together with their decreasing cost, one might expect that their collective ability to replace travel should by now be considerable and that measurable decreases in travel should have resulted.

Instead, we still see nothing of the sort. Aggregate measures of travel demand continue to demonstrate basically increasing trends worldwide (Giuliano and Small 1995; Salomon et al. 1993; Schafer and Victor 2000). In the United States, the repeated cross-sectional Nationwide Personal Transportation Survey is perhaps the best source of data on local passenger travel. Changes in data collection methodology make comparisons across all time periods problematic, but just comparing recent data shows an 11% increase in per capita person-miles traveled between 1990 and 1995 (Hu and Young 1999, calculated from table 1), a period of considerable development and adoption of new communication technologies.

The purpose of this article is to explore the reasons behind this observation. In the sections to follow, I respectively advance conceptual, theoretical, and empirical arguments in support of the claim that the *net* impact of telecommunications on travel will be to increase it rather than reduce it. Most of these arguments have appeared elsewhere, scattered throughout the writings of myself and other authors, but they are marshaled and amplified here specifically to make the case for complementarity in a more cogent and directed manner than before. The focus of the article is primarily on passenger travel, but goods movement is addressed briefly in the penultimate section.

Conceptual Considerations

To understand the impacts of telecommunications on travel (and vice versa), it is important to begin with understanding the conceptual relationships between the two. In Mokhtarian (1990), I observed that communication can be partitioned into three major modes, each requiring some form of travel to occur: face-to-face communication, involving passenger transportation; the transfer of an object containing information (a book, magazine, letter, or even diskette), involving freight transportation or goods movement (in the broadest sense); and telecommunications, now involving the transportation of electrons over cables or radio waves through the air.

As alternate modes of communication, then, a number of relationships are possible between physical travel and telecommunications. The literature (Claisse 1983; Mokhtarian and Salomon 2002; Niles 1994; Salomon 1985, 1986) identifies the following types of cross-mode relationships:

- *Substitution (also referred to as replacement or elimination)*. As indicated in the Introduction, the potential for alternative modes of communication to substitute for one another is clear to most people. Specific telecommunications applications that have been hypothesized to replace their travel-based counterparts include telecommuting, teleconferencing, tele-education or distance learning, telebanking, teleshopping, telemedicine, telejustice, and televoting and other government applications (see, e.g., U.S. DOT 1993, in which these applications are specifically referred to as "telesubstitutions").
- *Complementarity (also referred to as stimulation or generation)*. Use of one mode of communication can also increase the use of other modes, however. Salomon (1986) explicitly and Claisse (1983) implicitly subdivided this effect into two categories.

  1. *Enhancement* occurs when the use of one mode directly causes or facilitates the use of another mode. For example, the first words Alexander Graham Bell spoke over the telephone were "Mr. Watson, come here; I want you" (de Sola Pool 1977), generating a trip (in this case, only down the hallway) for his assistant. Numerous other examples of telecommunications stimulating travel can be produced, in which a phone call, letter, e-mail message, or fax prompts a trip, or in which individuals meet first over the Internet and then, finding common interests, arrange to meet in person. In general, the increased ease of communication expands the size of our contact sets (Niles 1994; Gaspar and Glaeser 1998; Couclelis 1999) and therefore increases the number of opportunities for face-to-face interaction. The increased availability of information about activities and locations of interest is also likely to whet the appetite to engage in such activities and visit such locations (Gottmann 1983). As an example in the reverse direction (of transportation enhancing the use of telecommunications), consider the impact of travel on mobile phone usage: The more one travels, the more useful, and used, a mobile phone becomes.
  2. *Efficiency* occurs when the use of one mode is a necessary accompaniment to or side effect of the use of another mode, thereby increasing the efficiency of the latter mode. The impact of mobile phones on travel is one example: at a person-to-person level, one of the main uses of mobile phones, according to some studies, is to schedule or modify face-to-face meetings (requiring travel) on the fly (Yim 2000). And in fact, taxi drivers are now using mobile phones as an adjunct to the traditional dispatching service, to more efficiently obtain real-time information on available fares from friends, relatives, and other drivers (Townsend 2000). A number of system-oriented telecommunications services also fall into this category; for example, the telegraph was used extensively to support train operations (e.g., Harlow

1936), and telegraph wires were often strung along railroad rights-of-way. Telecommunications are still indispensable to air traffic control, vehicle dispatching and tracking, and other logistics operations. Many intelligent transportation system (ITS) applications currently under development and in early stages of use provide further examples of efficiency, such as providing real-time traffic information and route guidance to a driver, enabling her to bypass congestion, or providing real-time arrival and travel time information to transit users, reducing the uncertainty associated with taking transit and thereby increasing its attractiveness.

- *Modification*. Sometimes, one communication mode modifies something about the use of another mode. For example, a phone call may alter the departure time or destination of a trip, or a real-time in-vehicle navigation device may alter the route of a trip. The trip still occurs, so it is not substituted, and it would have occurred anyway, so it is not generated by the communication; but it is changed. As Mokhtarian and Meenakshisundaram (1999) pointed out, depending on what the measure of interest is, modification effects may be reclassifiable as one of the other three effects. For example, if the measure is vehicle-kilometers traveled (VKT), then a change in departure time would be VKT neutral, and a change in route could constitute either generation or substitution, depending on whether the new route were longer or shorter, respectively, than the old one.
- *Neutrality*. In some circumstances, use of one mode may leave the use of other modes unaffected. A broadcast e-mail message may have no impact on travel; a routine trip to the grocery store may not create any phone calls. Although this seems self-evident, it is actually easy to overlook. In hoping for substitution effects, for example, it has sometimes been implicitly assumed that every telecommunication activity of a certain kind is replacing a travel-based version of the same kind of activity. In fact, quite often the alternative to the telecommunication activity is not the travel-based activity, but rather not conducting the activity at all. Thus, for example, not every distance-learning student would otherwise be enrolled on a conventional college campus, not every home-based worker would otherwise be commuting to a conventional job, not every participant in a teleconference would otherwise have traveled to a face-to-face meeting, and not every impulse purchase made over the Web (or from a home-shopping television channel or mail-order catalog, for that matter) would have otherwise involved a trip to a store to purchase the same (or similar) item.

Figure 1 offers a schematic portrayal of relationships among the three primary modes of communication over time, with the black lines depicting the situation at one point in time and the gray lines depicting a later point in time. The amount of communication occurring via a particular mode, represented by the size of the wedge for that mode, can increase over time as a result of three possible effects: (1) own-mode generation (increases in the given mode that occur independently of the other modes), (2) cross-mode substitution (in which the given mode replaces another mode, and hence the given mode increases while the other mode decreases), and (3) cross-mode complementarity (in which both the given mode and another mode increase). Of course, all three of these effects can occur to varying degrees for all three modes of communication, so the net outcome is a complex composite of multiple directions of causality and multiple directions of effect (positive or negative). The figure illustrates my view of reality, in which telecommunications is increasing in share with respect to the other modes, but use of all modes is increasing in absolute terms. This picture is consistent with simultaneous substitution and generation effects (Claisse 1983; Gautschi and Sabavala 1995; Couclelis 1999) and may help explain why both those who "believe" in substitution and those who "believe" in complementarity are right, up to a point. The ultimate question, however, is not the existence of, say,

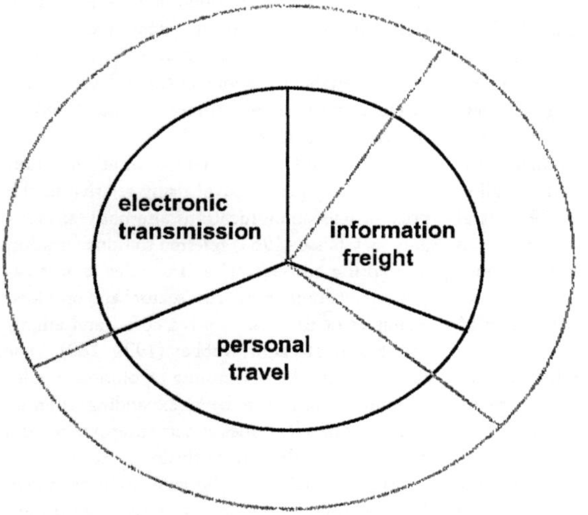

Figure 1 The black lines schematically represent total communications at one point in time, split among the three modes shown. The gray lines represent communications at a later point in time. The situation depicted is one in which the personal travel and information object transmission modes lose share to electronic communication over time, but in which the continued expansion of communication by all modes results in the absolute amounts of communication being greater for all modes at the later point in time than at the earlier point. *Source:* Mokhtarian (1990).[1]

substitution effects (which certainly do exist), but the net outcome of all the effects we have identified here (i.e., the *overall* change in demand for each mode).

As a side point, it can be commented that the measurement issues in actually developing an empirically realizable model that would account for all these effects are formidable (Salomon 1985). Mokhtarian and Meenakshisundaram (1999) discussed some of the challenges associated with finding and operationalizing a common metric for all modes of communication and implemented a first attempt at such a model, using the very crude metric of number of instances of each mode of communication. Their results are presented in Empirical Considerations below.

Theoretical Considerations

Price and Income Effects

What does economic theory suggest about the relationship between telecommunications and travel? At an elementary level, two effects come into play (this discussion is based on, but not identical to, a much briefer argument in Helling and Mokhtarian 2001). The *price effect* says that as the price of a good falls, the quantity demanded rises (and conversely for rising prices). The *income effect* says that as consumers' incomes increase, the quantity demanded rises. How do these two effects combine?

For telecommunications, both effects operate in the same direction: Rising consumer incomes and falling real prices both act to increase demand. Thus, it seems fairly safe to predict that the demand for telecommunications will continue to increase.

For transportation, the outcome is not as clear-cut. Consumers' incomes are increasing, but if the price of travel is also rising, as might be initially assumed, then the two effects will counteract each other, with an unknown net result. But is the price of travel in fact rising? The question immediately arises, Compared to what? At least two comparisons seem reasonable.

First, is the price of travel, as a proportion of real income, rising compared to its own historical price? Not really. Schafer and Victor (2000) presented data illustrating that the average proportion of per capita gross domestic product spent on travel is relatively stable across time within a variety of countries, albeit with predictable variations between countries based on transport policies and other factors. In the United States, they said, consumers compensated for relative rises in gasoline prices by purchasing more fuel-efficient cars, with the result that the travel money budget has remained at 8% to 9% of gross domestic product per capita for the 20-year period 1970–1990. Even those price increases were only relative, and temporary. According to the U.S. Department of Energy *Annual Energy Review* (www.eia.doe.gov/emeu/aer/contents.html), the average cost of gasoline (in 1996 dollars) was $1.17/gallon in 1999, compared to $1.31/gallon in 1966 and a high of $2.17/gallon in 1981. And according to the U.S. Bureau of Labor Statistics (www.bls.gov/cpihome.htm), the Consumer Price Index (CPI) for transportation was lower than the overall CPI in every year between 1966 and 1999, except for 1980–1982.

But what about congestion? Surely if we take into account the time costs of travel, those are increasing? Not necessarily: Although congestion appears to be increasing at an aggregate level (Lindley 1987; Hanks and Lomax 1991), it is the disaggregate impacts that govern an individual's choices. On a per capita basis in the United States, commute times appear to be remaining relatively stable even while commute lengths are increasing. This suggests that people are increasing their travel speeds by decentralizing to more peripheral areas, changing to faster modes and commuting more in the off peak; and, in some cases, reducing their commute frequencies through compressed work weeks and telecommuting (Pisarski 1992; Hu and Young 1999; Levinson and Kumar 1994; Gordon et al. 1990, 1991; Kumar 1990).

If we broaden the idea of "price" still further, beyond monetary and time costs to disutility in general, the argument of rising prices is even harder to maintain. It appears that individuals are quite adept at adopting "travel-maintaining" strategies that reduce the disutility of travel (Salomon and Mokhtarian 1997; Mokhtarian et al. 1997), or even at deriving positive utility from the travel itself, as well as from activities that can be conducted while traveling (Mokhtarian and Salomon 2001; Redmond and Mokhtarian 2001). A number of products and services, such as drive-up windows at fast food restaurants and increasingly multimedia in-vehicle entertainment systems, are oriented toward allowing travelers to engage in multitasking. Particularly ironic is that, as mentioned in the previous section, the proliferation of information technologies (such as mobile phones, laptops, in-vehicle special-purpose navigation systems and general-purpose computers, and, increasingly, wireless-Internet-enabled versions of these devices) acts to make travel time considerably more productive than ever before.

The second "compared to what" question asks, Is the price of travel rising relative to the price of telecommunications alternatives? Probably so. Claisse (1983) referred to transportation as a "rising cost sector" and to telecommunications as a "diminishing cost sector" and provided examples of the comparative costs (and energy utilization) of each. Webber (1991, 280) commented that "the continuing revolution in telecommunications has been expanding channel capacities of telephones at persisting exponential rates—and with constantly lowering costs. In contrast, capacities of the auto-highway system have been falling while costs have been rising." Comparison of the CPIs for transportation and for communication over time supports these assertions. Economic theory predicts that as the price of one good (travel) rises relative to the price of its substitute, the demand for the substitute good (telecommunications) rises. Thus, to the extent that telecommunications is a substitute for travel, the price comparison between the two will favor telecommunications. Two points should be considered, however.

First, how close a substitute is telecommunications for the travel alternative? Among others, I argue that it is often not a favorable substitute at all. For one thing, it is so far still the case that a telecommunications alternative is frequently simply not available or is prohibitively expensive. For another thing, the quality of communication that occurs by telecommunications cannot match the quality of face-to-face interaction. Although technological improvements are closing the gaps of ubiquity, cost effectiveness, and quality, complete parity to the "high touch" of face-to-face interaction can never be achieved (and as argued by Naisbitt and colleagues [1999], the greater the role of high tech, the greater the human need for high touch). Just as importantly, though, the direct communication that is the ostensible purpose of the interaction is only one reason, and sometimes not the primary reason, for making the trip (Day 1973). In the case of a business meeting or conference, "metamotivations" for traveling could include visiting an interesting location, visiting friends or relatives on the same trip, or escaping from the office or home (Button and Maggi 1994; Mokhtarian 1988). Shopping has been observed to serve entertainment, recreational, and social needs, as well as its direct utilitarian purpose (Gould and Golob 1997; Salomon and Koppelman 1988; Tauber 1972). Being at work has similarly been noted to fulfill metaneeds for social and professional interaction, and commuting to work offers a desired transition period between roles, as well as time for oneself (Albertson 1977; Edmonson 1998; Mokhtarian and Salomon 1997; Redmond and Mokhtarian 2001; Richter 1990; Salomon and Salomon 1984; Shamir 1991). In all these and other cases, the tele-alternative to the location-based activity may not provide an equally satisfactory, let alone superior, experience.

Second, viewing telecommunications as a substitute for travel is only one side of the coin. I have argued in the previous section that telecommunications and travel can also be complements, and economic theory suggests that a decrease in the price of one good (telecommunications) increases the demand for the complementary good (travel). Incidentally, the complementarity between telecommunications and travel seems asymmetric (although this does not affect the basic argument). It seems likely that a great deal of telecommunications takes place without generating travel. That is, much telecommunications (whether routine or nonroutine) is self-generated, with a neutral cross-mode impact; the alternative would have been not to communicate rather than to have sent a letter or gone in person. On the other hand, in today's society, it seems unlikely that much nonroutine travel takes place without telecommunications being an inevitable adjunct. It could be in facilitation of a vacation or business trip (prior phone calls, faxes, or e-mail messages coordinating time, place, and agenda; posterior communications relating to, for example, expense reimbursement, thank yous, and so on) or generated by the content of the face-to-face interaction (follow-up exchanges with people met there, new documents produced as a consequence of the trip and circulated to others)—the first case being an example of efficiency, and the second an example of enhancement.

The implications for the current discussion are (1) if travel continues to increase as a net outcome of the income and price effects, the telecommunications activities concomitant to that travel will increase accordingly. Therefore, again, (2) telecommunications are likely to increase both because of being substituted for travel in some cases and because of being complementary to travel in other cases. But (3) to the extent that telecommunications *are* a complement to travel, the relatively favorable price of telecommunications can still act to increase travel, even if indirectly (by making it easier to conduct activities, some activities that otherwise would have been too costly can now take place).

To summarize the evidence offered by microeconomic theory, then, it appears, first, that increases in telecommunications are virtually certain, because both income and price effects favor them, whether telecommunications is a substitute or complement for travel. Second, in my opinion, a continued increase in travel is highly likely: The income effect certainly favors it, the price effect favors it in many ways, and these combined effects probably outweigh the respects in which price effects are unfavorable. The outcome could be different, however, if the real price of travel were to increase considerably, for example through gasoline shortages or substantial congestion pricing.

Explanations of Observed Relationships

When measures of transportation and telecommunications are observed to increase together, three fundamental explanations are at work, to varying and typically unknown degrees. First, the observed relationship may be purely spurious: For example, any two series that increase over time appear to be correlated even if they are not related in any structural way whatsoever (e.g., Utts 1999). Second, each measure may be correlated with one or more other variables that are causing both transportation and telecommunications to increase separately. Third, there may be a genuine causal relationship in either or both directions, for the kinds of conceptual reasons discussed in Conceptual Considerations above.

Both the price and the income effects discussed in the previous subsection can contribute to observed relationships between telecommunications and travel that represent genuine causality, correlations with third-party variables that are structural but not directly causal, or entirely independent trends. For example, changes in the consumption of travel due to price changes will have one component based on the demand for travel *independently* of telecommunications activity (depending only on changes in the price of travel itself) and another component based on how travel and telecommunications relate as substitutes and complements (and depending on relative changes in the price of both goods), which are causal relationships. Similarly, increases in the consumption of travel due to rising income can have a component that is independent of increases in telecommunications due to the same rising income, and another component that is due to a causal relationship between the two for which income is serving as a marker or proxy. For example, higher-income occupations provide greater scope for one's expanding contact set (both business and personal) to directly generate new travel.

In interpreting the empirical evidence presented below, then, it should be kept in mind that (1) observing a strong empirical relationship between telecommunications and transportation *without* controlling for economic indicators and other important potential confounding factors tells us virtually nothing about causality between the two forces, but (2) observing a weak empirical relationship *after* controlling for such factors also tells us little about causality! Ultimately, here as in many other instances, causality must typically be inferred on the basis of external conceptual considerations together with statistical ones. On the other hand, more rigorous empirical (and potentially theoretical) studies can begin to assemble stronger statistical evidence for causality as well.

Empirical Considerations

What *does* the empirical evidence say? Mokhtarian and Salomon (2002) reviewed the empirical literature speaking to the impacts of telecommunications on travel and classified studies based on (1) whether they were focused on a single application (such as telecommuting or the telephone) or took a comprehensive approach to telecommunications and (2) whether they took an aggregate or disaggregate perspective. They commented that although the single-application studies often found substitution between telecommunications and travel, the comprehensive studies had a much stronger tendency to find complementarity. This is not surprising in view of figure 1, which illustrates that substitution can be happening at the margin, simultaneous with generation and complementarity happening overall. Focusing purely on a single application and looking for direct, short-term effects (in only one direction of causality), then, is likely to give an inaccurate picture of the probably complex, diffuse, multidirectional, and long-term relationships involved. Thus, in this article I have chosen to review only the studies attempting a more comprehensive measurement of telecommunications as well as travel. These studies can be classified as either aggregate or disaggregate.

The Aggregate Level

Perhaps the most elementary yet most dramatic empirical studies on this subject are those that simply plot measures of communication and travel in the same geographic area over the same time period. Such graphs are found in Grubler (1989) for France between 1800 and 1985, in Batten (1989) for Sweden between 1950 and 1985, and in Niles (1994) for the United States between 1980 and 1994; they inevitably show both telecommunications and travel generally rising together over time (also see Day 1973). The Grubler picture in particular seems to have struck a chord, having been reproduced in Batten (1989), Marchetti (1994), and Graham and Marvin (1996). Hojer and Mattson (2000) appropriately pointed out that the visual impact of the Grubler figure can be greatly altered by different choices of base point and scale. But the general trend of simultaneous increases in telecom and travel would remain, and it is likely that correlations for any of the referenced pairs of time series would be large and statistically significant.

As discussed in Explanations of Observed Relationships above, however, it is unknown to what extent simultaneous increases in telecommunications and travel are due to true causality between the two types of activities (perhaps partly mediated by third-party correlation of each time series with income and other indicators of economic activity), as opposed to independent correlations with third-party variables, or even just entirely independent increases over time.

A more sophisticated analysis of aggregate (nation-level) data was undertaken by Plaut (1997). She performed an input-output analysis of industrial consumption of transportation and communication services by nine countries of the European Community in 1980. She found strong evidence of complementarity, in the sense that use of transportation was strongly correlated with use of communications. Here too, however, the results do not speak to the degree of direct causality between the two sectors: The observed correlations may be due in some part to independent mechanisms that separately generate congruent transportation and communication demands.

Another aggregate study focused on per capita consumption expenditures on private transportation, public transportation, and communications. Using 1960–1986 time-series data from Australia and the United Kingdom, Selvanathan and Selvanathan (1994) estimated a simultaneous equation system of the consumer demand for these three kinds of goods separately, plus all others combined. Interestingly, this study found a pairwise substitution relationship among all three sectors. In reconciling these two studies, Mokhtarian and Salomon (2002) suggested that the enhancement and efficiency effects of complementarity may apply more cogently at this point to industry than to consumers, but that this may be changing (and may have already changed considerably from the 1986 endpoint of the data analyzed in this study). The difference in methodologies used in the two studies is also a confounding factor. In any case, Plaut (1997) pointed out that industrial expenditures on transportation and communications account for half to two-thirds of the total in Western countries, and hence the findings for industry are likely to dominate the overall relationships among these sectors of the economy.

The Disaggregate Level

A number of recent studies have examined relationships between information and communication technologies (ICTs) and travel at the individual level. Measures of travel are typically obtained through a travel or activity diary of some kind, and measures of ICTs are either somewhat general, or in the best cases also obtained through a diary or log (the remainder of this section draws heavily from Mokhtarian and Salomon 2002).

Zumkeller (1996) described a study in which 166 employees of the University of Karlsruhe, Germany, completed diaries recording information on all trips and contacts (communication activities) they made for 1 day in 1994. He concluded (p. 79) that "the complementary factor of the interrelationship between travel and communication is much stronger than the substitutional one" because high levels of trip making were found to be associated with high levels of communication activity. A similar observation was made more than a quarter century ago by Day (1973), citing an unpublished research proposal by James Kollen.

Johansson (1999) reported on the communications and travel behavior of a sample of about 2,000 respondents (ages 15 to 84) to the 1997 Swedish National Communication Survey, using a methodology close to Zumkeller's. She observed that the number of trips is positively correlated with the number of telecommunications contacts, but later commented that both are positively correlated with income. Neither study appears to have investigated the extent to which the observed results are due to the income effect.

Using time-use data, Harvey and Taylor (2000) analyzed the contact and travel behavior of nationally representative samples collected from Canada, Norway, and Sweden in 1990–1992 (total $N = 17{,}496$). They concluded (p. 53) that "there is a tendency for persons with low social interaction [specifically including those who work at home] to travel more. It is argued that individuals need, or want, social contact and if they cannot find it at the workplace they will seek it elsewhere thus generating travel. . . . [This suggests] that working in isolation at home will not necessarily diminish travel but rather may simply change its purpose."

Hjorthol (2002) studied the relationship between travel and home use of ICT for a sample of Norwegians in 1997–1998. The measures of ICT activity used in this study were general rather than based on a comprehensive diary of particular communication episodes. Using regression to model daily distance driven by car ($N = 786$), she found a small but significant positive effect of using a home computer for work, even after controlling for household income, number of cars in the household, gender, and age.

A Dutch study (KMPG 1997) looked at travel behavior by three categories of people: heavy information technology (IT) users (not specifically defined), a reference group of other people with sociodemographic characteristics similar to those of the heavy IT users, and the Dutch population as a whole. The study found that although the heavy IT users traveled more than the population as a whole, most of the difference was explained by sociodemographic distinctions because their overall trip frequency and distance traveled were similar to (although slightly higher than) that of the reference group. The heavy IT users, however, traveled considerably more frequently (47% more trips) and farther (53% more kilometers covered) for business than did the reference group.

I also directed a study of the relationships between telecommunications and travel across time (Mokhtarian and Meenakshisundaram 1999). Ninety-one adult residents of Davis, California, completed a communications/travel log in 1994–1995, in which they recorded instances of communication in each of several categories as well as trips and personal meetings. Logs were kept for four consecutive days at two points in time about six months apart. A system of structural equations was estimated, with the amount of activity in each communications/travel category at time 2 being modeled as a function of the amounts in all categories at time 1 and other explanatory variables (elapsed time between the two measurements, dummy variables representing seasons, age, household size, and occupation).

Results were as follows: (1) The elapsed time variable was positive in all equations and generally significant, meaning that each form of communication is generally increasing over time, all else equal. (2) The amount of communication by each mode in wave 2 was positively and generally significantly related to the amount by the same mode in wave 1. (3) Significant cross-mode relationships were mostly positive (the more communication by one mode in the first wave, the more communication by a different mode in the second wave), indicating the presence of complementary effects across modes. Taken together, these results suggest that self-generation and complementarity are the predominant impacts. Here too, however, a direct causality between the communications occurring in wave 1 and those occurring six months later in wave 2 cannot be claimed. The findings are arguably merely demonstrating associative tendencies, although it is clear that those tendencies are mainly complementary rather than substitutive.

The Impact of Telecommunications on Goods Movement

This article has focused primarily on the impact of telecommunications on the movement of people. Speculating briefly about the impact on the movement of other material goods, especially information objects, is of interest. It seems likely that similar relationships are at work. At the margin, telecommunications can certainly facilitate greater efficiencies in goods movement, through the timely exchange of information permitting more optimal load formation, dispatching, and routing. But the demand for goods movement is likely to increase in absolute terms, in part stimulated by those same telecommunications technologies. Consider the impact of Internet-based shopping, for example. Matthews and colleagues (2001) pointed out that the home delivery of millions of individual copies of books ordered on-line is less efficient than delivering those books to local stores in much larger batches, from the standpoint not only of transportation, but also of packaging (which will have domino effects on the transportation of packaging materials). Mokhtarian (2002) suggested a number of other ways in which on-line shopping could lead to increases in goods movement (such as increases in shopping frequency, obtaining goods from more distant sources, and increases in total per capita consumer spending stimulated by the convenience, variety, customization, information, and price advantages potentially offered by the Internet).

Also consider the demand for paper. Although the alleged "paperless office" may someday become a reality as technology progresses, it shows no signs of doing so yet (Salomon 1996). Huws (1999), for example, cited the statistic that paper consumption in the United Kingdom more than doubled in the period 1984–1995. Similarly, Giddens (1991, 25), citing Strawson (1980), commented, "It has been calculated that, on a global level, the amount of printed materials produced has doubled every fifteen years since the days of Gutenberg." Certainly, there has been substitution at the margin: Memos and other documents that used to be photocopied and circulated physically are now e-mailed electronically and deleted by many recipients without ever being printed. At the same time, however, the ease of access to reports and other documents on-line has resulted in a much greater volume of end-user printing than used to be the case. Production of information goods (including software and music, as well as printed media) is being increasingly decentralized toward the end user, which to some extent simply shifts the distribution of generic media (blank printer paper, electronic storage media) from the producer to the consumer. The elimination of the physical transportation between the producer and the consumer and the reduction of packaging when delivery is electronic will be counteracted (to a currently unknown extent) by increases in demand for information goods, stimulated by ICT.

Thus, the general principle that substitution is easier to detect when the focus is short-term and narrow, whereas complementarity emerges from a more comprehensive, longer-term focus, appears likely to hold for the relationship between telecommunications and the transportation of information objects as well. Bernardini and Galli (1993, 445) made a similar observation with respect to dematerialization in general:

> Even though there is abundant evidence of dematerialization trends in single products and systems, it is not at all clear that this effect is not being counterbalanced by an equivalent increase in the number of products used. Thus microelectronics and miniaturization have contributed to decreasing the minimum size and material content of television sets, but this has been accompanied by a strong decrease in prices which has favored the diffusion of multiple sets in households. Moreover, with declining prices households are expressing greater preference for large and extra large sets. These may be sufficiently strong effects to leave the overall material intensity associated with television receivers roughly unchanged.

In a parallel vein, in the context of analyzing trends in 1988–1998 U.S. energy consumption, Murtishaw and Schipper (2001) noted that improvements in the energy efficiency of a given mode (i.e., reductions in energy consumption per unit of activity conducted by that mode) are often more than counteracted by shifts from more efficient to less efficient modes and by increases in total activity levels. They specifically noted such effects with respect to freight transportation. Although both rail and trucking modes have become more energy efficient over time, trucking remains less efficient than rail, and shifts from rail to trucking, plus increases in total goods movement, have contributed to the increased energy consumption seen in this sector.

Conclusions

This article has examined the conceptual, theoretical, and empirical evidence with respect to the impact of telecommunications on travel. It argues that although direct, short-term studies of that impact focusing on a single application (such as telecommuting) have often found substitution effects, such studies are incomplete and likely to miss the more subtle, indirect, and longer-term complementarity effects that are typically observed in more comprehensive studies. From the comprehensive perspective, substitution, complementarity, modification, and neutrality within and across communication modes are all happening simultaneously. The net outcome of these partially counteracting effects, if current trends continue, is likely to be faster growth in telecommunications than in travel, resulting in an increasing share of interactions falling to telecommunications, but with continued growth in travel in absolute terms.

The caveat, "if current trends continue," is a nontrivial one. My expectations for the future are largely predicated on the assumption that the real price of travel will continue to decline or at least remain relatively stable. Should the price of travel escalate markedly—whether through natural shortages of hitherto "cheap" but nonrenewable energy resources, geopolitical events affecting the supply of petroleum, or domestic policies such as fuel taxes or congestion pricing—the substitutability of telecommunications will obviously become more attractive. Shifts toward telecommunications substitution may also occur for reasons such as an increasing societal commitment to more environmentally benign or sustainable communication modes, but experience suggests that such impacts will be modest at best.

All things considered, then, telecommunications and travel have risen together through many historical technological advancements and political events, and there is no compelling reason at this time to believe that current and future events will dramatically alter that relationship. If anything, complementarity may be reinforced, as the Internet permits an exchange of information about contacts, activities, and places on a hitherto unprecedented scale.

Several research needs have emerged through this discussion. One primary need is to improve our ability to quantify communications occurring by different modes with a common metric, and then to improve our ability to actually operationalize that metric and collect data from individuals on their communications activities by various modes. Although the difficulties are formidable, progress can be made at a conceptual level, and communication technologies themselves can assist at the operational level.

It would be extremely valuable to replicate the aggregate studies of Selvanathan and Selvanathan (1994) and Plaut (1997), to see if their seemingly contradictory findings of substitution in one case and complementarity in the other are sustained. Such a replication would preferably apply both of their methodologies (structural equation models on time-series data of consumption expenditures and industrial input-output analysis of cross-sectional data) to data for the same countries at the same, more recent, points in time. Although aggregate studies are far removed from a behavioral understanding of the phenomenon in question, they are indispensable in terms of offering a "big picture" perspective that is impossible for disaggregate studies to provide.

A final, paramount, question is how much the simultaneous increases observed in telecommunications and travel reflect true causal complementarity, and how much they are due to spurious third-party correlation with other variables. The issue is important because to the extent that the observed relationships are coincidental rather than structural, the more likely it is that changes in certain variables will affect the observed trends in unpredicted ways. The empirical evidence to date is quite limited in its ability to address this question. We can only claim that the conceptual and theoretical considerations discussed above suggest *some* causal influence, but it may or may not account for a substantial portion of the observed relationships. I believe that the conceptual and theoretical arguments for a considerable degree of structural causality are strong, but more can be done to control for confounding factors and to develop a more complete structural model of relationships. At this point, what we can say with confidence is that the empirical evidence for net complementarity is substantial although not definitive, and the empirical evidence for net substitution appears to be virtually nonexistent.

Acknowledgments

This article owes a considerable indirect debt to my extensive collaboration with Ilan Salomon on this subject. A number of conversations over the years with Hani Mahmassani and other colleagues such as John Niles have also been enlightening. In all those cases, predominantly remote interactions have facilitated and generated travel more than they have replaced it, and conversely, relatively few trips to meet face-to-face have prompted innumerable telecommunications. Entirely remote communications from anonymous referees and the Journal editors have also improved the article.

Note

1. Reprinted from *Transportation Research A 24A (3)*, Mokhtarian, P. L., A typology of relationships between telecommunications and transportation, pp. 231–242, copyright 1990, with permission from Elsevier Science.

References

Albertson, L. A. 1977. Telecommunications as a travel substitute: Some psychological, organizational, and social aspects. *Journal of Communication* 27(2): 32–43.

Albertson, L. A. 1980. Trying to eat an elephant. *Communications Research* 7(3): 387–400.

Batten, D. F. 1989. The future of transport and interface communication: Debating the scope for substitution growth. In *Transportation for the future*, edited by D. F. Batten and R. Thord. Berlin: Springer-Verlag.

Bernardini, O. and R. Galli. 1993. Dematerialization: Long-term trends in the intensity of use of materials and energy. *Futures* (May): 431–448.

Button, K. and R. Maggi. 1994. Videoconferencing and its implications for transport: An Anglo-Swiss perspective. *Transport Reviews* 15(1): 59–75.

Claisse, G. 1983. *Transport and telecommunications*. ECMT round table 59. Paris: European Conference of Ministers of Transport.

Couclelis, H. 1999. From sustainable transportation to sustainable accessibility: Can we avoid a new "tragedy of the commons"? In *Information, place, and cyberspace: Issues in accessibility*, edited by D. Janelle and D. Hodge. Berlin: Springer-Verlag.

Day, L. H. 1973. An assessment of travel/communications substitutability. *Futures* 5(6): 559–572.

de Sola Pool, I., ed. 1977. *The social impact of the telephone*. Cambridge, MA: MIT Press.

de Sola Pool, I., C. Decker, S. Dizard, K. Israel, P. Rubin, and B. Weinstein. 1977. Foresight and hindsight: The case of the telephone. In *The social impact of the telephone*, edited by I. de Sola Pool. Cambridge, MA: MIT Press.

Dilts, M. M. 1941. *The telephone in a changing world*. New York: Longman's Green.

Edmonson, B. 1998. In the driver's seat. *American Demographics* (March): 46–52.

Forster, E. M. 1928. The machine stops. 1909. In *The eternal moment and other stories*. New York: Harcourt, Brace.

Gaspar, J. and E. L. Glaeser. 1998. Information technology and the future of cities. *Journal of Urban Economics* 43: 136–156.

Gautschi, D. A. and D. J. Sabavala. 1995. The world that changed the machines: A marketing perspective on the early evolution of automobiles and telephony. *Technology in Society* 17(1): 55–84.

Giddens, A. 1991. *Modernity and self-identity: Self and society in the late modern age*. Stanford, CA: Stanford University Press.

Giuliano, G. and K. A. Small. 1995. Alternative strategies for coping with traffic congestion. In *Urban agglomeration and economic growth*, edited by H. Giersch. Berlin: Springer-Verlag.

Gordon, P., A. Kumar, and H. W. Richardson. 1990. Peak-spreading: How much? *Transportation Research* A 24(3): 165–175.

Gordon, P., H. W. Richardson, and M.-J. Jun. 1991. The commuting paradox: Evidence from the top twenty. *APA Journal* (Autumn): 416–420.

Gottmann, J. 1983. Urban settlements and telecommunications. *Ekistics* 50: 411–416.

Gould, J. and T. Golob. 1997. Shopping without travel or travel without shopping? An investigation of electronic home shopping. *Transport Reviews* 17(4): 355–376.

Graham, S. and S. Marvin. 1996. *Telecommunications and the city: Electronic spaces, urban places*. New York: Routledge.

Grubler, A. 1989. *The rise and fall of infrastructures: Dynamics of evolution and technological change in transport*. Heidelberg: Physica Verlag.

Hanks, J. W., Jr. and T. J. Lomax. 1991. Roadway congestion in major urban areas: 1982 to 1988. *Transportation Research Record* 1305: 177–189.

Harlow, A. F. 1936. *Old wires and new waves: The history of the telegraph, telephone, and wireless*. New York: D. Appleton-Century.

Harvey, A. S. and M. E. Taylor. 2000. Activity settings and travel behaviour: A social contact perspective. *Transportation* 27(1): 53–73.

Helling, A. and P. L. Mokhtarian. 2001. Worker telecommunication and mobility in transition: Consequences for planning. *Journal of Planning Literature* 15(4): 511–525.

Hjorthol, R. J. 2002. The relation between daily travel and use of the home computer. *Transportation Research* A 36: 437–452.

Hojer, M. and L.-G. Mattsson. 2000. Determinism and backcasting in future studies. *Futures* 32: 613–634.

Hu, P. S. and J. Young. 1999. *Summary of travel trends: 1995 Nationwide Personal Transportation Survey*. Report FHWA-PL-00-006. Washington, DC: U.S. Department of Transportation, Federal Highway Administration.

Huws, U. 1999. Material world: The myth of the "weightless economy." In *The socialist register*, edited by L. Panitch and C. Leys. www.yorku.ca/socreg/huws99.txt. Accessed December 2002.

Johansson, A. 1999. Transport in an era of communication. SIKA dokument 1999:1. Stockholm: Swedish Institute for Transport and Communications Analysis. Available from the author at anna.johansson@sika-institute.se.

KMPG (Bureau for Economic Research and Documentation.) 1997. *The influence of the information society on traffic and transportation*. Final report, commissioned by Ministry of Transport, Public Works, and Water Management of the Netherlands, Transport Research Centre (AVV), Strategic Studies Division (VMV), Rotterdam, The Netherlands. Hoofddorp: KMPG.

Kumar, A. 1990. Impact of technological developments on urban form and travel behavior. *Regional Studies* 24(2): 137–148.

Levinson, D. M. and A. Kumar. 1994. The rational locator: Why travel times have remained stable. *Journal of the American Planning Association* 60(3): 319–332.

Lindley, J. A. 1987. Urban freeway congestion: Quantification of the problem and effectiveness of potential solutions. *ITE Journal* (January): 27–32.

Marchetti, C. 1994. Anthropological invariants in travel behavior. *Technological Forecasting and Social Change* 47: 75–88.

Matthews, H. S., C. T. Hendrickson, and D. L. Soh. 2001. Environmental and economic effects of e-commerce: A case study of book publishing and retail logistics. *Transportation Research Record* 1763: 6–12.

Mokhtarian, P. L. 1988. An empirical evaluation of the travel impacts of teleconferencing. *Transportation Research* A 22A(4): 283–289.

Mokhtarian, P. L. 1990. A typology of relationships between telecommunications and transportation. *Transportation Research* A 24A(3): 231–242.

Mokhtarian, P. L. 2002. A conceptual analysis of the transportation impacts of B2C e-commerce. Submitted for publication (available from the author).

Mokhtarian, P. L. and R. Meenakshisundaram. 1999. Beyond tele-substitution: Disaggregate longitudinal structural equations modeling of communication impacts. *Transportation Research* C 7(1): 33–52.

Mokhtarian, P. L. and I. Salomon. 1997. Modeling the desire to telecommute: The importance of attitudinal factors in behavioral models. *Transportation Research* A 31(1): 35–50.

Mokhtarian, P. L. and I. Salomon. 2001. How derived is the demand for travel? Some conceptual and measurement considerations. *Transportation Research* A 35(8): 695–719.

Mokhtarian, P. L. and I. Salomon. 2002. Emerging travel patterns: Do telecommunications make a difference? In *In perpetual motion: Travel behaviour research opportunities and application challenges*, edited by H. S. Mahmassani. Oxford, UK: Pergamon Press/Elsevier.

Mokhtarian, P., E. A. Raney, and I. Salomon. 1997. Behavioral responses to congestion: Identifying patterns and socio-economic differences in adoption. *Transport Policy* 4(3): 147–160.

Murtishaw, S. and L. Schipper. 2001. Disaggregated analysis of U.S. energy consumption in the 1990s: Evidence of the effects of the Internet and rapid economic growth. *Energy Policy* 29: 1335–1356.

Naisbitt, J., N. Naisbitt, and D. Philips. 1999. *High tech, high touch: Technology and our search for meaning*. New York: Broadway Books.

Niles, J. 1994. *Beyond telecommuting: A new paradigm for the effect of telecommunications on travel*. Report DOE/ER-0626, September. Washington, DC: National Technical Information Service (NTIS). www.lbl.gov/ICSD/Niles. Accessed December 2002.

Pierce, J. R. 1977. The telephone and society in the past 100 years. In *The social impact of the telephone*, edited by I. de Sola Pool. Cambridge, MA: MIT Press.

Pisarski, A. E. 1992. *Travel behavior issues in the 90's*. Publication FHWA-PL-93-012. Washington, DC: U.S. Department of Transportation, Federal Highway Administration, Office of Highway Information Management.

Plaut, P. O. 1997. Transportation—Communication relationships in industry. *Transportation Research* A 31A(6): 419–429.

Redmond, L. S. and P. L. Mokhtarian. 2001. The positive utility of the commute: Modeling ideal commute time and relative desired commute amount. *Transportation* 28(2): 179–205.

Richter, J. 1990. Crossing boundaries between professional and private life. In *The experience and meaning of work in women's lives*, edited by H. Grossman and L. Chester. Hillsdale, NJ: Lawrence Erlbaum.

Salomon, I. 1985. Telecommunications and travel: Substitution or modified mobility? *Journal of Transport Economics and Policy* (September): 219–235.

Salomon, I. 1986. Telecommunications and travel relationships: A review. *Transportation Research A* 20A(3): 223–238.

Salomon, I. 1996. Telecommunications, cities and technological opportunism. *The Annals of Regional Science* 30(1): 75–90.

Salomon, I. and F. S. Koppelman. 1988. A framework for studying teleshopping versus store shopping. *Transportation Research A* 22(4): 247–255.

Salomon, I. and P. L. Mokhtarian. 1997. Coping with congestion: Understanding the gap between policy assumptions and behavior. *Transportation Research D* 2(2) (May): 107–123.

Salomon, I. and M. Salomon. 1984. Telecommuting: The employee's perspective. *Technological Forecasting and Social Change* 25(1): 15–28.

Salomon, I., P. Bovy, and J.-P. Orfeuil, eds. 1993. *A billion trips a day: Tradition and transition in European travel patterns*. Dordrecht, The Netherlands: Kluwer Academic.

Schafer, A. and D. G. Victor. 2000. The future mobility of the world population. *Transportation Research A* 34: 171–205.

Selvanathan, E. A. and S. Selvanathan. 1994. The demand for transport and communication in the United Kingdom and Australia. *Transportation Research B* 28B(1): 1–9.

Shamir, B. 1991. Home: The perfect workplace? In *Work and family*, edited by S. Zedeck. San Francisco: Jossey-Bass.

Strawson, J. M. 1980. Future methods and techniques. In *The future of the printed word*, edited by Philip Hills. London: Pinter.

Tauber, E. 1972. Why do people shop? *Journal of Marketing* 36: 46–49.

Townsend, A. 2000. Life in the real-time city: Mobile telephones and urban metabolism. *Journal of Urban Technology* 7(2): 85–104.

U.S. DOT (U.S. Department of Transportation). 1993. *Transportation implications of telecommuting*. Washington, DC: U.S. Department of Transportation.

Utts, J. M. 1999. *Seeing through statistics*. Second edition. Pacific Grove, CA: Duxbury Press.

Webber, M. M. 1991. The joys of automobility. In *The Car and the city*, edited by M. Wachs and M. Crawford. Ann Arbor, MI: University of Michigan Press.

Wells, H. G. 1968. When the sleeper wakes. 1899. In *The sleeper awakes*. 1954. London: Collins.

Yim, Y. 2000. Telecommunications and travel behavior: Would cellular communications generate more trips? Paper 00-0625 presented at the Annual Transportation Research Board Meeting, 9–13 January, Washington, DC.

Zumkeller, D. 1996. Communication as an element of the overall transport context: An empirical study. In *Proceedings, 4th international conference on survey methods in transport*, Vol. 1, 66–68. Leeds, UK: University of Leeds Institute for Transportation Studies.

About the Author

Patricia L. Mokhtarian is a professor in the Department of Civil and Environmental Engineering and associate director of the Institute of Transportation Studies at the University of California in Davis, CA, USA.

[25]

Sustainable Transportation

U.S. Dilemmas and European Experiences

Elizabeth Deakin

The approach to sustainable transportation issues in the United States was examined in light of findings from a study of sustainable transportation planning in Sweden, Germany, the Netherlands, and Scotland. In the European countries, reducing greenhouse gases has been the initial motivation for most sustainable transportation initiatives, but broader social, economic, and environmental concerns now figure into the idea of sustainability. In the United States, barriers to greenhouse gas reduction and planning for sustainability include uncertainty about the problem and the best ways to address it, uncertainties about public support, and the lack of a clear mandate for action. Nevertheless, efforts are under way locally in the United States to promote sustainable development, and transportation plays a central role in these plans. The European organizations visited are using many of the same strategies as are U.S. planners, but supporting their efforts are strong policy commitments, government incentives, and new planning processes emphasizing collaboration and performance measurement. Tracking the comparative success of these efforts would be an important next step.

Concerns about the sustainability of development practices are increasingly being voiced in the United States, Europe, and other developed countries as well as in the developing world. Early discussions of sustainable development focused largely on environmental damage (especially global warming) from land development and transportation, but increasingly, disparities in economic development and in the choices available to different nations and social groups for housing, jobs, services, and overall quality of life are among the matters of concern. Although the current administration has rejected the Kyoto Protocol, efforts are under way to find other ways to reduce greenhouse gases, and concern about broader quality-of-life issues remains strong.

Interest in sustainable development is motivated by a number of factors. There is a desire to reduce carbon loading of the atmosphere to avoid or reduce the impact of the greenhouse effect and its possibly catastrophic global consequences. There is also worry about a host of other environmental harms, ranging from the effects of deforestation on ecosystems and biodiversity to the impacts of fishery depletion on seacoast economies and ocean health. In both developed and developing countries, the effects of growing dependence on the automobile, rapid urbanization, and sprawl development are major motivations for considering sustainability. In addition, there is interest in maximizing the productivity of infrastructure investment and controlling costs. Desires to preserve farmland, wild land, and other open spaces and to improve urban amenity, redevelop brownfields, and make the suburbs more interesting are additional factors in current discussions. Finally, the social equity of current patterns of consumption domestically and internationally, the necessity of economic growth and development in the less-developed nations, and the relative contributions of the haves and the have-nots to environmental damage and to its potential restoration are additional issues that have entered into the debate over sustainable development.

University of California Transportation Center and Department of City and Regional Planning, 108 Naval Architecture Building, Berkeley, CA 94720-1782.

OVERVIEW OF PROPOSALS

Proposals that aim to promote sustainable development are as wide ranging as are the problems that motivate them. They include carbon taxes and fuel pricing strategies, new fuels and new vehicle technologies, strategies to phase out fossil fuel power generation, and industrial strategies focusing on efficiency gains. They also include local and regional growth management and urban revitalization efforts, transit investments, and traffic calming programs. The umbrella is wide enough to cover such U.S. federal and state programs as brownfields initiatives, Enterprise Zones, Location-Efficient Mortgage experiments, and Livable Communities programs, as well as programs to promote the use of recycled materials in construction and reconstruction, to implement intelligent transportation systems (ITS), and to formulate new processes of context-sensitive design.

In the United States and other developed countries, transportation is a major focus of sustainable development inquiries, for the simple reason that transportation is so prominent in the many problems that sustainable development aims to address. The United States produces some 30% of the world's total greenhouse gas emissions; transportation accounts for 25% of those emissions. Strategies that do not address transportation would fail to address a large share of total emissions.

Specific transportation measures that have been proposed in sustainable development discussions include the following:

- Electric and other alternative-fueled vehicles;
- High-efficiency gasoline, diesel, and hybrid vehicles;
- Traffic operations and logistics improvements using information technologies and ITS;
- Deployment and promotion of alternative modes such as rail freight transportation and transit, biking, and walking for personal transport;
- Land use strategies that reduce automobile dependence by focusing development on transit and in pedestrian-oriented areas;
- Land use strategies that shorten trips and encourage transit use, walking, and biking through compact, higher density, mixed-use development; and
- Telecommuting and other telecommunications substitutions for travel (*1*).

Although interest in these strategies is high, barriers to their successful implementation also are substantial. Technological alternatives are costly, take time to deploy, and do not always perform well.

Alternative modes are frequently neither as convenient nor as inexpensive—in time or dollars—as the automobile. Developers and lenders question the costs and profitability of infill and believe that high-density urban or town-center living is a specialized and limited market niche. Government officials note that funding for these strategies is limited, and they worry about public costs increasing along with densities. Institutional issues also are significant, with technological development, transportation infrastructure, and land use planning falling to different agencies with different approaches and missions [E. Deakin, *Combating Global Warming Through Sustainable Surface Transportation Policy*, paper prepared for TCRP, Washington, D.C., October 1998 (unpublished data)].

Nevertheless, efforts are under way to try to overcome the barriers to sustainable development, often focusing in particular on sustainable transportation and land use. State, regional, and local governments and agencies are increasingly working on sustainability issues, sometimes on their own and sometimes in cooperation with local businesses, developers, and community-based organizations. Many of the efforts rely on leveraged investments and involve both public and private partners. From the more successful experiences, a new planning style is beginning to emerge, emphasizing networks and coalitions of common interest, locally crafted solutions, experimentation, and rewards for accomplishment (E. Deakin, *Combating Global Warming Through Sustainable Surface Transportation Policy*, 1998).

For some, however, sustainability is best treated with skepticism. The concept seems too vague to be useful. To some, it seems antigrowth. Some states have rejected the approach outright; others find it controversial.

Despite these sensitivities, the U.S. transportation community has shown a growing interest in sustainable transportation and its linkages to land use and urban development patterns, economic growth, environmental impacts, and social equity. Federal, state, and local agencies as well as private organizations are working to translate the broad goals of sustainability into specific transportation policies, objectives, and programs. They are reexamining their policies, planning approaches, and evaluation methods. Also, they are considering changes to every aspect of practice, from the materials and designs used in construction to the kinds of alternatives considered for implementation.

Their efforts are known by a variety of names. Some are developing plans or programs for smart growth or livable communities. Some have conducted a visioning process. Others emphasize one or another aspect of sustainable development: infill, compact growth, transportation alternatives, habitat preservation, revitalization of established neighborhoods and shopping districts, and expansion of affordable housing opportunities. Still others have placed great emphasis on programs to improve public safety, access to jobs, and educational opportunities. Funding has sometimes been entirely local; sometimes it has been supported by grants obtained through the FTA's Livable Communities program, the Housing and Urban Development's Enterprise Zone community revitalization efforts, or the Environmental Protection Agency's Brownfields Redevelopment demonstrations. What all these efforts have in common is a desire to meet current needs in a fashion that improves social, economic, and environmental conditions while preserving good options for future generations (*2*).

Because of the growing U.S. interest in sustainability issues, a study was carried out to examine how other developed countries are addressing sustainable transportation issues. Sweden, Germany, the Netherlands, and the United Kingdom were identified as having several years of experience in addressing sustainable transportation issues. Visits to each country were arranged for a study group composed of representatives from the U.S. Department of Transportation, AASHTO, metropolitan transportation organizations, city and county governments, and the author. A set of questions was provided in advance to the European participants.

In each country, the U.S. group met for 2 to 3 days with officials engaged in sustainable transportation efforts. The topics covered included these:

- The context in which planning and decision making for transportation and development occurs,
- Definitions of sustainability,
- The policies and planning practices used in pursuit of sustainability,
- Sustainable transportation and sustainable development strategies, and
- Case studies and implementation examples.

Following the site visits, additional materials were provided by some of the organizations visited.

The study group noted differences in context that must be considered in assessing the potential for adoption of similar policies and practices in the United States: growth slower than that in many U.S. metropolitan areas, relatively homogeneous populations, higher development densities, and more extensive and more heavily used transit systems. Similarities include growing automobile ownership and use, suburban development, and public interest in community amenity and quality of life. The differences suggest that some translation for American settings will be necessary, whereas the shared concerns and objectives point to opportunities for mutual exchange and learning.

This paper examines the issues raised by sustainable development and sustainable transportation, drawing on U.S. experience—as identified through a literature review and interviews with 89 U.S. transportation officials and academics—as well on experience from the four European investigations. The paper briefly presents some background on greenhouse gases, the initial motivation for most sustainable transportation initiatives, and then examines the transportation strategies that have been proposed in the United States. Key issues or uncertainties that arise in U.S. discussions of sustainable transportation are identified; sustainable transportation as an emerging strategy in the United States is considered. Findings from the European study are then reported. The final section of the paper summarizes lessons learned.

BACKGROUND ON GREENHOUSE GASES

Scientists generally agree that increasing concentrations of greenhouse gases (water vapor, CO_2, methane, nitrous oxide, and halocarbons) in the atmosphere are causing the average temperature of the earth to rise. The timing, magnitude, and consequences of this temperature increase are not fully understood or agreed on, but most analyses have predicted that warming could be on the order of 1° to 4° C within a century. Average temperature increases of this magnitude could produce marked changes in precipitation patterns, with accompanying disruptions in other natural systems. It is also possible that the frequency and violence of storms could increase (*3*). The resulting changes could be so rapid that neither natural systems nor social systems would be able to adapt easily. Some system changes appear to be under way already, including increased global mean surface temperatures and rising sea levels (*4*).

In response to this potential threat to social, economic, and environmental well-being, a series of international conferences have been held to develop a plan of action. The Kyoto Protocol, hammered out in 1997, aimed to be step toward the reduction of greenhouse gas emissions over the next decades. The protocol sets out targets for industrialized nations, averaging out to about 5% below 1990 levels by the period of 2008 to 2012. For the United States, the target level was proposed to be a 7% reduction, a difficult target to meet considering that emissions have continued to grow each year. In contrast to most advanced industrial states, the U.S. Congress did not ratify the agreement, however, leaving the United States comparatively uncommitted to action. Subsequently, the current administration also rejected the protocol, while continuing to look for other ways to address greenhouse gas issues.

Finding ways to achieve significant reduction in greenhouse gas emissions in the United States is a major challenge. The largest energy user in the world, the United States is also the largest emitter of CO_2, currently accounting for almost one-quarter of the total. The CO_2 emissions in the United States come from transportation activities, residential and commercial activities, and industrial processes in roughly even shares. Transportation activities in the United States, which have been estimated to be the largest single source of greenhouse gas emissions in the world (*5*), include both motor vehicle emissions and other transportation emissions (e.g., from jet aircraft); however, surface transportation alone is 25% of the U.S. total. Three-quarters of that 25%, about 16%, are from motor vehicle use.

Debate continues over the range of options that might be used to reduce greenhouse gas emissions. Market approaches, regulation-mandated changes in technologies, and restraints on energy use are all under discussion. Reductions need not be targeted proportionally at the various sectors producing CO_2. Indeed, in the short run, developed countries might pay for technological improvements, or perhaps for forest preservation, in less-developed nations as a more cost-effective alternative than immediate technological change or demand suppression at home. Still, many analysts believe that a multipronged strategy will be necessary both to accomplish the needed reductions and to ensure a modicum of equity. Meanwhile, in the U.S. transportation sector, CO_2 emissions are expected to nearly double by the middle of the next century unless technological changes are vigorously introduced or demand is sharply curbed. Either type of reduction could have large economic, social, and environmental consequences far beyond its greenhouse gas effects; hence, research is needed on possible strategies, impacts, and implementation pathways (*6*).

TRANSPORTATION STRATEGIES FOR REDUCING GREENHOUSE GAS EMISSIONS

Table 1 presents a partial list of strategies for reducing greenhouse gas emissions in the United States, identified in the literature and in interviews with transportation officials and academics (2). The strategies are grouped into categories based on the component of the transport system addressed: vehicles, roadways, operations, or demand.

The first category of strategies would reduce greenhouse gas emissions through technological change in vehicles and fuels. In the short run, this would most likely be accomplished by improving the efficiency of conventional vehicles, and in the longer run through the introduction of new vehicle technologies and new fuels. Innovations emerging from manufacturer innovations could be put to use in reducing greenhouse gas emissions, for example. If, due to demand-side incentives or changes in public attitudes, consumers began to demand vehicles that are low emitters of greenhouse gases, suppliers would probably respond by offering consumers that choice. Alternatively, the government could provide incentives to manufacturers to produce low-emissions vehicles, or it could mandate the same through regulatory interventions. Corporate average fleet efficiency (CAFE) standards, research and development partnerships, taxes, rebates, and subsidies are specific options to consider, and they could apply to passenger vehicles and trucks both.

A second category of strategies to reduce greenhouse gas emissions would involve improvements to roadways and vehicle operations. Again, different approaches might be used in the short run and over the longer run. Conventional traffic flow improvements such as traffic signal timing, ramp metering, flow metering, and bottleneck removal all have the potential to reduce greenhouse gas emissions by smoothing the flow of traffic and reducing fuel-wasting stop-and-go travel. Driver education could reduce emissions by training drivers to avoid heavy accelerations and decelerations and to be mindful of the fuel consequences of high speeds. Scheduling trips outside of the peak periods could reduce congestion and thereby cut emissions.

Improved methods of accident and incident management and improved logistics and fleet management, both relying increasingly on advanced technologies for vehicle location and communication, also have substantial promise for increased efficiency of operations. Information technology–enhanced routing and scheduling can reduce the fuel needed for transport of both passengers and freight. Technological innovations currently under development offer the potential for significantly larger gains. They include the more advanced aspects of intelligent transportation system improvements such as smart highways and smart vehicles.

Demand management is a third category of strategies for reducing greenhouse gas emissions. Several subcategories of demand management are in use: modal substitution, telecommunications substitution, pricing, and land use strategies all can be thought of as forms of demand management.

Modal substitution means, for example, replacing car trips with transit, paratransit, ridesharing, biking, and walking for personal travel and substituting rail for truck and air freight. Such substitution can be accomplished by providing better modal options (offering services and improving their quality to attract travel to alternative modes) or through incentives for the use of the alternative modes (e.g., subsidies for users of preferred modes). Regulatory requirements (e.g., trip reduction ordinances requiring employers to obtain commute mode shares of no more than 50% by drive-alone) are also a possible way to induce modal substitution.

Telecommunications substitutions for travel also can be considered a form of demand management. Telecommuting, teleshopping, teleconferencing, and distance learning are varieties of telecommunications substitutes for travel.

Pricing incentives and disincentives could be used in the short run to reduce demand and encourage the use of alternative modes or the substitution of telecommunications for travel. In the longer run, vehicle technology improvements would likely be induced by the higher prices. Gas tax increases are the pricing strategy most commonly used in the United States and abroad; fees and taxes that affect vehicle ownership, such as sales taxes and registration fees, also are common. Variations that base taxes and fees on fuel efficiency, emissions, and expected vehicle life could specifically target the reduction of green-

TABLE 1 Strategies for Reducing Transportation Greenhouse Gas Emissions: A Partial List

Vehicle / Fuel Technological Changes:
Improved Efficiency of Conventional Vehicles
New Vehicle Technologies
New Fuels
Manufacturer Innovations / Supplier Offerings
Responses to Consumer Demand
Responses to Government Regulation and Incentives: CAFE Standards, R&D Partnerships, Taxes, Rebates, Subsidies
Road / Vehicle Operations Improvements:
Conventional Traffic Flow Improvements
Traffic Signal Timing
Ramp Metering
Flow Metering
Bottleneck Removal
Intelligent Transportation System Improvements
Smart Highways
Smart Vehicles
Accident / Incident Management
Routing and Scheduling Enhancements
Driver Education
Improved Logistics and Fleet Management
Demand Management:
Modal Substitution
Transit, Paratransit, Ridesharing, Walking, Biking Improvements and Incentives
Rail Substitutes for Truck
Telecommunications Substitutions
Telecommuting
Teleshopping
Teleconferencing
Distance Learning
Information Technology-Enhanced Routing and Scheduling (Passengers, Freight)
Pricing Incentives / Disincentives:
Gas Tax Increases
Vehicle Sales Tax Based on Fuel Efficiency and Expected Life
Vehicle Registration / License Fee Based on Fuel Efficiency, Use (Measured or Estimated)
Other Impact Fees Based on Use
Subsidies for Preferred Modes, Telecommunications Substitutes, Etc.
Land Use –Transportation Strategies:
Compact Development
Mixed Use Development
Higher Development Densities
Transit, Pedestrian, Bike-Friendly Development

house gases, as could "fee-bate" variations offering tax reductions for efficient, low-emissions vehicles, along with surcharges for high emitters. Or pricing strategies could base emissions or fuel surcharges on measured or estimated use (vehicle miles traveled/vehicle kilometers traveled). Finally, rather than use pricing to restrain emissions directly, pricing could take the form of subsidies for preferred modes or for telecommunications substitutes.

Land use and urban development strategies alter demand by reducing trip length (by providing a choice of close-by destinations) or by making alternatives to the automobile more competitive and cost-effective. (These strategies also may reduce emissions associated with building heating and cooling, service provision, and other factors.) For example, compact development, mixed-use development, and higher development densities can reduce trip lengths and make transit, pedestrian, and bike use practical and affordable. In some cases, compact development also may facilitate better management of urban freight transport (e.g., shipment consolidation, delivery scheduling).

This list of strategies will sound familiar to most transportation professionals. The same list has been used for many years in the search for strategies to reduce air pollution emissions and to manage traffic congestion. Also, many of the strategies have been used for decades. The federal government established CAFE standards to induce fuel efficiency in vehicles in the 1970s. Federal and state gas taxes are in place and are periodically increased. Highway departments and local traffic engineering offices routinely use traffic operations strategies to increase capacity and reduce environmental impact. Transit services and ridesharing programs are offered throughout most metropolitan areas and in rural ones as well. Land use strategies promoting infill, compact development, and mixed use are promoted in numerous cities and suburbs.

The widespread use of many of the listed strategies is both an advantage and a drawback. It is an advantage because it means that established programs are in place to offer evidence of efficacy and to potentially serve as a base for further expansion or innovation. It is a drawback because it means that many of the strategies will already be fully deployed where they are cost-effective, with their benefits already captured. Further deployment in these circumstances could produce limited results and in some circumstances could even produce disbenefits. As an example, consider the impacts of deploying fixed route bus service in low-density areas where the service is little used: emissions and fuel use per passenger carried can be higher than would occur by using automobiles or taxis to serve the trips. Hence, the issue is whether market niches cans still be found to which these strategies might effectively be applied.

For some of the strategies, many people would respond that there are indeed more markets to be served. For example, traffic signal timing and other operations improvements are in common use, but many localities have lacked the resources to upgrade equipment or to retime their signals on a regular basis. They could benefit from a funded traffic signal management program. Similarly, few localities have had the resources to fully implement bicycle networks, pedestrian improvements, and traffic-calming programs, and funds for such strategies are oversubscribed. Transit operators and ridesharing service providers often have lists of unfunded improvements. Few have even begun to explore the possibilities for shuttle services, subscription buses, and other innovations. Not all of these strategies would necessarily be cost-effective from a greenhouse gas reductions perspective, but some surely would be.

Other strategies remain in the early stages of deployment and a strategic effort to implement them might produce meaningful results. This is the case, for example, for many of the strategies involving advanced technologies for highway and transit operations. It also may be the case for certain vehicle and fuel technology strategies, in which the wider implementation of experiments and demonstration projects could be useful.

Land use strategies have recently begun to capture the attention of many interest groups, and studies and small programs to test these strategies' transportation effectiveness are under way. Here, too, wider experimentation and systematic evaluation could be useful.

Pricing strategies are still highly controversial in most parts of the country, although some local governments and private operators are successfully managing parking pricing. Also, projects with variable tolls and value pricing are under way in a handful of U.S. highway facilities. Proposed gas tax increases, fee-bates, emissions fees, and other measures have been evaluated in major studies, but so far implementation has not occurred, owing to concerns about equity and opposition to any strategy that looks like a tax. Nevertheless, there is enough interest in these strategies to consider a larger effort toward their implementation than has occurred to date.

How effective would the various transportation strategies be in reducing greenhouse gas emissions? TRB investigated this topic in a 1997 report (*1*). Four scenarios were tested. One emphasized demand management and land use planning, the second focused on improvements in vehicle efficiency, the third emphasized fuel price increases, and the fourth assumed the introduction of new vehicle technologies. Evidence was drawn from the literature on modeling studies and field experiments, and estimates of greenhouse gas reductions were produced for each scenario. The results, which are for the United States, were as follows:

- From aggressive demand management and land use planning strategies: 6% reduction by 2020 and 15% by 2040;
- From a 1.5% annual increase in average new vehicle fuel efficiency: 15% to 20% reduction by 2020 and 35% by 2040;
- From higher fuel prices amounting to a 3% increase per year: 20% reduction by 2020 and 40% by 2040; and
- From the introduction of new low-emissions vehicles (5% of fleet by 2020 and 35% by 2040): no significant change by 2020, and a 30% reduction by 2040.

Even if one assumes that these results are approximately correct, no one strategy by itself offers a silver bullet for the greenhouse gas emission problem. Furthermore, considerable uncertainty about implementation feasibility is connected to each of the scenarios, strengthening the conclusion that thorough consideration of the full range of options is a prudent approach.

KEY ISSUES ON TRANSPORTATION AND GREENHOUSE GAS REDUCTIONS

U.S. interviews revealed several key issues that may block or slow action on transportation strategies for greenhouse gas reductions: uncertainties about the nature and severity of the greenhouse gas problem, lack of agreement on the extent to which transportation strategies should be used in attacking greenhouse gas concerns, uncertainties about the effects of change in transportation technologies, and uncertain public support for intervention. Interview respondents noted that these issues reflect, and are reflected in, the lack of federal mandate and funding for greenhouse gas reduction activities.

Uncertainties About the Problem

Important uncertainties remain about the nature and severity of the greenhouse gas problem. Although most scientists agree that the Earth is warming, there is less agreement about specific mechanisms, including the role of the oceans and forests in carbon absorption and recycling. Furthermore, changes in activity in both developed and developing countries must be predicted to estimate carbon loading of the atmosphere. The pace of development and the choices made could greatly affect such forecasts. Uncertainties, coupled with the high stakes involved, make it hard to muster the political support needed for action.

Unresolved Responsibility for Reduction of Greenhouse Gas Emissions

Analyses often assume that transportation would aim to reduce greenhouse gases proportional to the transportation sector's contributions. However, this is not necessarily the most efficient or effective way to reduce the emissions. As noted earlier, reductions in greenhouse emissions could be obtained by helping developing countries limit or reduce their production of greenhouse gases (emissions trading). Or a higher than proportional share of U.S. reductions might be sought from other sectors of the economy, for example, power generation (accounting for about 36% of total U.S. emissions).

Either strategy raises its own set of concerns about equity, costs, and timing. There appears to be no simple solution that would alleviate all pressure for change in the transportation sector.

Technological Change and Its Implications

Technological advances in the automotive industry and other sectors of the economy have considerable potential to reduce greenhouse gas emissions and, for that matter, other externalities that are broader concerns about sustainability. Aggressive technology deployments, whether in the form of changes in conventional vehicles or through the introduction of radically different vehicle and fuel technologies, could produce substantial greenhouse gas reductions over the next several decades. To many, such invisible technological change would be more inviting than the prospect of higher prices or other demand reduction strategies. However, such technology deployments are by no means assured, and they may emerge in forms that are not as attractive as currently envisioned—considering price, availability, and performance characteristics. In addition, in the absence of public policy direction, the technological changes that do emerge may or may not be directed to greenhouse gas emissions. For example, at present many advances in automotive technology are being applied to increase acceleration and performance or to strengthen vehicle bodies, not to boost efficiency. A variety of interventions could change this situation directly or indirectly. Higher CAFE standards and higher gas taxes are but two of many possibilities but would certainly be controversial themselves.

Uncertain Public Support for Interventions

Reductions in transportation greenhouse gas emissions, as well as broader actions to move toward greater environmental, social, and economic sustainability, will probably depend at least in part on changes in consumer preferences. Although polls generally find widespread support for environmental protection and enhancement, observation suggests that many consumers are not yet ready to alter their travel behavior or consumer purchases because of the threat of global warming. For example, suburban utility vehicles and trucks remain popular despite their comparatively low fuel economy, and drive-alone mode shares are increasing.

Many analysts believe that only those measures that do not require significant changes in behavior have much chance of winning public support. Other analysts believe that pricing policy, such as higher fuel taxes or full-cost pricing for parking, could substantially change consumer choice. However, public opposition to such measures continues to make their implementation doubtful. Changes in travel behavior also could be instigated through changes in land use and location, modes offered and chosen, and overall activity patterns. The interest in livable communities and sustainable development suggests some sentiment in favor of land use and transport changes, but the vast majority of households and businesses continue to settle in conventional suburbs and to rely on motor vehicles as their main means of transport. Finally, changes in public attitudes and behavior might be forthcoming as public understanding of greenhouse gas issues increases, but so far there is little evidence that this is occurring.

Lack of Mandate and Funding for Action

A number of studies have developed and tested plausible scenarios for reducing transportation greenhouse gas emissions. However, in the United States, moving forward from scenario testing to the development and implementation of action plans has been a slow process hampered by two major barriers: the lack of a national (federal) mandate and the lack of a dedicated funding source. With neither requirements nor incentives for action, many agencies are reluctant to take on the greenhouse gas problem—especially when they consider that the topic is likely to be multifaceted and controversial and which may put them at a disadvantage compared with other jurisdictions that choose not to take action.

SUSTAINABLE TRANSPORTATION: AN EMERGING STRATEGY

Although the issues raised about transportation and greenhouse gases are difficult ones, a growing number of states and localities are developing programs for reducing greenhouse gases. Many of them have acted under the rubric of smart growth, or more broadly, sustainable development, of which sustainable transportation is a key part.

What exactly is sustainable development? A variety of definitions of have been put forward. There is growing consensus that sustainability must include economic betterment and social equity, not just a narrow technical focus on greenhouse gases or other aspects of the natural environment. Planning for sustainable development increasingly involves strategic coordination of efforts along all three dimensions. Furthermore, sustainability is increasingly understood as a collective process for considered decision making and action, not simply a particular end state or outcome.

In the last several years, sustainability initiatives have been undertaken in a number cities and regions here and abroad. Recent undertakings include the Maryland Smart Growth Initiatives; the Portland, Oregon, 2040 Plan; Sustainable San Francisco; Sustainable Toronto; Sustainable Seattle; and the Bay Area Alliance for Sustainable Development. Examples from the Netherlands, France, and Germany, as well as from Brazil, Argentina, and Indonesia also are well known. These efforts largely follow up on the 1987 report of the World Commission on Environment and Development (commonly called the Brundtland Commission report) and, in the United States, on the President's Council for Sustainable Development report, *Sustainable America—A New Consensus,* which argued that sustainable development can be achieved only by building sustainable communities.

Reflecting the recommendations and action items in those documents, the local and regional efforts typically focus on the interrelationships among transportation, housing, and employment trends and policies, and the resulting consequences for the environment (especially air quality), energy use, economic prosperity, and social equity. The efforts often are developed through a process involving a wide range of interests (e.g., business leaders, environmentalists, and social justice advocates, as well as public officials and agency staff members). Often they involve the negotiation of procedural agreements as well as the development of performance indicators and specific actions to be undertaken.

Table 2 lists some of the strategies that are proposed in sustainable development plans. As the table shows, the strategies range from land use planning to transportation, housing, and economic development; linkages and overlaps are strong.

Supporters of the strategies and initiatives believe there is considerable potential for important gains through sustainable development planning. Most admit, however, that the United States is behind many other countries in its efforts, and it is too early to tell about the U.S. experiments. Hence, a closer look at sustainability initiatives in other developed countries should provide valuable insights and examples.

EUROPEAN EXPERIENCES

The study tour to Sweden, Germany, the Netherlands, and Scotland was designed to examine how other developed countries are addressing sustainable transportation issues. Visits to each country provided the opportunity to conduct extended discussions with policy makers

TABLE 2 Strategies for Sustainable Development

Land Use and Community Development:	
	Preservation, Rehabilitation, and Redevelopment of Central Cities and High Density Inner Suburbs
	Infill in Cities and Suburbs – increased Density, Mixed Use
	Reusing Brownfields, Recycling Buildings
	TODs and PODs as the Paradigm for New Developments
	Quality of Life: Attention to Crime/Schools/Services/Amenities
	Recycling/Precycling/Composting Programs
	Transportation:
	Access vs. Mobility – Basic Concepts
	Bike- and Pedestrian-Friendly Cities
	Transit, Paratransit, Ridesharing
	Telecomuting/Teleconferencing
	New Technologies for Improved Efficiency: Electric Vehicles, Traffic Control Systems, Transportation Information Systems
	Prices and Subsidies Aligned with Sustainability
	Housing and Other Building Designs:
	A Range of Choices
	Energy-Efficient Buildings
	Edible Landscaping
	Natural/Indigenous Plants
	Business/Job Creation:
	Business Leadership
	Community Economic Development
	Clean/Safe Technologies
	Social Equity:
	Aligning Taxes and Subsidies with Sustainable Development
	Equitable Distribution of Resources

and practitioners and to more closely examine these countries' experiences. Each country is part of the European Union (EU), which provides a consistent overall framework for action; key features of EU environmental policy, enunciated in the Treaty of Rome and Single European Act, are as follows:

- Use a precautionary principle: better prevention than cure;
- Do early consideration of environmental effects;
- Avoid resource exploitation that causes significant damage: use but do not abuse;
- Avoid spillovers in other countries;
- Use a polluter-pays principle;
- Promote a worldwide environmental policy;
- Improve scientific knowledge on the environment;
- Share responsibilities for environmental actions;
- Assign responsibility for action to the appropriate level of government—as close to citizens as possible; and
- Conduct environmental education.

Within that framework, however, considerable variation among the countries was noted. The sections that follow discuss shared approaches in the four countries visited; Table 3 through Table 6 list specific strategies being used by each of the four countries.

Definitions of Sustainability

All of the countries visited use some variation of the Brundtland definition of sustainability—meeting the needs of the present without compromising the ability of future generations to meet their own needs—as the starting point for their efforts on sustainable development. The reduction of CO_2 is an important objective. However, sustainability is seen as a much broader concept having economic and social as well as environmental dimensions. Sustainable development is viewed as development that improves the standard of living and quality of life, while at the same time protecting and enhancing the natural environment and honoring local culture and history.

In this context, sustainable transportation is safe, of high quality, accessible to all, ecologically sound, and economical. It is also a positive contributor to regional development. Specific goals for sustainable transportation include improvements in service quality, safety, air quality, water quality, noise reduction, protection of habitat and open spaces, and historic preservation; the reduction of carbon emissions; and the attainment of local goals consistent with the overall objective.

Policies and Practices

The policies and practices used to pursue sustainability recognize the importance of collaboration, both as a means of reaching agreement on specific goals and objectives and as a way to pursue specific strategies. Each of the countries visited uses collaborative strategic planning to identify and evaluate ways to move toward sustainability and to devise performance measures for assessing progress. Collaborations involve the different levels of government, various agencies, citizens, and the private sector.

Both the EU and each country back the commitment to reducing CO_2 emissions through the use of policies for accomplishing that end. At the same time, they recognize that local governments and the general public typically have more immediate transportation concerns, including mobility needs as well as noise, speeding, and traf-

TABLE 3 European Strategies for Sustainability: Germany (Berlin)

Overall Strategies for Sustainability:

- Work within EU framework, apply local regulations (no overall package of measures for country)
- Qualitative vision and quantitative criteria – noise, air quality, acidification, CO_2 reduction, etc.
- Multimodal planning and least cost planning
- Avoid trips, shift to less damaging modes, optimize road capacity, improve vehicle technology, deploy telecommunications and ITS
- Land use strategies – German City Assn.: density and mixed use, corridor and wedge, reinforce existing centers, discourage or ban greenfield stand-alone malls; focus development at crossing of transit lines
- Education and research
- User fees to reflect full cost (not yet supported by public except perhaps for trucks)

Examples of Sustainable Development and Sustainable Transport:

- Truck impact management–vehicle taxes, fuel taxes, time restrictions
- Rail and sea emphasis for freight
- Logistics and ITS to manage freight movements
- Speed advisories to avoid congestion
- No new roads, but upgrade and some widening
- In longer term, highway management using ITS
- Traffic calming
- Federal regulations on urban development – compact growth, pedestrian, bike, and transit access
- Coordinated land development and transport improvements
- Emissions standards for vehicles
- Alternative fuels for vehicles
- Intermodal improvements including use of tunnels
- Transit emphasis
- Dedicated bus lanes
- Transit connection to airport
- Bicycle facilities well connected to rail
- Construction logistics to reduce adverse impacts
- Recycled materials in construction and reconstruction

TABLE 4 European Strategies for Sustainability: The Netherlands (the Hague, Rotterdam)

Overall Strategies for Sustainability:

- Get prices right
- Access as focus – land use as well as transport
- Support existing centers
- Provide multimodal transport in line with sustainable policies
- Quality services door to door
- Quality design – including undergrounding
- Public-private land use-transport solutions

Examples of Sustainable Development and Sustainable Transport:

- Manage demand by using pricing to appropriately reflect full costs
- ABC policy – focus development where there is greatest access
- Development contiguous to existing cities and towns, mixed use
- Infrastructure policies aligned to support sustainable development (including water, sewer, etc.)
- Limit stand-alone shopping malls, etc.
- Traffic calming
- Preserve Green Heart of the region
- Plan for whole trip chain not mode by mode (e.g., bike storage at apt., bike lane to train station, storage or on-board option, etc.)
- Truck logistics
- Plan for technological change in vehicles, ITS, etc. – focused on safety and congestion relief (getting 10% less congestion)
- Prohibit cell phone use while driving unless hands free
- Driverless people mover connecting office park and rail station
- Public-private partnership to help pay for high cost items
- Undergrounding roads, rail, parking

TABLE 5 European Strategies for Sustainability: Scotland (Edinburgh)

Overall Strategies for Sustainability:
- Regional strategies for development and transport; integrate transport and land use
- Central city vitality
- Public transport competitive and attractive
- Compact and contiguous suburban development
- Emphasis on exchange, not movement

Examples of Sustainable Development and Sustainable Transport:
- Travel Wise program – try to educate public – think before you travel, chain trips, walk or bike, etc.
- Extensive green lane system for bus priority
- Parking pricing
- Car club experiment
- Traffic calming
- Bike streets, bike ways
- Wider sidewalks (also good for business)
- Lower speeds in some zones
- Speed enforcement by camera
- Recycled materials in construction

fic problems. However, initiatives to address local concerns often reduce CO_2 emissions as well, thus contributing to a larger sustainability strategy. European policies therefore encourage and reward such local initiatives. Practices include the following:

- Offer local governments incentives for aligning their policies and practices with national objectives;
- Lead by example: show good practices in government first;
- Support local projects that move in the direction of greater sustainability, to build local understanding and support; and
- Try new ideas and see what works.

What makes sustainable transportation planning different from past practice is that social and environmental objectives are an integral part of sustainable transportation planning, rather than constraints or the focus of mitigation efforts. European policies on

TABLE 6 European Strategies for Sustainability: Sweden (Stockholm)

Overall Strategies for Sustainability:
- Strategy is "lots of small things" but done in collaboration and put together into an overall strategy
- Access, quality service, safety, good environment, economic development – all objectives for transportation plans
- Transportation providers must meet social and environmental objectives, are evaluated on social and environmental performance
- Collaborative efforts to identify and remove conflicts, pursue areas of agreement
- Strategic planning, performance measures, monitoring, evaluation and feedback to strategic plan
- Environmental goals integrated with planning processes
- Accelerate attainment rather than change direction
- Lead by example – show good practices in govt. first
- Try things out and see what works – e.g., fossil fuel-free community
- Recognize that general public is not so concerned or knowledgeable about global issues, but are concerned about local ones such as too much traffic – build upon local understandings, expand understanding and educate.

Examples of Sustainable Development and Sustainable Transport:
- Emphasis on making transit work – performance goals
- Subsidy reduced but more efficient service
- Customer orientation – market surveys, info systems at stops, remove barriers, etc.
- Quality architecture and landscape design in stations, stops
- New towns at walkable densities, near transit, etc.
- Redevelop centers – recognize cultural and social importance.
- Build and rebuild to reduce negative impacts, e.g., underground roads, traffic calming
- Biodiversity protected through good planning, design, and maintenance
- Remove barriers for animals
- Careful choice and use of road materials; recycled materials used
- Alt. fuels; hybrids for buses and govt. fleets; Zeus project
- Truck improvements being sought – incentives for cleanup
- Rail improvements for freight
- Zero deaths safety plan – grade separation, traffic calming, in-vehicle protection, education

sustainability have made transport agencies directly responsible for the social, economic, and environmental performance of their systems. This situation is leading to a changed set of priorities, emphasizing access and exchange rather than trips per se. It gives greater attention to the less environmentally damaging modes, optimizes the use of existing capacity, and encourages improvements in vehicle technology. Sustainability considerations are reflected in the types of projects pursued, in project location decisions, in the quality of design and landscaping, and in the choice of materials used.

Specific Strategies for Sustainable Transportation

A variety of strategies are being pursued to increase the sustainability of the transport system. The overall approach was described as doing a number of small things as part of a larger, strategic program.

Land use–transportation strategies are one component of the overall effort and are used to reduce trip lengths and facilitate the use of transit, biking, and walking. Specific actions include revitalization of existing centers; infill and brownfields redevelopment; focus on high-density development near transit; encouragement of development in and contiguous to existing centers already served by transit; and planning for compact, mixed-use suburban development that is both walkable and sufficiently dense to support transit services. Policies discouraging single-use, stand-alone developments such as shopping malls also exist or are being considered.

Transit improvements are a second element of the European strategy for sustainability. Subsidies have been reduced in recent years but are still provided as a matter of social and environmental policy. Specific strategies to improve transit service include the development of extensive systems of priority lanes for buses, high-quality architecture and landscaping at transit stations and stops, planning for door-to-door service (including walk and bike access planning as part of transit planning), improved intermodal transfers, and high-quality customer information services.

Biking and walking are a third key strategy for sustainability. Extensive systems of bikeways and bike parking exist in each of the countries we visited, and bikes can be taken on many transit systems. High-quality pedestrian spaces are plentiful, and agencies are creating more by widening sidewalks, calming traffic, establishing vehicle-free or vehicle-restricted zones, and redesigning intersections to facilitate pedestrian crossings.

Although considerable emphasis is given to alternative transport modes, automobiles and the highways that drivers use are recognized as essential to travel in every country, so most efforts focus on sound management rather than on disincentives. Efforts attempt to educate the public about the impacts of automobile use as well as to encourage trip chaining and trip scheduling to reduce harmful effects.

Streets and highways are being built and rebuilt to reduce negative impacts, in some cases by undergrounding the facilities. Parking also is underground in many instances, and it is priced according to cost. Joint development of air rights and partnerships with developers and owners of nearby properties are being used to help finance these costly projects.

Biodiversity is being protected through good planning, location, design, and maintenance practices, for example, by providing animal crossing corridors and maintaining shoulders and medians as habitat. Bioengineering is used to create environmentally sound, aesthetic structures, and recycled materials are chosen to reduce environmental impact.

Logistics and operations improvements and ITS technologies are increasingly being applied to maximize the capacity of existing facilities, thereby reducing congestion, improving safety, and cutting down on the need for new facilities. Traffic calming is widely used in residential areas and on major streets in shopping districts. Installations used for that purpose are made of high-quality materials and are well designed and landscaped. Both speed management and traffic calming are part of programs aimed at a goal of zero highway deaths.

Car-sharing programs are being tried to give households the convenience of occasional access to automobiles without necessitating ownership or costly rentals. Alternative fuels are being tested both for transit and for personal cars, to reduce pollution and carbon emissions. In addition, ITS technologies are being promoted to help drivers plan trips more effectively, avoid bottlenecks, and travel at speeds that reduce congestion and improve safety. Improvements in truck technology are being sought, and incentives for reducing truck emissions include both emissions pricing and restrictions on the use of dirty trucks in sensitive areas.

Road pricing is discussed as a way to properly reflect the social and environmental costs of automobile use but approached with considerable caution, because there is not much public support for it. Fuel taxes are already several times steeper than those in the United States, but they are not hypothecated; instead, funds are allocated to reflect government policies and priorities.

Despite the emphasis on higher densities, transit, biking and walking, automobile use is increasing in all of the countries, though it is still far lower than in the United States. Whether the combination of strategies—providing good modal alternatives and managing traffic more effectively—will produce the desired results is being tracked through monitoring studies using specific performance measures.

LESSONS LEARNED: SUMMARY AND IMPLICATIONS FOR THE UNITED STATES

In the United States, barriers to greenhouse gas reduction include uncertainty about the problem and the best ways to address it, uncertainties about public support, and the lack of a national mandate for action. However, a number of states and localities are promoting sustainable development. Their efforts typically take a broad view of sustainability, defining it to include economic growth, quality of life, and social equity as well as environmental and energy considerations specific to issues on climate change. Transportation is an important element in many of these plans and is often coordinated with housing and jobs strategies. Improvements for transit, biking, and walking; advanced traffic operations and logistics; and promotion of new fuels and vehicles are among the transportation strategies emphasized.

The European organizations that were visited apply similar strategies, but supporting their efforts are strong policy commitments, government incentives, and new planning processes emphasizing collaboration and performance measurement. Tracking the comparative success of these efforts would be an important next step.

U.S. agencies also might take a closer look at the European emphasis on policy consistency and cooperative problem solving among agencies with somewhat different objectives. Efforts to harmonize policies, undertaken at the EU, national, state, province, and local levels, show promise as a useful way to resolve transportation–environmental conflicts and to speed the attainment of environmental goals in the United States.

Planning approaches that might be adopted in the United States include visioning processes to develop shared goals, strategic planning for both the long term and the middle term, and backcasting to identify strategies and steps needed to achieve desired results. Another approach with high potential for the United States is the use of performance standards along with monitoring and reporting on progress. Such activity could be coupled, as it is in the countries visited, with fiscal incentives for actions supportive of adopted goals.

ACKNOWLEDGMENTS

The work presented here was supported in part through an international scanning tour funded by the U.S. Department of Transportation. The group that conducted the scanning tour contributed ideas throughout this study: team leader Susan Petty, FHWA; Peter Markle, FHWA, Charles Howard, Washington DOT; Ysela Llort, Florida DOT; Frances Banerjee, City of Los Angeles Transportation Department; Alex Taft, Association of Metropolitan Planning Organizations; David Pampu, Denver Regional Council of Governments; and Jean Jacobsen, National Association of County Officials. Additional support was provided by TCRP and by the University of California Transportation Center.

REFERENCES

1. *Special Report 251: Toward a Sustainable Future: Addressing the Long-Term Effects of Motor Vehicle Transportation on Climate and Ecology.* TRB, National Research Council, Washington, D.C., 1997.
2. Sustainable Transportation Project, University of California Transportation Center, www.uctc.net/sustrans.
3. *Climate Change Presentation Kit* (CD-ROM). EPA/NASA/NOAA Partnership, Washington, D.C., 1999.
4. Intergovernmental Panel on Climate Change. Climate Change 1995: The Science of Climate Change. Contributions of Working Group I to the Second Assessment Report. Cambridge University Press, Cambridge, England.
5. *Transportation and Global Climate Change—A Review and Analysis of the Literature*. Research and Special Programs Administration, U.S. Department of Transportation, 1998.
6. Deakin, E., and G. Harvey. CO_2 Emissions Reductions from Transportation and Land Use Strategies—A Case Study of the San Francisco Bay Area. In *Global Climate Change: European and American Policy Responses, Proceedings of the Peder Sather Symposium* (J. Trilling and S. Strom, eds.), Center for Western European Studies and Royal Norwegian Ministry of Foreign Affairs, Regents of the University of California, Berkeley, 1993, pp. 93–109.

Publication of this paper sponsored by Task Force on Transportation and Sustainability.

[26]

THE DEMAND FOR CARS IN DEVELOPING COUNTRIES

EDUARDO A. VASCONCELLOS
Rua República do Iraque 1605, 04611-003, São Paulo, Brazil

(Received 1 November 1994; in revised form 26 March 1996)

Abstract—The paper analyzes the misunderstandings that have occurred in dealing with the private vs public transportation issue in developing countries. Both the economic view of the car as just a "free consumer desire", and the psychological views of the automobile as symbol of "freedom", "status" and "power" are criticized. An alternative sociological approach to the automobile is proposed, based on transport technology as embedded in the contemporary pattern of social reproduction. It is argued that the demand for automobiles, in addition to its utility, has been induced by urban, economic and transportation policies directed towards selected social sectors — the middle classes — who in turn perceive the car as an essential tool for their social reproduction. The same policies keep transit alternatives impractical. Consequently, there are important political (and not psychological) obstacles to alternative, less auto-oriented urban transportation policies.

1. INTRODUCTION

In the last three decades, the rapid growth of large cities in the developing world has been accompanied by a growth in car ownership, profoundly changing the traffic circulation conditions within these cities.* Transportation planners seem to be headed down a dead-end street: the increasing use of the car seems unavoidable, as it appears as something people "naturally desire" to have and use, and as something attached to symbols of status, power and self-affirmation. This approach to the problem has been leading to either conservative policies reproducing automobile dominance or to efforts aimed at generating voluntary shifts to public transportation. In the first case, conservative policies are justified as answers to consumer desires. In the second case, transport planners initially believe the shift is convenient and viable, however when faced with the driver's refusal to use transit often blame the "insensitive people" who supposedly persist in their "selfish" behavior of traveling by car. In both cases, automobile dominance appears as inevitable and alternative transit-oriented policies seem unrealistic.

I argue that basic misunderstandings have occurred in analyzing this problem, with important impacts on transport policy decisions in the developing world. The misunderstandings relate to both the supposed psychological characteristic of car ownership, as a "desire for power and status," and the economic assumption of car purchase as resulting from a "free consumer choice." On one hand, the psychological approach is superficial, having a weak explanatory power with respect to the actual motives that induce people to commit themselves to buy and use automobiles. Conversely, the traditional economic view is inadequate to understand developing countries' context, where market failures abound and most people have no choice other than use public transportation. I argue that the effective demand for cars, in addition to its utility, is influenced by urban, transport and economic policies, which shape space and confine transport choices. Moreover, demand is socially determined, in the sense of being related to how social groups and classes — especially the middle class — see and interpret the process of economic modernization in contemporary capitalist societies. In this respect, the automobile is perceived by the middle classes as essential to perform their desired daily activities, that is, to ensure their social reproduction. Therefore, reactions against car restraint measures have a political nature, rather than a psychological one, posing important obstacles to alternative, less auto-oriented policies.

*The average annual population growth for 27 selected cities in developing countries in the 1970–80 period was 4.9%. The corresponding figure for car ownership increase was 8.9% (18 cities). See Banjo and Dimitriou (1990), Table 2.1.

Traditional approaches being inadequate, and actual transport choices being confined, conservative policies that intend to generate an automobile-oriented space as an answer to "free consumer desires" are unjustified in the developing world, on technical and social grounds. Similarly, attempts to reduce car usage without a proper understanding of its valuation turn out to be a waste of time. Therefore, transportation planners in developing countries should avoid misinterpretations and search for the actual factors that make the car "demanded." A better knowledge would help them develop better socially and environmentally sound transport policies.

The main objective of this paper is to search for a better explanation of the automobile's importance in contemporary developing societies and to help explain why there are limits to alternative, less auto-oriented urban transportation policies. As a complement to prevailing views of the automobile, I propose a more revealing sociological approach, which I believe goes furthest in explaining auto demand. This approach sees transport technology embedded in the contemporary pattern of urban social reproduction and, for several reasons that will be discussed later, associates the use of the automobile with the new middle class lifestyle that corresponds to capitalist modernization in developing countries. Therefore, middle class interests and political power with respect to urban circulation, and the role of urban, transport and economic policies in shaping accessibility and general travel conditions, are considered highly relevant to an understanding of the demand for cars in the developing world.

Section 2 proposes a sociological view of the automobile by analyzing the supply and use of the transport system as a tool for understanding the mobility of social groups and classes. Section 3 considers the specific conditions of urban and economic development in developing countries, with an emphasis on the Brazilian case. Section 4 analyzes the demand for cars in these countries using the proposed approach. Finally, Section 5 summarizes the main conclusions and briefly discusses the limits and possibilities for alternative urban transportation policies.

2. APPROACHES TO AUTOMOBILE DEMAND

2.1. Conventional views of the automobile

The meaning of the automobile in modern and contemporary societies is multifaceted: no single facet alone can explain why this technology has so profoundly influenced our lives. Roland Barthes, a French philosopher, wrote of the automobile's cultural position as "the Gothic cathedral of modern times" (quoted in Sachs, 1992). Just as the cathedral is not merely a shelter, the automobile is not only a means of transport. One has to consider several aspects of the car culture when analyzing its importance in our society. I believe four conventional views encompass most conceptions about the car. The first conventional view — anthropological — is that of the automobile as symbol of power, status, wealth, related to the linkage that it can create between possession, public appearance and the wealth of its owner. The second view — political — corresponds to symbols of freedom and privacy. The third vision — psychological — corresponds to the ideas of youth and athleticism, self-reliance, and personal pleasure. The fourth view — economic — relates to the utility of the car, as a technology that allows for an unprecedented amount of mobility and provides the most efficient trip-chaining capability. According to this view, the decision to buy the car — and the resulting rejection of public transportation — is seen as a natural consequence of a rational comparison between benefits and costs of several consumption possibilities. This vision is supported by a large economic literature (Button, 1982; Small, 1992). It represents an important change because it supersedes the drawbacks of the superficial approaches described before, by proposing the actual utility of the car as the main factor explaining its valuation.

The anthropological, political and psychological views are integrated in the discourse of daily life, whether in people's conversations as consumers, in industrial establishment communication, or in the general media. Within the latter, the advertising and marketing sectors are areas that best understand and manipulate these multiple views while motivating automobile purchases. The anthropological (group symbolism) and psychological (personal pleasure) views are the most superficial in explaining auto purchase and use, despite corresponding to actual values and expectations clearly extant in certain social sectors or age-groups in contemporary industrial society. The political view (mobility as freedom) is more powerful as an explanatory tool, related to the very essence of the ideology of capitalism itself, and particularly of capitalist modernization.

Finally, the economic approach is indeed the most powerful, connected to the nature of the consumer society and the corresponding rational decisions. It is largely used in the official discourse, as supporting automobile-oriented policy decisions.

Yet all these views are insufficient in explaining the degree of either automobile purchase or daily usage in the developing world. First, the expansion of the automobile industry and the large volume of automobile purchases are not solely a consequence of these marketing efforts or of the industry's communication ability.* One cannot imagine that this success was caused by people's "irresistible" attraction to automobiles as goods. People, as political beings inside social classes and groups of an industrial society, cannot be seen solely as potential "consumers" who throw themselves blindly into the purchase and use of an offered commodity. The decision to buy such an expensive commodity, often requiring years of monthly payments, cannot be compared to the decision to buy a shirt. There must be other reasons behind the decision to buy the automobile technology and they must relate to both the specific conditions within which this technology is offered and the way people imagine the car in addressing their needs.

Second, the use of the traditional economic approach is inappropriate in developing countries. First, it groups everybody into a homogeneous set of potential "consumers." However, when considering the rather long distances to be traveled in large towns, and the prevailing income distribution, most people have no choice at all other than to use public transportation. Therefore, consumer choice models are meaningless. In addition, the use of this modeling procedure to analyze the case of the selected groups which can afford to have cars is also inadequate, for all sorts of market failures are found as underlying their supposedly enlightened and free decisions: there is no perfect competition; transport prices are not defined at the marginal level; consumers have insufficient information about prices and services; and the costs of transport externalities (accidents, congestion, pollution) are not properly imposed on those who cause them (Bayliss, 1992).

Third, the statement about the "obvious utility" of the car as explaining purchase is a truism. It either closes the analysis or throws the analyst in the wrong direction, that of concluding that if people "demand" cars we should therefore just provide them with cars, and also streets and parking facilities, without considering the complex relationship among land use, social and economic factors, and transport demand.

Therefore, it is important to analyze other dimensions of these processes to understand the motives that shape the significance of the automobile. Who sees the car as a utility and under what conditions? How the built environment and the prevailing transport infrastructure influence the need to buy and use a car? How the automobile compares to alternative public transportation means? To accomplish this task, a sociological approach to the automobile within contemporary society is needed. It complements the economic approach by combining the idea of the utility of the car with the urban, social and political contexts where the automobile is offered as a transport technology.

2.2. The proposed approach

A sociological approach to transportation differs from the simple economic approach in three ways. First, transport technologies are analyzed in respect to a given urban structure (Castells, 1977), within specific social, economic and political contexts. Second, the sociological view uncovers the large, uniform set of "consumers" and reveals the internal social and economic differences among social groups and classes, to analyze who is purchasing cars, for what purposes and within which conditions. Finally, the proposed approach relates the demand for cars with the automotive industry interests and needs, as part of larger industrial and economic development projects.

The sociological approach to the automobile will therefore be developed by analyzing transport patterns in contemporary urban areas of developing countries, as influenced by their specific

*Sachs states that"Car commercials use these significations to do their best to keep them alive... They... do not draw their power of persuasion just from the inventiveness of graphic artists, rather they found the very melody that originated with the bicycle and that accompanied the use of automobiles and motorcycles: the joy of minor emancipation thanks to easy mobility" (Sachs, 1992, pp. 108).Conversely, Stokes and Haller place a high importance in the power of the media to create dependency on cars and hence to revert the situation if another transportation policy is designed; this seems to be exaggerated according to the views expressed in this paper. See Stokes and Haller (1992).

social, economic and political conditions. Considering the profound differences among these countries, I focus on those that already have a large industrial base and already use a large fleet of motorized vehicles, such as Brazil, Mexico, Venezuela, South Korea, Indonesia.

In addition, the proposed approach will be developed around three main concepts that will be explained below. The first is the "reproduction" of social classes, as those consumption activities that allow people to ensure living conditions for themselves and their families inside a particular economic system (e.g. work, education, health, leisure, shopping). The second is the "means of collective consumption," as those facilities (schools, hospitals, streets, parks) and tools (vehicles) which allow people to accomplish these activities (Castells, 1977; Preteceille, 1981). The third is the "transport (and consumption) strategy," representing the decisions taken by individuals and their families to organize their daily movements, and the corresponding "time and space budgets": they represent the amount of time and money that have to be consumed, and the distances that have to be traveled by people within the living space, in order to perform the desired activities (Hagerstrand, 1987).

2.3. Reproduction of social classes

To continue to live and participate in society, people have to develop several activities. They also have to support other people who, for several reasons, depend on them. This set of activities, which are believed necessary, ensure people their "reproduction", both in individual terms and with respect to their social class or group, when essential needs and values are preserved or enhanced.

Reproduction requires consumption activities that can be biological — food, water, clothes, shelter and health services; intellectual — education, political and cultural activities; and social— as social gathering, leisure, sports and recreation.* The range of these consumption activities, and their specific nature, varies in time and space. They also vary according to social, cultural, racial, religious, ethnic and economic characteristics. The most important characteristic of modern urban consumption is that it has been increasingly subjected to collective rules — e.g. water and electricity distribution, sewage and garbage disposal, housing and transport, entailing important equity and welfare issues. The most important feature of activities themselves is that they are socially determined, that is, they are felt, expressed and satisfied according to the specific conditions faced by people in society.† Therefore, in such complex contemporary urban environments people will establish strategies of consumption, which will depend on social, economic and physical (urban) conditions. The strategy will include selecting transport alternatives, and defining the way vehicles, streets and sidewalks will be used. Consequently, the final patterns of consumption and transport will present sharp differences among people, social groups and social classes.

In capitalist societies, complex class structures pose specific problems and constraints to the analysis of the reproduction process. Social mobility and migration further complicate the analysis. In addition, developing countries present three specific characteristics as influencing transportation policies: the state plays a prominent role in shaping economic development, as opposed to free, deregulated economies in the industrialized world (Martins, 1985); the decision-making process is far from democratic, with ruling sectors having a tight control over policy decisions (Cardoso, 1977); finally, profound social discrepancies, economic instability and widespread informal economic activities are often the rule (Banjo and Dimitriou, 1990). Therefore, transportation policies have to be analyzed in the face of these specific characteristics, and have to consider the relative interests and power resources of all relevant actors.

2.4. The use of the space

While circulating, the user develops wanted or needed activities interrelated by a time and space network. The daily operation of this network is done by people considering personal techniques to optimize time and cost, implying space and time "budgets" (Hagerstrand, 1987;

*The forms relate to each other but this simple division is considered adequate for the purposes of the paper.

†There are no "natural" needs other than biological. Consequently, nobody has the "natural" need to buy a car. It is obvious that to eat, to drink water and to use clothes against inclement weather are actual needs, regardless of the economic system. But the point is that the way these needs are felt, expressed and satisfied are themselves socially determined, having therefore specific characteristics in every case. For a detailed analysis see Preteceille and Terrail, 1985.

Goodwin, 1981). The decisions are highly constrained by the social and economic characteristics of the users and their families: a strong positive relationship exists between income and quantity and diversity of trips, and also between income and use of private transportation. The activities are also constrained by the physical structure of the city (inherited from the past), by the physical disposition and time of operation of urban equipment and facilities and by the available transportation means.

Different physical, social, and economic conditions combine to provide a set of travel possibilities that are found in contemporary societies. Considering the widespread urbanization process, urban activity networks are becoming increasingly larger and motorized transportation means are becoming dominant. Among them, the private automobile is by far the most efficient in optimizing network performance, given its flexibility.

Many authors have acknowledged this particular condition of the automobile, although in different ways than proposed in this paper. Buchanan (1963, p. 195) says that "the attractions of private cars are so great... there can be no denying the difficulties of providing public transportation services so intrinsically convenient." Stone (1971, p. 99) recognizes that the attractiveness of the car can be explained because "the random route system approximates a door-to-door transportation system." Orfeuil (1994, p. 39) stresses for developed countries that "suburbanization is rational at the individual level." Finally, Weber (1991, p. 274) says that "people everywhere are attracted to cars not because they are lovable nor because they are prestigeful, but because they offer better transport services." These views reinforce the economic approach to the valuation of the automobile, as related to its general utility. However, as previously emphasized, these views do not directly address the convenience of the car on sociological grounds, that is, relating the use of this technology — and therefore of alternate technologies — to social classes and groups, their needs and interests, and the context within which the technology is offered: the car does not run alone and one has to analyze who, within which conditions, has decided to purchase and use it.

Other authors have scrutinized the issue through a sociological approach. Whitelegg, for instance, states,(the) car... shapes the whole lifestyle (and)... creates a subtle dependence on itself (italics added)... in many parts... it is not possible... to reject car ownership... the car is the center of a complex web of lifestyle organization which sets it apart from many other consumer durables (Whitelegg, 1981).

Reichman (1983, p. 100) also acknowledges that "many factors support the idea that the car implies a whole lifestyle and mobility strategy." Finally, Bernard and Julien (1974, p. 100) stress that the use value of the car "is not explained by an abstract preference for this mode... The needs of transferring (between points) entails a need to effect certain trips by car, without alternatives. It is the form that urban space is organized which confers the automobile its use value."

These views get closer to a more delimited sociological approach to the car because they manage to relate its ownership and use to social and political characteristics of the economic development model in which they are immersed, and to the urban structure as well. To understand the distribution of accessibility in the urban areas of developing countries, and the role played by the automobile, it is necessary to analyze how cities were adapted by the capitalist modernization process of the postwar period. This will be done by using São Paulo as a representative case.

3. URBAN GROWTH AND TRANSPORT NEEDS IN DEVELOPING COUNTRIES

Economic modernization entails profound changes in the technology of production, with large impacts on land use, urban structure and travel patterns. In addition, modernization requires new educational and technical skills. Considering the prevailing social and economic discrepancies among people in the developing world, the closed nature of the political system, and the limited amount of investments, modernization is not open to everybody and will be limited (in its full impacts) to selected groups.* In this context, the new pattern of investment generates conditions to improve and sustain economically only selected social sectors, which can join the new development cycle. These social sectors are called "middle classes." Therefore, instead of an economic concept, strictly income-related, I am using here a broader concept of middle classes, as those sectors that

*After four decades of economic modernization in Brazil, 63.1% of people with 10 or more years earning money in jobs (excluding social security beneficiaries), receive no more than three minimum salaries a month (approximately US$ 180)(IBGE, 1990).

have cultural, social and economic conditions to commit themselves to and to benefit from the modernization process.*

How do these middle classes relate to the recent urban changes in the developing world? Large cities in developing countries, within the last three decades, experienced intense growth related to economic modernization and internal migration. In addition, they have experienced an intense process of physical reconstruction and adaptation of the road system as cities were adapted to cope with the increasing number of automobiles. As shown for São Paulo, transportation planning directed some of its most important actions to the enlargement and adaptation of the street and highway systems to guarantee comprehensive spatial interconnections, while traffic management used modern technologies and tools to yield high levels of fluidity. At the political side, these changes were supported by the ideological commitment between the technocracy and the middle classes concerning the process of modernization, and required the improvement of conditions to use private transport (Vasconcellos, 1993). Accordingly, these policies were carried out along with a rapid increase in the number of automobiles, that jumped from 160,000 in 1960 to 1.9 million in 1980. The availability of cars, despite being limited to selected sectors, was facilitated by economic policies in two ways: the financing of car acquisition and the organization of vehicle "consortiums," where people belonging to a limited group contribute for two or three years with monthly payments to receive a car. Besides economic policies, urban policies also worked to ease car use. New land use laws allowed the building of middle class housing and apartment complexes, in newly occupied areas where the state provided adequate infrastructure and appropriate traffic and parking conditions. Conversely, transit captive users (the majority) were subjected to poor transportation conditions.

The reshaping of the urban space corresponded to a new lifestyle, characterized by new, increasingly complex, patterns of consumption and social relations. People's daily activity networks have increased and diversified, with new destinations at often greater distances. This diversification occurred within all personal networks, as capitalist modernization has transformed all social and economic life: in São Paulo, this change led to a 122% increase in the average number of motorized daily trips between 1967 and 1977, as opposed to a 45% increase in population. However, the largest increase has occurred with auto trips, whose participation in the total number of trips rose dramatically: in this period, they almost tripled, compared to a 100% increase in bus trips (CMSP, 1978). Between 1977 and 1987, private transportation share again increased more than public transportation (36 vs 6%) (Table 1).

The large increase in auto trips by limited sectors was related to broader economic and urban changes that deeply affected middle class lifestyles: the daily activity network incorporated new trips, primarily related to private education, private health care, sports, leisure and shopping, with a profound impact on transportation needs. Before, these activities were not done at the same level of intensity, or they were performed free, often within walking distances: before modernization, most middle class children attended neighborhood public schools and used local public health services, and had their leisure playing on the street or on nearby, empty lots. Shopping was done on local small markets and stores, and distant out-of-town trips were limited to national and school holidays, most of them made by intercity bus or railway. Modernization increased

Table 1. São Paulo metropolitan area. Modal split changes, motorized trips, 1967–1987

	Daily motorized trips (1000)					
Mode	1967	%	1977	%	1987	%
Public	4894	68.1	9759	61.0	10343	55.0
Private	2293	31.9	6240	39.0	8473	45.0
Total	7187	100.0	15999	100.0	18816	100

Source: CMSP, 1988.

*In economic terms, there are no absolute figures to identify the middle class in countries such as Brazil. In addition, there are intermediate subgroups ("lower," "middle" and "upper" middle class), which further complicates the analysis. For the purposes of the paper, I will consider the middle classes as those social sectors which have family monthly incomes higher than 15 minimum salaries (approximately US$ 900 at the time of the Origin–Destination survey, in 1987). In the Sao Paulo Metropolitan Region, this accounts for about 25% of the population (see Table 1).

activities, distances and costs, both the direct costs (payments for the new services) and the indirect ones, related to the means of transportation needed to accomplish these new activities. Now middle class children go to private schools,* often located far away from their living place and requiring escorted automobile trips. Private medical services are also spread over the space, shopping is increasingly concentrated in large regional shopping centers,† and streets are closed to leisure activities, as parked and passing cars occupy all available space: leisure is provided in private clubs or in shopping centers, and in the few remaining large regional public parks. In addition, weekend trips are now a common form of leisure, supported by a modern highway system linking the São Paulo metropolitan region to other cities.‡ These new forms of consumption derive from the "commodification" of social relations proper to capitalist modernization.§ The most important consequence of this process in the Brazilian case was the definition of a dividing line between the middle classes and the working classes, to enhance their social and political differences: working classes continue to walk to nearby public schools and medical centers (no longer used by most of the middle class), use transit in longer trips and have most of their leisure playing on the street or on nearby empty lots. The differences in using the space are summarized by differences in mobility rates (Table 2) and trip diversity (Table 3).

Table 2 shows that mobility increases sharply with income. Mobility rates for higher income groups are twice as much as those for lower income groups (all trips), and more than four times higher when just motorized trips are considered. The data also show that in 49% of the families people make less than one motorized trip per day. This increased mobility translates into a more diversified travel pattern for the highest income levels, where frequent additional trips (leisure and

Table 2. São Paulo metropolitan area. Mobility rates, all trips, 1987

			Mobility rate (trips/person/day)	
Family income*	Autos/house	Share in pop. (%)	All trips	Motorized trips
< 240	0.13	20.8	1.51	0.59
240–480	0.29	28.1	1.85	0.87
480–900	0.57	26.0	2.22	1.24
900–1800	1.01	17.2	2.53	1.65
> 1800	1.61	7.9	3.02	2.28

*Estimated considering that one minimum wage was approx. US$ 60.
Source: CMSP, 1988.

Table 3. São Paulo metropolitan area. Mobility rates by purpose and income, 1987 (all trips)

			Mobility rate †	
Family income*	Autos/house	Share in pop. (%)	Work business school	Other‡
< 240	0.13	20.8	0.61	0.17
240–480	0.29	28.1	0.84	0.16
480–900	0.57	26.0	0.99	0.18
900–1800	1.01	17.2	1.14	0.27
> 1800	1.61	7.9	1.35	0.46

*Estimated considering that one minimum wage was approx. US$ 60.
†Trips per person per day; it does not include home returning trips.
‡Shopping, health and leisure.

*While public schools are free, private schools cost about US$ 400 per child, per month. While the number of children attending public schools in the Sao Paulo Metropolitan Region increased 38% between 1978 and 1990, the corresponding figure for private schools was 122%. (SEADE, 1991)

†From 1966, when the first shopping center was opened to the public, to 1990, 385,000 m^2. were built, offering 28,050 parking spaces, used by approx. 141,000 automobiles per day. See ABRASC, 1996.

‡For example, trips to the beach (50 miles long) are made by more than 100,000 automobiles on regular weekends and more than 250,000 on Carnival and 4–5 day summer holiday weekends. The automobile AADT increased 152% from 1972 to 1990, as opposed to a 100% increase in the population of the metropolitan area. See DERSA, 1992.

§The broadest economic changes characterize what may be called the 'commodification' of social relations: activities (e.g. leisure) and services (e.g. health care) that were free of charge enter the money circuit, being 'commodified', and people have to pay for them. This process radically changes the relationship among the state (as regulator), the private sector (as supplier) and the user (as consumers).

shopping) are possible. The different travel patterns are associated with different forms of using the space, as the use of transportation means is also highly differentiated (Table 4).

Table 4 shows that the three lowest income sectors travel predominantly by foot or public transportation. Conversely, people at the two upper income levels make motorized trips predominantly by car. These latter households account for approx. 25% of the total number of households in the metropolitan area and represent the social sectors for whom the car is essential.

Finally, journeys to school are also a clear demonstration of differences in using space (Table 5).

It is also important to compare travel conditions between modes. In this respect, car and buses (the leading public transportation alternative) have been remarkably different (Table 6). In 1987, average travel time by car was half that of the bus. In addition, access times to either buses or parked vehicles, which play an important role in defining actual accessibility conditions, were also remarkable different. Finally, the percentage of bus lines with unacceptable loading conditions in the peak-hour (more than 6.2 passengers/m^2) varied between 32 and 84% in the city (CET, 1984).

A portrait of differences among income groups can be seen by analyzing examples of daily activity networks. Shown below are daily trip patterns of two families with very different social and economic characteristics. The first is a working class family, with a monthly income of about US$ 360, and no car. The second is a middle class family, with monthly income of US$ 2400 and two cars. Despite the large diversity found in all families, these two are representative of overall patterns.

Table 7 shows the large diversity in activities and travel conditions. The middle class family made an average of 2.9 trips per person, compared to 1.6 by the working class family. Trips for the middle class family included escorting a child to attend a special class 2.9 km away, shopping

Table 4. São Paulo metropolitan area. Modal split and income, 1987

Family income*	Share in pop. (%)	Modal split (% of trips)		
		Public	Private	Foot
< 240	20.8	37.3	8.8	53.9
240–480	28.1	40.1	13.3	46.6
480–900	26.0	39.6	24.6	35.8
900–1800	17.2	33.3	41.4	25.3
> 1800	7.9	19.6	66.0	14.4

*Estimated considering that one minimum wage was approximately US$ 60.
Source: CMSP, 1988.

Table 5. São Paulo metropolitan area. Journey to school according to the highest and the lowest income levels, 1987, selected districts*

Monthly family income†	Transportation mode (5)			Average distance (km)
	Foot	Public	Private	
< 240	97	3	—	0.65
> 1800	18	—	82	2.94

*Three districts for every income level (35 households each). Children under 16 years old only.
†In US$.

Table 6. São Paulo metropolitan area.Access time and travel time by mode of transport, 1987

Mode	Access time (min)	Total travel time (min)
Auto	2	23
Taxi	3	25
Bus	13	55
Metro	14	70
Train	20	85

Source: CMSP, 1988.

away from home and carpooling with other middle class families to transport children to a private school 5.7 km away, plus leisure at night 3 km away from home. Trips for the working class family were restricted to a work trip (the father), a visit by foot to a neighborhood doctor (the mother) and walking to the nearby public school (the children). The availability of a housemaid in the middle class household allowed the mother to stay away while keeping a 9-yr old son at home. In the working class family, the 1-yr old daughter was left at home with an older brother (12-yr old) while the mother was away at the doctor's. In 82% of the time spent by the working class family, the middle class family covered three times as much distance. The corresponding average speed is about four times higher. The most important conclusion is that the attempt to replace auto trips by bus (or non-motorized) trips for the middle class family would make the daily schedule infeasible,* hampering the performance of activities which are perceived as necessary for their reproduction as middle class people: to accept such changes the father would take one hour to get by bus to the workplace, the children would have to attend local public schools, shopping would have to be done by foot nearby, night leisure would no longer be possible and the housemaid would probably be fired. By doing so, these people would no longer "be" middle class people, exactly what they will never accept.

The data analyzed so far demonstrate that activities and travel conditions in the city vary remarkably with income. The most important point with respect to Brazilian conditions (and other developing countries) is that actual conditions establish a dividing line between those with and without access to private transportation. The access to the automobile, limited to selected sectors, generates large differences in accessibility, convenience and comfort, as opposed to public transportation. Accordingly, any automobile users who attempted to replace auto trips by bus trips would face unreliability and discomfort, and would significantly increase total travel time required, rendering infeasible the current middle class daily schedules.

4. THE DEMAND FOR CARS

The analysis made so far helps us to understand why the restructuring of space and the insufficiency of public transportation supply placed the automobile in a unique position, as the sole means of transportation capable of guaranteeing a minimum level of efficiency with respect to existing public transportation. Therefore, the decision to use the car, for those who could afford it, was rational, but deeply influenced by prevailing conditions. In these specific contexts, the purchase and use of the car have therefore to be seen primarily as a class decision, not an individual one in the narrow sense of the term. Even if personal preferences are activated to select the product (e.g. color, size), the major decision is to buy (and use) the specific technology called automobile, as opposed to alternative technologies. This decision is socially determined, in the sense of fulfilling the perceived needs of a particular group, the new middle class created by modernization. If other urban, transport, social and political conditions were present, other purchasing decisions would be made. Consequently, in sociological terms, one can say that the automobile turned itself into a means of class reproduction, a vital tool for the very existence and reproduction of the new

Table 7. Daily travel patterns of two different families (examples)*

Data	Middle class family	Working class family
Persons	7 (one housemaid)	5
Autos	2	0
Income†	2400	360
Trips	20	8
Modes used	Car; carpool; foot	Bus; foot
Purposes	Work; school (private); shopping; leisure (at night)	Work; school (public); doctor
Distances (km)	51.8	16.3
Time (min)	210	255
Speed (km/hr)	14.8	3.8

*Examples drawn from the 1987 O–D survey (25,000 households).
†Monthly income in US$.

*The automobile chaining capability plays an important role: in selected districts, for the three highest income levels, automobiles account for a minimum of 40% and a maximum of 80% of chained trips.

middle classes generated by the income concentration process. To "be" a middle class citizen required the performance of a set of new (commodified) social, cultural and economic activities, whose time–space optimization relies on the car: as previously shown, the attempt to perform these activities by bus (or non-motorized means) would result in disaster. Therefore, for the middle classes described in this paper, the decision to buy the automobile is similar to decisions about enrolling children in private schools, paying private medical care, studying foreign languages, going to restaurants, making weekend trips and even travel abroad. These are all class decisions, to fulfill activities that are believed to maintain and reproduce people in the way they think they have to reproduce themselves and their children, to remain members of a class, or to achieve social mobility. Moreover, these activities are also believed necessary because they clearly separate the middle classes from the working classes. Considering prevailing spatial arrangements and the supply of public transportation, the automobile is the sole technology that can combine these activities efficiently, in a 'trip chaining way,' and the decision to buy the car follows logically.

In addition, the class nature of the choice is reinforced by the non choice condition imposed to most of the people: since its inception, the Brazilian automotive industry was never organized to be a mass industry, as the average income of most people is totally incompatible with automobile prices: the monthly minimum wage has been around US$ 60 in the last two decades, while the less expensive cars have been priced between US$ 6000 and US$ 10,000.* Therefore, most Brazilian people cannot even think about buying a car. Even if theoretically a 'desire' to buy an automobile may appear, it is almost as possible as to buy a boat or a plane, and impossible desires cannot be considered for policy analysis.

4.1. State, industry and middle class interests

The sociological approach to the automobile has to be taken also in face of the role of the automotive industry in contemporary capitalist economies, through a macro-economic approach to the car industry. Here the sociological and macro-economic approaches to the automobile get together: the expansion of the Brazilian automotive industry would not have happened without an actual market for its products. As stated by Whitelegg, consumer demand for cars in many different societies cannot be adequately understood outside of an appreciation of state support for motorized transport and changes in the built environment which generate further demand for car ownership.(Whitelegg, 1981, p. 155)

The market was then organized through the generation of conditions for the emergence of a middle class which corresponded to the ideological and economic project of the proposed development model: on political grounds, the new middle classes played a vital legitimizing political role, especially during the authoritarian period, from 1964 to 1982. Key to the objective of market generation was the income concentration process (Bacha and Klein, 1989). In addition, a built environment which enhances the need for car use was also generated: large investments in new urban and regional highway systems were provided, while railways were dismantled and public transportation systems were kept subjected to permanent crisis (Barat, 1991; Figueroa, 1991). The development project envisioned middle sectors committed to a new lifestyle who would then use all tools thought to be necessary to drive to a better future. In this sense, one can talk about a *symbiosis* between the middle class and the automobile, for one cannot survive without the other. It is the longest lasting and happiest marriage of our times. In addition, one can talk about a mutual sustaining relationship between the built environment and the automobile. It is like a *trap*, within which the middle class was born and guided from its inception to understand that the dream of social mobility would be possible only with the automobile.

4.2. Automobile demand in other countries

In developed countries, the same relationship between transport policies, urban structure and automobile dependence has been emphasized by several authors. For example, McShane and Koshi (1984, p. 104), analyzing the post — WWII increase in auto use in Sweden, stress that the publicly financed housing complexes were "located on the fringe of the cities, far enough from downtown centers to make the provision of other than public transport service expensive and

*Considering permanent inflationary processes, it is difficult to arrive to a precise figure. However, the mentioned price reflects actual market conditions during the 1970–90 period.

impractical. Consequently, car ownership became a necessity... (this) mirror what was happening in the U.S. at that time and likewise credited with much of the responsibility for accelerating American auto dependency." This example suggests that inferred consumer desires should be taken more carefully when analyzing policy decisions. As pointed out by Pucher, observed travel behavior is not simply the outcome of consumer sovereignty. Indeed, in some cases, policies are so extreme, and choices so restricted, that the resulting travel behavior — in particular, mode of travel — is practically preordained.*

In the US, the conjunction of economic prosperity, automobile subsidies (direct and indirect), urban residential policies and public transportation policies made transit impractical and automobile 'highly demanded'.† The built environment constrains the use of the car, as a survival tool, and relying on public transportation is irrational. As stressed by Rosenbloom when analyzing why North American families need cars, "...it is hard to see how any other option...[than the car]... could serve the complex travel needs of such families." (Rosenbloom, 1991, p. 39)‡

With capitalist European countries, the post WWII increase in auto ownership also led to pressures on more space availability for cars.§ Although the rate of automobiles per capita is, on average, about 50% less than that of the U.S., the use of the automobile is very different, in the face of traffic restrictions, higher gasoline prices, high taxes on car sales and high parking fees.¶ In addition, transportation policies share a more transit-oriented approach, that derives from both different historical and urban conditions, and from different political environments. Large public transportation systems — subways and railways — have long provided good transportation services. Moreover, governments have, for a long time, subsidized public transportation and supported non-motorized means such as the bicycle. In addition, the much higher urban land use densities historically found in Europe were maintained in new urban developments, for governments have been consistently supporting high-density occupation, served by public transportation means.|| These policies reflect a different approach to transportation policies, and can be said to be directly related to a more complex political arena, where radically different views of society struggle around policy outcomes.

Developing countries pursuing industrialization and economic modernization, in spite of different social and political conditions, have been experiencing the same process that occurred in Brazil, with similar results in respect to urban transportation systems.** In these countries, new middle sectors have been generated and the automobile technology has been replacing local transportation means. Car ownership has been confined to selected sectors and public transportation services have been subjected to permanent crisis (Figueroa, 1991). Sharp differences in trip patterns — with and without automobile — have been identified, reproducing the same differential forms of use of the space for circulation (Banjo and Dimitriou, 1990). A similar process can be identified in the former socialist countries (Pucher, 1993) as well as in China,†† where public (and non-motorized) transportation systems have been losing space to cars.

Regardless of their deep social, political and economic differences, automobile demand in these countries should not be seen as a consequence of a generalized 'natural consumer desire', or as an 'inevitable outcome of progress.' The demand for cars should instead be related to the specific social and political conditions found in the countries, and to the regional, urban and transport policies that shape space and transport supply. The utility of the car stressed by the economic approach has to be complemented by the sociological analysis of who is buying automobiles, for what purposes and within which social and economic conditions. As with the case of Brazil, the increase in automobile demand in such countries, as South Korea, Indonesia or China, cannot be attributed to a sudden 'desire' for this technology, divested from all the social, political

*Pucher, 1988, p.509.

†See, for example Barrett (1983) and Vuchic (1984). The former discusses the abandonment of public transportation systems and the latter stresses that automobile oriented policies pursued in the country were "by no means spontaneous desire of people...the choice (for the automobile) has been strongly influenced by biased policies..."(p.128)

‡She makes an interesting description of the U.S. case. However, she does not acknowledge that these families need the car because the built environment makes the auto indispensable and because there has been a powerful economic and political movement to support the automotive development model.

§Sachs, 1992; Dupuy, 1975.

¶Pucher, 1988.

||Pucher, 1988.

**For the Asian case, see Spencer and Madhavan (1989) and Dick and Rimmer (1986).

††The growth of cars in China can be viewed in the traffic safety study made by Navin *et al.* (1994).

and economic conditions within it is offered as an alternative. Particularly in the poorer countries, and in countries with highly unbalanced income distribution, the access to the automobile will be related to the creation of limited middle class sectors, who will be able to join and benefit from the economic modernization. They will see the automobile as vital for their reproduction as social classes, and they will make large efforts to purchase it, establishing a mutually reinforcing relationship with the automotive industry. Accordingly, they will try to use cars as extensively as possible, reacting against old, slower transport modes hampering the efficiency of the automobile, as well as against any restraint by traffic authorities. In these countries, modernization will change travel patterns and a new balance between modes will result, changing living and travel conditions for everybody. The final position of the automobile in the transport market will be related to how middle classes will influence policy decisions and to how excluded sectors will manage to have their travel needs attended to by existing non-motorized and motorized public transportation means. The costs and benefits of all these complex changes have to be analyzed in every case, for every particular condition. This challenge could be considered by transport analysts, while using the sociological approach suggested here to better understand recent urban and transport changes.

5. CONCLUSIONS

Misunderstandings about the factors that influence car demand in developing countries have been leading either to conservative automobile-promoting policies or to frustrated efforts to promote voluntary shifts to transit. The inadequacy of prevailing views requires that better explanations be suggested, to support improved policy decisions.

The paper proposes an alternative sociological approach, that is intended to better explain the demand for cars in developing countries. It complements the economic approach based on the utility of the car, by analyzing social, economic and political conditions that surround both the supply of this specific technology in developing countries and the individual decisions to purchase it. In this respect, the paper stresses that the valuation of the car in developing countries cannot be fully explained by psychological, political or anthropological approaches of 'status' and 'power', neither by the traditional economic approach of free consumer choices.

First, simple explanations of the success of the car related to issues of status, power, sexual strength and privacy are superficial.* Even considering that the car may be used as symbol of status in particular conditions, the decision to buy the car technology is socially determined and is seldom based just on the interest of demonstrating status or power: there are no psychological motives able to explain why so many people commit a significant share of their budgets, for such a long time, to buy such an expensive technology. These single factors are secondary and derive their explanatory power due to the large visibility from, and manipulation by, the media. Therefore, their use occurs in a parasitical manner in respect of the actual motives by which people in developing countries decide to purchase and use cars.

Second, the traditional economic approach neglects social differences and the actual political, economic and urban conditions in developing countries. Considering the long average distances found in large cities and the income distribution patterns, most people have no choice other than to use transit, making choice models meaningless. In addition, market failures abound in the transport sector and the option for the automobile is highly biased by the lack of proper alternatives. As shown in the case of São Paulo, the differences in efficiency and convenience between cars and buses are so great that the choice for the automobile appears inevitable for those who can afford it.

Therefore, a better explanation requires the analysis of specific social, economic and political conditions in developing countries which lie behind recent processes of capitalist modernization. In the contexts described in the paper, this economic modernization induced a new lifestyle for selected social sectors — the middle classes — that defined a clear dividing line with respect to working classes. This new lifestyle radically changed mobility needs, requiring increased incomes

*The ownership of a car can be seen as a symbol of status in some portions of the lower middle class; the make of the car can also be seen as a status symbol for high income sectors; but these features do not change the much higher importance of the automobile as a means of class reproduction for most middle class people.

and better transportation means, significantly increasing the use of the automobile by the middle classes. In this respect, the paper proposes that the valuation of the automobile is better explained by the sociological perception, by the middle class, of the car as a vital means to guarantee its reproduction as class, in a context where physical and social mobility are the prime objectives. The car is a tool whose use is deeply embedded in social, political and economic constraints: behind the wheel, rather than 'people' there are political beings with needs and interests, and with a definite view of society, as ideology. In addition, the valuation of the car is explained by the particular urban, economic and transport policies promoted in developing countries: these policies have been shaping the contemporary space in the developing world in a way that induces the need for the car, while making alternate public transportation means impractical. Considering the prevailing built environment, the ease to use automobiles and the poor supply of public transportation means, the middle class has no alternative but to purchase and use the car intensively. If other conditions prevailed, fewer people in the developing world would commit a substantial part of their incomes to 'demand' cars, and the use of the automobile would be much more selective.*

Hence, the decision to have and use cars is based on its utility, rather than on symbols of 'status', reinforcing the economic approach to the demand for the automobile. However, the demand is deeply constrained by the actual physical (urban space, transport supply), economic (transport costs and convenience), and political (class consciousness) conditions that surround the decision, enhancing the need for a broader sociological approach.

With respect to actual policy decisions, one can conclude that conservative policies designed to promote an automobile-oriented space as answers to 'free consumer choices' are unjustified in developing countries: the effective demand for cars is limited to selected sectors and highly influenced by urban, economic and transport policies. To interpret this biased demand from limited sectors as representing society's desire is technically mistaken and socially unacceptable, considering that prevailing urban and economic conditions transform most people in to public transportation captive users.

In respect to alternative transport policies, one can also conclude that the symbiosis between the middle class and the automobile in prevailing conditions presents enormous obstacles to changes in the current accessibility patterns in developing countries' cities. If the automobile is a means of reproduction of this class, there will be an immediate reaction against any limitation in its use. The reaction is therefore political, rather than psychological. Even when this resistance is exhibited on an individual basis — and reinforced by the prevailing superficial psychological and anthropological views — the underlying rationale represents class consciousness about the vital role of the car for its reproduction. In this respect, 'well intentioned' but wrongly approached policies to promote voluntary shifts to transit, or to increase auto occupancy, turn out to be a waste of time.† This reaction is reinforced by the middle class adherence to the ideology of increased social mobility, which translates into a desire for increased mobility in space and freedom to circulate. In addition, the accumulation rationale of the economic system set priorities for the automotive industry and its economic linkages. Consequently, powerful interests permanently induce an auto-oriented space and create obstacles to alternative technologies (Winner, 1977).

As a consequence, the actual social and environmental costs of cars seem very difficult to impose except in selected and limited sites, and the corresponding organization of a transit-friendly system, coupled with efficient non motorized transportation subsystems, will face enormous obstacles. There are plenty of traffic management alternatives to restrain the abuse of the automobile and to promote a more balanced urban traffic (May, 1986), however they will always confront reactions which are difficult to overcome. In the middle term, only the middle class concern with environmental issues appears as potentially encouraging significant changes, through the acceptance of travel pattern changes and automobile restriction measures intended to improve quality of life.

Finally, the sociological approach to car demand can be used to analyze similar processes in developed countries, in former socialist countries and in developing countries as well. The

*For instance, the positive relationship between urban density and use of public transportation means was analyzed in developed countries by Newman and Kenworthy (1989).

†In São Paulo, the attempt to shift auto trips to special privately organized bus services has just failed, after two other unsuccessful attempts made with public buses (the first in the 1970s and the second in the 1980s). Similar attempts to stimulate carpooling in worktrips in the 1980s have also failed. See CMTC (1992).

challenge for transport analysts is to reconsider existing interpretations of the demand for cars as expressing natural consumer desires, and search for better explanations on how the social, economic and political conditions in these countries influenced people's transport decisions.

REFERENCES

ABRASC — Associação Brasileira de Shopping Centers (1996) Dados sobre shopping centers no Brasil (internal report). São Paulo

Bacha, E. L. and Klein, H. S. (1989) *Social Change in Brazil: 1945–1985, The Incomplete Transition.* University of Mexico Press, U.S.A.

Banjo, G. A. and Dimitriou, H. (eds) (1990) *Transport Planning for Third World Cities.* Routhledge, U.S.A.

Barat, Josef (1991) *Transportes urbanos no Brasil: Diagnóstico e Perspectivas, Planejamento e Políticas Públicas*, no. 6, pp.75–96. Planejamento e Políticas Públicas, Brasilia.

Barrett, Paul (1983) *The Automobile and the Urban Transit.* Temple University Press, U.S.A.

Bayliss, Brian (1992) *Transport Policy and Planning — An Integrated Analytical Framework.* Economic Development Institute of the World Bank, Washington.

Bernard, Jean-Claude and Julien, Nicole (1974) Pour un analyse des transports urbains. *L'architecture d'aujourd'hui* **172**, 98–104.

Buchanan, Colin (1963) *Traffic in Towns.* HMSO, U.K.

Button, K. J. (1982) *Transport Economics.* Heinemann, London.

Cardoso, F. Henrique (1977) *O modelo politico Brasileiro.* Difel, Rio de Janeiro.

Castells, Manoel (1977) *The Urban Question.* University of California Press, U.S.A.

CET — Cia de Engenharia de Tráfego (1984) *Probus.* São Paulo.

CMSP — Companhia do Metropolitano (1978) *Pesquisa Origem-Destino 1977.* São Paulo.

— Companhia do Metropolitano (1988) *Pesquisa Origem- Destino 1987.* São Paulo. Author please complete

CMTC — Cia Municipal de Transportes Coletivos (1992) *Onibus Especial* (internal report). São Paulo.

DERSA—Desenvolvimento Rodoviario SA (1992) *Estatistica de Transito Rodoviario.* São Paulo.

Dick, H. W. and Rimmer, P. J. (1986) Urban transport planning in southeast Asia: a case study of technological imperialism? *International Journal of Transport Economics* **13**(2), 177–196.

Dupuy, G. (1975) *Une Technique de Planification au Service de l'Automobile: Les Modeles de Trafic Urbain.* Institut de Urbanisme de Paris, Paris.

Figueroa, Oscar (1991) La crise de court terme des transports en commun: l'experience de San Jose du Costa Rica. *Recherche Transports Securite* **31**, 47–56.

Goodwin, P. B. (1981) The usefulness of travel budgets. *Transportation Research A* **15**, 97–106.

Hagerstrand, T. (1987) Human interaction and spatial mobility:retrospect and prospect. In *Transportation Planning in a Changing World*, eds P. Nijkamp and S. Reichman. GOWER/European Science Poundation, Netherlands.

IBGE — Instituto Brasileiro de Geografia e Estatística (1990) *Pesquisa Nacional por Amostra Domiciliar.* Rio de Janeiro.

Martins, Luciano (1985) *Estado Capitalista e Burocracia no Brasil pós-64.* Zahar, Rio de Janeiro.

May, A. D. (1986) Traffic restraint: a review of the alternatives. *Transportation Research A* **20**(2), 109–121.

McShane, Mary P. and Koshi, Masaki (1984) Public policy toward the automobile: a comparative look at Japan and Sweden. *Transportation Research A* **18**, 97–109.

Navin, Francies, Bergan, Art and Qi, Jinsong (1994) Road safety in China. Paper presented at the 73rd TRB Annual Meeting, Washington.

Newman, Peter W. G. and Kenworthy, Jeffrey R. (1989) *Cities and Automobile Dependence: a Sourcebook.* Gower Technical, Australia.

Orfeuil, J. Pierre (1994) *Je suis l'automobile.* Editions de l'aube, Paris.

Preteceille, Edmond (1981) Collective consumption, the state and the crisis of capitalist society. In *City, Class and Capital*, eds M. Harloe and E. Lebas. Holmes and Meyer, 1–16, U.S.A.

Preteceille, E. and Terrail, J. P. (1985) *Capitalism, Consumption and Needs.* Blackwell, U.K.

Pucher, John (1988) Urban travel behavior as the outcome of public policy: the example of modal-split in Western Europe and North America. *APA Journal* **Autumm 1988**, 509–520.

Pucher, John (1993) The transport revolution in Central Europe. *Transportation Quarterly* **47**(1), 97–113.

Reichman, Shalom (1983) *Les Transports: Servitude ou Liberte*? Presses Universitaires des France, Paris.

Rosenbloom, Sandra (1991) Why working families need a car. In *The Car and the City — The Automobile, the Built Environment and Daily Urban Life*, eds Martin Wachs and Margareth Crawford. University of Michigan Press, U.S.A.

Sachs, Wolfgang (1992) *For Love of the Automobile.* University of California Press, U.S.A.

SEADE — Fundação Estadual de Análise de dados (1991) *Anuário estatístico do Estado de São Paulo.* São Paulo.

Small, Kenneth A. (1992) *Urban Transportation Economics.* Harwood Academic Publishers, U.S.A.

Spencer, A. H. and Madhavan, S. (1989) The car in southeast Asia. *Transportation Research A* **23**(6), 425–437.

Stokes, C. and Haller, S. (1992) The role of the advertising and the car. *Transport Reviews* **12**(2), 171–183.

Stone Talbot (1971) *Beyond the Automobile.* Prentice Hall, U.S.A.

Vasconcellos, Eduardo A. (1993) A cidade da classe média: estado e política de transporte em São Paulo. *Revista dos Transportes Públicos* **60**, 79–96.

Vuchic Vucan, R. (1984) The auto versus transit controversy: toward a rational synthesis for urban transportation policy. *Transportation Research A* **18**(2), 125–133.

Weber, Melvin (1991) The joy of the automobile. In *The Car and the City*, eds M. Wachs and M. Crawford. University of Michigan Press, U.S.A.

Whitelegg, John (1991) Road safety: defeat, complicity and the bankruptcy of science. *Accident Analysis and Prevention* **15**, 153–160.

Winner, Langdom (1977) *Autonomous Technology: Techniques Out of Control as a Theme in Political Thought.* M.I.T. Press, U.S.A.

[27]

Urban mobility in the developing world

Ralph Gakenheimer *

Department of Urban Studies and Planning, Massachusetts Institute of Technology, 10-403 Cambridge, MA 02139, USA

Abstract

Mobility and accessibility are declining rapidly in most of the developing world. The issues that affect levels of mobility and possibilities for its improvement are varied. They include the rapid pace of motorization, conditions of local demand that far exceed the capacity of facilities, the incompatibility of urban structure with increased motorization, a stronger transport–land use relationship than in developed cities, lack of adequate road maintenance and limited agreement among responsible officials as to appropriate forms of approach to the problem. The rapid rise of motorization presents the question: At what level will it begin to attenuate for given economic and regulatory conditions? Analysts have taken various approaches to this problem, but so far the results are not encouraging. Developing cities have shown significant leadership in vehicle use restrictions, new technologies, privatization, transit management, transit service innovation, transportation pricing and other actions. Only a few, however, have made important strides toward solving the problem. Developing cities have lessons to learn from developed cities as regards roles of new technologies, forms of institutional management and the long term consequences of different de facto policies toward the automobile. These experiences, however, especially in the last category, need to be interpreted very carefully in order to provide useful guidance to cities with, for he most part, entirely different historical experiences in transportation. Continued progress in meeting the needs of the mobility problem in developing cities will focus on: (a) highway building, hopefully used as an opportunity to rationalize access, (b) public transport management improvements, (c) pricing improvements, (d) traffic management, and (e) possibly an emphasis on rail rapid transit based on new revenue techniques.

1. Introduction

In the large cities of the developing world, travel times are generally high and increasing, and destinations accessible within limited time are decreasing. The average one-way commute in Rio de Janeiro is 107 min. In Bogota it is 90 min. The average vehicle speed in Manila is 7 miles per hour. The average car in Bangkok is stationary in traffic for the equivalent of 44 days a year.

* Corresponding author. Tel.: +1 617 253 1932; fax: +1 617 971 0421; e-mail: rgaken@mit.edu.

This is happening because vehicle registrations are growing fast on the basis of increased populations, increased wealth, increased commercial penetration, and probably an increasingly persuasive picture in the developing world of international lifestyle in which a car is an essential element. Accordingly, in much of the developing world the number of motor vehicles is increasing at more than 10% a year–the number of vehicles doubling in 7 yr. The countries include China (15%), Chile, Mexico, Korea, Thailand, Costa Rica, Syria, Taiwan, and many more.

What is the shape of increasing congestion and declining mobility? There are no widespread measures available for comparative purposes because decline in mobility is complicated. Congestion is always localized in time and space. A few things are nonetheless evident.

1. Congestion is reducing the mobility of auto users. It is clear by measures of traffic delay available and even by impressionistic evidence that in virtually all large cities of the developing (and developed) world congestion increasingly impedes mobility for auto users. The only exceptions are very poor metropolitan areas, some cities in the initial stages of relief from planned economy (e.g. Tashkent), and a very few with successful traffic management (of which the flagship example is Singapore).

2. Mobility is declining even more for public transport users. This is largely because transit routes characteristically follow the highest volume arteries, those most afflicted with congestion. Further, transit networks are usually dominantly radial, not permitting cross-town avoidance of congestion. Finally, transit users are not able to follow trip destinations that are displaced into the higher auto-accessible locations at the periphery because the transit network does not serve them. A policy emphasis on expanding the road system rather than improving transit often makes this quandary worse.

3. For the numerous individuals newly acquiring cars in the developing world, however, mobility is rising. This is simply because they are removing themselves from plight 2 above to plight 1, which is less severe.

The conflicting interest between group 3 above and the first two is, of course, one way to define the mobility problem. That is, no matter how bad congestion becomes, it is usually advantageous to use a car. It is a tragedy-of-the-commons situation.

This paper ranges among a series of topics that bear on mobility in cities of the developing economies. It is a search for elements of strategies to deal with rising motorization and falling mobility. In contrast to much of the literature, this paper focuses on personal mobility and access, rather than on their external impacts. The product will be not so much a list of recommendations as the threads of a process for thinking about this problem in the light of available evidence and candidate techniques from the developed economies.

2. Basic mobility issues

The first obligation of mobility in the developing city is to enhance the unique, essential functions of the large city. They are of special importance to a country whose central concern is economic development. Bangkok includes only 10% of the population of Thailand, but accounts for 86% of the country's GNP in banking, insurance and real estate, and 74% of it's GNP in manufacturing (Kasarda and Parnell, 1993). More broadly, large cities are sure to be centers of education, research, innovation of all sorts and the various aspects of globalization that are bringing the developing countries into the world-wide production system. The decline of mobility is damaging these roles significantly. Bangkok loses 35% of its gross city product in congestion.

The basic characteristics that differentiate the developing city from those of more advanced economies in regard to transportation are

1. Rapid pace of motorization. There is a significant portion of these cities where motorization is increasing at more than 10% a year. In China vehicles are increasing at 15% a year, automobiles at 25% a year. In Korea there was an annual increase averaging 23.7% for some 7 yr following 1985. Pace of motorization is important because related systems, such as transportation facility capacity and urban structure adjustments cannot keep up, resulting in enormous congestion. How else could Bangkok be more congested on a national average of 54 vehicles per 1000 population than American cities on national average 750 vehicles per 1000.

2. Travel demand that far exceeds the supply of facilities. High levels of congestion and high latent demand for travel is the result of motorization outstripping any possible expansion of highways. This condition exists in nearly all developing countries, except for a few very wealthy ones (c.f. some of the Gulf States) and some with such low initial motorization rates that the increase has not yet caught up with capacity. In some cases the prospects for privatization are sufficiently good and right-of-way acquisition obstacles so benign that the question of "how many highways to build?" is topical (viz. Guangdong, China). But there are not many areas in this situation.

3. High share of trips by public transit. Across much of the developing world urban vehicular trips are around 75% by transit. Exceptions include China, where a significant percentage are bicycle trips. This means that making public transport work has high priority and buses swamped in auto congestion is a difficult problem, but in most settings public transport is ridden with politics and institutional problems that make it very difficult to improve.

The split of spatial domain between the motorized population and the non-motorized, transit-using population is great. In cities ranging up to 250 persons per hectare (as in China) auto ownership and use are heavily constrained. As a result decentralization with motorization is explosive. Valued urban functions in the hands of the motorizing part of the population evacuate to the suburbs, leaving the city for only low income activities. Employment, increasingly in decentralized settlements is not accessible to non-motorized, lower income workers.

Under these circumstances the viability of public transport is particularly important. But public transport in every city is dominated by buses and they, as mentioned above, are generally more susceptible to increasing congestion than cars. There are possibilities of escaping this problem by assertive management of transit rights-of-way through such means as independent lanes or signalization that favors transit vehicles. The prospects for the improvement of mobility by this means are important, but few cities have been successful because under circumstances of increasing congestion the pressure to favor general use of the streets is high – automobile owners are a powerful lobby in most of the developing world.

4. Intense desire for auto ownership and use. According to government surveys, Chinese families are likely to be prepared to spend 2 yrs' income for a car that is expected to last for 10 yrs. (Americans spend about 27 weeks' salary.) Auto shows are thronged.

5. Urban structure incompatible with motorization. Residential densities in China are as high as 200 to 250 persons per gross hectare. (The Western European city is about 50 persons/gross hectare.) Street space is around 10% of the city surface (rather than 25% in the western city). Land

use is likely to be more mixed than in the western city and the average urban trip length much shorter (The average bike trip in Shanghai is 3.5–5 kms. in a city of 20 million people.) The lack of street space and parking results in explosive decentralization of land use.

6. Stronger land use/transportation relationship. Changes in the road system, such as through the construction of a new urban highway, has much more impact on urban development in a developing city simply because there are fewer high speed roads. The new one therefore provides more comparatively attractive access than in the developed city, where peripheral access is high in every direction. Also, more rapid urban growth (likely to be in the range of 5% per year) produces more rapid change in urban structure.

Further, in some parts of the developing world where motorization is rapid, governments have considerable influence, current or potential, to guide land use into mobility-friendly forms. This is partly because local government, as in China and Korea, is less divided within metropolitan areas. Unified metropolitan administration is important because small sub-metropolitan jurisdictions seldom take great interest in access.

7. Greater differences in vehicle performance. The wide variety of vehicle types on the streets presents difficult problems of efficiency and safety. Many cities have passenger vehicles ranging from human traction to high-speed sports cars, and every scale of freight vehicle. According to Darbera and Nicot, 1984 there are 16 modes of public transport on the streets of the cities of India. In China, while it is surely essential to assure the survival of adequate ways for bicycles, it is unquestionably inefficient for the street lanes to be divided into motor and non-motor lanes, especially because of difficulties of movements at intersections. All this does not deny the benefits of specialized vehicles for different roles, but it makes traffic management much more difficult than in developed economies.

8. Inadequate street and highway maintenance. Highways and arterials are often built by national agencies and maintained by local governments. No funding provisions are made for the maintenance, however, and the local government often has scarcely enough revenue to collect the trash. As a result, transport ways are often in very bad condition. Indeed, sometimes they are intentionally left that way because the local administration hopes the national agency will step in again when the deteriorated condition of the road is so bad that repair is in effect full reconstruction, once again recognized as a national responsibility.

9. Irregular response to impacts of new construction. In some countries new urban facilities are very difficult to build. Projects encounter strong resistance movements from impacted institutions and communities (especially in Latin America). In others, there is very little resistance (e.g. China). Air pollution is a matter of intense concern in certain cities (e.g. Bangkok) and very little in others (e.g. Cairo). There are indications that air quality is increasing as a concern in areas where it was not previously a major preoccupation. For example, at one point an issue of *India Today* (December 15, 1996) had a cover headline "Choking to Death: Polluted Cities," and a cover story titled "Gasping for Life," though the problem had gotten relatively little attention earlier. Though unexpected participatory politics occur everywhere, in the developing world they are especially unstable.

10. Fewer legal constraints on the use of new technologies. One of the strongest constraints on the introduction of new technologies in the West, for example, for driver advisory functions, is fear of legal suit. This concern is less problematic in most of the developing world, making innovation more feasible on this account.

11. Weak driver discipline in many countries. While driver discipline is equally strong or stronger in many East Asian countries than in the West, it is certainly weaker in most of the developing world. This is a problem for the implementation of many forms of traffic management. For example, transit-only lanes have been attempted in several cities where it was found that drivers would simply not respect them. But cities are learning. Bogota recently enacted counterflow lanes without barriers, resulting in generally satisfactory improvement.

12. Very limited agreement on planning approaches. Whereas the Western countries have cadres of engineers and planners with reasonably consistent perspectives on dealing with urban transportation problems (however much they may disagree on the details), the developing countries characteristically do not. They tend to borrow method and professional perspective from elsewhere and to have professional communities that are at crossroads of ideas, without stable commitments. This results in turbulence in the course of transportation problem solving, stalemates when trying to marshal strength to a particular solution and rapid change of strategies over time that keeps any strategy from having sound effect. This is a serious problem in transportation because there are so many alternative views. It presents an important need for professional education and leadership as a foundation for meaningful problem solving.

13. Capital is scarce and operating subsidies difficult to sustain. Cointreau-Levine points out that solid waste management consumes 20–50% of local expenditures in megacities (Cointreau-Levine, 1994). Some of those cities are virtually unsewered (e.g. Bangkok, Riyadh), and most of them have severe deficits of sewerage. These are circumstances that leave high net expenditures for urban transport in discouraging prospect.

14. Local transportation development is more centralized in the hands of a few elite players. This is because governments are more centralized in general, because much of the expenditure is central funds, and because high profile ideological differences among competing political parties may make organized community participation less viable.

3. Congestion and motorization in the developing world

There are no satisfying widely used measures that document the decline of mobility and serve to project it. To even casual observation, however, it is clear that congestion is increasing in most major cities. In the cities of China, India and Indonesia, rush hour speeds got slower through the 1980s, reaching speeds of less than 10 km an hour in major cities of those countries. In central Bangkok traffic speeds declined by 2% per year in the second half of the 1980s. These figures are believable, not only through intuitive observations, but because it is an expected consequence of rapid motorization.

There have been few useful citywide analyses of congestion anywhere. It is worth mentioning the work of the Texas Transportation Institute. They have developed a Roadway Congestion Index in which the independent variables are freeway vehicle-kilometers travelled/freeway lane-kilometers and arterial kilometers travelled/arterial-kilometers. This index has increased for cities across the US by roughly 20% during the period 1982–1991 (with a good deal of variation among cities.) TTI also estimated that during 1984–1991, for 50 large US cities total daily vehicle hours of delay increased by 21%. For a number of individual US cities it increased 30–50%. (Schrank et al., 1994 p. 31). (Note this is not per vehicle. It includes increases in number of vehicles and vehicle miles travelled.) These levels of increase under the circumstances of modest increase in motorization and urban population growth in US cities suggest the future consequences in the developing world because of much larger increases in both, and in many cases already more congested roads.

A somewhat similar measure has been attempted for the developing world for the UN Population Fund by the Institut d'Etudis Metropolitans de Barcelona (UNPF, 1988), but the effort is still in a primitive stage. There is no historical sequence of estimates and the survey appears to have included all roads (in part because freeways and arterials are hard to isolate in many developing cities). The measure uses vehicle registrations rather than vehicle kilometers travelled. Unsurprisingly, the ratio of vehicles per kilometer of road is much higher for the developed cities, even though congestion is worse in the developing cities. This tends to confirm the facts that (1) congestion is a condition localized to main ways that cannot be usefully averaged over a whole network and (2) that the developing, pre-motorized city has problems of adaptation to motor vehicles that are highly local in urbanized areas. We need further data to accumulate for developing cities (Table 1).

Perhaps the most telling data on mobility problems in the developing city is in journey-to-work travel times. It has been observed that travel times are remarkably similar from city to city. This was noted by Zahavi in the 1970s and recently concluded by Kenworthy et al., 1997a from survey data. On a world wide basis (excluding developing countries) the figure is roughly 30 min for a wide variety of different cities. In the developing world, on the other hand, in a set of data provided by UN Habitat there are several cities with average journeys to work (one way) around an hour for 1990. Those cities include Lilongwe (Malawi), Antananarivo (Madagascar), New Delhi, Harare, Quito, and Kingston in a list of 36. The top average work trip times were Rio de Janeiro at 107 min and Bogota at 90 min. Most megacities are for some reason missing from this list, but the cities with problematic commutes in general are not the larger ones included. If we isolated special suburban populations with long trip times it is probable that the set would include numerous fast-growing mid-sized cities. For example, commuting trips of 2 h occur from the suburbs of Kuala Lumpur – a metropolitan area of only 2 million (author's recent experience). Perhaps this is an indication that problems of urban mobility are not generic, but rather are special problems subject to correction.

4. Paths of motorization

Growth in the number of motor vehicles is at the base of mobility, on the one hand as an indication of increased motor mobility of the population and on the other as a force toward increased congestion. Although its significance to each is difficult to resolve, it is the best recorded variable. The work toward understanding future trends in motorization has been surprisingly limited, but there have been some recent interesting proposals.

One reason this analysis is of great interest is because it works toward understanding at what point rampant increase in motorization will approach saturation under given economic and regulatory conditions. This is an important question.

Table 1
Population, income, and journey to work, travel time, selected cities, 1990

Country	City	Metropol'n population	Cars per 1000 pop. (country)	Metro. fam. income US$ (1990)	Journey to work-(min. average)
Tanzania	Dal es Salaam	1 556 290	1.9	763	50
Malawi	Lilongwe	378 867	2.0	692	60
Bangladesh	Dhaka	5 225 000	0.4	1 352	45
Madagascar	Antananarivo	852 500	4.1	747	60
Nigeria	Ibadan	5 668 978	3.8	1 331	26
India	New Delhi	8 427 083	3.4	1 084	59
Kenya	Nairobi	1 413 300	5.5	1 500	24
China	Beijing	6 984 000	1.1	1 079	25
Pakistan	Karachi	8 160 000	6.4	1 622	NA
Ghana	Accra	1 387 873	5.5	1 241	35
Indonesia	Jakarta	8 222,515	7.2	1 975	40
Egypt	Cairo	6 068 695	19.8	1 345	40
Zimbabwe	Harare	1 474 500	30.2	2 538	56
Senegal	Dakar	1 630 000	8.6	2 714	35
Philippines	Manila	7 928 867	7.4	3 058	30
Cote d'Ivoire	Abidjan	1 934 398	12.9	3 418	38
Morocco	Rabat	1 050 700	26.7	4 158	25
Ecuador	Quito	5 345 900	15.4	2 843	56
Jordan	Amman	1 300 000	50.5	4 511	30
Colombia	Bogota	4 907 600	35.9	3 252	90
Thailand	Bangkok	6 019 055	21.4	4 132	91
Tunisia	Tunis	1 631 000	25.5	3 327	37
Jamaica	Kingston	587 798	28.3	3 696	60
Turkey	Istanbul	7 309 190	28.1	3 576	40
Poland	Warsaw	1 655 700	137.8	2 265	45
Chile	Santiago	4 767 638	53.9	3 433	51
Algeria	Algiers	1 826 617	29.0	7 335	30
Malaysia	Kuala Lumpur	1 232 900	103.3	6 539	34
Mexico	Monterrey	2 532 349	80.0	4 810	25
South Africa	Johannesburg	8 740 700	102.0	9 201	59
Venezuela	Caracas	3 775 897	80.2	5 123	39
Brazil	Rio de Janeiro	6 009 397	70.5	5 204	107
Hungary	Budapest	2 016 774	184.3	5 173	34
Czechoslovakia	Bratislava	441 000	207.0	3 677	40
South Korea	Seoul	10 618 500	32.1	19 400	37
Greece	Athena	3 075 000	172.9	14 229	40
Israel	Tel Aviv	1 318 000	174.5	16 680	32
Spain	Madrid	4 845 851	307.9	23 118	33
Singapore	Singapore	2 690 100	95.5	12 860	30
Hong Kong	Hong Kong	5 800 600	37.0	15 077	45
U.K.	London	6 760 000	363.5	18 764	30
Australia	Melbourne	3 035 758	435.6	26 080	25
Netherlands	Amsterdam	695 221	366.6	14 494	18
Austria	Vienna	1 503 194	387.9	22 537	25
France	Paris	10 650 600	417.3	32 319	40
Canada	Toronto	3 838 744	475.9	44 702	26

Table 1 (*continued*)

Country	City	Metropol'n population	Cars per 1000 pop. (country)	Metro. fam. income US$ (1990)	Journey to work-(min. average)
USA	Washington, DC	3 923 574	574.3	49 667	29
Germany	Munich	1 277 576	485.3	35 764	25
Norway	Oslo	462 000	380.0	34 375	20
Sweden	Stockholm	647 314	420.7	41 000	33

Source: United Nations Habitat, New York, 1993.

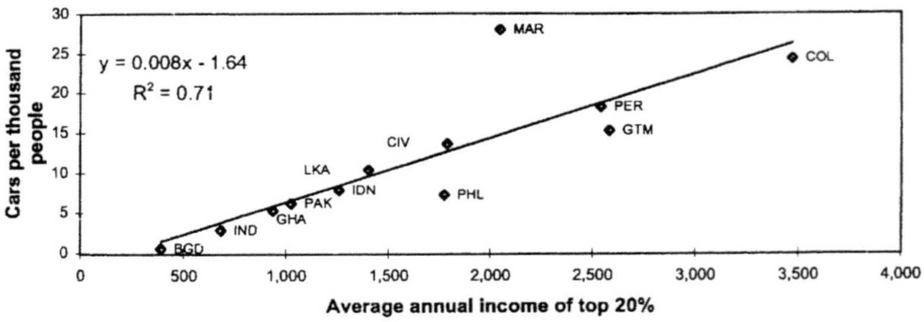

Fig. 1. Average annual income of top 20% of population vs. motorization (low income countries).

We have found that cars per 1000 population correlates very well with the annual income of the top 20% of population of the low income developing countries.[1] (See Fig. 1) Cars per 1000 also correlates well with percentage of the population in urban areas. To some extent, of course, percentage urban is a surrogate for income, since the vast majority of people in developing countries with incomes over the threshold of auto ownership live in cities (Gakenheimer and Steffes, 1995a) (see Fig. 2).

Other economic indicators perform very poorly in explaining patterns of motorization. We tried private consumption, industrial production (as a percent of GDP), openness of the economy (value of foreign trade/GDP), net current transfers (highlighting remittances from citizens overseas) and percentage of population in the labor force. None of these produced results of interest.

Several analyses have examined the income elasticities of motor vehicle ownership (based on GDP per capita), all yielding elasticities higher than one but with considerable variation (Stares and Zhi, 1996 p. 47). Most recently Kain and Liu found an elasticity of 1.44 for all motor vehicles and 1.58 for passenger cars, using a 52 country sample. Interpretation presents problems when considering that the part of the populations in the developing countries with incomes over the car-owning threshold is a very small part of the total, and their income growth probably does not vary with the average, particularly in transitional economies. One might expect better relationship with

[1] The low income developing countries are: Bangladesh, India, Pakistan, Ghana, Sri Lanka, Indonesia, Philippines, Ivory Coast, Guatemala, Morocco, Peru, and Colombia. (The lower middle income developing countries are Jamaica, Poland, Costa Rica, and Botswana. Upper middle are Malaysia, Venezuela and Brazil).

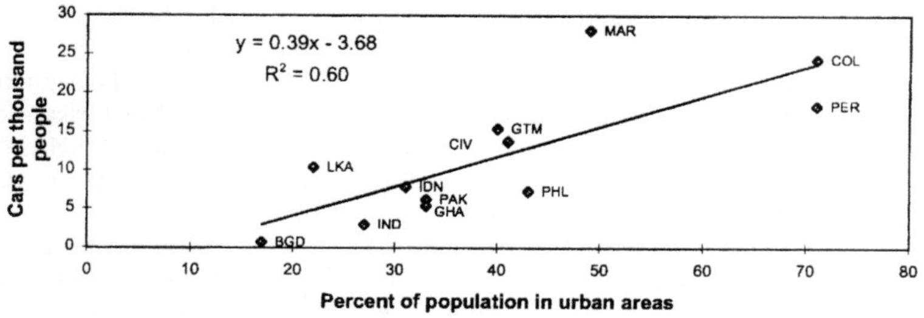

Fig. 2. Percent of population in urban areas vs. motorization (low income countries).

commercial vehicles, but their elasticity for commercial vehicles in the Kain and Liu study was only 1.15. At least, however, we can conclude that the elasticity is positive.

During the last 35 yrs most analysts have assumed that the variation conformed to a sigmoidal (logistic, "S" shaped) curve. This was established by J.C. Tanner in the early 60s (Tanner, 1962) and further developed in a number of papers, especially from the United Kingdom. The sigmoidal curve was originally introduced for biological, and epidemiological phenomena. It later became used for analysis of the diffusion of technological innovation. It is an intuitively satisfying curve for a process that begins slowly, matures into break-neck growth and must slow down at some point because of saturation. It has obvious special limitations, however, when applied to the developing countries where incomes are rising (or falling) and where motorization is far from reaching general saturation. In most of the developing world the decline toward the top of the "S" is not visible.

Some papers using the sigmoidal curve have interpreted the phenomenon as similar to technological innovation, where the process is one of increasing familiarity with the motor vehicle and adjustment of preferences to act on acquiring it (Jansson, 1989) Some have noted the need to shift the saturation rate upward repeatedly over time (Korver, 1993). Button et al., 1993 have proposed the use of a time variable to account for change in the relation between motorization and income. This may suggest the influence of increasing perception of a universal motorized life style, increasing market penetration of the industry, or similar effect. Mogridge, (1989 p. 55) points out that "automobile ownership and new car registrations show one of the largest year-to-year fluctuations of any economic variable."

We may conclude that the sigmoidal curve surely has some interpretation that illuminates motorization, but that it does not usefully project any slow-down phase of motorization in most of the developing countries.

An entirely new interpretation has been introduced by Debu Talukdar, 1997. He hypothesizes the relevance of the Kuznets curve. Originally introduced by economist Simon Kuznets to examine the relationship between economic development and income inequality, it has been used more recently to model the long term relationship between economic development and the environment. It projects relationships in an inverse "U" form that rises, peaks and then falls. In the case of environment it suggests rising damage from development, followed by investor, citizen and policy reactions that reduce such impacts as development continues. Its relation to motorization might suggest a rise in motor vehicles followed by a per capita (not necessarily absolute) decrease based on response to congestion, loss of novelty, and adjustments of public policy.

Talukdar presents this curve as a quadratic equation and tests it against the sigmoidal curve and the log-linear form, using a sample of 49 countries with substantial historical depth, and including 29 developing counties. He finds that the Kuznets curve provides a better statistical description of the long term relationship between economic development and per capita motorization than either of the other two.

His research was recently completed as a thesis at MIT. His data set indicates a peak in car ownership at the level of around $21 000 annual income. This does not mean that the developing countries need reach that level before experiencing some attenuation of motorization but it does suggest that significant decline of auto ownership growth in the developing world is in the distant future.

Another perspective on the future of motorized travel is offered by Andreas Schafer, 1997. There has been a belief sustained by evidence over the last 35 yr, initially by J.C. Tanner and later Yakov Zahavi, that personal travel occupies a constant budget of cost and time on average. The cost amounts to some 10% of personal consumption expenditure. The time in travel is somewhat more than one hour per day. This means that as incomes rise and time availability remains constant, people will spend more on travel per unit time (presumably using faster modes). Schafer reasons that at very high incomes people will proportionally reduce street travel (though not necessarily motorization) in favor of faster modes.

Accordingly, modal adjustments may be in store. In 1990 about 50% of global travel (in passenger kilometers) was by car, 30% by bus, 10% by rail and 10% by air. (Almost 80% of global bus traffic occurred in the developing world.) Schafer estimates by the year 2020 a rise in air traffic at about double the 1990 figure, to 20%, and a consequent only gradual decrease in the share of bus and car traffic. Auto traffic is projected at 43–53% percent of total passenger travel. Putting this into the perspective of an expected increase in total travel by a factor of three, this means that auto travel will grow by a factor of 2.5–3 while high-speed travel increases by a factor of 6. Based on expected kilometrage per automobile, this anticipates more than doubling the size of the automobile fleet by 2020, but the faster modes will increase more rapidly.

This perspective is another way of conceptualizing the constraints on auto mobility in the future. It does not isolate the situation of the developing world, but it infers that the balance in growth of motorization will increasingly be in the developing world. It does not suggest limits to growth.

In summary, the projection of motorization with reference to the developing world is a very difficult task at which the work has offered certain interesting insights but is a long way from confident estimates for the future at the level of 20 yrs and beyond. It has attempted to cope in various ways with the perception that such rapid increase has eventually to attenuate in some form, but it has presented as yet no persuasive view of altering trends.

5. Sparks of mobility leadership from the developing countries

Even though there is much less research and development on mobility and planning technique in the developing countries and public budgets are limited, they have certain important advantages in mobility innovation relative to developed economies. These include some cases in which there is

1. Stronger authority to increase mobility. There are countries in which urban governments have much more authority than in the developed world (often because they are single metropolitan governments rather than balkanized into a number of local administrative units, and sometimes on account of vested authority). Many of the countries have more power in central government guidance of local action. In some cases there are remarkable levels of charismatic leadership, such as former Mayors Jaime Lerner of Curitiba, Jamil Mahuad of Quito and Ronald McLean Albaroa of La Paz.
2. Lower personnel cost relative to capital costs. This simply results in different choices of actions, sometimes with consequences worth the attention of wealthier countries.
3. Fewer regulatory and legal barriers. These permit the introduction of guidance that would be halted in the developed world by fear of law suits in the case of malfunction.
4. Less convention in problem solving. In countries where transportation planning is a professional tradition, thinking is more conventional and there may be less scope for innovation. Innovation is sometimes easier in a less structured professional environments like those of the developing world.
5. A larger stake in solving mobility problems that better supports public action. This is because the problems are worse. The cities of the developing world have motorized faster, leaving urban structure further out of adjustment than in developed world cities.
6. Perceptibly growing problems. In many developing cities congestion is growing at a rate easily perceived year to year by even a casual observer. Any observer over 40 yrs of age remembers when central Miraflores outside of Lima, or Providencia outside of Santiago were quiet semi-commercialized areas with stores in former houses. Now they are occupied by 20 story buildings surrounded by massive congestion. This public awareness is leverage toward action in some cities.

With the help of the factors mentioned above to stimulate action, these innovations, existing or incipient, are from the developing world.

1. *High yield vehicle use restrictions*. Responsive to the severity of the problems, cities of the developing world often reach for higher achievement actions than developed cities. It is not unusual to have serious discussion or even attempts at implementation of actions that are almost patently impractical. For example, Bangkok made a recent serious effort to restrict all newly registered cars to use exclusively in non-rush hours.

Perhaps the most stringent restrictions have been imposed in China by municipalities concerned with mounting congestion. Some of them have limited the number of new motorcycle registrations each year. In Guangzhou it is not lawful to enter the city on a motorcycle registered anywhere but in Guangzhou. Many of them have limited the operation of commercial vehicles in unprecedentedly detailed ways (in terms of days, hours and localities). At last notice these restrictions had not been applied to private cars.

Some high yield restrictions have been associated with very high pollution levels. Restrictions in Santiago and Mexico City to limited days a week and limited parts of the city have emerged from this problem.

2. *New technologies*. Cities of the developed world have experimented with untried technologies. Brazil has been the first to build a substantial number of transitways. They have also been built in Instanbul, Ankara and Abidjan. The air propelled aeromovel has been introduced only in the

developing world (Sao Paulo, Porto Alegre and Jakarta). Whether successful or not, it is an example of adventuresome public innovation. In La Paz, Mayor McLean Albaroa advocated suspended cable car for new hilly transit routes. It may soon be implemented. Altogether, however, given current opportunities for private participation there is rather less innovation in developing cities than one might expect.

3. *Privatization of existing highways*. Certain countries have taken special initiative in the privatization of maintenance and extension of highways. It has been found a difficult job and there have been costly errors. These have included preparations that attracted insufficient bidders and excessively rapid pay-back schedules that produced very high tolls. But the efforts have generated valuable experience and may lead the way toward more general practice. Mexico, Argentina and Colombia have been particularly active in this matter. Other Latin American countries such as Chile, are following suit, in that case with the division of the Pan American Highway into several lengths for individual privatization. The Chilean government is adding the innovative dimension of contracting economic development services at the same time to convert the highway into a more significant development generator, at the same time, of course, creating market for travel on the highway.

In the construction of new highways there has been much activity engaging the private sector in the developing world for private toll facilities, build-operate-and-transfer (BOT) and other arrangements. While not originated in the developing world, the level of commitment to this form of new highway development may well exceed kilometrage in the developed world. Significant activity is taking place in India, Philippines, Indonesia, China (in Guangdong), Thailand, and elsewhere.

4. *Private non-unitary transit management*. The vast majority of public transport systems in the developing world are private and always have been. Most of them are made up of relatively small scale concessionaires each serving a limited number of routes and in some competition with one another. There is often a separate public transit authority serving a small portion of the demand. The management of transit is often a lively debate with cases of publicization of private systems as well as the privatization of public ones.

As a result, while not innovative, the competitive environment of privatized transit in much of the developing world provides a laboratory of experience in the management of concessions and other contractual arrangements for private service to the public under circumstances of competition among servers. Several cities have tried a number of alternatives, dramatically represented by the deregulation and reregulation of public transport in Santiago.

5. *Transit innovation*. There have been innovative ideas such as several from Curitiba: the platoon system of grouping buses, the boarding station enabling prepayment of the fare, and the practice of providing transit tokens for turning in a bag full of street trash. Brazil offers the experience of employer-provided transit passes for low income workers.

Perhaps the most useful experiences in this category have been those of flexible transit use under circumstances of permissive or sometimes unenforceable transit regulation. Routed vans and cars often switch to the role of taxis when business is slack and opportunity occurs. There is a variety of experiences with informally revised (i.e. unauthorized) transit routings, for example, that escape unprofitable congested streets at the city center through route terminations at the periphery of business districts. There have been informal resolutions of low volume service needs at urban peripheries and after hours requirements.

6. *Assertive Congestion Pricing and Other Ownership/Use Charges*. As means of controlling mounting congestion, high user charges are recurrently considered in many countries of the developing world. The examples of Hong Kong and of Singapore (where purchase taxes amount to some 300% of the price of the vehicle) are present examples. There have been temporary cases of high user charges in various countries, for example in Chile during the regime of the Unidad Popular in the early 1970s, when automotive imports were very heavily taxed, and in Korea during its period of rapid industrialization. Area licensing schemes resembling Singapore's have been repeatedly proposed, for example in Bangkok and in Kuala Lumpur. (Maybe replacing them eventually with broader congestion pricing, as in Singapore, will become a trend.) So far, no very assertive policy of pricing has appeared on a long-term basis in the lower income countries, but it remains a possibility as concerns rise and the dialogue continues.

7. *Rapid Transit Innovation*. There is incipient possibility of changing views on rapid transit in the developing world, especially suggested by recent discussions that now include its systems effects and other positive externalities. Up to now the position of the international lending institutions has been reluctant, or outrightly opposed, to nearly any investment in rail rapid transit on the grounds of its high capital cost and need for high operating subsidies. This is understandable since the only metros that currently recover their operating costs outside of Japan are Seoul, Santiago, and Hong Kong. Only Hong Kong covers full costs and is a very special case in various respects (e.g. 50 000 people live within 10 min of each stop, and a fare of over US$1 is feasible). Even the widely touted high volume of use of the Mexico City subway yields only 40% of operating costs from the fare box. In many cases metros cannot be self-financing because cities cannot charge high enough fares. Very few of the world's metros could be financed, even for the cost of operations, only by their users.

However, now that there are over 14 rail transit facilities in the developing world with some twenty years of record, it has become evident that cities with metros have better preserved downtowns than others. Further, it appears that a capacity up to some 70 000 passengers/ direction/hour in critical corridors has improved urban transportation systems in terms of overall system performance. These benefits are impossible to evaluate with any satisfying precision, but the visible evidence is persuasive.

This has led the World Bank to issue a surprising discussion paper, "Approaching Metros as Potential Development Projects" by Slobodan Mitric, 1997. Whereas World Bank transport policy up to now has been that metros are reasonable only in very exceptional cases when they are "likely to produce high rates of return," taken to mean very seldom, this new paper sustains the position that

> "…neither the state of the art of economic evaluation of metro projects nor its quality as practiced by consultants working in the developing countries are strong enough to justify treating the assessed economic rate of return as both a necessary and sufficient condition for project acceptance. It is simply too narrow, doing injustice to the complexity of the subject of cities in developing countries and their strategic decisions in the transport dimension."

The final section of the discussion paper reads like a design manual. It is difficult to say what impact this may have on the substantial number of cities in the developing world that recurrently debate the possibility of rail rapid transit, but it appears that encouragement might conceivably be in store. Successful management awaits resolution of the need to accommodate the contrasting concerns of public and private partners.

8. *Auto Cooperative Possibilities.* The movement toward car sharing (ubiquitous small scale rental locations) so far shows little evidence in the developing world, but there are grounds for regarding it as a hopeful possibility. Here are some reasons:

(a) In many of the more advanced developing countries there are significant populations with substantial incomes just under the threshold of auto ownership, with reasonable credit records and who share the world view that includes the personal need for auto mobility.

(b) The practice of sharing assets in general is a growing practice universally, and the developing countries are part of the trend. The trend may well be based, more than anything else, on institutional and advanced electronic developments. That is, it is now possible to negotiate and enforce more complex contractual agreements than formerly. Some countries are improving systems of these kinds. Further, in the developing countries' typically higher density cities there is considerable sharing of common building spaces and utilities connections. There is also sharing of vacation houses and work equipment (especially in fishing and agriculture). In some cases higher risks of breakdown and service interruptions have encouraged sharing agreements for back-up services (such as electric power generators).

(c) The issues of maintenance are surely a concern in auto cooperatives. This is a situation in which the relatively lower costs of labor in the developing world are significant. In many countries it would be practical to have a chauffeur capable of minor repairs permanently assigned to a particular car who would work in turn for its various users.

The actual existence of a market niche remains to be seen, but there are surely mobility behaviors that car share would fit in the developing world.

At the same time the environmental consequences should be considered. While car sharing among developed country car-owning environments normally reduces miles travelled (because people give up ownership and share), among developing country non-car-owning groups it would no doubt increase miles travelled with consequences in environmental and congestion impacts.

9. *Institutions for Credit Purchases.* The lack of credit to purchase vehicles has been a limitation throughout the developing world. There are, however, special institutions that have been used to overcome this problem. In the special case of vehicles in public transit, taxi and cargo service, there are several countries with special national funding for the replacement of vehicles in the public service. In Venezuela there has been a public Corporacion Financiera de Transportes making low interest loans for the replacement of buses, taxis and other vehicles that serve the public. A similar one has existed in Venezuela.

10. *Land Use Planning as a Mobility Tool.* Transportation and land use planning has a checkered history with limited achievement in the high income countries. This has been a consequence of limited metropolitan public powers, the need to accommodate varieties of stakeholders, and a limited need for such action in the eyes of responsible officials. There are indications of greater possibilities, however, in parts of the developing world. There are indications of this effect in the success of Korea in imposing development restrictions that have clustered demand around Seoul, special land assembly in Shanghai and Bombay, new cluster development in Bangkok, and other cases.

Reasons for the promise of transportation-friendly public action in land development include:

(a) High levels of public authority in the metropolitan areas of certain countries. In China the government owns the land, leasing it to private or public users through municipal district action. In Singapore government owns a large proportion of urban land (in fee simple). In Seoul, a mayor of the city reportedly created a scandal ultimately terminating him in office when he attempted simply to extend a dwelling he owned in a green belt. In Bangkok the governor was able to focus his authority on sites where transformation was desired to create a submetropolitan center for future development. This does not ignore, of course, that land development control in most developing countries is very weak.

(b) The speed of urban development promises significant effects in limited time. At typical rates of 5% per year, new urbanization that doubles the population of the existing city is created in only 14 yrs. Since much of this new population is at decentralized locations on new terrain, it is an opportunity for urban development planning.

All these items represent possibilities for coping with rapid motorization on the part of the developing world by means that are not directly borrowed from the developed countries. They may also be approaches that bear watching from the vantage of the high income countries. Probably the most promising initiatives are congestion pricing, other traffic management techniques and land development planning.

6. Borrowing from the developed countries

There is a pervasive belief in the developed world that it has much to teach the developing world about mobility and motorization. Holders of the this belief include people with contradictory different opinions. There can be little doubt that this assertion is certainly true in some sense, but its interpretation is bound to be controversial.

Let us divide lessons to be learned from the developed countries into three parts:

1. Lessons of technology. It is simply a fact that the vast majority of research and development funds are spent in the developed world (and for it). The developing countries are mostly borrowers of technology and some guidance would be in order.
2. Lessons of institutional management. The public and private sectors in the developed world have tried a number of things that don't work in administering public transport, managing vehicle use and so forth. This experience may be sufficiently basic that it could enable new managements to save costs and trouble. There have been success cases too.
3. Lessons of general experience. The urge to convey wisdom from past experience sometimes emerges from a belief that the developed countries have been along a path of mobility evolution on which the developing countries are coming along behind. Accordingly they should learn from the errors of the developed world, and its lost opportunities. This is the most complicated element of lessons to be learned. For one thing, it is not clear that the path is the same one. For example it was one thing to accompany the invention and industrial development of the automobile, and another to adopt it in later stages. Further, since these problems are often the by-product of a much-sought life style, recommendations sometimes bear the image of paternalism or even hypocrisy.

Let's look at some of these possibilities more closely. First, lessons of technology and technological loan possibilities. They include devices to reduce engine local pollutants, and global warming emissions. They include new low cost vehicle technologies, ITS equipment, and transport infrastructure designs such as transitways. The lessons and lending issues are straightforward in this category at the level of immediate workability. But there are questions to ask about the alternatives and the subsequent consequences. Some technologies would reduce congestion and pollution. Others would increase them.

The lessons of institutional management are also very good possibilities to facilitate mobility in the developing world. These are topics on which the developed countries have demonstrated capability, often learned through decades of trial and error. Items in this list are also restricted to those that are relatively non-controversial. Among the possibilities:

(a) Control of expenditures in the light of probable revenues and available budget. Transportation projects all over the world classically underestimate costs and overestimate receipts, resulting in serious financial problems. (This is particularly problematic in the liberalizing planned economies, where systematic concern for budgeting is not a strong part of their project administration background.) Learning on this problem is better characterized as a world-wide comparative experience, rather than a developed/developing country exchange. It is the case, however, that the problem is better documented in the developed world and refined techniques for cost and revenue estimation are more available from the developed world. This point is one way of introducing the whole subject of transportation systems planning technique, generally a useful contribution uncomplicated by controversy, and learnable as a set of skills. (Note that for the moment we ignore the *process* of transportation planning, which is in a different category.)

(b) Better privatization is a topic closely related to the last one. It is a second case of worldwide learning (with many of the important lessons coming from developing country experiences). It is nonetheless constructive for the developed community to convene the effort, bringing into play the considerable research that has been done on the subject.

(c) Traffic management techniques, including the institutionally complex issues of implementing ITS, is another potentially important contribution. Part of the challenge on this matter is assuring that the techniques installed respond to the serious needs of the developing world for high yield actions.

(d) Transit administration is an important possibility. While the developing countries have more public transport, their public management systems for transit are often poorly functioning concessionary systems that remain from the sector's early times and are not adaptive to contemporary scales of big city needs.

(e) Beyond this are numerous administrative practices ranging from vehicle registration systems to enforcement and educational needs in which the transfer or adaptation of management schemes would be very beneficial.

Lessons of general experience are the most complicated group, learning from the problems that have resulted from the whole overall prevalence of motor vehicles in the developed world. At one level we can pessimistically suggest that if the developed countries did not learn sufficiently to solve their problem while it was being created in their own environment, how can we expect a response from citizens of a country that has not experienced the problem yet? Further, in many cases the balance of advantages and disadvantages is such that final judgments about painful restraint behaviors during rapid motorization is subject to varying citizen values.

The emphasis here should be on descriptive analyzes of experience that are as value neutral as possible and which encourage independent decision on the part of developing country governments and private participants. They need to illustrate both the advantages and disadvantages. Such demonstrations have to relate to the experience of the listener. For example, it is one thing to study problems in cities where the impacts of rapid motorization are already taking place (Bangkok, Cairo...), and another to discuss them where such changes are only incipient (Colombo, Tashkent...).

7. Conclusions

As stated at the outset, the intent of this paper has been more to stimulate a conscious approach to the complexity of the problem than to produce a prescription.

A few things are certain. Though we don't know where the growth of motorization in developing countries will attenuate, it will surely continue to be rapid, outdistancing any efforts to accommodate it or adjust to it during the next number of years. Accordingly, actions to confront it must be high yield actions to avoid high economic and social impacts costs.

How do we choose these actions? Looking at issues in both the developing and developed economies, roughly the same suggestions arise in each with respect to the other. Urban transportation issues penetrate all aspects of urban culture and economy. In one's own culture one typically subsumes sufficient understanding of these connections to be able to act intuitively. (The adequacy of that assumption can be examined elsewhere.) At least with respect to a foreign culture (whether borrowing or lending techniques) it is necessary to be conscious about these connections.

One has to know more about details of the microeconomy. There are different traditions of exercising authority, different styles of public administration and of private management. What are the reputations of different kinds of public imposition on private prerogative? To whom is the individual morally responsible–family, society, an institution? What are the relations among social strata and the status of efforts to realign them? What are the definitions and expectations of social opportunity? What are the sensitivities to noise, light and air quality? What is the working definition of "democracy"–decisions that are best for the most people or a negotiation that begins with the assumptions that the vested interest of each stakeholder is as legitimate as each of the other's? All this conditions what works.

At the same time there is an empirical definition of what works. After all, there is a substantial resemblance among solutions in place around the world. Discovery and fashion in transportation solutions have world wide span–the land use/transportation movement, the traffic management movement, privatization, ITS, and so forth. They are all present across the world. Within each there are elemental discoveries–features that improve management of privatization that are being assembled from experience. The subtle evolution from road pricing to congestion pricing to value pricing makes multiple appearances world wide. The field is rich with invention that needs careful dialogue to evaluate for each case. The structure of change seems to be characterized more by solutions looking for application that by problems seeking new solutions.

The outcome will be different for each venue. Perhaps a pattern will emerge, especially in the middle to upper-income part of the developing world, where the variety of solutions in play is the greatest. Though detailed selections of technique and implementation style must be gauged to each application, the basic menu is:

1. *Highways.* The sure thing is that more highways will be built. There should be careful attention to maximizing their network effects and creating positive land use relationships.

Rationalizing land use patterns to go with them is an obvious very valuable step where possible. Certain venues are capable of this.

2. *Public transport management improvements.* These must, and will, be laboriously accomplished in spite of the inertia of existing institutions. Whether privatizing or publicizing, the important thing is to shake off the accumulation of inefficient practice through institutional change.

3. *Pricing to pay for infrastructure and rationalize the use of its capacity.* There are many ways to do this and some venues will do much more of a job of it than others.

4. *Traffic management to rationalize the use of infrastructure.* Again, there are dozens of options to choose from, and some cities can go much further than others.

5. *High capacity public transport.* Current trends across the world already suggest substantial increase in high volume modes. Advancing forms of contracting and financing to expand the set of beneficiaries to be called upon to support new systems will help.

As a background for advice it is useful to tease apart the alternative actions under these headings and the multiple forms of detailed application possible for each action.

References

Button, K., Ngoe, N. and Hine, J., 1993. Motor vehicle ownership and use in low income countries, Journal of Transport Economics and Policy, January, pp. 51–67.

Cointreau-Levine, S., 1994. Private sector participation in municipal solid waste services in developing countries, vol. 1, The Formal Sector UNDP/UNCHS/World Bank Urban Management Program.

Darbera, R. and Nicot, B.H., 1984. Le Planificateur et le Cyclopousse: Les Avateurs du Transport Urbain en Inde, Institut d'Urbanisme de Paris, Universite de Paris XII.

Gakenheimer, R. and Steffes, A., 1995a. A cross-sectional analysis of possible correlates of motorization in development countries, Working Paper, Cooperative Mobility Program, Center for Technology Policy and Industrial Development, MIT.

Jansson, J.O., 1989. Car demand modeling and forecasting: a new approach, Journal of Transport Economics and Policy, May, pp. 125–139.

Kasarda, J.D. and Parnell, A.M. (Eds.), 1993. Third World Cities: Problems, Policies and Prospects. Newbury Park, Sage Publications, Calif.

Kenworthy, J., Laube, F., Newman, P. and Barter, P., 1997. Indicators of Transport Efficiency in 37 Global Cities, a report to the World Bank, ms. February 1997.

Mitric, S., 1997. Approaching Metros as Potential Development Projects, TWU Papers, Discussion Paper, March 1997.

Mogridge, M.J.H., 1989. The prediction of car ownership and use revisited. Journal of Transport Economics and Policy XXIII (1), 55–74.

Schafer, A., 1997. The Global Demand for Motorized Mobility, ms., Cooperative Mobility Program, Center for Technology Policy and Industrial Development.

Schrank, D.L., Turner, S.M. and Lomax, T.J., 1994. Trends in Urban Roadway Congestion – 1982 to 1991, vol. 1: Annual Report. Texas Transportation Institute, College Station, September 1994.

Stares, S. and Zhi, L., 1996. Theme Paper 1: Motorization in Chinese Cities: Issues and Actions, In: Stares, S. and Zhi, L. (Eds.), China's Urban Transport Development Strategy World Bank Discussion Paper no. 352.

Talukdar, D., 1997. Economic Growth and Automobile Dependence: Is there a Kuznets curve for motorization? MCP Thesis, Department of Urban Studies and Planning, MIT.
Tanner, J.C., 1962. Forecasts of Future Numbers of Vehicles in Great Britain. Roads and Road Construction. XL, 263–274.
United Nations Population Fund, 1988. Cities: Statistical, Administrative and Graphical Information of the Major Urban Areas of the World. Published by the Institut d'Estudis Metropolitans de Barcelona.

[28]

An assessment of the political acceptability of congestion pricing

GENEVIEVE GIULIANO
University of California, Irvine, California CA 92717, USA

Key words: congestion pricing, public policy implementation, Los Angeles

Abstract. There is renewed interest in implementing congestion pricing in metropolitan areas throughout the US. This paper reviews changes in the transportation policy environment that have led to this renewed interest and identifies the major interest groups that support congestion pricing. A case study is used to demonstrate that significant barriers to congestion pricing implementation continue to exist. The paper concludes with some suggestions for developing politically acceptable pricing alternatives.

Introduction

There is growing support for congestion pricing in the United States. Frustrated by financial and environmental constraints, prodded by growing sectors of the public who are demanding that congestion problems be addressed, and encouraged by the many congestion pricing projects planned in cities outside the U.S., transportation planners and policy makers are considering congestion pricing more seriously than ever before. The Intermodal Surface Transportation Efficiency Act of 1991 (ISTEA), a blueprint for the next decade of transportation policy, identifies congestion pricing as a promising alternative for heavily congested urban areas and promises funding for pricing demonstration projects. In California a variety of public agencies are studying the feasibility and potential effectiveness of congestion pricing projects in response to legislative mandates for air quality improvements and congestion relief.[1]

Has congestion pricing's day come at last? Have planners and policy-makers finally accepted the logic of congestion pricing, and are they willing to take the political risk of actually testing the idea? Or, will the proponents of congestion pricing be defeated once again by public opposition? This paper presents an assessment of the feasibility of implementing congestion pricing. I will argue that despite apparent policy imperatives, it is unlikely that congestion pricing will be implemented to any significant extent in the U.S. Public skepticism regarding the effects of congestion pricing, resistance to high tolls, and pressures to divert toll revenues to new transportation facilities are barriers to effective congestion pricing programs. In addition, the concept lacks broad-based political support. More likely are tolls on new capacity, tolls for specific classes of users, and other less direct and less complex auto pricing strategies.

I begin this paper by providing a definition of congestion pricing and distinguishing it from other pricing policies. Second, I review changes in the transportation policy environment that have led to a renewed interest in congestion pricing and a broader base of support for pricing policies. In the third section I discuss the various interest groups that support congestion pricing. I use a case study example in the fourth section to illustrate acceptability problems associated with conventional congestion pricing alternatives. I conclude with a discussion of the attributes of more acceptable congestion pricing alternatives.

Congestion pricing in the context of transportation pricing strategies

The purpose of congestion pricing is to reduce congestion by reducing demand for peak travel on congested facilities. It is a pricing system aimed specifically at minimizing the total cost of travel over all travelers. Total cost is minimized by charging each user a fee equivalent to the incremental delay imposed on other users. By charging this fee, congestion is reduced, and thus the total cost of travel declines. Congestion pricing can be applied both in the short run and the long run. In the short run, highway facilities are fixed, and the toll establishes the most efficient use of the existing facilities. For the long run, congestion pricing can be combined with a highway investment model to achieve efficient levels of highway capacity and utilization, with the tolls providing revenue to achieve long run investment objectives.[2]

Singapore continues to be the only city in the world where congestion pricing exists. Implemented in 1975, it is a simple scheme: a flat toll is charged for entry into the central business district during the morning and afternoon peaks. Congestion pricing is to be distinguished from downtown area cordon fees, bridge and other toll fees that do not vary by level of demand, mileage-based vehicle fees, or other indirect pricing policies. Downtown area cordon fees, such as those existing in several Norwegian cities, are imposed to raise revenue or to discourage autos from downtown areas. They are flat fees unrelated to traffic congestion. Bridge and other flat toll fees are charged to cover the costs of construction, maintenance and operation of specific facilities. Mileage-based fees, such as the gasoline gallonage tax, are flat fees that apply to all vehicle travel and thus are independent of congestion. Such fees certainly affect auto travel demand, but they do not specifically affect peak demand, and thus are not examples of congestion pricing.

The transportation policy environment

Renewed interest in congestion pricing can best be understood in the context of changes in U.S. transportation policy that have taken place in the 1980s. Shrinking fiscal resources (in real dollar terms), rising construction costs, growing environmental concerns, and increasingly effective community resistance brought major infrastructure development to a virtual halt in most metropolitan areas. At the same time auto use continued to increase (Hu & Young 1992). The result of these changes is unprecedented (and still rising) levels of traffic congestion (U.S. General Accounting Office 1989; Hanks & Lomax 1991). Numerous surveys, as well as the visibility of the subject in the popular press, document growing public frustration with congestion problems. Public pressure to "do something" is increasing.

Evolution of U.S. Transportation policy

U.S. Transportation policy has historically sought to accommodate travel demand and enhance mobility and accessibility. The private automobile has been particularly favored, as demonstrated by the Interstate Highway Program, by state and local regulatory policies, and by auto-oriented land development policies (Altshuler, Womack & Pucher 1981; Dunn 1981; Cervero 1989). As environmental and energy concerns emerged in the decades of the 1960s and 1970s, public policy focussed on providing better alternatives to the private automobile by reconstructing and expanding the nation's mass transit systems, making operational improvements to the highway system, and promoting the use of vanpools and carpools. Glaringly absent from this list of policy choices was any attempt to restrict use of the private auto.

By the end of the 1980s, however, the fundamental purpose of transportation policy was beginning to be questioned, as traditional approaches to solving transportation problems appeared to be less viable. First, increasing the supply of transportation facilities to serve anticipated demand is financially prohibitive, particularly in densely developed urban areas. For example, estimates for the Los Angeles Region's planned additional capital facilities needs are in excess of $56 billion, simply to maintain *present* levels of congestion through 2010 under the currently adopted long range plan (Southern California Association of Governments 1988). Second, the correlation between highway system expansion and growth in vehicle travel has suggested to some analysts that such expansion is self-defeating. Providing more capacity, which is typically justified as a means for reducing congestion and air pollution, will induce more travel, which in turn will ultimately result in more congestion (Newman & Kenworthy 1988). If highway capacity is constrained, travelers will make other choices (e.g. use public transit, maker shorter trips or make fewer trips). Third, efforts made to reduce auto use by providing more public transit have been largely unsuccessful. Public transit's market share has continued to decline, despite massive investment in new vehicles and facilities; the 1990 National Personal Transportation Study data shows that public transportation modes account for just 2.5 percent of daily person trips (Hu & Young 1992). These factors have led to a growing consensus among policy-makers that "we can't build our way out" of urban congestion problems. Accommodating growing demands for auto travel may simply no longer be feasible, and coaxing auto users to mass transit has proven to be ineffective.

Additional pressure is being brought to bear on transportation policy by growth management advocates and environmentalists. Opposition to new development proposals often focuses on their traffic impacts (Giuliano & Wachs 1992). Major transportation projects are frequently opposed on the basis of their "growth inducing" effects; this growth is seen as causing more traffic congestion. Environmental advocates hold complementary views, and have been able to directly influence transportation policy through the Clean Air Act Amendments of 1990 and the conformity provisions contained in the 1991 ISTEA. Thus there is growing advocacy of policies to control or manage the private automobile.

Policy response

Policymakers have responded to these changing views by placing increased emphasis on managing the transportation system and increasing its efficiency. On the supply side this emphasis is more obviously manifested in the myriad new technology programs currently underway (e.g. Intelligent Vehicle-Highway Systems) that seek to increase highway vehicle capacity. On the demand side, there has been a rapid proliferation of trip reduction ordinances and regulations. These "demand management" programs mandate the provision of incentives to attract more commuters to ridesharing and transit modes.

Favorable results to these system management efforts have been slow to materialize, however. The new technology approach requires a long lead time and may be very costly, and more conventional system management techniques (e.g. ramp metering, signal timing) have already been extensively implemented. Moreover, recent studies of various trip reduction programs reveal that they have had only mixed success in reducing solo-driving (Giuliano 1992; Giuliano & Wachs 1992). It is in this context of weakening consensus regarding the goals of transportation policy, growing barriers to traditional solutions for transportation problems, the linking of transportation and air quality policy, and continued pressure to reduce traffic congestion that congestion pricing has once again surfaced as a possible strategy.

Other perspectives

Several recent papers and presentations have addressed the question of why congestion pricing seems more likely to be implemented now than it has been in the past. Reasons given are listed in Table 1. These reasons are generally complementary to the discussion presented here, with two exceptions. The availability of new technology is frequently identified as making congestion pricing easier to implement. While it is certainly true that new technology can eliminate toll booths, make more complex pricing schedules possible and make tolls more convenient to pay, technology has never prevented pricing from being implemented. The Singapore scheme uses a decal license that must be bought; bridge and road tolls have been collected manually for several centuries.[3]

Table 1. Why congestion pricing is more likely to be implemented in the 1990s.

Reason	# of Times cited
Environmental concerns	5
Traffic congestion	4
Technology availability	4
Need for new facilities	2
European examples	2

Sources: Bhatt 1991; Gomez-Ibanez 1992; Kirby 1991; Poole 1992; Reinhardt 1990; Siracusa 1991; Small, Winston & Evans 1989.

Two papers cited the presence of European examples, such as the cordon fees in Norway, or the planned congestion pricing projects in Great Britain and The Netherlands. I do not find this a persuasive argument, since U.S. transportation policy has rarely, if ever, looked to European models. Vehicle and fuel taxes, transit subsidies, and vehicle safety are just a few examples of areas of major differences between U.S. and European policy (Dunn 1981; McShane, Koshi & Lunden 1984; Pucher 1988). Rather, the absence of congestion pricing in Europe is indicative of the difficulties involved in implementing it, as I will further discuss in a later section.

Sources of support for congestion pricing

Serious discussion of congestion pricing alternatives has gone far beyond its traditional proponents, transport economists and free market advocates. This section identifies the major interest groups that have become supporters (at varying levels of enthusiasm), and provides some possible explanations for their support. I focus mainly on California examples.

Environmentalists and transit advocates

Environmental advocates have influenced public policy in a number of areas in recent years.[4] Air pollution has been a major focus of lobbying activities. In California, a lawsuit brought by the Coalition for Clean Air was the impetus for Southern California's 1988 Air Quality Management Plan. The Plan is significant in that it contains a set of specific strategies for meeting clean air standards within the required time period (by 2010). The Plan has given rise to a series of new regulations on mobile sources, and includes options such as restrictions on new vehicle registrations and drivers licenses should air quality improvement targets not be met (South Coast Air Quality Management District 1988). Following the Coalition's success in Southern California, the Sierra Club filed a similar suit in the San Francisco Bay Area, and a corresponding planning effort has resulted.

Environmental advocates see road pricing as one of the few effective means for reducing the use of the automobile, and thus reducing air pollution. It is argued that the pollution costs of the auto are not covered by existing user fees, thus higher fees are justified. A recent Southern California study sponsored in part by the Environmental Defense Fund, for example, advocates a set of five "market-based" strategies that include congestion pricing, mandatory parking fees, and "smog-based" auto registration fees, as the most efficient way to reduce congestion and air pollution in Southern California (Cameron 1991).

Environmentalists are also often strong supporters of mass transit. Transit advocates argue that increased transit use is necessary to achieve more efficient land use patterns and reduced energy consumption (Newman & Kenworthy 1988). They argue that efforts to re-establish a market for mass transit cannot succeed as long as the mispricing of the auto continues. In addition, congestion tolls could provide a source of revenues for transit subsidies.

It is important to note that environment and transit advocates focus on restricting the use of the auto. Their interest is not necessarily in achieving a socially optimal investment in highway facilities, as is implied by long run congestion pricing and investment policy. Rather, congestion pricing is viewed as a legitimate rationing device that can help to achieve environmental goals.

Real estate and development industry

A second significant source of support for congestion pricing is the real estate and development industry. An increasingly powerful anti-growth political movement, generated by public frustration with rapid development and traffic congestion, has made new development more difficult and costly in rapidly growing areas. In some cases, land is simply downzoned in an effort to control traffic growth (Wachs 1990). In other cases, new development is held responsible for funding local infrastructure improvements to mitigate expected traffic problems. Major projects are also often held to specific "trip reduction" requirements, or to the development and funding of ongoing transportation management programs as a means of reducing traffic impacts (Giuliano & Wachs 1992).

Under these circumstances, congestion pricing provides several potential benefits for development interests. First, it shifts the burden of reducing congestion back to the public sector (e.g. the operators of road facilities). Monitoring and enforcement become the responsibility of the public agency, rather than of individual business or project managers. Congestion pricing thus does not introduce any competitive disadvantage to new development, in contrast to impact fees or trip reduction requirements. Second, congestion pricing can reduce the demand for new facilities and provide a revenue source for financing them, thus potentially reducing the fees imposed on new development. Finally, existing facilities could support more intensive development if congestion pricing were employed, and consequently may be perceived as an alternative to downzoning and other development restrictions.

Other business interests

Business interests in California, including major service and manufacturing firms, have also emerged as supporters of so-called market-based strategies such as congestion pricing mainly in reaction to the rapid proliferation of air pollution regulations since 1988. In Southern California, the South Coast Air Quality Management District (SCAQMD) has emphasized "command and control" strategies that impose specific standards that must be met. Critics claim that this stringent regulatory environment is harmful to the long-term economic health of the region, and that appropriate market incentives and disincentives could more efficiently achieve air quality goals.

One example of the "command and control" approach is SCAQMD's Regulation XV, which requires all employment sites with 100 or more workers to develop and implement a plan to achieve specified target levels of ridesharing. Compliance with the Regulation involves substantial cost and effort on the part of employers. Opposition to the Regulation has grown as a result of labor-management conflicts over plan incentives, slow progress in meeting ridesharing goals, and rising program costs (Giuliano, Hwang & Wachs 1992). Some employers consider region-wide pricing strategies to be a favorable alternative to Regulation XV, as employers would be relieved of the responsibility for influencing their employee's commuting behavior.

Perhaps the best example of pricing advocacy is the Bay Area Economic Forum. The Forum, a public/private interest group, was organized to develop a market-based program for air quality compliance in the San Francisco Bay Area. The program includes peak-period bridge tolls, parking fees and VMT fees. A major impetus for the Bay Area Forum was the "command and control" approach of SCAQMD, which was perceived by the Forum as harmful to the regional economy and ultimately ineffective. The Forum achieved national recognition as a model for establishing a constituency of support for auto pricing policies (Gomez-Ibanez 1992).

The diverse interest groups that have emerged in support of auto pricing strategies suggest that possibilities for implementing pricing policies are better than they have ever been. Notably absent from the above list, however, are elected officials and the general public – the constituency required for implementing congestion pricing.

Problems of traditional applications of congestion pricing

Many attempts have been made to implement congestion pricing since the mid 1960s, yet, with the lone exception of Singapore, all have failed as a result of political opposition. These include the Smeed Committee proposal for peak tolls on congested arterials in Greater London in the 1960s, the U.S. Urban Mass Transportation Administration's efforts to conduct downtown area pricing demonstration projects in the 1970s, and California's long-range plan of 1976 that included a congestion pricing component (Morrison 1986; Higgins 1979; Elliot 1986). More recently, the Hong Kong central area pricing experiment was abandoned after a 15 month trial of the technology, but prior to actual implementation (Borins 1988). In the 1990s, plans for central area congestion fees in Stockholm have been deferred indefinitely; plans for central area congestion tolls in Cambridge and other cities in Great Britain have been stalled by political opposition; and in the Netherlands, a regional congestion pricing plan was shelved after it caused political problems within the ruling party (Hau 1992; Jones 1991; Jones & Harvey 1991). Congestion pricing apparently remains a political impossibility, even in countries where auto restraint policies are widely accepted and employed, and where transportation policy decisions are made with less direct public participation than is the case in the U.S. (See May, 1992, for more information about international experience.)

The obvious question is, why is there so much public opposition to congestion pricing? This question is particularly perplexing to congestion pricing proponents, who cite the extensive research that has been devoted to demonstrating its economic and welfare benefits. I use the example of Los Angeles to show why conventional applications of congestion pricing encounter strong opposition.

Traffic congestion in the Los Angeles region

The duration and geographic extent of congestion in major metropolitan areas makes congestion pricing increasingly attractive. Economic theory suggests that the net benefits of congestion pricing will increase with the initial level of congestion; a simulation study based on San Francisco Bay Area data provides corroborating evidence (Small 1983).

Congestion is no longer restricted to the downtowns of metropolitan areas. Rather, congestion occurs around major activity centers and on the freeways and arterials connecting them (Hanks & Lomax 1991). This is certainly the case in the Los Angeles region, where over half of all the region's freeways are congested during peak periods (Southern California Association of Governments 1987). The extent of freeway congestion in the region is illustrated in Fig. 1. Freeways are categorized by level of service (LOS) for a typical weekday. The usual LOS categories of A through F are extended to levels of F which are based on the daily duration of congestion.[5] Figure 1 shows that much of the system is congested for several hours each day.

A closer examination of conditions in the Los Angeles region shows that demand far exceeds available capacity, and that travelers have responded in several ways. First, growing traffic demand has resulted in extensive peak spreading, as trips are shifted to other time periods. In many places the "peak" as conventionally defined has disappeared. Figure 2, for example, gives freeway traffic volumes by time of day for the I-405 freeway, just south of the I-10 freeway. This location is categorized as LOS F-1, and so does not constitute a worst case example. There are five lanes in each direction. The volumes shown are the average of three consecutive Wednesdays in May 1991. Traffic volumes build rapidly in the early morning, then drop around 8 AM as congestion reduces travel speeds and throughput. Volumes remain near capacity throughout the day, then decline relatively slowly as the afternoon peak stretches to 8 PM or later. Schedule adjustments (shifts of trips to non-peak time periods) are considered to be an alternative to paying a congestion toll. However, when the "peak" extends from 6 AM to 7 PM there are few opportunities to shift one's travel to a non-peak time period.

Second, discretionary trips have been largely eliminated from congested facilities. National travel data show that the journey to work accounts for only about half of all peak period trips, suggesting that many peak trips are discretionary, and therefore could easily be eliminated or shifted to other

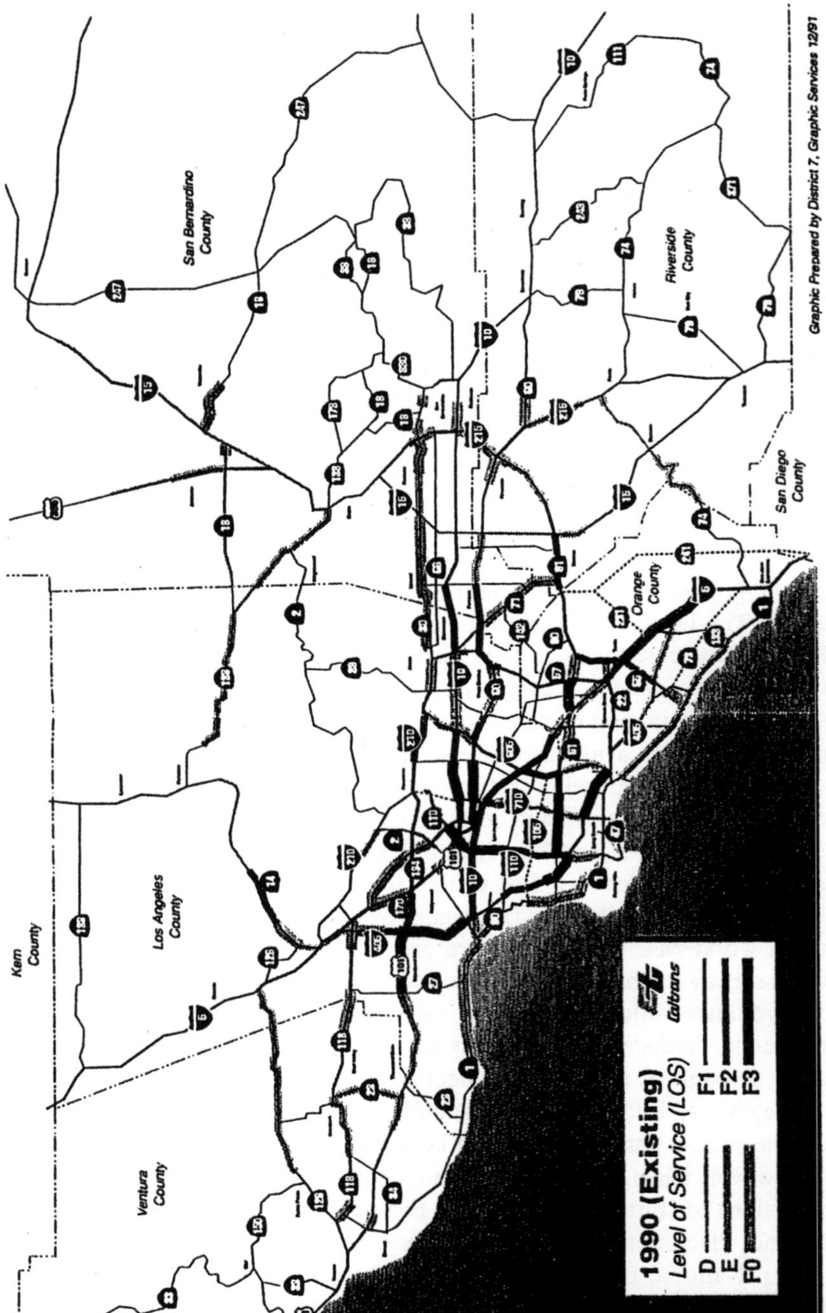

Fig. 1. 1990 Level of service on the Los Angeles region freeway system.

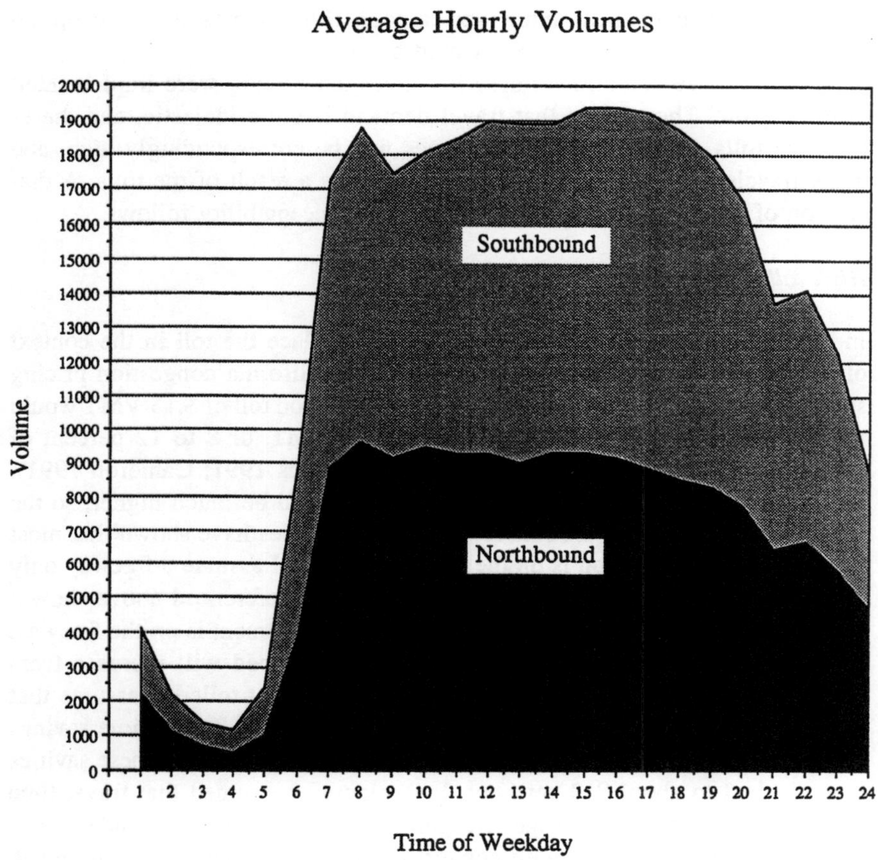

Fig. 2. Average hourly volumes on a Los Angeles freeway.

time periods (Richardson & Gordon 1989). This is not the case on congested freeways in the Los Angeles region. Traveler surveys of three different Southern California freeways revealed work and work-related travel to account for 80 to 90 percent of all peak trips, with the peak defined as 6 to 9 AM and 3 to 6 PM. In contrast, shopping trips averaged less than 1 percent and school trips averaged 2 percent of trips during these periods.[6] These findings make sense: travelers will avoid congestion to the extent that they find it possible to do so.

Finally, travel in the Los Angeles region is overwhelmingly oriented to the auto. Outside of downtown Los Angeles, nearly 80 percent of all workers drive alone and about 13 percent carpool. The transit share is just 2 percent (Giuliano, Hwang & Wachs 1992). Massive investments in new transit facilities and services are underway, but for most commuters, the only viable option to driving alone in the near future is to carpool. Thus the most available option for responding to congestion tolls would be carpooling.

What do these conditions imply if congestion pricing were implemented in the region? They imply that travel demand is quite inelastic, and therefore that tolls will be high, toll revenues will be correspondingly high, and many travelers will be substantially worse off as a result of the tolls. A discussion of how these outcomes affect political acceptability follows.

High tolls, winners and losers

In considering the issue of tolls, it is useful to place the toll in the context of a typical commute. Two different Southern California congestion pricing studies estimated that a regionwide average congestion toll of $.15/VMT would result in a 4 to 6 percent reduction of annual VMT, or 8 to 12 percent of congested VMT (The Urban Institute/KT Analytics 1991; Cameron 1991). Of course, tolls on the most congested routes would be much higher, so the numbers presented here are conservative. Since studies have shown that most of the region's congestion is on the freeway system, I assume a freeway only congestion toll of $.15/VMT. The freeway surveys referenced above show a median freeway work trip of 20 miles, of which 30 percent is on the freeway. If the toll eliminates freeway congestion, travel speed will increase from about 35 mph to 60 mph. Since surface streets are not tolled, I assume that the non-freeway portion of the trip remains unchanged.[7] Travel time savings of 4.3 minutes/trip would result at a toll of $.90/trip. The cost of these savings is equivalent to about $13 per hour. It is generally assumed that travel time is valued at about half the wage rate (Small 1992). In order to consider themselves at least as well off with congestion pricing as they were without it, commuters paying the toll would have to earn an hourly wage of about $26. However, the 1989 average hourly wage (the most recent year for which data are available) in the region is about $12 (U.S. Department of Commerce 1991).

It is important to note that congestion theory *requires* that peak period commuters on average be made worse off as a result of the tolls. If this were not the case, insufficient numbers of commuters would be tolled off (shift to other alternatives), and the expected reduction in congestion would not occur.[8] My point here is that the higher the toll, the greater the number of commuters (all else equal) made worse off by the toll, despite the fact that positive net social benefits may result overall.

Who are likely winners and losers when congestion tolls are imposed on existing facilities? Gomez-Ibanez (1992) presents the list shown in Table 2. The size of group 1 depends on the toll; the higher the toll, the greater the value of time required to offset the toll. Group 1 will consist mainly of higher income commuters. Group 2 is the "captive" HOV market, e.g. lower income

Table 2. Winners and losers when an existing facility is tolled.

Winners	
Group 1:	Those who continue to drive alone and whose value of time saved exceed the toll
Group 2:	Those who used HOV services and continue to do so
Group 3:	Recipients of toll revenues
Losers	
Group 4:	Those who continue to drive alone, but whose value of time saved is less than the toll
Group 5:	Those who shift to a less convenient untolled facility
Group 6:	Other (pre-existing) users of untolled facilities
Unknown	
Group 7:	Those who shift from driving to HOV modes (depends on relative benefits of avoiding toll vs inconvenience of HOV modes)

Source: Gomez-Ibanez 1992.

commuters, or those without access to a car. The size of group 3 depends on the extent and how tolls are redistributed.

Loser groups are potentially large and represent a wide range of income levels. Group 4 again depends on the amount of the toll, as well as the availability of alternatives to paying the toll. In heavily congested areas with few alternatives to paying the toll, this group would be large. The extent to which commuters divert to untolled facilities (or off-peak times) determines the size of groups 5 and 6. Group 7 commuters are more likely also losers, at least in the short run, because of the inconvenience of HOV modes for most trips. Although transit and HOV services will expand in response to the increased demand over a period of time, there would be an interim period of adjustment.

Any individual's assessment of the potential impact of congestion tolls depends on her particular circumstances. It is easy to see that many commuters would view themselves as captives of the tolls and consider any alternative to paying the tolls to be unacceptable. These individuals would strongly oppose the tolls, because they would not personally benefit from them.

Equity and fairness

Closely related to the discussion of winners and losers is the issue of fairness. The distributional fairness of congestion tolls has been a subject of concern from the time they were first proposed (Foster 1974; Richardson 1974; Vickrey 1955). Equity is widely considered to be a major barrier, and several studies have been devoted to analyzing the distributional effects of congestion pricing (e.g. Else 1986; Cohen 1987; Small 1983).

The economic assessment of equity depends both on the relative propensity of various income groups to drive to work on congested facilities and the disposition of toll revenue. If it is assumed that revenues are not redistributed in any way, congestion tolls will generally result in gains for upper income groups and losses for lower income groups (Else 1986; Cohen 1987).[9] Net benefits to all income classes is possible if revenues are redistributed among income classes at least in proportion to each class' toll contribution, for example by commensurately reducing other taxes (Small 1983). The incidence of net benefits across income groups also depends on the level of congestion, which determines the level of the toll. The higher the toll, the greater will be the differences in net benefits between income classes.

The studies on which these observation are based employ many simplifying assumptions that limit their applicability to real world situations. These include holding location choices, land values and total travel demand fixed, limited or no changes to transit service, no shifts in travel schedules, and of course correct tolls. Empirical analysis of actual pricing experiments would be necessary to determine distributional impacts more accurately. The only available potential case study is that of Singapore, and the distributional impacts of that program have not yet been studied.

Results of prior simulation studies suggest that it is certainly possible to offset distributional impacts of congestion pricing by allocating toll revenues in some compensatory way. For example, revenues could be used to replace existing regressive taxes, to provide direct user subsidies, or to subsidize alternative modes (Small 1993). Would such policies be sufficient to persuade the public that equity concerns have been addressed?

No matter how revenues may be redistributed, some individuals may still be made worse off, since congestion tolls do not lead to Pareto improvements. Harvey (1991) shows, for example, that workers in areas poorly served by transit would likely pay a peak bridge toll despite their lower income profile because of the lack of suitable alternatives. Second, the compensatory effects of revenue redistribution may be difficult to communicate or accomplish. If revenues were used to support transit facilities, for example, transit users would reap most of the benefits. If revenues were to replace existing taxes, distributional impacts would depend on differences in incidence and would be difficult to predict.

Although equity is frequently identified as a major barrier to implementing congestion pricing, it can be argued that the distributional impacts of congestion tolls are not particularly important with respect to public acceptance. Equity does not appear to be a major public policy issue. For more than a decade, Federal tax and revenue policy has become less equitable as redistributive programs have been reduced or eliminated. States and local governments have also shifted emphasis to more regressive tax instruments such as sales taxes and state lotteries. Revenue sources have followed the same trend in transportation financing: contributions from general sales taxes have increased while contributions from income taxes have decreased (Rock 1990).

Nor have distributional considerations been a definitive consideration in other areas of transportation policy. The case of public transit is illustrative. Several studies have shown that transit fare policy favors middle class commuters, yet only a few transit agencies have restructured fares in response (Pucher 1981; Cervero et al. 1980). Moreover, transit investment policy is oriented to the middle class commuter. Mass transit projects are justified on their expected contribution to congestion relief, as commuters shift from their cars to trains or trolleys. Such policies ignore the reality that most transit users are low income household members with limited access to an auto. Indeed, current investment policies often have the perverse effect of reducing the quality and availability of transit for those who use it the most, as the high costs of operating new rail lines forces cutbacks and fare increases in other parts of the system (Wachs 1989).

Given the evidence from other areas of public policy, then, it seems at least possible that distributional considerations are not a primary concern of congestion pricing opponents. Rather, distributional equity may present an apparently legitimate basis for opposition that is actually motivated by other reasons.

The problems of toll revenue

Implementation of congestion pricing in any circumstance where congestion is extensive will lead to a political mixed blessing: a large amount of toll revenue. The Southern California studies are again illustrative. If a program of freeway congestion pricing were implemented with an average toll of $.15/VMT on all congested freeway VMT in the Los Angeles region, annual revenues would be in the range of $2.6 to $2.8 billion (The Urban Institute/KT Analytics 1991). These are large numbers: more than twice what a 1 percent regional sales tax would generate, and equivalent to an annual tax of $530–$570 per household. In a context of severe budget constraints and limited opportunities for new revenue sources, the attraction of congestion tolls from the point of view of public agencies is clear. Not only would congestion problems be reduced, but financial resources for improving the system would also be increased.

Would tolls of this magnitude be politically acceptable? At present, there are no local option tax programs of comparable magnitude in existence. Moreover, it has proven to be impossible to increase fuel taxes more than five cents per gallon at a time – which amounts to only $0.7 billion per year from the same 5 county area. It is at least questionable that high tolls and their corresponding revenues would have broad political support.

Existing transportation programs are informative when considering how toll revenue disposition might be structured. Efforts to raise taxes for special purposes in the U.S. have shown that revenue disposition must be clearly established in order to gain voter approval. Successful transportation measures typically contain specific lists of projects to be funded by the tax revenue. Recent surveys conducted in Great Britain showed public attitudes to congestion pricing to be quite similar. Congestion pricing was found to be most acceptable when linked with a program of expanded public transit, lower transit fares, and selected expansion of the road system (Jones & Harvey 1991). It seems reasonable to expect that similar revenue disposition programs would be proposed in the U.S., e.g. capital intensive, transit oriented programs. Programs of revenue redistribution to offset toll impacts on lower income groups are far less likely, given the tradition in the U.S. of restricting transportation revenues to related uses and political pressures to expand mass transit options in many U.S. metropolitan regions.

Although the issues of toll impacts and revenue disposition are common to all congestion pricing proposals, conditions of heavy congestion make these issues more critical because of the magnitude of the market clearing toll and the extent to which demand must be reduced to achieve congestion reduction benefits.

Skepticism and uncertainty regarding congestion pricing outcome

In addition to concerns related to tolls and revenues, other barriers to public acceptance exist. Prior analyses of public responses to congestion pricing reveal that skepticism or misunderstanding of the concept is a major source of opposition (see Table 3). This may be somewhat surprising to researchers, given the extensive evidence of substantial net benefits provided by simula-

Table 3. Reasons for opposing congestion pricing.

Reason	# of Times cited
Skepticism/misunderstanding	5
Equity/fairness	4
Implementation problems	2
Right to travel	2
Pricing what was free	2
Tax resistance	1
Harmful to business	1
Privacy (AVI)	1
Restriction of choice	1

Sources: Button 1984; Cohen 1987; Higgins 1979; Jones & Harvey 1991; Kirby 1991; Siracusa 1991.

tion studies (for summaries, see Button 1984; Morrison 1986). The logic of congestion pricing rests on the assumption that time has a monetary value, and that therefore individual travelers are willing to pay a fee in order to save time. Examples of such behavior abound, and it appears that the trade-off between money and time is well understood by most consumers. This is apparently not the case for congestion pricing. Button (1984) speculates that the problem is one of perceptions: the congestion toll is immediate and tangible while time savings are not. Jones and Harvey (1991) report that many respondents simply don't believe that tolls will cause people to shift to other modes of travel times; others claim that tolls could not be set high enough to substantially reduce congestion.

These comments reveal concerns that congestion pricing would not work as economists claim it would. Some analysts view public skepticism as a public relations problem: the public has not been adequately informed or educated on the concept (Higgins 1986; Button 1984). Others blame established interest groups (e.g. elected officials, highway lobbies) for misinforming the public (Elliot 1986). In either case, these claims suggest that once the concept is properly communicated, the public will support congestion pricing. Skepticism may be a deeper problem, however, and may reflect a more fundamental distrust of untested policy alternatives.

Impacts on business

Fear of negative impacts on business within areas subject to congestion tolls is another illustration of how the concept is misunderstood. Congestion pricing theory suggests positive effects: congestion tolls reduce the full cost of travel; thus the targeted area would have a competitive advantage relative to areas that remain subject to congestion, as long as toll revenues remain within the targeted area. If congestion pricing were imposed in a downtown, this would mean that the toll revenues be used either to improve transportation services within the area or to somehow reduce the cost of doing business within the area. For example, free shuttle services, discounted taxi services or other partransit could be supported by toll revenues. Moreover, the little evidence available from Singapore suggests that business is not likely to be adversely affected.

It also bears noting that heavily congested downtowns and other major employment centers are competitively disadvantaged now because of congestion. Economic theories of growth posit that the clustering of activities occurs as a result of the benefits of agglomeration economies. Eventually, however, these agglomeration benefits are offset by congestion, causing new activities to seek out lower cost locations. In U.S. metropolitan areas, this process is illustrated by the historical decline in the proportion of metropolitan employment located in downtown (Linneman & Summers 1991). If access costs are reduced, it seems reasonable to expect that these areas would benefit. The fear that congestion tolls would harm downtowns seems to be based on consideration of the tolls only, and not of the time savings that would result from the tolls.

Right to travel and pricing what was free

When the focus of congestion pricing is on the tolls rather than on its net effect, other reasons for opposition are more easily understood. Congestion pricing conflicts with a fundamental and highly valued belief held by many Americans, namely that mobility is a right. Mobility provides opportunities to engage in spatially dispersed activities, and many consider these opportunities important. As pointed out earlier, U.S. transportation policy has historically reflected this belief and fostered high levels of mobility by making travel as easy and cheap as possible. Congestion pricing may be viewed as restricting the right to travel to the extent that some travelers are priced off the road and into less preferred modes, and may consequently be seen as a reversal of traditional policy, despite the fact that it would generate benefits to society as a whole.

Opposition to congestion tolls on previously "free" facilities is similarly based on considering the tolls in isolation. Use of congested facilities is of course not free: users are paying with time rather than money. Sometimes the toll is considered another tax. In this case, the argument is that user fees already exist to support the highway system and therefore additional tolls are not justified. Travelers assert that they are being asked to pay twice for the same facility when tolls are added. In either case, the arguments do not consider the time costs of travel. Many of the criticisms of congestion pricing, then, are related to a misunderstanding of the concept itself.

On the other hand, skepticism of congestion pricing is understandable, given the uncertainty involved in attempting to predict the results of any specific pricing proposal. Outcomes would depend on the level of tolls, how they vary by time of day, where they are imposed, what travel substitutes are available, and how tolls are spent.

Pitfalls in implementation

There are at least three potential pitfalls to implementing effective congestion pricing strategies. First, setting the correct toll level is difficult, given the absence of empirical data on responses to tolls. If set too high, too many travelers will be diverted to other modes, routes or schedules, and time savings benefits will not be sufficient to offset toll costs. If set too low, congestion will continue, and travel time savings will not be realized.[10] In a real world setting, policymakers could ill afford to err substantially in either direction. Given the history of other user fees in the U.S., it seems most likely that tolls would be set too low, because tolls are perceived as taxes, and therefore attempts to impose high tolls will be opposed regardless of their promised benefits.

Second, any limited implementation of congestion pricing (e.g., on a single freeway) could have significant spillover effects if large numbers of trips are diverted to alternate untolled routes. In such cases, the outcome of congestion tolls may be a redistribution of congestion, rather than any net reduction.

Finally, although congestion toll revenues are likely to be more than sufficient to offset any negative distributional impact, there is no guarantee that the funds will be used for this purpose. Rather, as explained earlier, funds are likely to be restricted to transportation uses, and thus equity problems will not be redressed. Whether or not such funds are used efficiently depends on how they are distributed among investment alternatives. If they are used to fund costly transit capital facilities, for example, only a small portion of travel demand will benefit. If they are used to increase highway capacity, the political coalition supporting congestion pricing may fall apart. As one observer has explained, environmental advocates see tolls as a means to force carpooling and mass transit use, thus reducing the need for new roads. Business interests, on the other hand, see tolls as funding resources for new highways and continued economic development (Reinhardt 1990).

Towards more acceptable congestion pricing options

Reasons for opposing congestion pricing are numerous and varied, and the circumstances in which it might be most effective have additional potential pitfalls. Given the depth of opposition and skepticism, it is very unlikely that congestion pricing could be implemented to any extent on existing highway facilities or in metropolitan employment centers. Nor is it likely that tolls could be set high enough to achieve congestion reduction goals in heavily congested areas. Rather, toll will be more acceptable to the public when linked with new facilities or some form of added highway capacity. These may include privately financed highways or highway improvements, peak tolled express lanes in existing freeway medians, or even allowing tolled access on existing HOV facilities for single occupant vehicles.

An understanding of the reasons why congestion pricing has been opposed provides a basis for identifying some characteristics of pricing strategies that are more likely to be found acceptable and implementable. The benefits of a congestion pricing policy must be significant, obvious, and easily understood. This is easiest to accomplish in the case of added capacity, particularly if tolls are needed to finance the new capacity. New tolled facilities provide another alternative for commuters while preserving existing untolled options. Moreover, the added capacity should benefit all travelers by improving average system speeds.

Prior studies have shown that the concept of congestion pricing is difficult to communicate. As a result, it is perceived as a tax with no obvious associated benefit. Thus it is criticized as unfair or harmful to business. If congestion tolls are perceived as taxes, they must be justified as taxes. Special purpose taxes are more acceptable when linked to a specific investment program. Linking the toll with a revenue disposition program identifies program benefits and focuses public debate on the tangible item (tolls) rather than the intangible item (time savings). If tolls are used to finance new facilities, the revenue linkage is direct and easy to justify.

Tolls on added capacity also effectively eliminates the loser groups listed in Table 2. Those whose value of time is less than the toll can continue to drive alone and still benefit from some reduction of congestion as a result of other travelers with higher time values shifting to the faster tolled route. Since traffic conditions will improve overall with the introduction of added capacity, all travelers receive some benefit. Moreover, when added capacity is restricted to existing rights of way, environmental impacts on adjacent properties are minimal.

The difference in responses to changing policies on existing facilities versus introducing the identical policy on new facilities is well documented by U.S. experiences with high occupancy vehicle (HOV) lanes. Two of the most visible policy failures of recent years include the I-10 (Santa Monica Freeway) "Diamond Lane" in Los Angeles and Boston's Southeast Expressway HOV lane, where existing general purpose travel lanes were restricted to high occupancy vehicles. Both were abandoned as a result of public pressure (Fisher & Sunkowity 1978). In contrast, new HOV facilities have been successfully introduced in several U.S. metropolitan areas.

In addition to restricting tolls to added capacity, potentially successful congestion pricing projects must be easy to use and understand, enforceable, and responsive to privacy concerns. There are a wide array of toll collection alternatives available, and the key to acceptability is ease of use. A toll system using electronic collection technology has clear benefits in terms of convenience. It is simply one more credit card and monthly bill. Since travelers would not pay tolls directly (at the time the trip is taken), there may be less opposition to the tolls. It would also be convenient for business: the toll bill would provide documentation for reimbursements and tax deductions. Debit "smart cards" or other technologies can be used to protect travelers' privacy. The toll itself must also be comprehensible. Thus a simple peak/off-peak pricing differential (marketed as an off-peak discount) is much preferred to a congestion-based variable toll. Variable tolls are conceptually elegant and technically feasible, but potentially confusing to the consumer.

Finally, any implementation of congestion pricing must overcome the skepticism that exists toward the concept. Planners and policy makers need to learn more about its sources and how it might be ameliorated. They also need to convince the public of the potential benefits of congestion pricing. This is of course a much easier task if pricing is linked with added capacity, and another reason why public acceptance is more likely for such proposals.

Political barriers to congestion pricing are formidable. Congestion pricing represents a fundamental change in transportation policy orientation. Although there seems to be a growing realization that transportation resources must be used more efficiently, public resistance to policies which appear to restrict or raise the price of using an auto continues to be strong. If congestion pricing projects are to be more successful in the future, they will have to be designed with more sensitivity to public acceptance, even if some of the economic efficiency benefits are lost in the process.

Acknowledgement

Los Angeles area freeway data were provided by the District 7 office of the California Depatment of Transportation.

Comments on an earlier draft from Peter Gordon, James Moore, Kenneth Small and Martin Wachs, as well as anonymous referees, are gratefully acknowledged.

Notes

1. State Proposition 111, passed on June 5, 1990, requires each county to implement a Congestion Management Plan if traffic congestion exceeds specified standards as a condition of receiving revenues from the Proposition's authorized gasoline gallonage tax increase. The California Clean Air Act of 1988 mandates the implementation of Transportation Control Measures as part of a comprehensive strategy to improve air quality in non-attainment areas.
2. Numerous reviews of congestion pricing are available, e.g. Morrison 1986; Small 1992.
3. Gomez-Ibanez (1992) makes a similar argument.
4. Some California examples include state mandated trash recycling requirements, public notification requirements on the use of hazardous or carcinogenic materials, and the defeat of several off-shore drilling proposals.
5. Categories are as follows: F-0, up to 1 hour of travel speeds of 30 mph or less; F-1, 1 to 2 hours at 30 mph or less; F-2, 2 to 3 hours at 30 mph or less; F-3, 3 or more hours at 30 mph or less.
6. Source: compiled by the author from survey data collected in 1985 and 1987 in Orange County, CA, and Orange County Transit District (1985).

7. Reduced vehicle trips suggest less congestion on surface streets, but freeway tolls would divert some trips to surface streets, suggesting more congestion. For simplicity, I assume these effects are offsetting.
8. The disposition of toll revenue has not been considered here; revenue disposition will be discussed in a later section.
9. No redistribution is equivalent to the toll agency putting the funds in a safe and not using them for any useful purpose.
10. Note that AVI-based tolls many reinforce this effect, since travelers will be less sensitive to indirect forms of toll payment.

References

Altshuler A, Womack J & Pucher J (1981) *The Urban Transportation System*. Cambridge, MA: MIT Press.

Bhatt K (1991) Congestion pricing of urban street network in Los Angeles Region. Paper presented at the University of California Transportation Center Conference on Congestion Pricing, San Diego CA.

Borins S (1988) Electronic road pricing: An idea whose time may never come. *Transportation Research A*, 22A(1): 37–44.

Button K (1984) Road pricing – An outsider's view of American experiences. *Transport Review* 4: 73–98.

Cameron M (1991) *Transportation Efficiency: Tackling Southern California's Air Pollution and Congestion*. Los Angeles, CA: Environmental Defense Fund and Regional Institute of Southern California.

Cervero R (1989) *America's Suburban Centers: The Land Use – Transportation Link*. Boston MA: Unwin Hyman.

Cervero R, Wachs M, Berlin R & Gephart R (1980) *Efficiency and Equity Implication of Alternate Transit Fare Policies*. Report DOT-CA-11-0019. Washington, D.C.: U.S. Department of Transportation.

Cohen Y (1987) Commuter welfare under peak period congestion tolls: Who gains and who loses? *International Journal of Transport Economies* XIV(3): 239–266.

Dunn J (1981) *Miles to Go: European and American Transportation Policies*. Cambridge, MA: The MIT Press.

Elliot W (1986) Fumbling toward the edge of history: California's quest for a road pricing experiment. *Transportation Research A* 20A(2): 151–156.

Else P (1986) No entry for congestion taxes? *Transportation Research A* 20A(2): 99–107.

Fisher R & Simkowitz H (1978) Priority treatment for high occupancy vehicles in the United States: A review of recent and forthcoming projects. Final Report No. UMTA-MA-0049-78-11. Urban Mass Transportation Administration, U.S. Department of Transportation.

Foster C (1974) The regressiveness of road pricing. *International Journal of Transport Economies* 1: 133–141.

Giuliano G & Wachs M (1992) Transportation demand management as part of growth management. In: Stein J (ed) *Growth Management: The Planning Challenge of the 1990's*. Newbury Park, CA: Sage Publications.

Giuliano G, Hwang K & Wachs M (1992) Mandatory trip reduction in Southern California: First year results. *Transportation Research A* (forthcoming).

Giuliano G (1992) Transportation demand management and urban traffic congestion: Promise or panacea? *Journal of the American Planning Association* 58(3): 327–335.

Giuliano G, Levine D & Teal R (1990) Impact of high occupancy vehicle lanes on carpooling behavior. *Transportation* 17: 159–177.

Gomez-Ibanez JA (1992) The political economy of highway tolls and congestion pricing. In: Exploring the Role of Pricing as a Congestion Management Tool: Summary of Proceedings of the Seminar on the Application of Pricing Principles to Congestion Management (pp. 5–9). Searching for Solutions: A Policy Discussion Series, No. 1. Washington DC: U.S. Federal Highway Admnistration.

Hanks J & Lomax T (1991) Roadway congestion in major urban areas: 1982–1988. *Transportation Research Record* No. 1305: 177–189.

Harvey G (1991) The suitability of Bay Area toll bridges for a congestion pricing experiment. Paper presented at the University of California Transportation Center Conference on Congestion Pricing, San Diego, CA.

Hau T (1992) *Congestion Charging Mechanisms: An Evaluation of Current Practice.* Washington DC: The World Bank.

Higgins T (1986) Road-pricing attempts in the United States. *Transportation Research A* 20A(2): 145–150.

Higgins T (1979) Roadpricing – Should and might it happen? *Transportation* 8: 99–113.

Hu P & J Young (1992) *Summary of Travel Trends: 1990 National Personal Transportation Study.* Report FHWA-PL-92-027. Washington DC: Federal Highway Administration.

Jones P & Harvey S (1991) Urban road pricing: Dealing with the issue of public acceptability – A U.K. perspective. Working Paper TSU669, Transport Studies Unit, University of Oxford, U.K.

Jones P (1991) Assessing traveler responses to urban road pricing. Working Paper TSU677, Transport Studies Unit, University of Oxford, U.K.

Kirby R (1991) Implementation issues – The local government view. Presented at the Seminar on the Application of Pricing Principles to Congestion Management, Washington, D.C.

Linneman P & Summers A (1991) Patterns and processes of employment and population decentralization in the U.S., 1970–1987. Paper presented at the Regional Science Association North American Meeting, New Orleans, LA.

May AD (1992) Road pricing: An international perspective. *Transportation* 19(4): 313–333 (this issue).

McShane M, Koshi M & Lundin O (1984) Public policy toward the automobile: A comparative look at Japan and Sweden. *Transportation Research A* 18A(2): 97–109.

Morrison S (1986) A survey of road pricing. *Transportation Research A* 20A(2): 87–96.

Newman P & Kenworthy J (1988) The transport-energy trade-off: Fuel efficient traffic vs fuel efficient cities. *Transportation Research A* 27A(3): 163–174.

Orange County Transit District (1985) *Commuter Market Study.* Garden Grove, CA: Orange County Transit District.

Poole R (1992) Introducing congestion pricing on a new toll road. *Transportation* 19(4): 383–396 (this issue).

Pucher J (1981) Equity in transit finance: Distribution of transit subsidy benefit and costs among income classes. *Journal of the American Planning Association* 47: 387–407.

Pucher J (1988) Urban travel behavior as the outcome of public policy. *Journal of the American Planning Association,* Summer: 509–519.

Reinhardt W (1990) Congestion pricing is coming: The question is where and when. *Public Works Financing,* June: 7–14.

Richardson HW (1974) A note on the distributional effects of road pricing. *Journal of Transport Economics and Policy* VIII(1): 82–85.

Richardson H & Gordon P (1989) Counting nonwork trips: The missing link in transportation, land use and urban policy. *Urban Land,* September: 6–18.

Rock S (1990) Equity of local option taxes. *Transportation* 44(3): 405–418.

Siracusa A (1991) Implementation issues – The business view. Presented at the Seminar on the Application of Pricing Principles to Congestion Management, Washington, D.C.

Small K (1983) The incidence of congestion tolls on urban highways. *Journal of Urban Economics* 13: 90–111.

Small K (1992) *Urban Transportation Economics*. Philadelphia, PA: Harwood Academic Publishers.

Small K (1992) Using the revenues from congestion pricing. *Transportation*, 19(4): 359–381 (this issue).

Small K, Winston C, Evans C (1989) *Road Work*. Washington, D.C.: Brookings Institution.

South Coast Air Quality Management District (1988) *Air Quality Management Plan*. Los Angeles, CA: South Coast Air Quality Management District and Southern California Association of Governments.

Southern California Association of Governments (1987) *Cost of Congestion*. SCAG Economic Development Program, Los Angeles, CA.

Southern California Association of Government (1988) *Regional Mobility Plan*. Los Angeles, CA: Southern California Association of Governments.

The Urban Institute/KT Analytics (1991) *Congestion Pricing Study*. Final Report. Los Angeles, CA: Southern California Association of Governments.

U.S. Department of Commerce (1991) *County Business Patterns: 1989 California*. Washington, D.C.: Economics and Statistics Administration, U.S. Department of Commerce.

U.S. General Accounting Office (1989) Traffic congestion: Trends, measures and effects. GAO/PEMD-90-1. Washington, D.C.

Vickrey W (1955) Some implications of marginal cost pricing for public utilities. *American Economic Review* 45: 605–620.

Wachs M (1990) Regulating traffic by controlling land use: The Southern California experience. *Transportation* 9(1 & 2): 241–256.

Wachs M (1989) U.S. transit subsidy policy: In need of reform. *Science* 244: 1545–1549.

Name Index

www.ingramcontent.com/pod-product-compliance
Ingram Content Group UK Ltd.
Pitfield, Milton Keynes, MK11 3LW, UK
UKHW050233200926
475702UK00020B/3017
* 9 7 8 1 1 3 8 3 7 8 4 6 9 *